DONALD L. JANSSEN is Director of Veterinary Services at the Zoological Society of San Diego.

SUSIE ELLIS is the former Senior Program Officer of the IUCN Species Survival Commission's Conservation Breeding Specialist Group. She now is Vice President of Conservation International's Indonesia and Philippines Program.

D1266299

Giant Pandas: *Biology, Veterinary Medicine and Managen*

"The giant panda is one of the world's most recognized anima
biology of this threatened species has been a mystery. For ex:
female giant panda is sexually receptive for only 2 to 3 days o
and, if pregnant, can produce twins, one of which inevitably
China undergoing unprecedented changes at a rapid and acc
rate, can such a highly specialized species survive? Giant par
in zoos are a favorite of the public, but more importantly are a
source of new biological information that can be applied to ι
ing and protecting the species in nature. This book is the first
summarize the present state-of-knowledge about giant pand:
the topics of reproduction, behavior, nutrition, genetics and
also offers the latest advances in neonatal care, preventative n
use of assisted breeding and recent progress in population bi
etically manage the worldwide 'insurance' population of gia
living in zoos and breeding centers. The exciting result is a gr
lation of giant pandas in captivity that, in turn, is allowing ani
zoos that produce funding to conserve the species in nature,
habitat protection. This book is an example of how zoos, thro
and awareness, contribute to the scientific understanding an
of one of the planet's most beloved animals, the giant panda

DAVID E. WILDT is Senior Scientist and Head of the Departm
Reproductive Sciences at the Smithsonian's National Zoolo

ANJU ZHANG is the former Director of the Giant Panda Tech
Committee in Chengdu, China under the auspices of the C
Association of Zoological Gardens.

HEMIN ZHANG is Director of the Wolong Nature Reserve in !
Province, China. Both Anju Zhang and Hemin Zhang are an
leading authorities on giant panda reproduction and heal:

Giant Pandas

Biology, Veterinary Medicine and Management

Edited by

DAVID E. WILDT
Senior Scientist and Head of the Department of Reproductive Sciences at the Smithsonian's National Zoological Park

ANJU ZHANG
the Former Director of the Giant Panda Technical Committee in Chengdu, China under the auspices of the Chinese Association of Zoological Gardens

HEMIN ZHANG
Director of Wolong Nature Reserve in Sichuan Province, China. Both Anju Zhang and Hemin Zhang are among China's leading authorities on giant panda reproduction and health

DONALD L. JANSSEN
Director of Veterinary Services at the Zoological Society of San Diego

SUSIE ELLIS
Former Senior Program Officer of the IUCN Species Survival Commission's Conservation Breeding Specialist Group. She now is Vice President of Conservation International's Indonesia and Philippines Program

CAMBRIDGE UNIVERSITY PRESS

Cambridge, New York, Melbourne, Madrid, Cape Town, Singapore, São Paulo

Cambridge University Press
The Edinburgh Building, Cambridge CB2 2RU, UK

Published in the United States of America by Cambridge University Press, New York

www.cambridge.org
Information on this title: www.cambridge.org/9780521832953

First published 2006

Printed in the United Kingdom at the University Press, Cambridge

A catalogue record for this publication is available from the British Library

ISBN-13 978-0-21-83295-3 hardback
ISBN-10 0-521-83295-0 hardback

Contents

Contributors

Autumn Anderson
Conservation and Research for Endangered Species, Zoological
Society of San Diego, 15600 San Pasqual Valley Road, Escondido,
CA 92027, USA

Tomas W. Baker
Department of Surgical and Radiological Sciences, School of
Veterinary Medicine, Veterinary Medical Teaching Hospital/Small
Animal Clinic, 1 Shields Avenue, University of California, Davis,
CA 95616, USA

Jonathan D. Ballou
National Zoological Park, Smithsonian Institution, 3001
Connecticut Avenue N. W., Washington, DC 20008, USA

Mollie A. Bloomsmith
Zoo Atlanta, 800 Cherokee Avenue S. E., Atlanta, GA 30315, USA

Janine L. Brown
National Zoological Park, Conservation & Research Center,
Smithsonian Institution, 1500 Remount Road, Front Royal, VA
22630, USA

Nancy M. Czekala
Conservation and Research for Endangered Species, Zoological
Society of San Diego, 15600 San Pasqual Valley Road, Escondido,
CA 92027, USA

Victor A. David
Laboratory of Genomic Diversity, National Cancer Institute,
National Institutes of Health, Frederick, MD 21702, USA

Autumn P. Davidson
Department of Medicine and Epidemiology, School of Veterinary
Medicine, Veterinary Medical Teaching Hospital/Small Animal
Clinic, 1 Shields Avenue, University of California, Davis, CA 95616,
USA

Barbara S. Durrant
Conservation and Research for Endangered Species, Zoological
Society of San Diego, 15600 San Pasqual Valley Road, Escondido,
CA 92027, USA

Mark S. Edwards
San Diego Zoo, Zoological Society of San Diego, P.O. Box 120551,
San Diego, CA 92112, USA

Susie Ellis
Conservation International, 1919 M Street, Suite 600,
Washington, DC 20036, USA

Lisong Fei
Chengdu Zoo, Northern Suburb, Chengdu, Sichuan Province
610081, People's Republic of China

Frank Göritz
Institute for Zoo Biology and Wildlife Research Berlin,
Alfred-Kowalke-Strasse 17, D-10315 Berlin, Germany

Mark Greenberg
San Diego Medical Center, 350 Dickinson Street, University of
California, San Diego, CA 92103, USA

Fernando Gual-Sil
Zoológico de Chapultepec, Paseo de la Reforma s/n, Mexico City
11000, Mexico

Janet Hawes
San Diego Zoo, Zoological Society of San Diego, P.O. Box 120551,
San Diego, CA 92112, USA

Kathy Hawk
San Diego Zoo, Zoological Society of San Diego, P.O. Box 120551,
San Diego, CA 92112, USA

Guangxin He
Chengdu Research Base of Giant Panda Breeding, 26 Panda Road,
Northern Suburb, Fu Tou Shan, Chengdu, Sichuan Province
610081, People's Republic of China

Thomas B. Hildebrandt
Institute for Zoo Biology and Wildlife Research Berlin,
Alfred-Kowalke-Strasse 17, D-10315 Berlin, Germany

Rong Hou
Chengdu Research Base of Giant Panda Breeding, 26 Panda Road,
Northern Suburb, Fu Tou Shan, Chengdu, Sichuan Province
610081, People's Republic of China

JoGayle Howard
National Zoological Park, Smithsonian Institution, 3001
Connecticut Avenue N. W., Washington, DC 20008, USA

Daming Hu
China Conservation and Research Center for the Giant Panda,
Wolong Nature Reserve, Wenchuan, Sichuan Province 623006,
People's Republic of China

Shiqiang Huang
Beijing Zoo, 137 Xiwai Dajie, Xicheng District, Beijing 100044,
People's Republic of China

Yan Huang
China Conservation and Research Center for the Giant Panda,
Wolong Nature Reserve, Wenchuan, Sichuan Province 623006,
People's Republic of China

Donald L. Janssen
San Diego Zoo, Zoological Society of San Diego, P.O. Box 120551,
San Diego, CA 92112, USA

David C. Kersey
National Zoological Park, Conservation & Research Center,
Smithsonian Institution, 1500 Remount Road, Front Royal,
VA 22630, USA

Mabel Lam
M. L. Associates, LLC, 123 Feldspar Ridge, Glastonbury, CT 06033,
USA

Desheng Li
China Conservation and Research Center for the Giant Panda,
Wolong Nature Reserve, Wenchuan, Sichuan Province 623006,
People's Republic of China

Guanghan Li
Chengdu Giant Panda Breeding Research Foundation, Chengdu
Research Base of Giant Panda Breeding, 26 Panda Road, Northern
Suburb, Fu Tou Shan, Chengdu, Sichuan Province 610081,
People's Republic of China

Xuanzhen Liu
Chengdu Research Base of Giant Panda Breeding, 26 Panda Road,
Northern Suburb, Fu Tou Shan, Chengdu, Sichuan Province
610081, People's Republic of China

I. Kati Loeffler
National Zoological Park, Conservation & Research Center,
Smithsonian Institution, 1500 Remount Road, Front Royal,
VA 22630, USA

Li Lou
Chengdu Research Base of Giant Panda Breeding, 26 Panda Road,
Northern Suburb, Fu Tou Shan, Chengdu, Sichuan Province
610081, People's Republic of China

Xiaoping Lu
CITES Management Authority of China, 18 Hepingli East Street, Beijing 100714, People's Republic of China

Terry L. Maple
Center for Conservation & Behaviour, School of Psychology, Georgia Institute of Technology, Atlanta, GA 30332, USA

Nathalie Mauroo
Ocean Park, Aberdeen, Hong Kong

Laura McGeehan
Conservation and Research for Endangered Species, San Diego Zoo, Zoological Society of San Diego, P.O. Box 120551, San Diego, CA 92112 and Department of Biology, University of California, Riverside, CA 92521, USA

Rita McManamon
Zoo Atlanta, 800 Cherokee Avenue S. E., Atlanta, GA 30315, USA

Philip S. Miller
Conservation Breeding Specialist Group, IUCN–World Conservation Union's Species Survival Commisson, 12101 Johnny Cake Ridge Road, Apple Valley, MN 55124, USA

R. Eric Miller
Saint Louis Zoo, WildCare Institute, 1 Government Drive, St Louis, MO 63110, USA

Steven L. Monfort
National Zoological Park, Conservation & Research Center, Smithsonian Institution, 1500 Remount Road, Front Royal, VA 22630, USA

Richard J. Montali
National Zoological Park, Smithsonian Institution, 3001 Connecticut Avenue N. W., Washington, DC 20008, USA

Patrick J. Morris
San Diego Zoo, Zoological Society of San Diego, P.O. Box 120551,
San Diego, CA 92112, USA

Tatsuko Nakao
Adventure World, Shirahama, Nishimuro-gun, Wakayama 649-22,
Japan

Stephen J. O'Brien
Laboratory of Genomic Diversity, National Cancer
Institute, National Institutes of Health, Frederick,
MD 21702, USA

Andreas Ochs
Zoological Garden Berlin AG, Hardenbergplatz 8, D-10787 Berlin,
Germany

Mary Ann Olson
Conservation and Research for Endangered Species, Zoological
Society of San Diego, 15600 San Pasqual Valley Road, Escondido,
CA 92027, USA

Megan A. Owen
San Diego Zoo, Zoological Society of San Diego, P.O. Box 120551,
San Diego, CA 92112, USA

Wenshi Pan
College of Life Sciences, Peking University, Beijing 100871,
People's Republic of China

Lyndsay G. Phillips
School of Veterinary Medicine, 1 Shields Avenue, University of
California, Davis, CA 95616, USA

Bruce A. Rideout
Conservation and Research for Endangered Species, San Diego
Zoo, Zoological Society of San Diego, P.O. Box 120551, San Diego,
CA 92112, USA

Oliver A. Ryder
Conservation and Research for Endangered Species, Zoological
Society of San Diego, 15600 San Pasqual Valley Road, Escondido,
CA 92027, USA

Ulysses S. Seal (deceased)
Conservation Breeding Specialist Group, IUCN–World
Conservation Union's Species Survival Commisson, 12101 Johnny
Cake Ridge Road, Apple Valley, MN 55124, USA

Fujun Shen
Key Laboratory for Reproduction and Conservation Genetics,
Chengdu Research Base of Giant Panda Breeding, 26 Panda Road,
Northern Suburb, Fu Tou Shan, Chengdu, Sichuan Province
610081, People's Republic of China

Robert Sims
Department of Applied & Engineering Statistics, Fairfax, VA, and
George Mason University, 4400 University Drive, MS 4A7 22030,
USA

Rebecca J. Snyder
Zoo Atlanta, 800 Cherokee Avenue S. E., Atlanta, GA 30315, USA

Lucy Spelman
National Zoological Park, Smithsonian Institution, 3001
Connecticut Avenue N. W., Washington, DC 20008, USA

Rebecca E. Spindler
Toronto Zoo, 361A Old Finch Avenue, Scarborough, M1B 5KY,
Canada

Karen J. Steinman
National Zoological Park, Conservation & Research Center,
Smithsonian Institution, 1500 Remount Road, Front Royal,
VA 22630, USA

Shan Sun
Laboratory of Genomic Diversity, National Cancer Institute,
National Institutes of Health, Frederick, MD 21702, USA

Meg Sutherland-Smith
San Diego Zoo, Zoological Society of San Diego, P.O. Box 120551,
San Diego, CA 92112, USA

Ronald R. Swaisgood
Conservation and Research for Endangered Species, San Diego
Zoo, Zoological Society of San Diego, P.O. Box 120551, San Diego,
CA 92112, USA

Chunxiang Tang
China Conservation and Research Center for the Giant Panda,
Wolong Nature Reserve, Wenchuan, Sichuan Province 623006,
People's Republic of China

Jason C. L. Tang
Ocean Park, Aberdeen, Hong Kong

Kathy Traylor-Holzer
Conservation Breeding Specialist Group, IUCN–World
Conservation Union's Species Survival Commisson, 12101 Johnny
Cake Ridge Road, Apple Valley, MN 55124, USA

Chengdong Wang
Chengdu Research Base of Giant Panda Breeding, 26 Panda Road,
Northern Suburb, Fu Tou Shan, Chengdu, Sichuan Province
610081, People's Republic of China

Jishan Wang
Chengdu Research Base of Giant Panda Breeding, 26 Panda Road,
Northern Suburb, Fu Tou Shan, Chengdu, Sichuan Province
610081, People's Republic of China

Pengyan Wang
China Research and Conservation Center for the Giant Panda,
Wolong Nature Reserve, Wenchuan, Sichuan Province 623006,
People's Republic of China

Qiang Wang
Chengdu Zoo, Northern Suburb, Chengdu, Sichuan Province
610081, People's Republic of China

Rongping Wei
China Conservation and Research Center for the Giant Panda,
Wolong Nature Reserve, Wenchuan, Sichuan Province 623006,
People's Republic of China

David E. Wildt
National Zoological Park, Conservation & Research Center,
Smithsonian Institution, 1500 Remount Road, Front Royal,
VA 22630, USA

Zhong Xie
Chinese Association of Zoological Gardens, Room C-6009,
13 Sanlihe Avenue, Beijing 100037, People's Republic of China

Zhiyong Ye
Chengdu Research Base of Giant Panda Breeding, 26 Panda Road,
Northern Suburb, Fu Tou Shan, Chengdu, Sichuan Province
610081, People's Republic of China

Jianqiu Yu
Chengdu Research Base for Giant Panda Breeding, Northern
Suburb, Fu Tou Shan, Chengdu, Sichuan Province 610081,
People's Republic of China

Anju Zhang
Chengdu Giant Panda Breeding Research Foundation, Chengdu
Research Base of Giant Panda Breeding, 26 Panda Road, Northern
Suburb, Fu Tou Shan, Chengdu, Sichuan Province 610081,
People's Republic of China

Cheng Lin Zhang
Beijing Zoo, 137 Xiwai Dajie, Xicheng District, Beijing 100044,
People's Republic of China

Guiquan Zhang
China Research and Conservation Center for the Giant Panda,
Wolong Nature Reserve, Wenchuan, Sichuan Province 623006,
People's Republic of China

Hemin Zhang
China Conservation and Research Center for the Giant Panda,
Wolong Nature Reserve, Wenchuan, Sichuan Province 623006,
People's Republic of China

Jinguo Zhang
Beijing Zoo, 137 Xiwai Dajie, Xicheng District, Beijing 100044,
People's Republic of China

Meijia Zhang
Chengdu Research Base of Giant Panda Breeding, 26 Panda Road,
Northern Suburb, Fu Tou Shan, Chengdu, Sichuan Province
610081, People's Republic of China

Ya-Ping Zhang
Key Laboratory of Cellular and Molecular Evolution, Kunming
Institute of Zoology, Chinese Academy of Sciences, Kunming
650223, People's Republic of China

Zhihe Zhang
Chengdu Research Base of Giant Panda Breeding, 26 Panda Road,
Northern Suburb, Fu Tou Shan, Chengdu, Sichuan Province
610081, People's Republic of China

Wei Zhong
Chengdu Research Base of Giant Panda Breeding, 26 Panda Road,
Northern Suburb, Fu Tou Shan, Chengdu, Sichuan Province
610081, People's Republic of China

Xiaoping Zhou
China Conservation and Research Center for the Giant Panda,
Wolong Nature Reserve, Wenchuan, Sichuan Province 623006,
People's Republic of China

Foreword

Conserving biodiversity is a daunting and complex task. Perhaps no species presents a greater challenge than the giant panda – one of the most recognized and threatened animals on the planet. Its difficult-to-traverse, mountainous habitat in China makes quantifying population numbers in the wild exceedingly difficult. Despite a recent survey suggesting that the wild population may be growing, there is no disagreement that the primary threat is severely fragmented habitat. There now are more than 40 isolated populations, many too small or containing too few giant pandas to be demographically and genetically viable for much longer.

Seminal studies have been conducted on wild giant panda ecology by pioneers such as Wenshi Pan, Zhi Lu and George Schaller. However, we still have only touched on the full complement of information necessary for integrated and robust conservation initiatives. One threat to overall giant panda conservation is simply the lack of broad-based knowledge about its biology. This is particularly important for such an evolutionarily distinct species. Its biological systems are unconventional: distinctive from bears, but a derivative of the ursine lineage; a bear-like, monogastric animal that largely survives on grass (bamboo); and a species that has somehow survived to modern times despite an extraordinarily short (three-day) window of sexual receptivity for the female. Surely, a more detailed understanding of such phenomena is critical, both from a scholarly perspective as well as to provide data that can inform wise management decisions. This requires coordination and collaboration among numerous stakeholder groups, including governments, academia, conservation organisations, zoos and breeding centres and local communities.

In addition to the estimated 1500 giant pandas now thought to inhabit the wild, a second population also exists – in Chinese zoos and breeding centres. This population, accessible to the public and to scientists, represents a valuable resource for answering many of the research questions that, for various reasons, cannot be addressed in the wild. Maintaining the captive population should never be viewed as a substitute for conserving giant pandas and their wild habitat. Nevertheless, data gleaned from studies in this environment can help to round out the available information that will contribute to informed decision-making as comprehensive management and recovery plans are developed.

Conservation breeding of giant pandas in China is a fascinating story. From a futile beginning only a few decades ago, the Chinese have made remarkable progress in developing a healthy population of giant pandas that not only provides a valuable research resource but also an insurance policy against extinction. Many of the steps forward have been made by the Chinese themselves, but advancements accelerated even more quickly with a request for advice from the Chinese Association of Zoological Gardens to the IUCN/SSC Conservation Breeding Specialist Group. The result was a stakeholder workshop followed by a unique Biomedical Survey – a multidisciplinary, cross-cultural effort that started as an experiment and blossomed into a relationship that has generated massive new information on the giant panda, most of which is summarised in this book. This project produced a host of new research questions, partnerships and capacity building initiatives. Additionally, data generated from this project have taken giant panda management in new directions, from enhancing reproduction in previously infertile individuals, to developing a global cooperative breeding programme to maintain genetic diversity.

This book is the first-ever compendium on the biology and management of giant pandas, and provides a summary of contemporary scientific information derived from studying more than 60 giant pandas living in zoos and breeding centres in China. It adds data to our fragmented knowledge concerning giant panda biology, including health issues, behaviour, nutrition, reproductive physiology/endocrinology, assisted breeding, early development and social competence, behavioural enrichment and medical and genetic management.

As importantly, these new data have been gleaned from a multi-institutional, multidisciplinary approach involving partnerships and coordinated teams of western and Chinese scientists. In addition to the scientific data presented, this text tells an appealing story of part-nerships and collaboration, describing how people from diverse cultures worked hand-in-hand to resolve the health and reproductive problems facing giant pandas living in Chinese zoos and breeding centres. Each chapter also describes what we do not yet know while offering explicit recommendations for future studies. I heartily concur with the conclusion of virtually all authors that the highest priority is to continue to build capacity, generating a cadre of young, enthusiastic scientists prepared to tackle the difficult issues facing the rich biodiversity of China (far beyond giant pandas). Extending this capacity, most of all, is the legacy of this ground-breaking endeavour.

Russell A. Mittermeier
President
Conservation International

Acknowledgements

Initially, we planned to focus this book entirely on the substantial results of the Biomedical Survey organised under the umbrella of the Conservation Breeding Specialist Group (CBSG) of the IUCN–World Conservation Union's Species Survival Commission. However, as our work continued within China, we discovered a wealth of additional studies being conducted by both western and Chinese colleagues. Thus, the book rapidly expanded to cover a number of related topics of interest to anyone concerned about giant pandas or fascinated with their biology, husbandry, medical care and management. We are most grateful for the dedicated efforts of each of the authors who contributed to this book. Kathy Carlstead, Lori Eggert, JoGayle Howard, Olav Oftedal, Jesus Maldonado, Jill Mellen, Suzan Murray, Amanda Pickard, Kathy Traylor-Holzer, Rebecca Spindler, Karen Terio and Duane Ullrey generously assisted the editors in providing reviewer comments to chapters.

This endeavour would not have been possible without the incredible trust and accessibility to expertise, physical resources and especially the animals offered by Chinese colleagues. Their confidence, friendship, enthusiasm and hospitality were inspiring. The CBSG Biomedical Survey was possible first because of the enormous amounts of time dedicated by all members of the team – more than 65 active investigators. Anju Zhang, Zhihe Zhang, Guangxin He, Hemin Zhang, Jinquo Zhang, Don Janssen and JoGayle Howard played particularly important leadership roles in the myriad of survey responsibilities, especially in planning, interpretation and follow-up. We thank the directors of the Chengdu Research Base of Giant Panda Breeding, Chengdu Zoo, Chongqing Zoo, Beijing Zoo and the China Conservation and Research Centre for providing physical resources and animals, as well as the staff at each of these institutions for their kind and willing assistance. The Survey also required significant amounts of funding and in-kind support. We are deeply appreciative of financial support provided by the Giant Panda Conservation Foundation (GPCF) of the American Zoo and Aquarium Association (AZA), the Zoological Society of San Diego, the Smithsonian's National Zoological Park, Columbus Zoo, Zoo Atlanta and the Saint Louis Zoo. David Towne (GPCF) and Kris Vehrs (AZA) deserve special credit for encouraging North American zoo directors to invest in the future of giant pandas by supporting the Biomedical Survey and other studies described in this book. British Airways generously donated many of the tickets used by USA-based scientists to travel to China. Our other corporate sponsors – Nellcor Puritan Bennett, Heska, Sensory Devices Inc., InfoPet Identification Systems, Air-Gas Inc., Ohaus and Olympus America Inc. – provided in-kind donations of survey equipment now in use in Chinese institutions.

The Chinese Association of Zoological Gardens, especially Menghu Wang and Zhong Xie, resolved numerous political challenges that continue to facilitate studies to this day in China. The Biomedical Survey and the many spin-off studies would never have emerged without the initial invitation of Madam Shuling Zheng, of the Ministry of Construction, that prompted CBSG's involvement with giant pandas. The Department of Wildlife Conservation and the Giant Panda Office paved the way for expanding the Survey to include giant pandas under State Forest Administration purview. More recently, the China Wildlife Conservation Association (in particular Rusheng Cheng and Shanning Zhang) have assisted, especially in the organisation of training courses directed to *in situ* conservation. During the first year of the Survey, Wei Zhong's translation helped avoid what could have been numerous misunderstandings as both the China- and USA-based teams learned to work together. Throughout our efforts in China, Xiaoping Lu and Mabel Lam (formerly of the Zoological Society of San Diego and now with M. L. Associates, LLC) have been relentless problem-solvers and amazingly capable of understanding and resolving cultural challenges and the occasional miscommunication. Their efforts to build bridges between the two cultures have been tireless, and it is safe to say that everyone associated with every study in this book extends their heartfelt appreciation to these two wonderfully dedicated people.

David Wildt thanks the staff of the Department of Reproductive Sciences and the administration of the National Zoo for their patience in understanding the amount of time taken from normal duties to complete this book. Laura Walker of the National Zoo's Conservation and Research Centre provided invaluable help in the final phases of manuscript preparation. All of the co-editors are grateful to Alan Crowden, Mike Meakin, Ward Cooper and especially Clare Georgy of Cambridge University Press for believing in the need for this book, for answering countless questions and facilitating the assemblage of a high-quality product.

Finally, during the course of this work, we (and the entire conservation community) lost two colleagues who were intimately associated with the CBSG Biomedical Survey. Arlene Kumamoto (Zoological Society of San Diego) was the geneticist/laboratory technical specialist during the first year of the Survey; she died of pancreatic cancer in 2000. Her diligence, good humour, friendship and strong belief in the importance of collaborative science are sorely missed. Ulie Seal (former Chairman of CBSG) was one-of-a-kind – a charismatic scientific leader, Renaissance Man and a person who believed that people and collaborative problem-solving were the keys to successful conservation; he died of lung cancer in 2003. Ulie made us all believe that it wasn't an option to give anything but our very best – and then some. Arlene and Ulie would have been delighted with what has been accomplished but at the same time would have said 'Do more'. We dedicate this book to their memory and hope that it is a useful step in generating more scholarly information through collaborative science, both of which are needed to conserve this Earth's increasingly threatened biodiversity.

I

The giant panda as a social, biological and conservation phenomenon

SUSIE ELLIS, WENSHI PAN, ZHONG XIE, DAVID E. WILDT

INTRODUCTION

The giant panda has captured the world's imagination. Its seemingly harmless, playful nature, velvety black and white fur, flat face, softly rounded body and soulful black eye patches combine to make it resemble an oversized and loveable teddy bear (Fig. 1.1). Its upright posture and famous 'panda's thumb' – an elongation of the wrist bone that allows it to grasp bamboo and other food much like people do – further adds to its widespread appeal. From the most prominent government authorities to young children, people are passionate about protecting the giant panda. This fervent interest has caused the panda to emerge as the most highly visible of all endangered species, even though few people have actually ever seen one in the wild. Furthermore, this single species has become a worldwide icon for the need to conserve animals, plants and habitats. Therefore, it is ironic that the giant panda, which evokes so much attention by the public, scientific and conservation communities, still remains such a mystery with so many pieces still missing from a biological jigsaw puzzle that, if solved, could improve species management, welfare and conservation. The purpose of this book is to provide, and then assemble, a few more pieces of this enormous puzzle.

Giant Pandas: Biology, Veterinary Medicine and Management, ed. David E. Wildt, Anju Zhang, Hemin Zhang, Donald L. Janssen and Susie Ellis. Published by Cambridge University Press. © Cambridge University Press 2006.

Figure 1.1. The giant panda (photograph by Jesse Cohen).

WHY THE GIANT PANDA IS UNIQUE AMONG SPECIES, ESPECIALLY BEARS

Within China, the giant panda often is called *daxiongmao* by local people, literally 'large bear-cat' in Chinese (Schaller *et al.*, 1985). Its scientific name *Ailuropoda melanoleuca* actually means black and white cat-footed bear. The black and white colouration of the panda allows it to blend in with its high mountain forest surroundings, which often are blanketed with thick snow. When threatened, pandas climb the nearest tree, where this coloration renders them almost undetectable.

The giant panda is indeed a type of bear, of the subfamily Ailuropodinae in the family Ursidae. During evolution, it diverged from the main bear lineage (comprised of seven other species) 15 to 25 million years ago (Lumpkin & Seidensticker, 2002). Interestingly, the giant panda's nearest relative (genetically speaking) is the spectacled bear (*Tremarctos ornatus*) which inhabits the mountainous regions of South America. Unlike its ursid counterparts, which are principally omnivores, the giant panda is a 'grass-eating' bear with 99% of its diet as bamboo. This, in part, explains some of its unique morphology, including the skull's expanded zygomatic arches and the associated powerful muscles for mastication (Nowak & Paradiso, 1983). The giant panda's dentition is also different from, for example, a similarly sized black bear because of broad, flattened premolars and molars designed to break and grind bamboo. As a 'hypo carnivore', the giant panda is also partial to meat, with its teeth specially suited to crush bones (Lumpkin & Seidensticker, 2002). Its forefoot, with the modified and well-described sixth toe (Gould, 1982) or wrist bone, is considered unique among species, allowing it to grasp its food more securely.

During winter, the giant panda's survival in its cold, wet mountainous habitat is enhanced by the superb insulation provided by short, thick fur. It has no tolerance for heat, in part because of its lack of capacity for passive heat loss or evaporative cooling (Lumpkin & Seidensticker, 2002). Unlike its bear counterparts, the giant panda does not hibernate, probably because of the need to forage throughout the year for its low-energy diet of bamboo. One of its most unique features is its adaptation from carnivory to herbivory while amazingly retaining the digestive system of the former. The result is the need to spend 14 hours of each day searching, selecting and consuming bamboo (Lumpkin & Seidensticker, 2002).

Perhaps most interesting (and one of the incentives for the work in this text) is the overall low reproductive rate of the giant panda. Note the use of the word 'low' and not 'poor'. There has been much embellishment by the popular press about 'poor reproduction' in this species. This misperception is derived from the well-known challenges of breeding pandas in artificial conditions in captivity. However, there has never been any systematic study of reproductive efficiency in wild giant pandas in nature. Obviously, it cannot be true that the giant panda is normally poor at reproduction or it would never have evolved or survived to modern times. Nonetheless, the species has developed some fascinating and rather illogical characteristics that are less than ideal for ensuring reproductive success. It is a seasonal breeder with the female entering oestrus (heat) in late winter/early spring. This trait in itself is nothing special, and the environmental stimulus inducing oestrus is likely to be increasing day length, although no one is sure. However, unlike other bears, the giant panda is monoestrus, displaying sexual receptivity once per year for only 2 to 3 consecutive days. In turn, the male produces prodigious numbers of motile spermatozoa, probably because of the need to ensure conception if given the chance to mate with a female, who normally is sexually 'turned off' for more than 360 days per year. Further evidence for the physiological reproductive prowess of the male giant panda includes the species' comparatively short and repeated copulations, each 1 to 8 minutes in length (Zhang et al., 2004), unlike in other bears. Resulting embryos are

free-floating in the uterine horns for an undetermined interval (a phenomenon called delayed implantation), which is common in bears, as is
the eventual production of one or two small, comparatively immature
cubs. Enigmatically, however, in the case of giant panda twins, one
offspring is usually rejected by the dam and dies soon after birth. The
giant panda cub is relatively slow-growing, although the species as a
whole achieves sexual maturity at a time comparable to other bear
species.

STATUS IN NATURE AND THREATS

The giant panda is endemic to the mountains of Sichuan, Gansu and
Shaanxi Provinces in China. The species is now found in only six mountain ranges at the eastern edge of the Tibetan plateau, distributed in
as many as 30 to 40 distinctive populations (Fig. 1.2; Plate I). The
Min Shan Mountains are the heart of panda numbers and activities,
probably sustaining half the remaining wild individuals (Lumpkin &
Seidensticker, 2002). Historically, the species was widely distributed and
may have numbered 100 000 animals, but has declined to likely no more
than 1500 animals in total. In reality, this number is only a broad
estimate – even recent surveys have been unable to produce an absolute
number of giant pandas living *in situ*. This is largely because these are
extreme habitats with steeply ascending ridges that plummet into deep
and narrow valleys. It is exceedingly difficult to traverse this terrain,
let alone see elusive giant pandas or their signs. Historically, these
rugged landscapes have protected the region's biodiversity. However,
as China's human population continues to grow, human settlements
are expanding into these remote areas.

As with virtually all endangered species, the giant panda has been
most affected by human forces, especially overall habitat loss as a result
of logging and farming operations. More than half of this habitat was
destroyed from the mid 1970s through the 1980s, a time when there
was enormous concern and publicity about conserving the species. The
magnitude of this destructive impact has been effectively illustrated
by Lumpkin and Seidensticker (2002) who have pointed out that the
resulting ecospace for all giant pandas became 5000 square miles,
which is *less than 25%* of the size of the Greater Yellowstone Ecosystem.
Giant pandas became the ecological losers in terms of total habitat
available. Compounding this problem was habitat fragmentation, the
breaking apart of existing forest into small patches with no corridors for

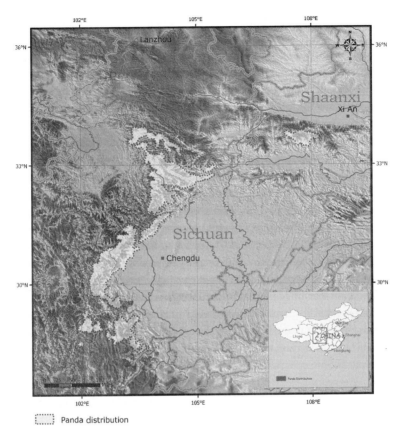

Panda distribution

Figure 1.2. Range map for remaining fragmented populations of giant pandas living in nature. (See also Plate I.)

genetic exchange. Although no one is sure of the number of individual pandas in each of these isolated areas, it is highly probable that some populations are not self-sustaining. As human demands escalate, many nature reserves are being heavily used for economic purposes (Liu *et al.*, 1997). Furthermore, many of the official protected areas (currently more than 40 reserves) are severely under-resourced, lacking the infrastructure (roads, buildings), personnel (managers, field staff) and equipment (ranging from vehicles to binoculars) to attend properly to daily and routine activities, let alone conservation priorities. And, of course, not all giant pandas live inside protected areas.

Historic dangers for the wild giant panda included hunting as trophies (mostly by westerners), museums and zoos. Hunting was officially banned in 1963 for any purpose. Poaching still occasionally causes mortality, although most of these are probably incidental deaths in snares targeting other species rather than deliberate acts directed at giant pandas. Until recently, it was common practice to 'rescue' giant pandas from the wild to support zoo breeding programmes. As described in Chapter 2, the *ex situ* breeding community committed to abandoning the practice of taking giant pandas from nature in 1996.

Adequate supply of appropriate food sources has been debated as a potential threat, especially given the significance by the popular press to the flowering die-offs of bamboo. Lumpkin and Seidensticker (2002) indicated that this impact is probably less significant than once believed because most habitats contain at least two bamboo species that do not flower in tandem. Thus, the panda simply switches bamboo species, if necessary. Total available bamboo also is not likely a significant factor because, although quality generally is marginal, supply is usually generous and rather consistent. There is growing concern, however, about panda–human competition for wild bamboo, including shoots (a dietary favourite of both species) and stems that have many uses by people ranging from basket weaving to tools to fencing.

Certainly, a threat to giant pandas is the lack of broad-based knowledge about their biology and numbers in nature. It is impossible to manage any habitat or species without understanding its status through systematic and continuous studies. Pioneering studies that methodically monitored life history, behaviour, mating and foraging were conducted by Schaller *et al.* (1989), Reid *et al.* (1989), Pan & Lu (1993), Pan (1995), Pan *et al.* (1998) and Lu *et al.* (2000, 2001). However, given all of the unknowns about contemporary panda activities (including how many pandas are out there), a continued lack of basic information certainly hinders appropriate decision-making to best manage wild populations.

Finally, some have asserted that the *ex situ* (captive) population threatens giant pandas living *in situ*. Essentially, the argument is that if too much attention is directed at pandas living in zoos, then the wild population is 'out of sight, out of mind and out of luck' – the distraction paradigm. The concern is that because there are healthy, reproductively fit pandas in zoos, there would be no urgency, or even a real need, to protect wild counterparts or their habitats. In our opinion, this theory is not valid, especially considering the intense worldwide interest in the species. We fully realise, however, that this theory *could* have validity, but only if we failed to clearly articulate *and* demonstrate the value of individuals managed *ex situ*, especially their potential in contributing to the conservation of the wild giant pandas. Much of this book is dedicated to this goal.

GIANT PANDAS IN CAPTIVITY IN CHINA

Unlike other prominent species (e.g. the tiger and crane), the giant panda has never been entrenched in historical Chinese culture, including the arts and literature. The earliest recorded giant pandas in captivity were held in the Emperor's garden during the Han Dynasty (206 BC to AD 226) in the then-capital of Xian. In more modern times (mid-20th century), the species was held by more western than Chinese zoos. The first serious interest in exhibiting the species in China occurred in Chongqing in 1941, but it was 10 years later when pandas began appearing regularly in Chinese zoos.

By the early 1960s there was evidence of targeted management, largely on the basis of reproductive success, albeit with inconsistency (see Chapter 19). The first ever birth by natural mating in captivity occurred at the Beijing Zoo in 1963. This same institution produced the first cub from artificial insemination (AI) with fresh sperm in 1978. The Chengdu Zoo was the first to produce a cub by AI with frozen-thawed semen in 1980. Through 1989, giant pandas were successively bred at zoos in Kunming, Shanghai, Hangzhou, Chengdu, Chongqing, Fuzhou and Xian, and at the Wolong Nature Reserve's breeding centre. From 1990 to 2002, 179 cubs were born from 126 pregnancies, with 71% of neonates surviving (see Chapter 19). And, interestingly, dedicated captive breeding activities complemented parallel efforts at protecting giant panda habitat as the first three giant panda reserves were established in 1963, growing to 13 by 1989 and to more than 40 today.

GIANT PANDAS IN THE WESTERN WORLD

The giant panda was virtually unknown outside China until the 1800s when the declining Qing Dynasty opened China to western trade. The species was first described in the western world by the missionary naturalist and explorer Père Armand David who described a giant panda specimen shot by Chinese hunters in Baoxin County, Sichuan Province in 1869 (Hu & Qiu, 1990). It was not until 1916 that the first westerner, Hugo Weigold, saw a live giant panda, and then it was another 14 years until the next sighting was reported. In the years following its discovery, killing of giant pandas became a goal of western museum collectors and hunters, beginning with Kermit and Theodore Roosevelt, Jr, sons of Teddy Roosevelt, who shot a specimen on an expedition sponsored by the Chicago Field Museum (Sheldon, 1975).

The first live giant panda was exported to the USA by Ruth Harkness, widow of the wealthy adventurer William Harkness, who 'rescued' a cub in Sichuan Province. In late 1936, after trouble with customs, Mrs Harkness took the cub out of China with a customs voucher that said 'one dog, $20.00' (Sheldon, 1975; Schaller *et al.*, 1985). This animal, Su Lin, had been destined for the New York Zoological Society, but the zoo refused it because of perceived health problems (Schaller *et al.*, 1985). The National Zoological Park in Washington, DC also declined to accept it, due to a rather extraordinary asking price (Lumpkin & Seidensticker, 2002). After a whirlwind tour of San Francisco, Chicago and New York, Su Lin ended up at Chicago's Brookfield Zoo, where she died of pneumonia in April 1938. The 'pandamania' spawned by Harkness and others' 'bring 'em back alive' approach led to the export of at least 16 giant pandas to western zoos over the next 15 years. Without readily available fresh bamboo or husbandry expertise, western zoos were ill-equipped to care for these animals, and none survived beyond 10 years of age.

The further exportation of giant pandas from China stopped with the Cultural Revolution and the formation of the People's Republic of China in 1949. A handful of animals were sent to zoos in Europe and North Korea. Then, the re-initiation of diplomatic relations between China and the USA (spearheaded by Mao Zedong and Richard Nixon) resulted in a 1972 gift of two giant pandas to the Smithsonian's National Zoological Park. This was followed by similar state gifts to Japan, France, the UK, Mexico, Spain and Germany. Only three of these pairs produced surviving young. The pairs in Japan and Mexico still have surviving offspring. The pair in Spain had two cubs; one survived for 4 years, but all offspring are now deceased.

Species charisma, relentless media coverage and parallel explosions in visitation at holding zoos in the west provoked the 'rent-a-panda' programme of the 1980s. This involved short-term loans from only weeks to a few months duration in exchange for substantial amounts of cash. Because these activities had no clear benefits for the species, it did not take long to attract the attention of conservationists as well as the USA Government which quickly saw the programme as strictly exploitative. The giant panda was placed on the USA Endangered Species List in 1984, which was followed by an all-out importation ban in 1988. Through a loophole, the Columbus Zoo arranged a short-term loan of giant pandas in 1992. This controversial loan set the stage for the future, in that funds raised as a result of the loan were used to establish new reserves in wild panda ranges in China. To buy time, the US Fish and Wildlife Service enacted a moratorium on any further giant panda importations. The goal was to formulate a policy ensuring that any further trade in giant pandas would not be detrimental to the species in nature. In fact, the most important part of the guidelines mandated that any loan be connected to *enhancement of conservation* of giant pandas in nature and not linked to commercial gain.

The result was that zoos in the USA were forced to develop highly organised scientific and management plans before being considered as candidates for importing giant pandas from China. There were also substantial financial costs to each loan, generally about $1 million annually for the loan plus additional costs to support the home institution's research and training programmes in the USA as well as in China (see Chapter 22). Even given these challenges, to date four institutions in the USA currently maintain giant pandas, including the San Diego Zoo (beginning 1998), Zoo Atlanta (1999), the Smithsonian's National Zoological Park (2002) and Memphis Zoo (2003). The San Diego Zoo also achieved the first milestone in North America, the production of a surviving cub by AI (Hua Mei, studbook number 487, born in 1999) who was subsequently returned to China and reproduced in 2004. Most recently (July 2005), the National Zoo produced a cub (Tai Shan, SB 595) by AI which survives at the time of writing.

CURRENT STATUS OF THE WORLD'S *EX SITU* GIANT PANDA POPULATION, INCLUDING THREATS

The notion of 'conservation breeding' of giant pandas is not new – the Chinese have long recognised this need and produced the first cub in captivity almost 40 years ago. Births in Mexico, Japan and the USA (often following complicated behavioural and reproductive monitoring as well as sophisticated assisted breeding technologies) also demonstrate global interest and dedication to propagating the species. But through-out history, what is apparent and common to all giant panda-holding institutions is sporadic, inconsistent success at reproduction followed by survival to adulthood. Lu and colleagues (2000) correctly pointed out some of the problems that have plagued panda-breeding programmes, including the enormous amount of funds expended on captive breed-ing; the high failure rate of reproduction (by 1997, 74% of adults had not bred); and the lack of appropriate *ex situ* environments for this specialised species.

From our overview here, it is probably apparent that nothing is simple about giant panda conservation, biology or politics. It is a species under enormous pressure by people, and yet it relies on people to ensure its ultimate survival. Nonetheless, progress is being made. In 1996, when the activities associated with this book began, there were about 124 giant pandas living in captivity worldwide. Today, there are more than 160 living individuals (Xie & Gipps, 2003) with the majority under the management authority of the Chinese Ministry of Construc-tion and its Chinese Association of Zoological Gardens. A counterpart Chinese agency, the State Forestry Administration, manages all pandas in the wild plus a captive population at its China Conservation and Research Centre for the Giant Panda in the Wolong Nature Reserve and a more recent collection in Ya'an (Ya'an Bifengxia Base of China Conser-vation and Research Centre for Giant Pandas). There now are approxi-mately 29 pandas living in zoos in North America, Europe, Japan and Thailand.

Despite the charisma, controversies, money and politics swirling around the species, improvements in captive management are being made. This is largely for two reasons: the application of an integrative, multidisciplinary scientific approach (see Chapter 2) and the develop-ment of partnerships, including training and the emergence of trusting relationships, across often complex cultural and agency boundaries (see Chapter 22). Before 1996, the primary threats to a sustainable captive panda population were lack of knowledge and no coordinated way to address routine problems encountered in management and husbandry. In fact, the challenges had never been clearly defined, and zoo man-agers encountering the same health, behavioural, genetic and repro-ductive problems rarely cooperated scientifically. However, now (as hopefully will become clear throughout this book) there is much new information on the specific factors that limit giant panda reproduction and survival in captivity. Furthermore, there have been many positive efforts to resolve issues by applying the scientific method. However, this in turn requires the consistent provision of adequate resources by governmental and non-governmental agencies and partnering organisations. Training the next generation of scientists is also imperative. And all of this can be accomplished because of the intense international interest in exhibiting the species, which can be translated into cooperation, financial support and more basic and applied research (Zheng et al., 1997).

VALUE OF GIANT PANDAS *EX SITU*

If giant pandas should be maintained in captivity then the role of that population in conservation needs to be clearly articulated. Given that we can adhere to a goal whereby the *ex situ* population is not a detriment to the wild, but rather an advantage for 'enhancement' of giant pandas in nature (Zheng et al., 1997), then we recognise at least six ways in which the captive population of giant pandas is of conservation value.

1. *Ambassadorial value.* Few people have been fortunate enough to see a giant panda in the wild. Even so, this rarely glimpsed creature has become a worldwide ambassador for the need to conserve threatened habitats and diverse species. What happens to the giant panda also happens to other species sharing the same habitat – the 'umbrella' effect. Saving the mountain regions in which giant pandas live means the protection of the golden monkey, takin, serow, muntjac, tufted deer, red panda, golden pheasant, giant salamander and thousands of other species, including rare plants and invertebrates. Because of the precarious status of wild populations and the difficulty in viewing them in nature, giant pandas in zoos and breeding centres play a crucial role in educating the public. Giant pandas 'up close and personal' are commanding emissaries for their wild counterparts and a tangible reminder of why so much effort needs to be directed at saving wild places.

2. *Educational value.* In a similar fashion, there is a need to educate the general public about the precarious status of wild populations. Those facilities exhibiting pandas have the responsibility to provide visitors with synthesised lessons about

animal anatomy, physiology, ecology and behaviour, ultimately instilling an appreciation of the species and its particular adaptations to the natural environment. Most importantly, zoos and breeding centres must emphasise the imperilled status of wild giant pandas and send the message that captive management is not a substitute for intensive efforts to conserve the species and its habitat in nature. And, finally, given the rapid progress made from systematic studies, we would suggest that interest in giant pandas and the stories emanating from research could become a model to 'turn on' the general public (especially children) to science by demonstrating its value in managing and conserving one of the world's most beloved species.

3.   *Insurance value.* The status of wild giant panda populations is uncertain at best. Although logging operations have ceased, Chinese forests remain fragmented, corridors among habitats have not been established, and new reserves are not yet capable of optimal management. Humans often encroach upon and economically exploit existing reserves, reducing the quality and quantity of habitat (Liu *et al.*, 1997). Most worrisome is the lack of reliable knowledge about numbers, demography and genetic viability of giant pandas in each of these isolated populations. A fragmented population is highly vulnerable to unpredictable events, for example, a disease epidemic or natural catastrophe such as a bamboo die-off. Thus it makes sense that any species facing such a precarious future be 'insured'; a captive programme provides an insurance policy. However, part of the dividend payment by zoos and the public they serve must be dedicated to protecting pandas in nature, thereby avoiding the need to ever 'cash in' the policy.

4.   *Funding value.* Whether we like it or not, the ability to 'experience' giant pandas can have a profound impact on our ability to raise funds – in no other case is it routine to generate $1 million per year to import a wildlife species. Panda appeal translates into serious funding for conservation, not just benefiting giant pandas but many other species sharing the same habitats. Under present conditions set in place by the US Fish and Wildlife Service, the $1 million per year from each panda-holding zoo in the USA becomes available for building capacity, whether it involves building roads and ranger stations at newly developed

protected areas or training the next generation of Chinese field biologists and zoo scientists (see Chapter 22). Access to giant pandas held in zoos and breeding centres helps to convince politicians, corporations and the private sector to give money that, in turn, will help to ensure resources for conservation now and long into the future.

5. *Value for scholarly knowledge.* An *ex situ* giant panda population serves as an invaluable resource for basic and applied biological research. Overall, there has been little detailed, integrated knowledge about giant panda biology, especially in the life sciences. Yet our descriptions above, about species uniqueness explain the need for many more systematic studies. How, for example, can one study disease susceptibility, digestion dynamics or sperm biology in a species that lives in remote and thick, mountainous bamboo forests? One of the most exciting progressions in panda biology in the last few years is agreement among holders that the captive population must be used to better understand the species from a scholarly perspective. Buy-in to this concept is assisted by the realisation that the resulting information will vastly improve *ex situ* management and eventually may contribute to more enhanced *in situ* conservation. This book is a testament to the advantage of having accessibility to giant pandas living in controlled environmental conditions for research.

6. *Unknown value for the future.* There is a sixth undefined reason for maintaining giant pandas *ex situ*, and that involves unpredictable future advantages of maintaining a genetically viable population. Certainly from an applied conservation perspective there have been recurrent discussions about reintroducing giant pandas into nature – adding new individuals to existing or new reserves (Mainka, 1997). In an ideal world, wild individuals would serve as the source for these movements. However, we must also consider that, realistically, captive populations may be the most reasonable source for these individuals (despite our current vast lack of knowledge about exactly *how* to reintroduce captive-produced pandas into wild habitats). And, finally, from a scholarly angle, one never knows how basic studies of one species will benefit another. For example, how indeed can a species evolve and survive to modern times when the female is sexually actively for less than 1% of an entire year? Perhaps there are

lessons here for other mammals (including humans) in what controls reproductive success. Thus who knows what can be learned from the biologically mysterious giant panda that will benefit other living things?

PRIORITIES FOR THE FUTURE

It is fortunate that now there is so much intensive interest and action in place for the giant panda and *before* demographic and genetic instability has set in. In numerous other species, experience has shown that it is essential to develop comprehensive management and/or recovery plans well before species numbers become critical. In such cases, early intervention (which may include captive breeding) can provide a timely and cost-effective, integrated approach that allows problems to be addressed before there is a crisis and no time for research or errors (Ellis & Seal, 1995). For many threatened species, such as the black-footed ferret, California condor and Hawaiian crow, captive breeding options were resisted until the wild populations crashed (often to fewer than 20 individuals), genetic erosion had begun, and the species was at maximum risk.

In contrast, the Chinese approach has been bold as well as visionary – acting now while there is time and adequate genetic diversity (see Chapter 10 and 21). A dream to develop a self-sustaining captive population of giant pandas in China was initiated only in 1996. As this book explains, Chinese efforts to achieve this goal while simultaneously contributing to the protection of giant pandas in nature are well on their way. Certainly many obstacles remain but the purpose of this text is to demonstrate the value of taking many small and integrated steps. The priority is to be absolutely resolute, not faltering or becoming frustrated by the political complexities that generally accompany studying the world's most high-profile species. Rather, for both the wild and captive populations, there is a single priority: to continue to work together in intensive partnerships to create more biological knowledge that will ensure a genetically stable and viable population – in perpetuity.

ACKNOWLEDGEMENT

We thank Hao Wang for providing the map of giant panda distribution. The authors were inspired by the thorough documentation of giant panda 'history' provided in the superb book by Susan Lumpkin and John Seidensticker (*Smithsonian Book of Giant Pandas*).

REFERENCES

Ellis, S. and Seal, U. S. (1995). Tools of the trade to aid decision-making for species survival. *Biological Conservation*, **4**, 553–72.

Gould, S. J. (1982). *The Panda's Thumb: More Reflections in Natural History*. New York, NY: W. W. Norton & Co.

Hu, J. and Qiu, X. (1990). History and progress of breeding and rearing giant pandas in captivity outside China. In *Research and Progress in the Biology of the Giant Panda*, ed. J. Hu, F. Wei, C. Yuan and Y. Wu. Chengdu, China: Sichuan Publishing House of Science and Technology, pp. 326–33.

Liu, J., Ouyang, Z., Yang, Z. *et al.* (1997). Human factors and panda habitat change in the Wolong Nature Reserve. *Proceedings, Ecological Society of America*, 1997 Annual Meeting, Albuquerque.

Lu, Z., Pan, W., Zhu, X., Wang, D. and Wang, H. (2000). What has the panda taught us? In *Priorities for the Conservation of Mammalian Diversity: Has the Panda Had Its Day?*, ed. A. Entwistle and N. Dunstone: Cambridge: Cambridge University Press, pp. 325–34.

Lu, Z., Johnson, W., Menotti-Raymond, M. *et al.* (2001). Patterns of genetic diversity in remaining giant panda populations. *Conservation Biology*, **15**, 1596–607.

Lumpkin, S. and Seidensticker, J. (2002). *Smithsonian Book of Giant Pandas*. Washington, DC: Smithsonian Press.

Mainka, S., Ed., (1997). *Proceedings of the International Workshop on the Feasibility of Giant Panda Re-introduction, Wolong Nature Reserve, Sichuan, China, 25–29 September*. Beijing: China Forestry Publishing House (in Chinese and English).

Nowak, R. M. and Paradiso, J. L. (1983). *Walker's Mammals of the World*, 4th edn. Baltimore, MD: Johns Hopkins University Press, pp. 976–7.

Pan, W. (1995). New hope for China's giant pandas. *National Geographic*, **187**, 100–15.

Pan, W. and Lu, Z. (1993). The giant panda. In *Bears: A Complete Guide to Every Species. Majestic Creatures of the Wild*, ed. I. Stirling. London: Harper Collins, pp. 140–5.

Pan, W., Oftedal, O. T., Zhu, X., *et al.* (1998). Milk composition and nursing in a giant panda (*Ailuropoda melanoleuca*). *Acta-Scientiarum-Naturalium- Universitatis-Pekinensis*, **34**, 350–1.

Reid, D., Hu, J., Dong, S., Wang, W. and Huang, Y. (1989). Giant panda *Ailuropoda melanoleuca* behavior and carrying capacity following a bamboo die-off. *Biological Conservation*, **49**, 85–104.

Schaller, G., Hu, J., Pan, W. and Zhu, J. (1985). *The Giant Pandas of Wolong*. Chicago, IL: University of Chicago Press.

Schaller, G., Tang, Q., Johnson, K. *et al.* (1989). The feeding ecology of giant panda and Asiatic black bear in the Tangjiahe Reserve, China. In *Carnivore Behavior, Ecology and Evolution*, ed. J. Gittleman. Ithaca, NY: Cornell University Press, pp. 212–41.

Sheldon, W.G. (1975). *The Wilderness Home of the Giant Panda*. Amherst, MA: University of Massachusetts Press.

Xie, Z. and Gipps, J. (2003). *The 2003 International Studbook for Giant Panda (Ailuropoda melanoleuca)*. Beijing: Chinese Association of Zoological Gardens.

Zhang, G., Swaisgood, R. R. and Zhang H. (2004). An evaluation of the behavioral factors influencing reproductive success and failure in captive giant pandas. *Zoo Biology*, **23**, 15–31.

Zheng, S., Zhao, Q., Xie, Z., Wildt, D. E. and Seal U. S. (1997). *Report of the Giant Panda Captive Management Planning Workshop*. Apple Valley, MN: IUCN–World Conservation Union/SSC Conservation Breeding Specialist Group.

2

The Giant Panda Biomedical Survey: how it began and the value of people working together across cultures and disciplines

DAVID E. WILDT, ANJU ZHANG, HEMIN ZHANG, ZHONG XIE, DONALD
L. JANSSEN, SUSIE ELLIS

INTRODUCTION

This book deals mostly with new biological knowledge and the *use* of that knowledge to benefit the giant panda by enhancing health, reproduction and management. It is an important strategy for modern-day zoo scientists, conducting 'basic research' to learn as much as possible about previously unstudied phenomena in any species, especially those that have received little, if any, attention.

In many ways, a scientist affiliated with a zoo is no different than a university research professor – both study mechanisms by using the scientific method to test hypotheses (Wildt, 2004). What is different about zoo science is the growing emphasis on results having practical uses – addressing issues that are relevant to allowing an animal to be better maintained in captivity, to allow it to thrive, reproduce and help sustain its species. In a perfect world that new knowledge will have duality of purpose, being useful to improving the conservation of *in situ* as well as *ex situ* populations. In fact, there now are many examples of 'captive' studies that have been useful for re-invigorating

Giant Pandas: Biology, Veterinary Medicine and Management, ed. David E. Wildt, Anju Zhang, Hemin Zhang, Donald L. Janssen and Susie Ellis. Published by Cambridge University Press. © Cambridge University Press 2006.

17

or re-establishing wild populations (e.g. golden lion tamarin, scimitar horned oryx, Florida panther, black-footed ferret, red wolf, California condor, among others). But having an *in situ* benefit is not an essential prerequisite to studying a species because the primary target is always the production of new knowledge – intellectual capital that improves our understanding of the wonders of biology and the natural world.

University- and zoo-based science also differ because zoo scientists still have the opportunity, if not the mandate, to work with whole living animals. By contrast, university investigators are relying more and more on cellular and molecular technologies, largely because knowledge in common species is so advanced that the next logical step is investigating inside cells and genomes. And while these approaches generate valuable information, there is something lost by not working with a whole living animal – a definite advantage to investigators intrigued with zoo-held species.

A third distinctive trait of zoo- and conservation-based research is the importance of an integrative, multidisciplinary approach (Wildt *et al.*, 2003; Wildt, 2004). Research in the biological sciences historically has been rather unidimensional. While in training, a scientist learns a set of scientific skills and then throughout a career applies those tools to test hypotheses in one or several species. While the approach is classic, it limits the value of the resulting knowledge for wildlife management and conservation purposes, the applied targets of good basic research. The successful breeding, management and preservation of any species *ex situ* (as well as *in situ*) are enormously complex and not easily achieved using any single discipline (e.g. behaviour versus nutrition versus veterinary medicine). Focus on one discipline and you surely can (and will) miss some other cause-and-effect. Rather, the task of achieving high-quality animal care management can only be achieved by collecting and linking data from many biological angles, relying on a multidisciplinary strategy that exploits various scientific fields simultaneously in ways to address problems most effectively. The paradigm then becomes adding the knowledge of parallel studies in behaviour *plus* nutrition *plus* veterinary medicine (as well as the other life sciences) rather than being singularly focused.

This chapter addresses the value and need for integrative approaches in wildlife science, including sharing one example (experience with the cheetah) that was used to develop a foundation for the giant panda. More importantly, this chapter shares the origin of multidisciplinary, cooperative research in China through a vision initiated by Chinese managers of giant pandas and facilitated by one of the most active groups of the IUCN–World Conservation Union's Species Survival Commission – the Conservation Breeding Specialist Group (CBSG).

VALUE OF INTEGRATIVE RESEARCH APPROACHES IN ZOOS: THE CHEETAH BIOMEDICAL SURVEY AS AN EXAMPLE

While the cheetah is best known for high-speed pursuit of prey on the African plains, zoo managers best recognise the species as historically difficult to reproduce (Fig. 2.1). Although maintained in zoos for millennia, the cheetah has never been consistently propagated. An early review suggested that only about 15% of all wild-caught cheetahs have ever reproduced in captivity, with an even lower success rate for zoo-born individuals (Marker & O'Brien, 1989). By the late 1980s, these failures were being openly discussed by zoo managers who were distressed because it appeared that the only way of sustaining a captive cheetah population in North America was to continue extracting animals from Africa.

Figure 2.1. The cheetah, the first species studied through a coordinated Biomedical Survey and a model for the later giant panda survey (photograph reproduced with permission from Alexandra von Knorring).

In 1989, zoo managers involved in the North America Cheetah Species Survival Plan (SSP, a consortium under the umbrella of the American Zoo and Aquarium Association, AZA) came to a rather radical conclusion: the cheetah population held in AZA-accredited zoos should be designated for research, to be studied to understand how to better manage and reproduce the species in captivity. The first step was to understand the well-being and physiological status of the existing North American cheetah population. Because empirical causes of poor reproductive efficiency were unknown, it was logical to establish first the reproductive and health status of the extant population, information that would be critical for making subsequent research and management decisions. In a crucial meeting, the managers agreed to evaluate every available cheetah to learn as much as possible about the health and reproductive status of the existing population. Thus was born the first large-scale, *ex situ* Biomedical Survey for an endangered species.

More than 125 cheetahs distributed across 18 institutions throughout the continental USA were offered by zoos for examination. There was a need for a specialist team that could commit substantial time (sometimes consecutive weeks) to travel, collect and interpret data at cheetah-holding facilities. These scientists were required to create standardised protocols and mobile laboratory methods that allowed data to be gathered in a single fashion. Only by using consistent methods would it be possible to compare findings among institutions and across age groups of animals. The scientific team also had to excel at communication, making sure their hosts were comfortable with plans and procedures that would involve risky manipulations of valuable specimens. The survey would only be successful with the full cooperation of the partners – more than 100 scientists, curators, animal care and veterinary personnel who would be required to dedicate serious time to the project while simultaneously ensuring animal (and people) safety. And, of course, there were substantial costs associated with travel, the shipping of mobile laboratory equipment, the actual animal evaluations and subsequent data analysis.

The dedication of cheetah aficionados made this first interdisciplinary Biomedical Survey a success (Wildt *et al.*, 1993). Thousands of dollars were generated by financial donations to the SSP and in-kind support from local institutions. One hundred and twenty-eight cheetahs (60 males and 68 females) were evaluated, of which only 21% had ever reproduced.

Curators and animal-care staff were interviewed on topics ranging from husbandry practices to each animal's unique behavioural predilections and health conditions. Each animal was then anaesthetised for collecting morphometric and biomedical data. The results were no mortalities and no adverse, post-survey events, despite using rigorous manipulatory protocols which included serial blood sampling, semen collection in all males and laparoscopy of every female. Fourteen animals reproduced within less than a year after the Survey. Most striking was a 6-year-old male and 3-year-old female which had never mated or produced young. Within 17 days of evaluation these cheetahs mated with each other, and the female produced seven cubs. In another case, the team laparoscopically examined a female, which unbeknown to the host zoo staff was pregnant. Despite anaesthesia and the surgical manipulations, this female delivered three live cubs 66 days later.

Most importantly, this multidisciplinary, multi-institutional survey helped to sort out which biological and management factors needed further study and which required no further attention. For example, there were consistent abnormalities such as a very high incidence of ovarian inactivity – half of the adult female cheetahs were not reproductively active. Such findings provided a blueprint for designing later research projects. In the case of quiescent cheetah ovaries, this finding was incentive for Brown et al. (1996) to develop faecal hormone monitoring techniques that eventually proved that most female cheetahs were only sporadically reproductively active. This discovery, in turn, motivated a study by Wielebnowski et al. (2002) which linked a shutdown in ovarian activity to reproductive suppression due to inadequate management. Without the original Biomedical Survey there would have been no guideposts for Brown and Weilebnowski to develop their experimental designs.

This is only one example of the many subsequent research and management actions that emerged from the Cheetah Biomedical Survey, all of which involved one or more related disciplines – infectious disease, pathology, nutrition, behaviour, genetics and reproduction, efforts that continue to the present day (see review in Wildt et al., 2001). The result is that our collective wisdom about the life science of the cheetah is more extensive than for any other rare wildlife species. The Biomedical Survey was the initial driving force, but was spurred on by the satisfaction of people working together across diverse disciplines – both scientists and managers – and institutions. There was something special and exciting about combining forces to solve a management/

conservation challenge. The Cheetah Biomedical Survey demonstrated, for the first time, that it could be feasible and practical to conduct a continent-wide, highly manipulatory survey of an endangered species. As you will see, lessons learned from the cheetah were important in creating a plan of action to assist the equally enigmatic giant panda.

THE MANAGEMENT OF GIANT PANDAS IN CHINA AND THE SITUATION IN 1996

Giant pandas in China are managed under the authority of two federal agencies. One is the Ministry of Construction (MoC), which regulates giant panda activities in Chinese zoos under the umbrella of its Chinese Association of Zoological Gardens (CAZG). The CAZG (based in Beijing) has the enormous task of monitoring more than 80 zoos scattered throughout China. In 1996, there were 104 giant pandas distributed across 27 zoological parks throughout China, with the majority held in two facilities, the Chengdu Research Base of Giant Panda Breeding (and its allied Chengdu Zoo) and Beijing Zoo. The other management authority is the State Forestry Administration (SFA), which is responsible for all giant pandas living in nature plus a major facility (China Conservation and Research Centre for the Giant Panda) in the famous Wolong Nature Reserve (Sichuan Province) located about 130 km northwest of Chengdu. In 1996, this facility held 29 pandas.

By the mid 1990s, there were three large and serious breeding programmes for giant pandas:

1. China Conservation and Research Centre for the Giant Panda (within the Wolong Reserve).
2. Chengdu Research Base of Giant Panda Breeding (on the outskirts of Chengdu City).
3. Beijing Zoo (in the heart of the national capital).

These efforts were complemented by a few zoos (for example in Chongqing, Fuzhou and Shanghai) which maintained a handful of pandas that occasionally reproduced. Lastly, there were approximately 23 zoos that held only one to three individuals, with pairs rarely producing offspring.

Historically, communication among the giant panda stakeholders was less than ideal with hesitancy to share information mostly related to competition for resources, specifically funding. The giant panda is a national treasure for China, so there is naturally much prestige and favour associated with the ability to be the best at managing captive populations. Nonetheless, the giant panda world in China – in terms of total people involved – is relatively small. Regardless of home institution or federal affiliation, the various captive breeding managers and staff generally knew and respected one another despite having few opportunities to meet. One exception was what was called the Annual Technical Meeting for Giant Pandas, which was sponsored by the CAZG and generally held in Chengdu. This three-day event was usually designed as a 'reporting meeting', an opportunity for giant panda managers to share information from the previous breeding season and hopefully boast about the number of cubs produced at their institutions.

However, by 1995 there was concern that little overall progress was being made. While indeed some individual giant pandas were reproducing, most were not. There were reports of mortalities and illness. Occasionally, giant pandas were taken from the wild to be incorporated into breeding programmes – 18 individuals from 1991 through 1996. These challenges, although unsettling, were motivation for action. The key instigator was Shuling Zheng, then Vice-Director of the Department of Urban Construction of MoC. Madam Zheng directed her staff within the CAZG to put in place a masterplan for captive giant pandas. It was her vision that prompted the beginning of an exciting era in zoo research and management in China.

ORIGINS OF ACTION AND THE ROLE OF CBSG

When it came to identifying an organisation that could assist Chinese zoos and breeding centres with the contemporary challenges in *ex situ* panda management, the natural choice for CAZG was the Conservation Breeding Specialist Group, better known as CBSG. This small organisation, with only six permanent staff in 1996, is based in Apple Valley, MN, at the Minnesota Zoo. CBSG's organizational parent is the prestigious IUCN–World Conservation Union of Gland, Switzerland. Through the IUCN's Species Survival Commission, CBSG stormed into the conservation world under the dynamic leadership of Dr Ulysses S. Seal (Fig. 2.2). Better known as Ulie (from zoo directors to animal keepers alike), Seal had worked tirelessly since 1980 to build CBSG into a force to benefit conservation. His arrival on the scene was perfectly timed as it coincided with a rapidly emerging attitude within the zoological community at large – zoos needed to be more than simply amusement parks for the public – they must contribute to conservation.

Figure 2.2. Dr U. S. Seal, Chairman of the Conservation Breeding Specialist Group of the IUCN–World Conservation Union's Species Survival Commission.

Ulie Seal was trained as a biomedical scientist and for decades had conducted human and animal-related research at the Veterans Administration Hospital in Minneapolis. But his real interest was always in zoos, wildlife and how people could make a difference in preserving and better managing species. Of his many strengths, Seal was best known for his uncompromising philosophy that people can, and should, work together to solve problems. He was relentless in his dedication to breaking down political territorialities and jealousies, which are rife in the wildlife world and, on occasion, have contributed to the demise of entire species. During the 1980s, CBSG had become involved in some of the most difficult recovery programmes ever undertaken for endangered wildlife, including the Florida panther, black-footed ferret and Sumatran tiger, among others. Seal was a master of provoking action by bringing people together, even diverse characters with strong distastes for one another. His dynamism, folksy mannerisms, good humour and cajoling ability to convert even the most cantankerous person into an obliging partner were legendary. Most of all, CBSG developed a reputation as being 'agenda-less' – a neutral organisation with its roots in strong science that could enter any difficult fray to generate positive energy – and action.

CBSG was not new to China. In 1993, along with the IUCN Cetacean Specialist Group, it had been invited by the MoC to conduct a Population and Habitat Viability Assessment (a risk-evaluation workshop) for the baiji or Yangtze river dolphin. Despite this species being on the edge of extinction, the Chinese were impressed with Ulie Seal's objective, respectful and dedicated attempts to find potential options for saving this critically endangered species. This effort was followed by a 1995 invitation to develop a Captive Breeding Management Plan for the South China tiger. Again, this was no small task since this unique tiger subspecies is extremely rare in zoos (fewer than 65 individuals total), and the population is poor at reproduction, probably due to inbreeding depression. Again, the Chinese appreciated CBSG's equitable and proactive approach.

The ability to organise effective workshops, fairness in ensuring that all stakeholders were 'heard' and direct assistance in producing written documents (guidebooks for the future to help each species) solidified CBSG's credibility in China. As a result, the MoC issued an invitation to CBSG in early 1996 to assist in developing a 'masterplan for captive giant pandas'. More specifically, the invitation called for help in developing a 'scientifically based management programme that would result in a healthy, growing population of giant pandas in China'. How this eventually could emerge was left to the devices of CBSG.

Ulie Seal immediately responded to China's invitation, but added that CBSG's participation came with a condition – that the visiting team be comprised only of specialists from institutions that were not interested in loans of giant pandas. This caveat was related to Seal's insistence that the process be neutral. He suggested that a workshop of four days be held at a time and location most convenient to the Chinese. Seal at once formulated his specialist team which included Phil Miller (population biologist on the CBSG staff), Jill Mellen (a specialist in animal behaviour and enrichment from the Washington Park Zoo in Portland), Lyndsay Phillips (a veterinarian on the faculty of the University of California-Davis) and David Wildt (a reproductive biologist from the Smithsonian's National Zoological Park).

Beginning on 10 December 1996 the CBSG team met their 30 Chinese colleagues in a cold and drafty conference room in a public park building in downtown Chengdu. The only other foreigner present was Don Lindburg, an animal behaviourist from the Zoological Society of San Diego. The San Diego Zoo had recently finalised a giant panda loan agreement with the SFA and, thus, already was deeply involved in *ex situ* studies in China. Because he was a guest of the Chinese as well as a renowned scientist, Seal agreed that Lindburg's presence did not compromise CBSG's impartiality. Meanwhile, the Annual Technical Meeting for Giant Pandas was being held in parallel at another Chengdu site with additional Americans attending, but not allowed to attend the CBSG workshop (Fig. 2.3).

A CBSG workshop is a loosely structured process with minimal formality and maximal opportunities for open discussion, sometimes in plenary but most often in small working groups of three to six people. Every significant point is recorded on flip chart paper and/or in laptop computers to ensure that everyone is 'heard' and that there is a permanent record of progress. In the case of this first giant panda workshop, there also was the challenge of language differences. There was only one Chinese delegate, a young biologist by the name of Wei Zhong, who spoke fluent English; none of the Americans spoke a word of Chinese. A professional translator had been hired but she spent much of her time stymied by the foreign (to her) biological terms being tossed about by the CBSG team and the Chinese delegation. No doubt many of the participants were (at least internally) beginning to panic during the early hours of the workshop as everyone struggled to communicate while wondering what would emerge from this cross-cultural experience.

Ulie Seal persevered though, as he had done so many times before. He stood before all the participants on the afternoon of day one and, with young Wei Zhong translating, took control. After complimenting the Chinese on their advances in captive breeding and their foresight in developing a plan for the future, he declared in his booming Georgia drawl that 'Y'all are a black hole for giant pandas.' The American team was stunned, and poor Wei Zhong, who had been so proudly standing next to the famous Professor Seal, looked dumbfounded about how to translate such potentially inflammatory words. So as not to miss this one opportunity to get everyone's attention, Seal looked directly at the young Chinese biologist, repeated his words and politely asked him to translate. A long stream of Chinese words emerged from Zhong while all the nervous Americans intently watched the faces of their Chinese colleagues. Although we will never know for sure what Zhong said, it was heartening that gradually

Figure 2.3. Participants who attended the masterplanning meeting for captive giant pandas in Chengdu in 1996 plus attendees of the parallel Annual Technical Meeting for Giant Pandas.

throughout the room we saw the nodding up and down of heads in agreement.

Seal then ask the delegates the question: what is your reason for having giant pandas in zoos in China? What is your goal? This provoked more than 15 minutes of frenetic discussion, all of which, of course, was a mystery to the American contingent. Then, Mr Anju Zhang, the most senior scientist at the workshop and a respected authority on giant pandas, gave the group's consensus answer:

> The goal is to develop a self-sustaining population of giant pandas that will assist supporting a long-term, viable population in the wild.

This statement could not have been more appropriate or profound for the situation. It reflected two major points. First, it was a public declaration that the *ex situ* giant panda population was not self-perpetuating, and that it was a mistake to continue to support zoo breeding programmes by removing more pandas from nature. Second, that there was a need to articulate clearly the value of these special animals in captivity and that their presence needed to contribute somehow to conserving giant pandas in nature.

The setting of a goal clearly provided the guideposts for the remainder of the workshop. The participants were asked to use their knowledge and experience to generate a list of their concerns which might prohibit them from reaching their stated goal. A long list of issues emerged, which was condensed into three categories:

1. demographics and history of the current population;
2. reproduction, behaviour and management;
3. mortality, veterinary issues and nutrition.

Each of these was then addressed in small working groups which were facilitated by the CBSG representatives. While the many details emanating from these discussions are beyond the scope of this book (see Zheng *et al.*, 1997 for details), a few highlights will allow the reader to understand how the participants eventually reached the conclusion for a much-needed Biomedical Survey.

Highlights from the demographics working group

The Giant Panda Studbook (containing information on every individual dating back to Su Lin, the original animal imported to the USA in 1936) was used with computer simulations to better understand population status. Laptop modelling revealed that a self-sustaining captive giant panda population, with no augmentation from the wild, could increase at 5 to 6% per year and, therefore, *theoretically double within only 12 to 14 years*. However, this would depend on identifying and then resolving all the limitations to reproductive success. The records also showed that the annual growth rate of the captive population had fluctuated widely. There was a burst of growth from 1984 to 1986, a flat, no-growth period from 1987 to 1990, another spurt in growth from 1991 to 1995, followed by no net growth in 1996. More detailed examination revealed additional findings, one being that some of the growth was not due to births but rather animals being captured from the wild. Second, regardless of these extractions, reproductive success was inconsistent with most young being produced by only a few animals. In theory, breeding-age females (generally 6 to 20 years old) can produce, on average, two litters every three years or about 10 litters in a lifetime. But by 1996 many eligible females (almost 65%), including founders (previously captured from the wild), had produced no offspring. There were two results, the first being that about 10 of the females were responsible for half of all births and, second, valuable genes were being lost from founders who never produced offspring.

The situation was even more ominous for males. In 1996, there were 33 males (6 to 26 years of age and presumably able to produce sperm) in captivity in China. Only five (15.2%) had descendants with a whopping 22 animals being valuable (wild-born) founders that had never reproduced.

In summary, this working group acknowledged that there was a relatively high retention of gene diversity within the living captive population. But there were many animals that were unrepresented due to poor reproduction. Thus the highest priority was to identify the reason(s) for both male and female propagation failure so that these genetically valuable pandas could be reproductively recruited into the population. The group concluded that, if this could be achieved, then there was more than adequate gene diversity in the founder stock to meet a programme goal of retaining 95% of existing genetic variation for the next 100 years (a common target in zoo breeding programmes). And there would never be the need to remove another giant panda from the wild.

Highlights from the reproduction, behaviour and management working group

This group was comprised of a large contingent of managers who had a wealth of information on giant pandas, most of which was in their heads, having never been written down. Again, common problems among institutions surfaced, especially personal experiences in failed reproduction for both male and female pandas. Frequent comments referred to animals that were too aggressive, too meek or completely lacking interest in sex, frequent dystocia (difficult birth) and spontaneous abortion. In an attempt to determine if there was any common factor across the diverse organisations, a large paper matrix was created on one of the walls of the meeting room and the group recorded their experiences on the basis of such factors as age of weaning and later reproductive success, ability to reproduce in captive-born versus wild-caught individuals and the influence of differing diets or enclosure enrichment schemes. Nothing significant emerged from this rough and nonsystematic analysis. Nonetheless, a pattern of results was being revealed.

This working group concluded that, even if natural reproduction could be improved, there remained a need to develop assisted breeding procedures. Reproductive technologies normally span a wide range of procedures from straightforward AI to controversial cloning. Although the Chinese expressed interest in the potential of embryo transfer for more rapidly increasing offspring production, the group did not get carried away with high technology. The predominant needs involved developing techniques for monitoring reproductive status by measuring hormones in urine or faeces and improving artificial insemination with fresh and frozen sperm. The latter also stimulated a recommendation for eventually creating a genome resource bank, a frozen repository of giant panda sperm that could be used to move genetic material from one institution to another.

Highlights of the mortality, veterinary and nutrition working group

Participants interested in nutrition rapidly found common areas of concern, especially being suspicious that a suboptimal diet for captive-held giant pandas contributed to poor growth, health and reproduction.

Each of the holding institutions relied on widely variant feeding pro-grammes with no standardised protocol to address the question: what is the best dietary protocol for zoo-held giant pandas? This working group also believed that disease was a major threat to the captive population, especially for infants and subadults five years or less in age. Although there were few quantitative data, the diseases or conditions listed as occurring in giant pandas in Chinese institutions were chronic gastro-intestinal distress, haemorrhagic enteritis, epilepsy, infectious viruses (canine distemper and parvovirus) and demodectic mange.

No other working group emphasised more the importance of building capacity to assist in ensuring the health and reproduction of giant pandas in the future. These participants recognised a need for significant amounts of information sharing, especially in diagnosing and treating risky health conditions. They realised that much of their information was based on anecdotal evidence, in part because of lack of specialised training, inadequate equipment, no computerised record-keeping and too little reliance on the field of pathology to understand the root causes of many medical problems.

REACHING CONSENSUS ON NEXT STEPS FOR GIANT PANDAS IN CAPTIVITY

Every Chinese and American participant had spent four intensive days in this Chengdu meeting room hearing the many problems encoun-tered in day-to-day (and year-to-year) management of giant pandas *ex situ* – sexual incompatibilities, behavioural aggression, questionable nutrition, too few breeding males, females that mated but failed to produce offspring, neonatal deaths, poor growing animals, health issues in all age classes, pandas with unknown paternity, inadequate infrastructure (exhibits, enclosures, veterinary hospitals and research laboratories) and the need for more technology transfer, from veterin-ary medicine to nutrition to reproductive monitoring and the use of assisted breeding.

Despite the workshop's intensive efforts, the overall task of get-ting this animal population organised now seemed almost overwhelm-ing. It was beginning to settle upon everyone (Chinese and CBSG included) that it would be impossible to solve any of these challenges immediately. If anything, the picture was more complicated because there was an entire menu of potential culprits contributing to the nonviability of the *ex situ* giant panda population in China.

As the workshop began to wind down, two conclusions were drawn. First, it would be impossible to develop any sort of genetic management plan in the near future. But the good news was that, despite the long list of concerns articulated by the participants, there were 104 giant pandas in captivity in China in 1996, a population that held significant amounts of genetic diversity. This was a huge advantage as there were plenty of animals to conduct the necessary studies to solve the problems, and without taking more animals from the wild. The second conclusion emanated from the first – before being genetically managed, the population had to be understood, to be studied intensively by collecting new biological information, especially pertaining to health and reproductive status. The target was identifying the factor(s) that were limiting reproductive success followed, of course, by implementing remediation.

CBSG presented an idea to the Chinese participants. A Biomedical Survey approach had been used effectively across a diverse array of American zoos to identify factors that were contributing to reproductive inefficiency and health concerns in the cheetah. Could this strategy be considered for the giant panda? If so, then it would be necessary to form an international partnership involving multidisciplinary teams that, in turn, would require Chinese holding institutions to also work together. Everyone, sensing a huge opportunity, immediately embraced the idea. The only Chinese caveat was an insistence that CBSG be willing to serve as the organising institution; its neutrality was seen as a prerequisite to success.

The Chengdu workshop ended on a high note – the participants had identified the magnitude of the challenge, concluded that it was exponentially bigger than any single person or institution and had decided unanimously that the next step would depend on a partnership that would scientifically target the factors limiting captive breeding success.

IMPLEMENTING A BIOMEDICAL SURVEY PLAN

Concluding the need for a systematic assessment of as many giant pandas as possible was simple compared to all the other steps required to implement such a bold plan. The potential impediments were substantial, perhaps the most uncertain being our ability to secure appropriate official approvals. Recall that the 1996 workshop was sanctioned and sponsored by MoC with no SFA participation. Thus, more than 25% of the giant panda population (mostly housed at the breeding centre in Wolong) could perhaps be excluded from the Survey due to federal agency boundaries. There were also, of course, the issues of identifying leaders, teams, funding, study times and locations.

Because of his extensive involvement in the Cheetah Biomedical Survey, David Wildt was invited by Ulie Seal to lead the eventual visiting team to China. Wildt, in turn, invited Susie Ellis, then the Senior Program Officer for CBSG, to co-lead the operation. They travelled to China as representatives of CBSG in September 1997 to negotiate a final implementation plan with Shuling Zheng and the CAZG staff. During this meeting, it was decided that the Survey would commence in February 1998 (at breeding season onset so that data were collected from animals that hopefully were approaching their physiological peak). CAZG would coordinate negotiations with holding zoos, and it was decided that most emphasis would be placed on institutions holding significant numbers of giant pandas. For this first visit, that meant that the target institutions would be the Chengdu Research Base of Giant Panda Breeding, nearby Chengdu Zoo, Chongqing Zoo and Beijing Zoo. There was agreement that as much data as possible would be collected from every animal, including proven breeders as well as individuals that were ill. There was also consensus that animals would be permanently marked via both tattoos and electronic (chip) transponders implanted under the skin. CBSG agreed to be responsible for assembling the team with the appropriate expertise and securing the funds for equipment purchase, travel and daily allowances.

IMPLEMENTING THE BIOMEDICAL SURVEY

As will become apparent in Chapter 3, the 1998 Biomedical Survey was successful far beyond everyone's expectations, allowing it to continue into 1999 and 2000 and resulting in data from 61 giant pandas. This included extensive cooperation from the China Conservation and Research Centre for the Giant Panda (of the Wolong Nature Reserve under the auspices of SFA). By 1998 and, in part, because of its already existing relationship with the Zoological Society of San Diego, the centre at Wolong had generously invited CBSG to include its giant pandas in the Survey.

During the three-year course of study, we adhered to six rules that may be useful for others interested in integrating scientific disciplines in a cross-cultural fashion to address species-based problems.

1. *Be diverse in choosing team members.* It would have been feasible to assemble a US-based team originating solely from a single institution. We chose an opposite approach that, although causing more logistical challenges, produced a compatible, eclectic team with varied expertise.

2. *Be respectful of the host institution.* The animals are the responsibility of the home institution and its director, curator and/or veterinarian. These individuals have the ultimate responsibility for making all final decisions, and visitors must remain deferential.

3. *Write and sign a memorandum of understanding (MoU) with each host institution.* To avoid every possible misunderstanding, it is wise to develop a MoU to ensure that everyone is comfortable with the approach to be used and expectations. These written agreements are a convention in Chinese culture (see Chapter 3 for an example).

4. *Develop methods for learning from each other and across cultures.* A 'missionary' approach will fail every time when working in a foreign culture. Our Biomedical Survey strategy was to work hand in hand across the two cultures. Progress was substantial because both teams learned to work side by side and to learn from each other.

5. *Conduct all evaluations in-country with minimal export of biomaterials.* There are growing issues worldwide with proprietary ownership of materials and data. From the onset, we agreed to avoid any controversy by insisting that all biomaterials be left and analysed within China. In some cases, this caused complexity, including the need to develop a molecular biology laboratory and provide intensive training (i.e. for sorting out giant panda paternity; see Chapter 10). Yet adhering to this rule helped build capacity within China (Chapter 22).

6. *Share all results verbally and in writing immediately and identify the next step(s).* From a home institution perspective, nothing must be quite so frustrating as experiencing an exciting opportunity to collect data from one of the world's rarest species and then be left without a written or verbal report by the visiting scientific team.

At the end of each Survey visit, at least half a day was reserved for questions, case discussions and identifying next steps for follow-up. Each institution also received a written report and a PowerPoint presentation on all findings from each evaluated individual.

On the surface, these guidelines appear to be common sense. However, too often they are easily ignored. We believe they were the foundation for our success.

PRIORITIES FOR THE FUTURE

There are hundreds of species worldwide that could benefit from a thoughtful, systematic approach to health, reproduction and management challenges faced in an *ex situ* environment. Although we promote zoos as safe animal havens and as resources for science, we actually do amazingly few experimental investigations to enhance species health and reproduction.

In a way, the giant panda (and the cheetah for that matter) was 'low-hanging fruit' – it was reasonably easy to find interest and funding to elicit the actions described in this and subsequent chapters. The challenge for the future is creating strategies and evoking passion for less charismatic species. Certainly, the giant panda and the cheetah are not the only recalcitrant species maintained in the world's zoos. The main lesson from this chapter is that there are now positive examples of people working together, setting aside personal agendas and dedicating their expertise, time and resources to formulate action plans by consensus, including across cultures. The following chapters will illustrate how the actual science was conducted and how it has contributed to enhanced knowledge and improved giant panda health, reproduction and management.

ACKNOWLEDGEMENTS

Events described in this chapter would not have occurred without the vision of Shuling Zheng and the wisdom, leadership and encouragement of Ulysses Seal. The sponsors of the 1996 workshop in Chengdu deserve special recognition: the Chinese Association of Zoological Gardens; Ministry of Construction; the Columbus Zoological Gardens; and the American Zoo and Aquarium Association Giant Panda Program.

REFERENCES

Brown, J. L., Wildt, D. E., Wielebnowski, N. *et al.* (1996). Reproductive activity in captive female cheetahs (*Acinonyx jubatus*) assessed by faecal steroids. *Journal of Reproduction and Fertility*, **106**, 337–46.

Marker, L. and O'Brien, S. J. (1989). Captive breeding of the cheetah (*Acinonyx jubatus*) in North American zoos (1871–1986). *Zoo Biology*, **8**, 3–16.

Wielebnowski, N. C., Ziegler, K., Wildt, D. E., Lukas, J. and Brown, J. L. (2002). Impact of social management on reproduction, adrenal and behavioral activity in the cheetah (*Acinonyx jubatus*). *Animal Conservation*, **5**, 291–301.

Wildt, D. E. (2004). More meaningful wildlife research by prioritizing science, linking disciplines and building capacity. In *Experimental Approaches to Conservation Biology*, ed. M. Gordon and S. M. Bartol. Davis, CA: University of California Press, pp. 282–97.

Wildt, D. E., Brown, J. L., Bush, M., *et al.* (1993). Reproductive status of cheetahs (*Acinonyx jubatus*) in North American zoos: the benefits of physiological surveys for strategic planning. *Zoo Biology*, **12**, 45–80.

Wildt, D. E., Ellis, S. and Howard, J. G. (2001). Linkage of reproductive sciences: from 'quick fix' to 'integrated' conservation. In *Advances in Reproduction in Dogs, Cats and Exotic Carnivores*, ed. P. W. Concannon, G. C. W. England, W. Farstad *et al.* Colchester, Essex: Journals of Reproduction & Fertility Ltd, pp. 295–307.

Wildt, D. E., Ellis, S., Janssen, D. and Buff, J. (2003). Toward more effective reproductive science for conservation. In *Reproductive Sciences and Integrated Conservation*, ed. W. V. Holt, A. R. Pickard, J. C. Rodger and D. E. Wildt. Cambridge: Cambridge University Press, pp. 2–23.

Zheng, S., Zhao, Q., Xie, Z., Wildt, D. E. and Seal, U. S. (1997). *Report of the Giant Panda Captive Management Planning Workshop*. Apple Valley, MN: IUCN–World Conservation Union/SSC Conservation Breeding Specialist Group.

3

Factors limiting reproductive success in the giant panda as revealed by a Biomedical Survey

SUSIE ELLIS, DONALD L. JANSSEN, MARK S. EDWARDS, JOGAYLE HOWARD,
GUANGXIN HE, JIANQIU YU, GUIQUAN ZHANG, RONGPING WEI, R. ERIC MILLER,
DAVID E. WILDT

INTRODUCTION

There is surprisingly little published information about giant panda biology, especially in the life sciences. This poor quantity (and quality) of data has been due primarily to too few individual animals available for study and a traditional hands-off policy towards hands-on research in such a rare and high-profile species. However, recent changes (see Chapter 2) have created important, new opportunities for giant panda investigations. People responsible for ensuring that the species survives now realise that giant pandas living in zoos and breeding centres are a valuable research resource (see Chapter 1). It also has been recognised that this population must be intensively managed if it is truly to support giant pandas that are surviving precariously in nature. The intended result will be an ever-increasing amount of new, scholarly information *and* sufficient panda numbers to continue educating the public, helping to raise conservation funding, serving as a hedge against extinction, and even as a source of animals for potential re-introductions. However, these laudable goals can only be achieved by

Giant Pandas: Biology, Veterinary Medicine and Management, ed. David E. Wildt, Anju Zhang, Hemin Zhang, Donald L. Janssen and Susie Ellis. Published by Cambridge University Press. © Cambridge University Press 2006.

first understanding and then rigorously managing the captive population so that it becomes demographically and genetically stable. This, in fact, has become the mantra of Chinese managers of the *ex situ* population: 'to develop a self-sustaining, captive population of giant pandas that will assist supporting a long-term, viable population in the wild' (see Chapter 2).

When this goal was articulated in 1996 (Zheng *et al.*, 1997) it was realised that it would be impossible to achieve without first learning what was prohibiting consistent reproduction. This prerequisite involved using a consistent array of modern interdisciplinary techniques to conduct a thorough examination of as many giant pandas as possible. It was believed that this multidimensional approach would allow the quick identification of limitations that, in turn, could be resolved. As each issue was addressed, the hope was that the population would become healthier and more reproductively fit, eventually becoming self-sustaining.

This chapter summarises the findings from examining 61 giant pandas over the three-year Biomedical Survey conducted at four institutions (Chengdu Research Base of Giant Panda Breeding and its associated Chengdu Zoo, Beijing Zoo, Chongqing Zoo and the China Conservation and Research Centre for the Giant Panda in the Wolong Nature Reserve). General methods and results are presented to lay a foundation for later chapters which deal with more specific and, in some cases, remediation efforts.

BIOMEDICAL SURVEY EXPERTISE AND GENERAL METHODS

The Biomedical Survey was successful because of three key elements: effective communication; multi-skilled participants; and adequate funding. Constant translation was provided by two team members who were fluent in both languages, had lived in both cultures and were comfortable with biological terms and laboratory/animal procedures – these people probably were the most valuable of all the participants. Diversity in skills was achieved by selecting team members representing a wide range of scientific disciplines, including veterinary medicine, animal behaviour, reproductive physiology, genetics, nutrition and pathology. Each person had also demonstrated a history of successfully working in teams. Having people from different home organisations (Box 3.1) enriched the team with assorted philosophies while increasing overall problem-solving capacity. Numerous donors (Box 3.2) ensured

Box 3.1. Participating institutions and investigators

Beijing Zoo
Jinguo Zhang
Zheng Xin Peng
Cheng Lin Zhang
Shi Quang Huang
Yi Luo
Yan Lu
Ming Hai Yang
Tian Chun Pu
Yan Ping Lu
Wanmin Wang
Jiang Jun Peng
Fei Bing Zhu
Qiming Hou

**Chengdu Research
Base of Giant Panda
Breeding and
Chengdu Zoo**
Anju Zhang
Yuezhong Tian
Guangxin He
Guanghan Li
Yunfang Song
Jianqiu Yu
Zhiyong Ye
Qiang Wang
Zhihe Zhang
Zi Yang
Shunlong Zhong
Hongwei Chen
Xuebing Li
Mingxi Li
Xiangming Huang
Jingchao Lan
Meijia Zhang
Shurong Yu
Jishan Wang
Rong Hou
Wei Zhong

**China Conservation and
Research Centre for the
Giant Panda, Wolong
Nature Reserve**
Hemin Zhang
Peng Yan Wang
Guiquan Zhang
Chun Xiang Tang
Quan Chen Li
Jian Yiang
Ping Tan Xian
Yan Wang Pong
Rongping Wei
Yan Huang
Desheng Li
Jun Du

**Chinese Association of
Zoological Gardens**
Zhong Xie
Menghu Wang

Chongqing Zoo
Ximu Zhou
Youxin Xie
Wei Guo
Xiancum Zheng
Aiping Wang
Denfu Wu

Columbus Zoo
Ray Wack

**Conservation Breeding
Specialist Group, IUCN/SSC**
Susie Ellis (team leader)

St Louis Zoo
R. Eric Miller

**Smithsonian's
National
Zoological Park**
David Wildt (team
    leader)
JoGayle Howard
Richard Montali
Rebecca Spindler

**University of
California at
Davis**
Lyndsay Philips

Zoo Atlanta
Rebecca Snyder

**Zoological Society
of San Diego**
Barbara Durrant
Mark Edwards
Donald Janssen
Arlene Kumamoto
Mabel Lam
Mary Ann Olson
Bruce Rideout
Sandra Skrobot
Meg Sutherland-
    Smith
Jeff Turnage
Lee Young

Box 3.2. Donors to the Biomedical Survey

Air-Gas, Inc.
British Airways
Columbus Zoo
Giant Panda Conservation Foundation, American Zoo and Aquarium
    Association
Heska
InfoPet Identification Systems
Smithsonian's National Zoological Park
Nellcor Puritan Bennett
Ohaus
Olympus America, Inc.
St Louis Zoo
Sensory Devices, Inc.
Zoo Atlanta
Zoological Society of San Diego

---

the final element – funding to support the Survey, including monies needed for travel, daily allowances, equipment, supplies and data analysis/distribution.

The Survey was conducted in February and March (the onset of the giant panda breeding season) in each of the years 1998, 1999 and 2000. During the first year of collaborating within a given institution, a memorandum of understanding (MoU) was signed between each Chinese breeding centre or zoo and the CBSG team. This is a common practice in China, which greatly facilitates understanding mutual expectations while avoiding misunderstandings (see the sample MoU Appendix 3.A).

Biomedical Survey methods were consistent throughout all three years to allow data comparisons over time. Anaesthesia was induced in all giant pandas through the intramuscular (i.m.) administration of ketamine hydrochloride (see Chapter 4). The primary advantage of this anaesthetic was ease of administration (usually via blow dart or pole syringe) and rapid onset and recovery. Once each animal was tractable, the following procedures were performed:

(1)   Using a needle inserter, an electronic transponder chip (Trovan, Eidap, Inc., Sherwood Park, Alberta, Canada) was inserted

subcutaneously in the interscapular area at the dorsal midline at the cranial aspect of the black and white hair interface of each animal (Fig. 3.1). A Trovan transponder reader confirmed that each chip was working, and that the correct number was recorded.

(2)   Each animal was tattooed in the mucosa of the upper left lip with its studbook (SB) number (Fig. 3.2).

(3)   A 0.5 × 0.5 cm skin biopsy was incised from the inner thigh of a hind limb, and 20 hairs and 10 ml of heparinised blood were collected for preparing samples for genetic analysis (see Chapter 10). Each skin biopsy was minced, immersed in 1 ml of cryopreservation medium and placed in cryovials containing 1 ml of freeze medium. The medium consisted of alpha minimum essential medium (MEM, Irvine Scientific, Santa Ana, CA) supplemented with 10% fetal bovine serum (Irvine Scientific), 1% glutamine (Sigma-Aldrich, St Louis, MO) and 1% penicillin–streptomycin with 10% dimethyl sulphoxide (Sigma-Aldrich). Cryovials were frozen by being placed directly into a primed liquid nitrogen vapour shipper. The blood samples were mixed 1:1 with storage buffer consisting of 0.2 M NaCl, 0.1 M EDTA (ethylenediaminetetra-acetic acid) and 2% sodium dodecyl suphate and frozen at $-30°C$. Hair samples were stored at ambient temperature in labelled plastic bags. All biosamples were maintained at each of the respective sites until later genetic analysis in China.

(4)   Selected, healthy males were subjected to an approximately 20 minute electroejaculation procedure (see Chapter 7). Ejaculate volume was recorded, and fresh semen was evaluated for sperm motility, normal/abnormal morphology and sperm acrosomal integrity. Fresh semen was also mixed in various diluents used historically by the Chinese and USA team members (see Chapters 7 and 20). Sperm were then evaluated for the ability to retain viability or to survive cryopreservation in pellets or straw containers. Aliquots from each sample were thawed and evaluated by both teams together to determine optimal freezing and thawing methods, and to assess the ability of sperm from various males to survive a cryopreservation stress (see Chapter 7). All cryopreserved sperm samples were stored on site at the respective collection locations.

(5)   A general physical examination was performed which included body weight, body measurements, oral/dental examinations, and assessment of limbs and external genitalia; 25 to 30 ml of blood

Figure 3.1. Using Trovan injection device (arrow) (photograph by S. Ellis).

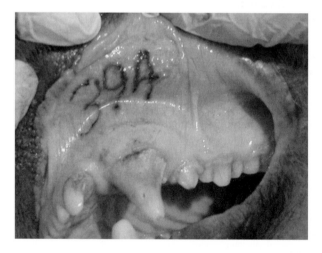

Figure 3.2. Tattooed mucosa of the inside upper lip for identification purposes (photograph by D. Janssen).

(including that used for the genetic analysis) were collected from the jugular vein (Chengdu institutions and the Wolong facility) or from the cephalic vein (Beijing Zoo) of each panda. Portions were used for haematology and serum chemistry analysis. Testes were measured and palpated for tone and consistency. A vaginal smear for cytology was prepared from each female and a faecal smear for cytology and Gram stain from many individuals of both sexes. An ultrasound examination of the abdominal cavity was conducted. During anaesthesia, each animal was monitored by noninvasive blood pressure, transcutaneous pulse oximetry, body temperature and direct observation. Some blood parameters (total number of white blood cells and percentage values for neutrophils, band neutrophils, lymphocytes, monocytes, eosinophils and basophils) were assessed onsite using a portable analyser (I-Stat, I-STAT Corporation, East Windsor, NJ) or haemocytometer. Other blood traits (e.g. sodium, potassium, chloride, pH venous, partial carbon dioxide, blood urea nitrogen and total and serum protein) were evaluated at a local human hospital (see Chapter 4). Remaining blood products were stored on site at the respective locations.

(6)   A nutrition survey form was distributed at each location and then evaluated (see Chapter 6). The focus here was on:

    a.   identifying and quantifying nutrient content of all offered food items;

    b.   evaluating offered diets, especially mass;

    c.   generating data on food intake, especially mass consumed;

    d.   measuring feeding frequency of various diet components;

    e.   characterising food distribution within enclosures and sites of food presentation;

    f.   determining seasonal variations in provided diets;

    g.   recording information on food storage and preparation facilities.

Diets were evaluated using Zoo Diet Analysis Program software (Allen & Baer Associates, Silver Spring, MD). Faecal volume and consistency traits also were described, including typical faecal output over 24 hours and mucous stool frequency, with and without blood.

(7)   Historical and behavioural data were collected, specifically origin, date of birth, health and reproductive history, past

reproductive success and opportunities to breed. Information included evaluations provided by 38 keepers on more than 20 behavioural characteristics (e.g. calmness, shyness or aggressiveness), which then were statistically analysed (see Chapter 5).

At each location, all results were discussed between the CBSG and various Chinese teams to reach a consensus on findings and interpretation. All data and resulting management recommendations were provided to all participants in electronic and hard-copy format.

SURVEY RESULTS

Early in the process, the combined CBSG and China teams agreed that each giant panda should be classified into a category that would identify its probable future value to the *ex situ* population. Four categories were defined and used.

An animal in good-to-excellent weight and health, displaying 'normal' behaviours and with a history of either successful reproduction or no reproductive deficiencies was categorised as a *Prime Breeder*.

A young, prepubertal, vigorously healthy animal of normal weight was classified as a *Potential Breeder*.

A *Questionable Breeder* was a giant panda that either was near reproductive senescence or was demonstrating a modest health problem that perhaps could be successfully treated; if resolved, the animal could potentially enter (or re-enter) the breeding population.

A giant panda in the *Poor Breeding Prospect* category was reproductively senescent or seriously ill and/or experiencing developmental problems that left no hope for future reproduction.

Twenty-four males and 37 female giant pandas were evaluated (Table 3.1). Of these, approximately 38% were Prime Breeders, 39% Potential Breeders, 11% Questionable Breeders and 8% Poor Breeding Prospects (see Table 3.1). Although most (>75%) met the criteria indicating high value to the captive management programme, at least 20% experienced one or more problems that likely prohibited reproductive success. There were two animals, both adult males, which did not fit cleanly into any of the four categories. Studbook 323 and 345 experienced unilateral testicular hypoplasia or atrophy (see Chapter 7), with the former having produced many offspring. Thus, while this animal truly met our Prime Breeder definition, at the same time he was

Table 3.1. Assigned categories of reproductive potential by institution and origin (male:female) based on the results of the three-year Biomedical Survey

Institution	Total no. animals	Origin	Total no. animals	Prime Breeder	Potential Breeder	Questionable Breeder	Poor Breeding Prospect	Not Classified
Beijing Zoo	4.5	Wild born	0.1	0.1	–	–	–	–
		Captive born	4.4	1.1	0.2	1.0	0.1	2.0
Chengdu Research Base and Zoo[a]	6.14	Wild born	2.0	2.0	–	–	–	–
		Captive born	4.14	0.5	3.6	0.1	1.2	–
Chongqing Zoo	0.3	Wild born	0.1	0.1	–	–	–	–
		Captive born	0.2	0.1	–	0.1	–	–
Wolong Centre[b]	14.15	Wild born	5.6	4.4	–	1.1	0.1	–
		Captive born	9.9	2.1	7.6	0.2	–	–
Total	24.37		24.37	9.14	10.14	2.5	1.4	2.0

[a]Chengdu Research Base for Giant Panda Breeding and the Chengdu Zoo were considered as one institution; [b]China Conservation and Research Centre for the Giant Panda

'undesirable' because this abnormality may have heritable origins or be genetically transmittable. These two males were listed in Table 3.1 as 'Not classified'.

The teams had begun the Biomedical Survey knowing intuitively (from the 1996 masterplanning discussions in Chengdu) that certain factors were adversely influencing health and reproduction of the contemporary panda population. Now, however, there were explicit data that allowed descriptive and quantitative details for each animal. The historical information that had been gathered also allowed us to determine if past events somehow influenced an animal's current status. Taken together, all data could help develop a blueprint for remediation and research action.

Of the variables examined, the most influential were assembled under six broad areas:

- unknown paternity;
- genetic over-representation by certain individuals;
- behavioural deficiencies;
- suboptimal nutrition;
- 'Stunted Development Syndrome';
- males with testicular hypoplasia (or atrophy).

UNKNOWN PATERNITY

Developing a self-sustaining *ex situ* population of giant pandas is strongly dependent on successful reproduction by a significant proportion of available individuals. This is attributable to the need to maintain as much of the original (founder) gene diversity as possible, avoiding inbreeding depression, which occurs with the loss of valuable 'wild' genes. The objective is not only to produce lots of giant panda offspring but rather young that represent *all the valuable genotypes* within the population (see Chapter 21). The genetic and demographic assessments from the 1996 masterplanning workshop revealed that there was excellent gene diversity in the captive population, with no need for more founders from nature (see Chapter 2).

However, the Biomedical Survey revealed that the genotypes of many captive-born young in the extant population were unknown. Only a few pandas were behaviourally capable of mating naturally due to widespread sexual incompatibility. At the onset of the Survey, virtually every breeding facility had only one or two natural breeders. The extraordinarily short and tricky window of opportunity (the two- to three-day oestrus period once annually) for each female exacerbated the challenge. The standard protocol was that managers allowed a female access to that facility's one available breeder each day she was in oestrus. Then, to maximise the chance of pregnancy, each mated female was immediately anaesthetized for AI with sperm from a non-breeding male. This scenario often occurred on sequential days of oestrus. It was not unusual to use one or more sires for breeding and two or more others as sperm donors: mating roulette. The impact of this practice became apparent when the Survey teams constructed first-cut pedigrees using studbook data (Xie & Gipps, 1999, 2001). Figure 3.3 represents a typical pedigree from one institution showing many individuals with unknown paternity. Without explicit sire identification, it is impossible to implement a valid genetic management programme – managers could be unknowingly mating related animals. As discussed by David *et al.* (in Chapter 10), this issue was tackled by developing a molecular genetics laboratory in China to sort out many panda

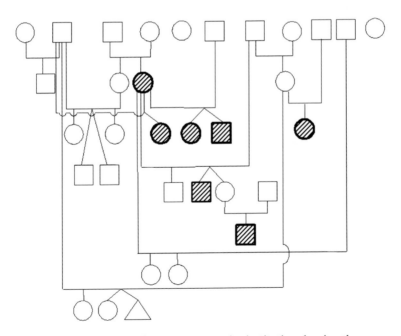

Figure 3.3. Pedigree from a representative institution showing the prevalence of giant pandas with unknown paternity (shaded). □, male; ○, female; △, cub that died without gender being determined.

paternities. Although 'missing links' remain (because of dead animals and the lack of DNA samples), these new data have allowed addressing the need for genetic management (see Chapter 10).

GENETIC OVER-REPRESENTATION

The Biomedical Survey confirmed that not only were few individuals reproducing, but also some pandas were over-represented genetically. The sample pedigree from one institution (Fig. 3.4) illustrates this problem. The first impression is that an impressive number of offspring has been produced. Closer examination, however, reveals that most of the young are derived from a single male and two females. Other founders in this subpopulation are under-represented or have never reproduced. This issue has been addressed through training in modern genetic theory along with beginning to manage giant pandas gen-etically in China (and abroad) (see Chapter 21). Nonetheless, discovering the prevalence of unknown paternities and over-represented indi-viduals emphasised that these challenges will be ongoing and that continuous genetic monitoring will be essential to maintaining a heterozygous, healthy, reproductively fit giant panda population in captivity.

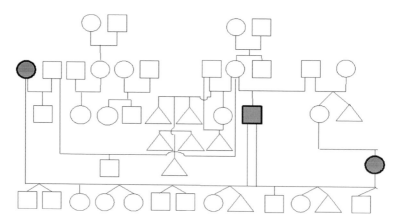

Figure 3.4. Partial pedigree from a representative institution showing over-representation by one male and two females (shaded). □, male; ○, female; △, cub that died without gender being determined.

BEHAVIOURAL DEFICIENCIES

A review of animal histories indicated a wide range of behavioural irregularities throughout the population. Many potential breeding opportunities had been derailed because of aggression displayed by males toward females, including injuries that prevented safe introduction for mating. We focused on learning the history of each panda, as well as determining whether behavioural or 'personality' traits may have contributed to an individual's reproductive success or failure. To secure the best information possible, we surveyed the people who knew the giant pandas best – the curators and keepers.

Table 3.1 shows that the survey included seven wild-born males and eight wild-born females, along with 17 males and 29 females born in captivity. Origin (wild-born) and, ironically, aggressiveness (for both sexes combined) contributed most significantly to successful production of offspring (see Chapter 5). Wild-born animals, whether male or female, were more effective at mating and rearing offspring. While aggressiveness to other pandas was not associated with breeding success for either males or females, when data for both males and females were combined, aggressive animals tended to have greater reproductive success. Obviously, there is a fine line between 'healthy' aggressive behavioural characteristics, which contribute to successful reproduction and inappropriate aggressive behaviours that thwart successful mating (see Chapter 5).

SUBOPTIMAL NUTRITION

Nutrition and nutrient status can obviously impact on any individual's health, including growth, reproduction and disease resistance. Our approach was to secure detailed information at each institution on feeding, dietary husbandry and nutrient contents, while relating these findings to parallel metrics of body weight, anatomical measurements and body condition scores (see Chapters 4 and 6). During the Survey, it became apparent that individual institutions were exploring dietary alternatives to the usual feeding of bamboo. The reason was primarily one of logistics and convenience, especially for large centres holding many giant pandas. An adult individual can consume 6–10 kg of the 10–18 kg of bamboo offered per day; finding, cutting, transporting and feeding this amount of vegetation is a challenge. Chinese managers have used substitute diets (high starch bread or gruel) which help to reduce labour and feeding costs.

Our most significant finding was that the use of these alternative diets caused lower-than-expected dry matter and fibre intakes. These substitutes provided a high-energy, highly digestible diet, which, in turn, resulted in lower faecal output than observed in giant pandas fed strictly with bamboo. Abnormal mucous stool output tended to be frequent in some individuals, which may have been related to the high starch and low fibre in the alternative diets. In turn, reproductive success was lower in pandas fed an insufficient bamboo diet, which perhaps was related to inadequate fibre intake.

STUNTED DEVELOPMENT SYNDROME

During the Survey, the teams were struck by the poor condition of almost 15% of the evaluated population (9 of 61 individuals). These giant pandas were small in stature (Fig. 3.5; Plate II), usually with a distended abdomen (from ascites), a coarse hair coat and worn and stained teeth (see Chapter 4). Some animals appeared chronically ill, with purulent mucous streaming from the nares, weepy eyes and other symptoms. Due to the comparatively small size of these animals (which, generally, was only two-thirds of their healthy counterparts), this con- dition was labelled Stunted Development Syndrome. A more detailed

Figure 3.5. Giant panda with Stunted Development Syndrome (SB 325) (photograph by D. Janssen). (See also Plate II.)

examination of the health histories revealed that these pandas generally had experienced multiple chronic diseases (especially gastrointestinal distress) with an absence of sexual development and activity. For
example, of the four females with this syndrome, only one (a wild-
caught female, SB 358, with shortened limbs and frequent mucoid
stools) had ever demonstrated oestrual activity.

The aetiology of this syndrome remains unknown (see Chapter 4).
Five of the affected animals originated from one of the panda centres,
indicating perhaps a strong relationship to a location-specific cause.
Many of the symptoms related to this condition, particularly ascites,
did not appear associated with clinical illnesses related to cardiac or
hepatic failure. Additionally, there was no consistent correlation to
abnormal serum chemistry values such as hypoproteinaemia (Zhang
et al., 2000; B. Rideout, pers. comm). Speculative potential causes include suboptimal nutrition, infectious disease or an environmental
toxicant.

TESTICULAR HYPOPLASIA OR ATROPHY

Abnormally small testes were discovered in three males, one of which
(SB 356) was an individual with Stunted Development Syndrome (see
Chapter 7). However, two normal-sized adult males (SB 323 and 345) had
only a single, normal-sized testis in the scrotum. The contralateral testis
was hypoplastic or atrophied (see Chapters 4 and 7). The retained testis
was situated adjacent or perhaps within the inguinal canal and could
be palpated for measuring. An examination of the pedigree as well as
photographic records revealed another male (SB 181, now deceased) also
with a single hypoplastic testis.

There was no obvious aetiology to this condition – the only
common factor being that SB 181, 323 and 345 had been housed at one
time at a single breeding facility. For example, at the time of the Biomedical Survey, SB 323 was a robust 12.5-year-old male and a consistently successful natural breeder. SB 181 (deceased) was the sibling of SB
323's dam, suggesting, perhaps, a heritable, genetic defect, similar to
the phenomenon of cryptorchidism observed in other mammals, including in wildlife (Thomas & Howard, 1975; Burton & Ramsey, 1986;
Roelke et al., 1993). Curatorial records also curiously indicated that
both testes in SB 345 were at one time descended (although one testis
was always smaller) with the onset of unilateral testicular atrophy
occurring at 7.5 years of age. Until this condition's aetiology is well

understood, males from this genotypic line are not recommended for breeding.

CONCLUSIONS

Perhaps the most important revelation of the Giant Panda Biomedical Survey was that no one factor was impeding reproductive success in the *ex situ* population. In fact, some of the variables (e.g. poor nutrition leading to compromised health that directly or indirectly decreased reproduction or offspring survivorship) were so interlinked that tackling this problem by relying on only one discipline would have failed. Our multidisciplinary approach was invaluable.

The resulting compendium of detailed information is now offered in the rest of this book. Many of the findings have also been applied in improving captive management. Examples include:

1. setting up a molecular laboratory to determine panda paternities (see Chapter 10) to facilitate implementing genetic management (see Chapter 21);
2. testing hypotheses to understand why so few captive-born males reproduce (see Chapter 14);
3. developing new artificial diets that provide adequate fibre (see Chapter 6);
4. formulating new technologies in semen cryopreservation, assisted breeding (see Chapters 7 and 20) and reproductive monitoring (see Chapter 8).

PRIORITIES FOR THE FUTURE

The Giant Panda Biomedical Survey began with a request from Chinese colleagues for advice about how to better manage disparate groups of giant pandas living in zoos scattered throughout China. This led to identifying a need to systematically assess not just a few of these unique creatures, but more than half the existing captive animals in China. This was not carried out in a unidimensional way or by a handful of people, but rather by an eclectic mix of experts across two diverse cultures. Working hand in hand, the results exceeded our expectations in two ways, first, in creating copious amounts of new biological information for the species, and second by translating the data into a blueprint for continued action to increase health and reproduction.

We believe that these cooperative efforts have resulted in a remark-ably quick turnaround in giant panda cub production and survival in China. However, this is not a one-time solution or 'quick fix'. Managers and scientists need to remain vigilant while actively following up on high priorities (offered at the end of each chapter in this book). Much remains to be done with perhaps the highest priority being further implementation of a long-range genetic management plan (see Chapter 21) to ensure that all wild-born founders and genetically valuable individuals reproduce.

There is also a need to determine how best the rapidly growing *ex situ* population might someday contribute to reversing the dire situation facing giant pandas living in nature. For example, how can new genetic technologies developed to sort out paternity (see chapter 10) be used to facilitate better surveys of wild population numbers? How can the non-invasive monitoring of adrenal hormones (see chapter 8) be used to measure stress in free-ranging pandas under pressure from local encroachment? What diseases are present in the wild population, how do they impact mortality and how can we mitigate their effects? And, if it becomes relevant, how might suitable panda candidates for reintroduction be identified, and how might a reintroduction protocol be tested based on experience with other carnivores?

Such questions highlight the broad knowledge gaps that remain for wild giant pandas, including inadequate ecological information. One of the most serious impediments is the ban by the Central Govern-ment of China for using radio-collars to study this species in nature (Mainka *et al.*, 2004). This has a harmful impact on the need to imple-ment long-term projects to examine social behaviour and community structure, diseases and the causes of mortality, population trends and habitat preferences. Solid knowledge of all these topics will be essential to planning and implementing a successful metapopulation manage-ment strategy that someday will link wild and captive giant panda populations.

Finally, it is important to re-emphasise that priority needs for all giant pandas and the subsequent implementation of actions have been, and will continue to be, the responsibility of Chinese professionals. Although the species is a beloved worldwide icon, successful conser-vation in the wild and in zoos will ultimately be the result of decisions and actions of the Chinese and not westerners. The Biomedical Survey adhered to this philosophy and proved the value of working across cultural and disciplinary boundaries while sharing scientific expertise and information, honouring previous work, and appreciating our differences and similarities. The result has been a true international and sustained mutual commitment to tackling important issues in giant panda biology, management and conservation.

ACKNOWLEDGEMENTS

We are grateful to Arlene Kumamoto for her contributions to shaping the Biomedical Survey methods and to her unwavering enthusiasm and commitment to scientific excellence, and to Mabel Lam and Wei Zhong, whose patient translations made this work flow smoothly.

APPENDIX 3.A

A sample 'Memorandum of Understanding' document used by the Chinese and CBSG teams for institutions in the Chinese Association of Zoological Gardens:

MEMORANDUM OF UNDERSTANDING
[INSTITUTION NAME]
and the
Conservation Breeding Specialist Group

WHEREAS, the Chinese Association of Zoological Gardens (CAZG), [NAME OF INSTITUTION] and the Conservation Breeding Specialist Group (CBSG) of the IUCN–World Conservation Union's Species Survival Commission are concerned about the grave situation of the giant panda (*Ailuropoda melanoleuca*), an endemic species of China, and:

WHEREAS, in December 1996 a Captive Management Plan was developed by CAZG in the city of Chengdu in partnership with its breeding centre and zoo and facilitated by CBSG, and:

WHEREAS, one of the high priority recommendations of the Plan was to conduct a Biomedical Survey of giant pandas held in Chinese zoos for the purpose of determining each animal's health, genetics and reproductive status and ensuring that each was appropriately and permanently marked for identification, and:

WHEREAS, the CAZG has invited CBSG to assemble a team of scientists to work in partnership with Chinese colleagues to develop a long-term approach on behalf of the continued breeding and successful conservation of the giant panda.

THEREFORE, the parties wish to document the agreed terms and conditions for this cooperation:

(1)   The parties agree to work together to conduct a Biomedical Survey of male and female giant pandas at the institutions of the Chinese Partners.

(2)   The evaluations will involve a total of *x* male and *y* female giant pandas at the designated facility, and the Chinese partners will select these animals.

(3)   The evaluations will be conducted in partnership by teams of scientists and managers from both institutions and CBSG.
   •   The CBSG Team will consist of [NAMES, ROLES AND AFFILIATIONS OF EACH TEAM MEMBER].
   •   The Chinese Team will consist of [NAMES, ROLES AND AFFILIATIONS OF EACH TEAM MEMBER].

(4)   All travel, housing and food expenses for the CBSG team will be paid by CBSG. The [NAME OF CHINESE INSTITUTION] will provide local transportation and working lunches.

(5)   CBSG agrees to provide the [NAME OF CHINESE INSTITUTION] [LIST OF EQUIPMENT TO BE DONATED]. The [NAME OF CHINESE INSTITUTION] agrees to loan necessary equipment for the length of the project.

(6)   Veterinarians at the [CHINESE INSTITUTION] will be responsible for anaesthetising each animal. The CBSG Team veterinarians will provide assistance when needed. The safety of the animals is the highest importance.

(7)   Once a safe level of anaesthesia has been achieved, the evaluation procedures will be as follows:
   a.   Identification transponders will be placed. Each animal will be implanted subcutaneously with an electronic transponder chip and tattooed. The chip will be placed in the scapular region at the anterior edge of the black and white juncture. Each animal will be tattooed in the left upper lip. CBSG Team veterinarians will demonstrate and provide training on the use of the transponders, reader and tattooing techniques. Extra transponder chips will be provided for future use by the Chinese partners.
   b.   A health evaluation will be made by the CBSG and Chinese veterinary teams. A thorough physical examination will be performed that will include:

      i.   weighing, body measurements and oral/dental exams (the CBSG Team will bring a portable scale);

      ii.   blood samples (25 ml per animal) collected for haematology, serum chemistry, serum banking and genetics;

      iii. biopsies of abnormal tissue growths, if accessible;

      iv. an ultrasound examination of the normality of the reproductive system using a rectal probe (other body systems in the abdomen and thorax will be scanned, if time is available);

      v.   careful monitoring of all anaesthetised animals with pulse oximetry, indirect blood pressure, body temperature and direct observation.

c.   The CBSG Team will bring equipment that will allow the electronic monitoring and some of the clinical blood assessments. The remaining assessments will be done by a local human hospital. CBSG will pay for the latter analysis. The CBSG Team will provide instruction and training on the appropriate use of all equipment.

d.   Samples for genetic studies will be collected from each animal. These will include a small skin biopsy (0.5 × 0.5 cm), 10 ml of heparinised blood (part of the original 25 ml of collected blood) and 20 pulled hairs. The skin sample will be processed by the geneticists from both teams and then will be stored in liquid nitrogen at the [CHINESE INSTITUTION OR DESIGNATED HOLDING FACILITY]. Blood and hair samples will also be catalogued and stored at these same sites. All samples will be maintained in [CHINESE INSTITUTION OR DESIGNATED HOLDING FACILITY] for use in future genetic studies.

e.   X [NUMBER] of males (studbook numbers X, Y, Z, etc.) will be subjected to a standard electroejaculation procedure while under anaesthesia. This procedure will require 20 minutes and will be done in partnership between the CBSG and Chinese scientific teams. Semen will be examined for volume and sperm concentration, motility and percentage of normal sperm forms (after fixing in glutaraldehyde), including assessing the incidence of intact sperm acrosomes. Good quality semen will be cryopreserved by splitting the semen sample and freezing half of it using the Chinese partner's method and half using a method developed in the USA. A small sample of cryopreserved sperm can then be thawed to determine the method that results in the best post-thaw survival. The CBSG Reproductive Physiologist will provide training on the new electroejaculator and other western laboratory procedures. All semen samples will remain in [CHINESE INSTITUTION OR DESIGNATED HOLDING FACILITY].

f.   Before or after each anaesthesia episode, historical and behavioural data will be collected and stored on written data sheets and in a laptop computer. This information will include origin, date of birth, reproductive history, and method of oestrus detection, including past opportunities to breed and reproductive success. Information will also include an evaluation by keepers of the animal's behaviour, including such characteristics as shyness or aggression that may be correlated with reproductive success. This will be accomplished by questionnaires in Chinese.

g.   The Chinese Team will provide information on diet, which will be forwarded to [NAME AND INSTITUTION OF CBSG TEAM NUTRITIONIST] for further analysis.

(8)   All collected information will be the property of the [CHINESE INSTITUTION] and the CAZG. A report on data collected on each animal will be provided. Further, on the last day of the project at [CHINESE INSTITUTION], a meeting will be held to review findings and progress and to exchange ideas. Additionally, copies of all written and computerised data sheets will be shared among the Chinese institution, CAZG office and CBSG.

(9)   Resulting publications will be done in full partnership between the Chinese and CBSG Teams. If scientific papers are published in USA journals, CBSG Team members will be primary authors. If papers are published in Chinese journals, Chinese Team members will be primary authors.

This Memorandum of Understanding will be written in Chinese and English and will be legally valid. Other matters not considered in this Memorandum of Understanding will be decided by consensus of the team members.

This Memorandum of Understanding is signed on [DATE].

By: _____   _____

             [NAME]   [NAME]

             [POSITION, INSTITUTION]   [POSITION, INSTITUTION]

REFERENCES

Burton, M. and Ramsey, E. (1986). Cryptorchidism in maned wolves. *Journal of Zoo Animal Medicine*, **17**, 133–5.

Mainka, S., Pan, W., Kleiman, D. and Lu, Z. (2004). Reintroduction of giant pandas: an update. In *Giant Panda Biology and Conservation*, ed. D. Lindburg and K. Baragona. Berkeley, CA: University of California Press, pp. 246–9.

Roelke, M. E., Martenson, J. S. and O'Brien, S. J. (1993). The consequences of demographic reduction and genetic depletion in the endangered Florida panther. *Current Biology*, **3**, 340–50.

Thomas, W. P. and Howard, M. H. (1975). Cryptorchidism and related defects in dogs: epidemiological comparisons with man. *Teratology*, **12**, 51–6.

Xie, Z. and Gipps, J. (1999). *The 1999 International Studbook for Giant Panda (Ailuropoda melanoleuca)*. Beijing: Chinese Association of Zoological Gardens.

Xie, Z. and Gipps, J. (2001). *Giant Panda International Studbook*. Beijing: Chinese Association of Zoological Gardens.

Zhang, A., Zhang, H., Zhang, Z. et al. (2000). *1998–2000 CBSG Biomedical Survey of Giant Pandas in Captivity in China*. Apple Valley, MN: Conservation Breeding Specialist Group, IUCN Species Survival Commission.

Zheng, S., Zhao, Q., Xie, Z., Wildt, D. E., and Seal, U. S. (1997). *Report of the Giant Panda Captive Management Planning Workshop*. Apple Valley, MN: Conservation Breeding Specialist Group, IUCN Species Survival Commission.

4

Significant medical issues and biological reference values for giant pandas from the Biomedical Survey

DONALD L. JANSSEN, MARK S. EDWARDS, MEG SUTHERLAND-SMITH, JIANQIU
YU, DESHENG LI, GUIQUAN ZHANG, RONGPING WEI, CHENG LIN ZHANG, R. ERIC
MILLER, LYNDSAY G. PHILLIPS, DAMING HU, CHUNXIANG TANG

INTRODUCTION

The Giant Panda Biomedical Survey sought to establish a baseline of scientific information on giant pandas living in Chinese zoos and breeding centres as a first step towards establishing a self-sustaining captive population (Zheng *et al.*, 1997; see also Chapter 2). To produce the most information that would allow an understanding of the health and reproductive status of the extant population, we chose an interdisciplinary approach to examine as many health and reproductive traits as possible. What was crucial was the trusting relationship that developed early in the process between the Chinese and American teams which led to a thorough understanding of giant panda biology – information that not only was fascinating from a scholarly perspective but also valuable to improving *ex situ* management.

This chapter provides detailed methods and medical findings following the assessment of more than 60% of the living Chinese population of giant pandas (as existed in 1996 when the need for a Biomedical Survey was recognised). The results in this chapter address issues

Giant Pandas: Biology, Veterinary Medicine and Management, ed. David E. Wildt, Anju
Zhang, Hemin Zhang, Donald L. Janssen and Susie Ellis. Published by Cambridge
University Press. © Cambridge University Press 2006.

ranging from disease conditions to reproductive compromise, all of which ultimately allowed classifying each animal as to its usefulness in achieving the goal of population self-sustainability. The practices and reference values described here will also be useful to those who are interested in closely studying and managing giant pandas in the future.

Animals

Sixty-one animals were available for the Biomedical Survey. To allow age comparisons, we divided the population into three groups: juveniles (animals 1 to 18 months of age); subadults (19 months to 4.5 years); and adults (>4.5 years). The rationale for the age division between the juvenile and subadult category was largely based on knowledge that young appear to be weaned in nature at about 1.5 years of age (Zhu et al., 2001). For the other division, our Biomedical Survey indicated that some males and most females were pubertal (had reached sexual maturity) by 4.5 years of age (see Chapter 7). Of the surveyed population, nine individuals were juveniles, 12 were subadults and 40 were adults. Six adults were examined twice, so our data set includes a total of 67 examinations (from 61 individuals) over a three-year period.

Anaesthesia

Each of the 61 pandas was anaesthetised using a ketamine hydrochloride (HCl)-based protocol. Including the six animals evaluated twice, our findings were based on a total of 67 anaesthetic episodes. Each individual was fasted for 0 to 12 hours before anaesthetic administration by a remote delivery darting system (Daninject Pistol System; Wildlife Pharmaceuticals, Fort Collins, CO), usually in the animal's home enclosure. Induction time was defined as the time from drug administration to recumbency (head and body contact with the ground). Procedure length was considered the time from first physical contact with the animal to the point of last physical contact. Most entire examinations required less than 60 minutes.

In addition to using ketamine HCl, on 22 occasions sedative drugs, including chlorpromazine (mean 0.40 mg kg^{-1}; range 0.26–0.59 mg kg^{-1}; n = 12), xylazine HCl (mean 0.35 mg kg^{-1}; range 0.21–0.48

mg kg^{-1}; n = 7) and/or diazepam (mean 0.13 mg kg^{-1}; range 0.09–0.21 mg kg^{-1}; n = 4), were given in an attempt to smooth the body rigidity normally incurred with ketamine HCl alone. With the exception of one procedure, supplementary ketamine HCl (administered intramuscularly or intravenously) or the use of isoflurane gas was always required to provide the needed time to complete the medical examination. Supplemental ketamine HCl was given intramuscularly or intravenously.

In 12 cases where the procedure was longer than normal (e.g. to conduct abdominal laparoscopy), we used inhalation anaesthesia delivered by face-mask or endotracheal tube. To accomplish the latter, isoflurane and oxygen were administered via a face-mask using a precision isoflurane vaporiser set at 4–5% and a circle-system anaesthetic machine. The animal was placed in lateral recumbency. Once jaw relaxation was achieved, the animal was intubated using a 12-, 13- or 14-mm internal diameter endotracheal tube. The head was extended dorsally as much as possible, and the mouth was held open with ropes held by assistants. The larynx was directly visualised with a laryngoscope and then intubated, or the endotracheal tube was gently guided through the mouth and into the larynx without viewing. In adult animals, laryngoscopy required an extra-long (35-cm) blade to view the vocal folds. Once intubated, the animal was maintained on the appropriate concentration of isoflurane to maintain a surgical plane of anaesthesia.

All pandas were given supplementary oxygen by a face-mask covering the mouth and nares. Animals were initially placed in right lateral recumbency and were monitored at frequent intervals with one or more pulse oximeters (N-20 and N-40, Mallinckrodt, Hazelwood, MO) with clip-type sensors located on the ear tip, tongue, vulva or prepuce (Fig. 4.1; Plate III). Temperature, heart rate and respiratory rate were tracked regularly. Additionally, indirect blood-pressure was monitored (Criticon Dinamap, Mallinckrodt) using an appropriately sized blood-pressure cuff (width approximately 40% of limb diameter) placed on the left foreleg.

Medical procedure techniques

We chose to use a systematic approach to ensure complete and comparable data sets for all animals and all facilities. The goal of each anaesthetic procedure was *always to safeguard the individual animal* while

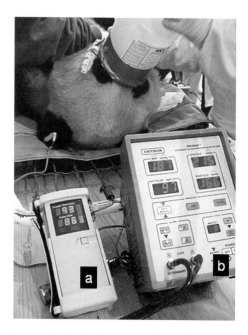

Figure 4.1. Anaesthetised giant panda illustrating supplemental oxygen administration. (a) Portable pulse oximeter; (b) indirect blood pressure monitor. (See also Plate III.)

collecting information that could benefit the entire captive population. Multiple medical procedures were chosen for the purpose of most effectively collecting baseline biomedical data, providing unique permanent identification and conducting diagnostic evaluations for individuals with reproductive or medical problems. The medical procedures performed included a physical and reproductive examination, body measurements, body weight and blood sampling. A small (approximately 0.5 × 0.5 cm) skin tissue sample was collected from the medial tibial area for genetic evaluation (see Chapter 10). Biological samples were also obtained for vaginal cytology, urinalysis, faecal parasite examination and faecal cytology. Photographs were taken of dentition, general body condition and detectable lesions.

Both males and females were subjected to an ultrasound scan at the end of the medical evaluation while each animal was in dorsal recumbency. An Aloka SSD-500 portable ultrasound machine (Aloka Co., Ltd., Wallingford, CT) with 3.5-MHz transducer and printer (or similar portable units available locally) were used to obtain ultrasound images of the abdomen. In some cases, we used a 5-MHz rectal transducer to obtain more detailed views of the uterine body. To minimise hair removal, we clipped small windows ventral to the urinary bladder and in an area caudal to the xyphoid. We concentrated on examining the reproductive tract for normality but we also visualised other abdominal organs, generally obtaining good views of all, or portions, of the liver, gall bladder, spleen, urinary bladder and occasionally the kidneys.

Three adult females were also subjected to laparoscopy (Olympus America, Inc., Melville, NY). In brief, this involved clipping most of the abdominal region free of hair before placing each female in a head-down (35°) angle on a surgical table. This area was surgically prepared and draped, the abdominal cavity filled with room air via a Verres needle and then a sterilised trocar-cannula inserted in the umbilical area through a 3 cm-long skin incision. The trocar was replaced with a fibre optic telescope. Internal organs were manipulated through a laterally placed, secondary trocar-cannula using an accessory grasping forceps. This technique allowed clear viewing of most aspects of the reproductive tract, including both ovaries as well as the spleen and liver. No anomalies were observed in any of the females undergoing laparoscopy. These examinations demonstrated the feasibility of this more invasive approach for detailed assessment of the abdominal cavity content.

During each anaesthetic episode, one team member recorded all detailed findings on a comprehensive data form. We then compiled the data into a spreadsheet for later analysis. Results (and in fact all findings and analyses) were recorded on raw and summarised data sheets which were shared with collaborators before leaving each facility.

Permanent identification methods

Individual permanent identification for animals is critical for cooperative breeding and management programmes, largely to avoid confusion and inaccuracy. In our study, each animal was given two forms of physical identification. Each was tattooed on the mucosal surface of the upper lip using the unique studbook (SB) number (Spaulding Electric Tattoo Marker Kit, Voorheesville, NY). Each animal received a microchip transponder placed subcutaneously along the dorsal midline between the scapulae (ID 100 transponder, Trovan, Eidap, Inc., Sherwood Park, Alberta, Canada). This identifier could be read through the skin with a hand-held reader (Trovan, Eidap, Inc.). All pandas tolerated both identification procedures well. For example, animals were seen eating bamboo later the same day after tattooing.

Morphometric methods

At the onset of each evaluation, a digital, load-bar weighing device (Model 500 with RL 15 load bars, Reliable Scale, Calgary, Alberta, Canada) with 0.1-kg resolution was used to determine individual animal body mass. A nine-point subjective scoring system, which relied on manual palpation to assess distribution of muscle mass and subcutaneous adipose tissue, was used to generate a body condition score (BCS). Individuals having a marginal or poor body condition were assigned a BCS of 1–3. Animals with moderate body condition were given a BCS of 4–6 (5 being ideal) whereas pandas with excessive body condition received a BCS of 7–9.

During each medical procedure, simple external measurements of defined anatomical points were also collected. The actual measures recorded were those modified from recommendations for the red panda, with additional metrics collected on the basis of earlier ursid studies (Roberts, 1994; Lundrigan, 1996). Unless specified otherwise, the following metrics were obtained using a standard, flexible measuring tape (data collected in 1999 and 2000, but not in 1998).

- *Head length.* From the tip of the animal's nose along the curves to the base of the skull's occipital ridge.
- *Body length.* From the base of the skull's occipital ridge along the curves to the root of the tail.
- *Tail length.* With the tail held at 30–45° above the dorsal aspect of the body, from the root of the tail to the tip of the tail (excluding hair extending beyond the tip).
- *Nose-to-tail body length.* From the tip of the animal's nose along the curves to the root of the tail.
- *Neck girth.* The circumference of the neck immediately behind the base of the skull.
- *Axillary girth.* The circumference of the body immediately behind the forelegs (at the chest).
- *Abdominal girth.* The circumference of the abdomen around the widest portion of the abdomen, across the ilium.

- *Fore foot.* With the right foot positioned with toes straightened: the length from the calcaneum (heel) to the tip of the longest toe, *sine unguis* (excluding the tip of the claw).
- *Hind foot.* With the right foot positioned with toes straightened: the length from the calcaneum (heel) to the tip of the longest toe, *sine unguis* (excluding the tip of the claw).
- *Elbow width.* The distance across the condyles of the right elbow.
- *Knee width.* The distance across the condyles of the right knee.
- *Axillary skin fold.* The thickness of the skin fold in the right axillary region (using callipers).
- *Inguinal skin fold.* The thickness of the skin fold in the right inguinal region (using callipers).
- *Wrist skin fold.* With the right fore foot perpendicular to the fore leg in a plantigrade position: the thickness of the lateral skin fold at the wrist (using callipers).
- *Ankle skin fold.* With the right hind foot perpendicular to the hind leg in a plantigrade position: the thickness of the lateral skin fold at the ankle (using callipers).

Animals were grouped by age based on dentition and onset of reproductive activity. The mean, standard deviation and range for each age class were calculated, and data for males and females were combined (because there were no gender differences; $p > 0.05$).

Clinical pathology methods

Blood was collected from the medial cephalic vein using a Vacutainer system (Vacutainer; Becton, Dickinson and Co., Franklin Lakes, NJ) within 15 minutes of handling an anaesthetized animal. Blood analyses included complete cell blood count (from EDTA-treated blood), blood-gas analyses (from whole heparinised blood) and selected chemistries (on whole heparinised blood and serum). After sample collection, blood-gas and chemistry analysis was performed within 15 and 30 minutes, respectively, using a portable analyser (iSTAT; Heska Corporation, Ft Collins, CO). A complete blood cell count was performed using manual methodology. A total white blood cell count was calculated using a Unopette (Becton, Dickinson & Co., Franklin Lakes, NJ) technique and a haemocytometer. A differential cell count was calculated using JorVet Dip Quick (Jorgensen Laboratories, Inc., Loveland, CO) stained blood smears. Total serum protein was measured using a hand-held temperature compensated refractometer (Leica, Inc., Buffalo, NY). Blood pH, partial carbon dioxide, total carbon dioxide, bicarbonate and base excess were included in the venous blood-gas analyses. Glucose, blood urea nitrogen, haematocrit and haemoglobin values were included in the chemistry analyses.

Semen collection and evaluation

Every adult male was subjected to a standardised electroejaculation procedure while under anaesthesia. The purpose was

1.   to determine the quantity and quality of spermatozoa as an index of male fertility and
2.   for research to understand male gamete biology and to enhance the ability to cool, cryopreserve or culture sperm *in vitro.*

All details associated with this process and related findings are presented by Howard *et al.* (in Chapter 7).

RESULTS AND DISCUSSION

Anaesthesia and monitoring data

Ketamine HCl was used without any additional sedative drugs on 44 occasions (mean 5.7 mg kg^{-1}; range 4.1–8.8 mg kg^{-1}; mean time to induction 10.3 minutes; range 1–58 minutes). The total procedure length averaged 43 minutes (range 28–144 minutes) with a total accumulated ketamine dose averaging 9.1 mg kg^{-1} (range 4.2–16.7 mg kg^{-1}). In the remainder, ketamine was the primary agent but chlorpromazine, xylazine or diazepam was also used (see p. 60–61). In general, relaxation was poor with ketamine HCl alone, with no noticeable improvement with the additional injectable sedatives at the described dosages. Regardless, all animals were adequately immobilised for transport from the animal's home enclosure to the examination area to begin the procedure. Overall induction time averaged about 10 minutes with the total dosage used averaging about 1.5 times the initial dose. In most cases, oral examination and tattooing were difficult due to poor jaw relaxation and spontaneous head movement, often requiring ketamine HCl supplements of 50–100 milligrams (i.v. or i.m.) per animal or the administration of isoflurane. Despite this, virtually all individuals remained in a light plane of anaesthesia throughout the procedure, with some animals attempting to roll over or stand by the end of the examination. In contrast to other bears, the giant panda in this light anaesthetic plane rarely appeared dangerous to personnel.

Of the 67 anaesthetic episodes, there were no serious events during examination or recovery, although two individuals retched or vomited during induction or near the onset of recovery. Both of these animals had inadvertently received food a few hours before being anaesthetized. Of particular interest was the lack of untoward effects with ketamine HCl alone in the giant panda in contrast to more severe impacts in other carnivores, such as severe muscle rigidity and seizures (Ilkiw, 2002). Only one animal had seizure activity, which lasted less than 1 minute. For this reason, we concluded that ketamine HCl was a safe and effective anaesthetic choice for short, non-invasive procedures such as those used in the Biomedical Survey. Because of its high margin of safety, we suspect that, for all ages, a more rapid and smoother induction with less need for supplementation could be achieved by increasing the initial ketamine HCl dosage to 8–10 mg kg^{-1}.

The various small, portable devices described in the 'Methods for medical evaluations' (from p. 65) were effective for monitoring multiple physiological parameters. This enabled closely tracking the plane of anaesthesia while collecting relevant data to evaluate the efficacy of the anaesthetic regimen. Virtually all pandas responded similarly and predictably to the ketamine HCl. The average systolic blood pressure was 155 mmHg (individual value range 86–237 mmHg with an average of eight measures per individual). Higher systolic pressures tended to occur in the larger adult males and were often sustained throughout the anaesthetic interval. For example, mean systolic pressure was higher ($p < 0.05$) in adult males (175 mmHg) compared to juveniles and adult females (157 and 149 mmHg, respectively).

Acute hypertension (>200 mmHg) was pronounced in six adult males, two females and one subadult (from the 1999 data subset only). In one case, hypertension developed coincidentally with electroejaculation. Although rarely used alone in the domestic dog, ketamine HCl is known to cause increased blood pressure in that species (Ilkiw, 2002). No giant panda appeared to suffer any ill effects from the hypertension episodes. However, complications such as cerebral haemorrhage would be a concern in an anaesthetized giant panda where hypertension is sustained over the course of the procedure (see Chapter 15).

Pulse oximetry revealed that relative oxygen saturation averaged 92% (individual value range 74–100% with a mean of 11 measures per individual). A value of <85% was usually a single datum point that represented a transient fluctuation or a sensor clip placement problem. The generally high and favourable oxygen saturation values were probably related to the relatively light anaesthesia plane provided consistently by ketamine HCl with or without other drug supplementations. Meanwhile, body temperature remained stable throughout the examination, averaging 37.2°C (range 34.2–38.8°C). Heart rate averaged 102 beats per minute (range 62–145) whereas average respiratory rate was 28 breaths per minute (range 12–60).

Body weights, body condition scores and morphometrics

Table 4.1 lists the body weights, BCS ratings and the morphometry measures for all giant pandas that did not meet the criteria of Stunted Development Syndrome (see below) during the 1999 and 2000 survey. Table 4.2 lists those measurements from adult pandas that did meet Stunted Development Syndrome criteria during those two years. In general, most giant pandas were not excessively lean or overweight on the basis of absolute weight or our subjective categorisation of body condition. The greatest variation in BCS occurred in the adult group (Table 4.1). The various morphometry measures found in the remainder of this table can be considered 'normal' for animals meeting our criteria for an adult, subadult or juvenile. There was no impact ($p > 0.05$) of gender on these variables. Thus, a male and female at a given age generally have comparable weights, body conditions and morphometries.

Clinical pathology

Data on blood values for all giant pandas examined (with stunted and other abnormal animals excluded) are illustrated in Table 4.3; the blood values for just those pandas affected by Stunted Development Syndrome are in Table 4.4. For the former, we have provided these 'normative' data on the basis of our three age classifications: adult; subadult; and juvenile. There were no differences on the basis of age and sex. Further, the values in general appeared consistent with published haematology and chemistry values for the free-living black bear (Storm *et al.*, 1988; DelGiudice *et al.*, 1991) and the giant panda (Mainka, 1999). Because of sample size differences, it was not possible to compare the normal size and stunted groups statistically but no major differences were apparent.

Table 4.1. Age, body weight and selected physical measurements of giant pandas without Stunted Development Syndrome[a]

	Adult					Subadult					Juvenile				
	n	Average	S. D.	Min	Max	n	Average	S. D.	Min	Max	n	Average	S. D.	Min	Max
Age at collection (years)	28	10.5	3.9	5.3	16.5	7	3.2	0.74	2.5	4.5	7	1.5	0.1	1.4	1.6
Body weight (kg)	28	100.1	16.5	67.1	130.6	7	94.4	14.9	80.0	121.1	7	54.6	11.9	43.7	73.0
Body condition score[b]	24	4.8	1.0	2.5	7.0	7	4.9	0.2	4.5	5.0	7	4.9	0.6	4.0	6.0
Head length (cm)	25	37.4	3.9	29.5	44.3	7	37.9	2.7	34.0	41.5	7	35.3	3.0	31.5	40.2
Body length (cm)	25	123.4	10.3	98.0	145.0	7	112.1	9.8	97.0	129.5	7	101.4	9.2	92.5	114.7
Tail length (cm)	25	15.4	2.6	11.0	20.0	7	15.7	2.1	12.0	18.5	7	13.4	2.7	10.0	18.4
Nose–tail body length (cm)	25	176.2	12.8	146.5	199.8	7	165.8	11.6	154.0	188.5	7	150.2	13.8	135.5	172.3
Neck girth (cm)	25	73.3	6.2	58.0	83.0	7	71.5	5.3	63.3	77.8	7	61.9	6.0	55.3	72.5
Axillary girth (cm)	25	106.6	7.4	93.3	124.0	7	104.4	7.9	96.5	116.0	7	88.2	4.4	82.5	92.2
Abdominal girth (cm)	25	116.1	7.8	100.5	132.0	7	109.1	9.6	96.8	122.0	7	90.1	8.6	78.8	99.0
Fore foot (cm)	25	17.3	1.3	16.0	21.5	7	18.5	1.4	16.4	20.5	7	15.9	0.8	15.0	17.0
Hind foot (cm)	25	22.2	2.1	17.0	26.5	7	22.2	2.3	19.0	25.5	7	19.8	3.5	14.5	24.4
Elbow width (cm)	24	7.0	1.2	5.0	8.9	7	6.8	1.2	5.2	8.6	7	6.2	0.9	4.6	7.2
Knee width (cm)	25	7.3	0.8	6.5	8.8	7	8.0	0.6	7.2	8.9	7	6.7	1.1	5.2	8.1
Axillary skin fold (cm)	25	1.4	0.4	0.8	2.4	7	1.6	0.4	1.1	2.3	7	1.4	0.4	1.0	1.9
Inguinal skin fold (cm)	25	1.5	0.4	0.9	2.1	7	1.5	0.2	1.2	1.9	7	1.4	0.2	1.2	1.8
Wrist skin fold (cm)	24	0.9	0.4	0.1	2.2	6	1.3	0.3	0.7	1.6	7	0.9	0.4	0.2	1.4
Ankle skin fold (cm)	25	1.0	0.4	0.3	1.8	7	1.0	0.2	0.7	1.2	7	0.9	0.5	0.2	1.8

[a] Data collected in 1999 and 2000; n = total number of individuals; S. D. = standard deviation; min = minimum value among all individuals; max = maximum value; [b] Body condition score based on ratings of 1 to 9 (ideal = 5).

Table 4.2. *Age, body weight and selected physical measurements of adult giant pandas with Stunted Development Syndrome*[a]

	n	Average	S. D.	Min	Max
Age at collection (years)	4	9.7	4.4	5.5	14.5
Body weight (kg)	4	75.4	11.7	60.0	88.2
Body condition score[b]	4	4	1	3	5
Head length (cm)	4	36.8	3.3	33.6	41.0
Body length (cm)	4	115.0	8.9	103.0	124.3
Tail length (cm)	4	17.3	2.1	14.2	19.0
Nose–tail body length (cm)	4	169.1	11.6	156.0	184.3
Neck girth (cm)	4	56.1	14.3	35.5	68.3
Axillary girth (cm)	4	91.3	8.9	78.0	97.2
Abdominal girth (cm)	4	111.9	8.0	100.0	117.5
Fore foot (cm)	4	16.1	0.8	15.0	16.8
Hind foot (cm)	4	20.8	2.4	19.2	24.3
Elbow width (cm)	4	6.1	1.5	4.2	7.3
Knee width (cm)	4	7.6	0.7	6.6	8.1
Axillary skin fold (cm)	4	1.0	0.2	0.7	1.1
Inguinal skin fold (cm)	4	1.0	0.4	0.6	1.5
Wrist skin fold (cm)	4	0.6	0.3	0.2	0.9
Ankle skin fold (cm)	4	0.6	0.3	0.2	0.8

[a] Data collected in 1999 only; measurements not taken in 1998; n = total number of individuals; S. D. = standard deviation; min = minimum value among all individuals; max = maximum value; [b] Body condition score based on ratings of 1 to 9 (ideal = 5).

Stunted development syndrome

Nine of the 61 giant pandas (14.8%; seven adults, two juveniles; three males, five females) had a stunted stature (comparatively small size with shortened body conformation) with various associated medical problems (Table 4.5). A common characteristic was an unthrifty appearance and rough hair coat. All (except SB 358) were captive-born, none had reproduced, and only one of five females had demonstrated oestrual activity. Five had moderate-to-severe ascites as indicated by ultrasound, and three of these (SB 325, 388, 393) had visible abdominal enlargement. Five had significant dental disease typified by excessively stained, pitted and worn teeth, far advanced for their age. In the younger cohort, the enamel was pitted, thin or even missing along some of the tooth surface (enamel dysplasia). All nine giant pandas

Table 4.3. *Clinical pathology values for the giant panda*[a]

	Adult					Subadult					Juvenile				
	n	Average	S.D.	Min	Max	n	Average	S.D.	Min	Max	n	Average	S.D.	Min	Max
White blood cells (mm³)	40	9411	3668	5000	18100	12	9652	2869	5600	15500	6	8335	1443	6400	9778
Neutrophils (mm⁻³)	40	6630	3083	2850	14046	12	7743	2392	4672	12090	6	5165	1476	3360	6985
Band neutrophils (mm⁻³)	40	3	21	0	126	12	45	156	0	540	6	0	0	0	0
Lymphocytes (mm⁻³)	40	1644	931	504	4511	12	2040	1054	648	4704	6	1620	959	18	2542
Monocytes (mm⁻³)	40	671	434	0	1629	12	555	392	0	1113	6	773	225	573	1088
Eosinophils (mm⁻³)	40	464	379	0	1333	12	363	415	0	1339	6	382	207	192	770
Basophils (mm⁻³)	40	0	0	0	0	12	1	3	0	9	6	0	0	0	0
Sodium (mmol l⁻¹)	40	130	16	122	139	12	131	4	120	136	6	131	2	129	135
Potassium (mmol l⁻¹)	40	4.4	6	3	6	7	4.6	0.3	3	5	6	4.6	0.8	3.3	5.5
Chloride (mmol l⁻¹)	40	103	21	93	198	7	97	4	91	106	6	98	3	95	102
pH venous	39	7.328	0.038	7.258	7.392	7	7.334	0.061	7.246	7.420	6	7.317	0.046	7.276	7.387
Partial carbon dioxide, venous (mmHg)	39	36	6	27	50	7	36	4	29	43	6	38	5	32.4	45.9
Blood urea nitrogen (mg dl⁻¹)	40	10	6	5	16	7	12	3	8	17	6	9	2	7	13
Glucose (mg dl⁻¹)	40	89	22	56	156	7	76	4	70	96	6	83	8	73	94
Haematocrit (%)	40	39	5	31	50	7	40	3	27	43	6	37	3	31	39
Haemoglobin (g l⁻¹)	40	13	5	10	17	7	14	1	9	15	6	13	1	11	13

Table 4.3. (cont.)

	Adult					Subadult					Juvenile				
	n	Average	S. D.	Min	Max	n	Average	S. D.	Min	Max	n	Average	S. D.	Min	Max
Total carbon dioxide, venous (mmol l^{-1})	40	20	5	15	31	7	20	1	15	21	6	21	2	18	23
Bicarbonate, venous (mmol l^{-1})	40	18	5	14	27	7	19	1	17	20	6	20	2	17	22
Base excess, venous (mmol l^{-1})	40	−7	9	−13	1	7	−7	2	−9	−5	6	−2	10	−9	17
Serum protein (refractometer) (gm dl^{-1})	8	7.0	1.2	6.0	9.9	3	6.9	1.4	6.0	8.5	0				

[a] Data collected in 1998, 1999 and 2000 with stunted and other 'abnormal' individuals excluded; n, total number of individuals; S. D., standard deviation; min, minimum value among all individuals; max, maximum value.

Table 4.4. *Clinical pathology values for giant pandas with Stunted Development Syndrome*[a]

	n	Average	S. D.	Min	Max
White blood cells (mm^{-3})	9	9836	2088	6494	14100
Neutrophils (mm^{-3})	9	5681	1782	3415	9588
Band neutrophils (mm^{-3})	9	0	0	0	0
Lymphocytes (mm^{-3})	9	2610	1053	940	4410
Monocytes (mm^{-3})	9	1261	2293	0	7622
Eosinophils (mm^{-3})	9	522	464	112	1410
Basophils (mm^{-3})	9	23	60	0	180
Sodium (mmol l^{-1})	9	126	4	119	131
Potassium (mmol l^{-1})	9	4.6	0.6	3.5	5.5
Chloride (mmol l^{-1})	9	98	5	88	104
pH venous	8	7.301	0.033	7.266	7.373
Partial pressure carbon dioxide, venous (mmHg)	8	42	5	37.9	53
Blood urea nitrogen (mg dl^{-1})	9	16	6	8	29
Glucose (mg dl^{-1})	9	85	6	73	92
Haematocrit (%)	9	36	6	30	49
Haemoglobin (g l^{-1})	9	12	2	10	17
Total carbon dioxide, venous (mmol l^{-1})	8	21	2	19	26
Bicarbonate, venous (mmol l^{-1})	8	20	2	17	24
Base excess, venous (mmol l^{-1})	8	−6	2	−9	−3

[a] Data collected in 1998 and 1999; n = total number of individuals; S. D. = standard deviation; min = minimum value among all individuals; max = maximum value.

with Stunted Development Syndrome had a history of chronic gastrointestinal disease, some beginning during juvenile development, characterised by frequent mucous stools and chronic diarrhoea.

To begin to establish formal criteria for what constituted this condition, an animal with Stunted Development Syndrome was defined as an individual with comparatively small body stature for its age and with at least three of the following:

- dental developmental anomalies;
- chronic gastrointestinal disease;
- moderate-to-severe ascites;
- lack of sexual maturity for its age;
- failure to reproduce; and/or
- poor body condition.

Table 4.5. *Information on nine giant pandas with Stunted Development Syndrome and associated medical problems*[a]

Studbook number	Sex	Year examined	Age (years)	Weight (kg)	Associated medical problems
393	Female	1998	5.5	73	Small stature; no oestrus historically; poor condition; severe ascites; excessive tooth wear; poor hair coat; demodecosis
373	Female	1998	6.5	62	Small stature; no oestrus historically; chronic gastrointestinal disease
356	Male	1998	8.5	75	Small stature; emaciated; asymmetrical dental wear; chronic nasal discharge; juvenile genitalia
325	Female	1999	12.5	75	Small stature; non-reproductive; worn teeth; enamel dysplasia; moderate to severe ascites
400	Male	1999	5.5	87	Normal size; chronic gastrointestinal disease; immature genitalia; delayed development
358	Female	1999	14	78	Shortened limbs; good reproduction record; frequent mucous stools; moderate ascites
453	Female	1999	1.5	56	Small size for age; chronic gastrointestinal disease; tooth surfaces stained and pitted
454	Male	1999	1.5	52	Small size for age; chronic gastrointestinal disease; tooth surfaces heavily stained and pitted; moderate ascites
388	Female	1999	6.5	60	Small stature; multiple medical problems; severe ascites; worn teeth; conjunctival chemosis; eosinophilia

[a] All captive born with the exception of SB 358.

The aetiology of this syndrome is far from clear. Detailed examinations of juvenile pandas with this condition have consistently revealed one or more systemic illnesses during critical growth and developmental stages. Any sickness may lead to an interval of suboptimal nutrition which results in malnutrition, enamel dysplasia and, eventually, stunted growth. We also have anecdotal and unsubstantiated evidence that some animals may recover and develop into reproductively successful individuals, with only minor residual effects, such as dental abnormalities. Others are unable to recover, continue to develop serious chronic medical problems and never reproduce. Although a definitive cause has not been established, infectious disease (e.g. canine distemper, canine parvoviral enteritis or *E. coli* enteritis) may play a role in inciting this syndrome in juveniles. *Baylisascaris* infections are generally considered non-pathogenic, but may also influence the health of young, developing giant pandas. Dietary excesses during growth, nutritional deficiencies (including premature weaning) or diet-induced gastroenteritis may cause growth and developmental challenges. Specific dietary problems, such as lactose intolerance or intestinal dysbiosis, may be examples. Causes specific to the dental abnormalities, such as tetracycline administration or fluorosis, should also be considered.

Aside from failing to understand its aetiology, the major concern with Stunted Development Syndrome is its influence on captive population viability. As discussed by Ballou *et al.* (see Chapter 21) it is essential that every valuable individual reproduce with its most genetically appropriate mate. The finding of a developmental anomaly that occurs *after* the critical neonatal period illustrates the importance of being vigilant about the health of juveniles. We now know that ensuring that young cubs survive is not enough – rather, the appropriate management of juveniles is essential to assuring long-term health and eventual reproductive success. Although it appears possible to recover the health of some younger animals that are on track for becoming 'stunted', years of reproductive success can be lost, sometimes permanently.

Abdominal effusions (ascites)

An abdominal effusion (fluid accumulation) was detected by ultrasound in 15 of 61 (24.6%) animals examined. All of these were adults (three males, 12 females), and nine had been born in captivity. Amounts measured were generally considered mild to moderate but three pandas (SB 325, 388 and 393) were experiencing severe fluid accumulation surrounding the abdominal organs (Fig. 4.2). Each of these individuals had a visibly distended abdomen and other signs of chronic disease, including decreased body condition, generalised oedema and

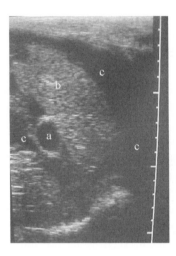

Figure 4.2. Abdominal ultrasound image of giant panda SB 388 showing significant intra-abdominal fluid. This 6.5-year-old female had a distended abdomen and multiple medical problems consistent with Stunted Development Syndrome. a, Gall bladder; b, liver; c, abdominal fluid pockets.

abnormal dentition. The cause of the abdominal fluid has not been determined, and ascites has not been reported as a significant medical problem in this species, although occasional ascites episodes have been noted, including in a giant panda maintained outside China (see Chapter 17). Regardless, it should be emphasised that small amounts of ascites appeared unrelated to clinical disease or to reproductive failure.

Dental disease

Dental disease, characterised by broken canine teeth or excessive wear with apparent enamel dysplasia, was not uncommon in the surveyed population. Fractured canine teeth with exposed root canals occurred in 10 of 61 (16.4%) giant pandas (two males, eight females, all adults, with half of these born in captivity). Although radiographs were not taken to assess extent of tooth-related disease, fistulations were absent upon physical examination. However, one individual had a firm swelling at the base of a fractured tooth (Fig. 4.3). For the long-term health of these animals, it will be important in the future to perform endodontic procedures to restore these teeth.

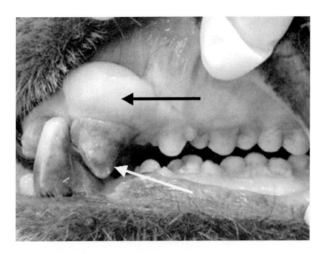

Figure 4.3. Lateral view of the dentition of a 16-year-old male giant panda (SB 305; wild born). The upper left canine is fractured exposing the root canal (white arrow). Above the fractured canine is a hard, bulbous swelling possibly related to the damaged tooth (black arrow). An X-ray unit was unavailable to help determine the nature of this swelling.

Another nine giant pandas (14.8%) were experiencing excessive tooth wear, pitting and/or staining compared to counterparts of similar age and sex (Figs. 4.4 and 4.5; Plates IV and V). The most extreme cases involved pandas with enamel loss so severe that the underlying dentin was clearly exposed. This finding led us to suspect that this condition has its origin during early development (see 'Stunted Development Syndrome', p. 70),

Demodecosis

Six of 61 giant pandas (9.8%) had clinical disease associated with *Demodex* sp. skin mites (Fig. 4.6). Diagnosis was made by performing a skin scraping of the affected area followed by microscopic examination (Fig. 4.7). All but one affected panda was an adult, and four had been captive born. The most typical lesions were mild alopecia associated with crusting and swelling of the eyelids. In one case, a 5.5-year-old male (SB 392) had severe, generalised demodecosis with widespread alopecia, erythema, pyoderma and lichenification of the skin.

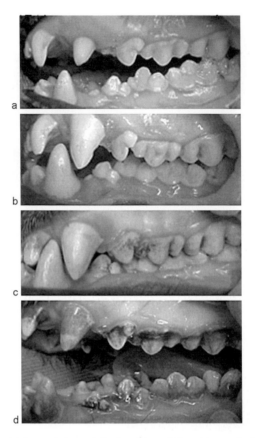

Figure 4.4. Lateral view of the dentition of four 18-month-old, juvenile giant pandas. (a) Male SB 461 and (b) female SB 452 are 'normal' compared to (c) female SB 453 and (d) male SB 454 who have enamel pitting and excessive teeth staining (along with chronic gastrointestinal disease and lower body weight for their ages). (See also Plate IV.)

Ascarid infection

Baylisacaris (Ascaridia) schroederii is a well-known parasite of the giant panda (Qiu & Mainka, 1993; Mainka, 1999). Although it is not uncommon for pandas to pass whole adult worms in the faeces or vomit, corresponding ova are infrequently detected in routine microscopic examinations of faecal samples. Using routine faecal flotation techniques, we detected ascarid ova in only a single juvenile female (SB 477; 1.5 years old). We are confident that ascarids are present in the captive environment of the giant panda. However, the prevalence and, more

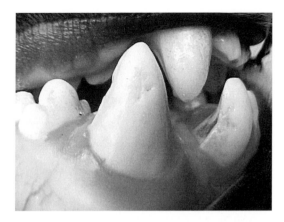

Figure 4.5. Close-up of newly erupted canine teeth and lower third incisor of an 18-month-old giant panda male (SB 454). Defects in the enamel (including staining and pitting) can be observed.) (See also Plate V.)

Figure 4.6. Giant panda male (SB 287) with hair loss, skin crusting and swelling around the eyelid margins caused by *Demodex* sp. skin mites.

importantly, the impact of this parasite on growth and development of the young giant panda remain unclear and deserve more research attention.

Testicular hypoplasia

Two cases of asymmetrically sized testes were encountered in normal-sized giant pandas during the Survey (SB 323 and 345). From histories, it was possible that the testicles developed normally and then decreased in size (atrophy) or at least in one case (SB 323) became hypoplastic (see Chapters 7 and 16) (Fig. 4.8; Plate VI). In both cases, the testes

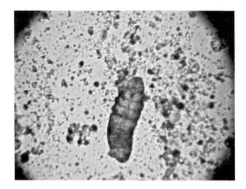

Figure 4.7. Microscopic view of a *Demodex* sp. skin mite obtained by a skin scraping from the eyelid margins of giant panda SB 287.

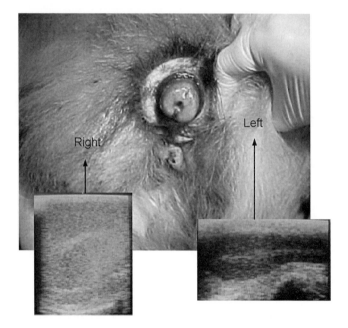

Figure 4.8. A 12.5-year-old male giant panda (SB 323) with a hypoplastic left testicle. The ultrasound images show the normal architecture and size of the right testis versus the hypoechoic condition of the smaller left testis. (See also Plate VI.)

were completely descended and in the normal location in the scrotum. Another male, with Stunted Development Syndrome (SB 356), had bilaterally juvenile testes despite being 8.5 years old (see Chapter 7).

Squamous cell carcinoma

Studbook 305, an adult male (originally wild born) was examined for the Survey at the China Conservation and Research Centre for the Giant Panda (Wolong Nature Reserve) in 2000. This individual had recently been moved from another institution, in part because of its serious health condition – an extremely large (approximately 50-cm diameter) open wound on the central back area. The veterinary team suspected skin cancer as the cause, and biopsies from the site confirmed a locally invasive squamous cell carcinoma. Prognosis for such an extensive lesion was deemed poor. This giant panda was treated supportively but died several months later. Despite the illness, semen quality was excellent at the time of the Biomedical Survey and, prior to his death, sperm from this male were used to artificially inseminate SB 432 who later produced a surviving female cub (SB 512) (see Chapter 20).

Uterine/cervical infection

During the ultrasound examination of each female, the focus was largely on identifying and evaluating reproductive tract status. In most cases the uterine body was readily identified through the bladder 'window' or directly by rectal probe (Fig. 4.9). Giant pandas SB 374 and SB 404 had evidence of fluid present within the uterine body as revealed by a hypoechoic line in the uterine lumen when viewed longitudinally (Fig. 4.10). The 6.5-year-old SB 404 also had a history of 'weak' oestrus and had never conceived despite having mated and being artificially inseminated. Direct vaginal examination indicated the presence of a cloudy discharge at the cervical os. Cytology and biopsy of the cervix revealed a suppurative cervicitis. The uterus was flushed with a solution of saline and gentamicin and the animal treated with a three-week course of oral ampicillin. That same breeding season, she displayed normal oestrus, mated and produced a male cub (SB 518).

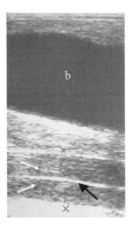

Figure 4.9. Ultrasound image obtained with a rectal transducer of a normal urinary bladder (b) and uterine body (white arrows). The uterus is free of luminal fluid as indicated by the hyperechoic endometrial line (black arrow). The diameter of this portion of the uterine body is ~2 cm.

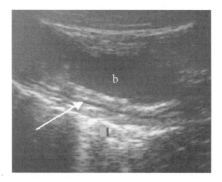

Figure 4.10. Ultrasound image obtained with an abdominal transducer of the bladder (b) and uterus showing a hypoechoic line representing uterine fluid (arrow). This 6.5-year-old female (SB 404) had a purulent cervicitis, which was presumed to be related to endometritis. She was treated with injectable antibiotics and successfully conceived shortly after this image was taken.

PRIORITIES FOR THE FUTURE

Sixty-one animals, ranging from 18 months to 16 years of age, were anaesthetised using ketamine HCl as the primary drug. All animals were weighed, measured and subjected to a thorough and systematic biomedical examination. From this effort, we generated substantial data on animals that appeared 'normal' versus those with various medical and developmental challenges. No significant problems were encountered using the anaesthetic regimens described. Ketamine HCl alone was safe and effective for such short-term evaluative procedures. Standard monitoring techniques (including pulse oximetry, ultrasound and laparoscopy) proved useful and effective. There was a significant prevalence of abdominal effusions, and severe ascites was detected in three individuals. Fractured canine teeth with pulp exposures were observed in 16% of the population with excessive tooth wear for age observed in another 15%. Mild hair loss on the eyelid border attributed to *Demodex* sp. mites was seen in five animals with one additional individual experiencing severe demodecosis with extensive hair loss and pyoderma. Asymmetrical testicles were observed in two adult males, and in both cases the smaller testis appeared hypoplastic. There was one case of an extensive squamous cell carcinoma and another of a treatable uterine/cervical infection. Of most interest was the discovery of nine giant pandas with stunted development associated with multiple chronic medical problems. We speculate that a systemic illness during critical growth and developmental stages may be occurring, which results in malnutrition, enamel dysplasia, stunted growth and ultimately failed reproduction.

A priority to define the cause of Stunted Development Syndrome

Determining the cause of this syndrome is one of the highest priorities for the future of a healthy, reproductively sound giant panda population in captivity. The order of action should first focus on better defining the syndrome in the context of risk factors, case definition and differential diagnoses. This should be followed by applying new knowledge and the best contemporary diagnostic modalities to address this severe anomaly. Strategic directions for future study should include improved parasite monitoring, serosurveys for understanding the prevalence of infectious diseases, advances in molecular diagnostics for detecting specific pathogens and endoscopic evaluation of the gastrointestinal tract of affected versus unaffected individuals.

A priority to continue routine examinations of the ex situ population

The Biomedical Survey of the giant panda revealed health, reproductive and developmental findings which were significant for today or could be sentinels for emerging trends in this valuable *ex situ* collection. Therefore, it makes sense to continue monitoring this population actively to determine if adverse observations (severe ascites, infections, Stunted Development Syndrome) are becoming less common or if new medical challenges are materialising. Furthermore, the examination of all available animals is essential as the sporadic testing of only the occasional individual will fail to reveal important population tendencies. In short, without a continued emphasis on biomedically surveying the giant panda population in (and outside) China, the present effort becomes only a 'snapshot' of population status at one time in history, quickly becoming outdated and of little use to genetic management. Therefore, this type of intensive monitoring of this important flagship species is vital.

A priority to develop a centralised database of biomedical information

A key to effectively managing and using all the information collected during the Biomedical Survey (and thereafter) is the availability of a centralised database. Such 'information repositories' are already available (e.g. a software program known as MedARKS; International Species Information System, Minneapolis, MN) and need to be developed cooperatively by all giant panda holders (within and outside China). A biomedical database would allow managers to recover medical information quickly to determine if an animal is healthy or to select a potential therapeutic treatment. These information systems also allow identifying or tracking trends in population health and reproductive status to permit changing management more effectively to prevent a problem or to take advantage of improvements occurring at another facility. Finally, such databases allow the *scientific* study and interpretation of data to improve our overall understanding of a species.

A priority for more sharing of information and technology transfer

As explained in Chapter 22, one of the most productive outcomes of the Biomedical Survey was realising the need for and value of information sharing among the partners. In terms of improved management of the *ex situ* giant panda population, there continues to be a strong desire for extensive capacity development in the biomedical sciences, especially in veterinary medicine. One priority in particular is the need for training in disease identification and management. Certainly, the disease issues recognized and discussed in this chapter and later by Loeffler *et al.* (in Chapter 16) reflect the complexity and dynamism of the situation in the extant giant panda population in China. As illustrated in Chapter 22, highly successful workshops have already been conducted in veterinary diagnostics and pathology. But these efforts are only a modest beginning in addressing the extensive need for training that will allow diseases to be effectively diagnosed, managed and then eliminated in captive populations within China. Thus, although listed as the 'last' priority in this section, the need for capacity building can by no means be delayed. For those willing to provide training within China, such efforts are absolutely essential to achieving the targeted goal of developing a 'self-sustaining captive population of giant pandas within China' (see Chapter 2).

ACKNOWLEDGEMENTS

The authors thank Arlene Kumamoto, Kim Williams RVT, Jeff Turnage RVT and Sandi Skrobot RVT for technical assistance in planning and performing the procedures described in this chapter. We also thank the directors and veterinarians at each of the institutions participating in the Biomedical Survey. Special thanks are due to Mabel Lam for coordinating and facilitating our activities.

REFERENCES

DelGiudice, G. D., Rogers, L. L., Allen, A. W. and Seal, U. S. (1991). Weights and hematology of wild black bears during hibernation. *Journal of Wildlife Disease*, **27**, 637–41.

Ilkiw, J. E. (2002). Injectable anesthesia in dogs. Part 2. Comparative pharmacology. In *Recent Advances in Veterinary Anesthesia and Analgesia: Companion Animals*, ed. R. D. Gleed and J. W. Ludders. Ithaca, NY: International Veterinary Information Service, www.ivis.org, A1415.0702.

Lundrigan, B. (1996). Standard methods for measuring mammals. In *Wild Mammals in Captivity*, ed. D. G. Kleiman, M. E. Allen, K. V. Thompson and S. Lumpkin. Chicago: University of Chicago Press, pp. 566–70.

Mainka, S. A. (1999). Giant panda management and medicine in China. In *Zoo and Wild Animal Medicine V*, ed. M. E. Fowler and R. E. Miller. Philadelphia, PA: W. B. Saunders Co., pp. 410–14.

Qiu, X. and Mainka, S. A. (1993). Review of mortality of the giant panda (*Ailuropoda melanoleuca*). *Journal of Zoo and Wildlife Medicine*, **24**, 425–9.

Roberts, M. (1994). *Red Panda Species Survival Plan Fact Sheet*. Washington, DC: Smithsonian's National Zoological Park.

Storm, G. L., Alt, G. L., Matula, G. J. and Nelson, R. A. (1988). Blood chemistry of black bears from Pennsylvania during winter dormancy. *Journal of Wildlife Disease*, **24**, 515–21.

Zheng, S., Zhao, Q., Xie, Z., Wildt, D. E. and Seal, U. S. (1997). *Report of the Giant Panda Captive Management Planning Workshop*. Apple Valley, MN: IUCN–World Conservation Union/SSC Conservation Breeding Specialist Group.

Zhu, X., Lindburg, D. G., Pan, W., Forney, K. A. and Wang, D. (2001). The reproductive strategy of giant pandas: infant growth and development and mother–infant relationships. *Journal of Zoology (London)*, **253**, 141–55.

5

Life histories and behavioural traits as predictors of breeding success

SUSIE ELLIS, REBECCA J. SNYDER, GUIQUAN ZHANG, RONGPING WEI,
WEI ZHONG, MABEL LAM, ROBERT SIMS

INTRODUCTION

Among mammals, the giant panda is reproductively unique. The female is a seasonal, monoestrual breeder, experiencing a single- two to three-day period of sexual receptivity once per year, presumably triggered by increasing day length. In the wild, male giant pandas compete with conspecifics for access to oestrous females (Schaller et al., 1985). Giant pandas produce copious sperm numbers (see Chapter 7), presumably as 'insurance' to ensure conception and the perpetuation of the male's genes if given the opportunity to mate during a female's brief window of fertility. Although the extraordinarily short oestrus is a fascinating biological trait, it does not appear to limit reproductive success in captivity given that a sexually compatible male is available and breeding occurs. It does, however, present challenges for captive management for cub production.

The wild-born giant panda cub stays with its mother for 1.5 to 2.5 years (Schaller et al., 1985). This almost always is *not* the case in Chinese zoos and breeding centres, because of the practice of promoting annual cub production by early weaning, usually before six months of age (see Chapter 14). The consequences of this short-term gain on long-term development remain a question, and studies are continuing on the

Giant Pandas: Biology, Veterinary Medicine and Management, ed. David E. Wildt, Anju Zhang, Hemin Zhang, Donald L. Janssen and Susie Ellis. Published by Cambridge University Press. © Cambridge University Press 2006.

impact of disrupted early rearing on adverse behaviours, including inappropriate aggression, inadequate sexual behaviour and/or incompetent maternal behaviour (see Chapter 14). These anomalies are rather common in the *ex situ* giant panda world. Many males tend to show aggressive rather than affiliative behaviours, even to females demonstrating strong oestrus. Some females also display 'weak' periods of sexual receptivity (low intensity interest in males and/or absent or feeble lordotic behaviour).

The manifestation of behavioural characteristics or 'personality' traits can be influenced by many different and possibly interacting factors, from rearing history to housing conditions. Genetic predispositions and interactions with the environment offer yet another layer of complexity, the 'nature versus nurture' paradigm (Lorenz, 1965; Wilson, 1975; Lewontin *et al.*, 1984).

The Giant Panda Biomedical Survey afforded an important opportunity to determine if and how rearing history and behavioural (i.e. personality) characteristics influence an individual's ability to successfully reproduce. As each animal was evaluated, we collected extensive historical data from curators and senior keepers, and administered surveys to those people who knew the pandas best – the animal keepers. This chapter reviews trends and behavioural patterns in the giant panda that may (or may not) play a role in reproductive success.

METHODS

Survey of historical information

The behavioural specialist on the Biomedical Survey team used a standardised form (Table 5.1) to collect relevant health and reproductive histories on every giant panda from senior keepers and/or managers at all four institutions. These data were relevant for establishing a baseline on health and reproductive history for each individual as well as ascertaining early rearing information. To ensure consistency, one person (with a Chinese translator) collected all information.

Keeper survey of behavioural traits

As a complementary tool, we also administered a detailed survey to animal keepers, in Chinese, which allowed a subjective assessment of specific behaviours. This method was based on earlier work of Gold and Maple (1994; gorilla), Fagen and Fagen (1996; brown bear), Wielebnowski

Table 5.1. *Giant Panda Biomedical Survey individual animal historical data form*

Year	19– 19 – 19 – 19– 19–
If female, did oestrus take place?	
Placed with a male when in oestrus?	
Identification number of male	
Did copulation take place?	
If not, why did pairing fail?	
Did pregnancy take place?	
Abortion, live or dead birth?	
Identification number of offspring, if survived	
Length of offspring survival	
Artificially inseminated?	
Identification number of male sperm donor	
Was sperm fresh or frozen?	
Did pregnancy take place from artificial insemination?	
Abortion, live or dead birth?	
Identification number of offspring, if survived	
Length of each offspring's survival	
Housed with males in non-breeding season?	
Health problems or surgical procedures?	

Studbook number/name _____

Facility _____

Date or year of birth _____

Wild caught or captive born? _____

How long has the animal been at the facility? _____

How many times has the animal been moved between facilities? _____

Other comments _____

_____

_____

(1999; cheetah) and was identical to that used by Feaver *et al.* (1986; domestic cat).

Twenty-three behavioural adjectives (Box 5.1) were measured on a form by 38 keepers at four institutions. In brief, each form listed the 23 adjectives next to a calibrated horizontal line. Each line was 100 centimetres long and was a continuous scale for a particular adjective. The minimum score (0) was placed at the left terminus to the line, and the maximum score (100) at the right terminus. Keepers scored each panda that they personally cared for at their home institution, marking every

---

Box 5.1. Behavioural definitions of adjectives used in the keeper survey

(1) Active: moves around frequently

(2) Aggressive to other pandas: frequently reacts with hostility and threats to conspecifics

(3) Alert: pays attention to surroundings and changes in environment

(4) Anxious: seems worried and apprehensive

(5) Amiable: pleasant and good natured

(6) Calm: not easily disturbed by changes in the environment

(7) Curious: readily explores new situations, environments or objects

(8) Eccentric: shows stereotypic behaviour or unusual behaviour

(9) Oestrus strength (females only): intensity of demonstrated oestrual behaviours

(10) Friendly to other pandas: approaches and seeks contact with conspecifics

(11) Friendly to people: approaches people readily and in a friendly manner

(12) Fearful of other pandas: reacts to conspecifics by moving away

(13) Fearful of people: reacts to people by moving away

(14) Insecure: seems scared or threatened easily

(15) Interest in other pandas during oestrus: reacts with positive interest in opposite sex during oestrus

(16) Irritable: reacts excessively to events and situations

(17) Oblivious: unresponsive to, and seemingly unaware of, significant events and situations

(18) Playful: engages in play with other pandas or objects in its surroundings

(19) Secure: shows confidence and calmness when dealing with a variety of situations

(20) Shy: reluctant to engage in social situations

(21) Solitary: spends time alone

(22) Spirited: abundant physical and mental energy

(23) Tense: shows restraint in movement and posture; carries its body stiffly

---

individual's score by placing an X along the line next to each behavioural adjective. The distance from 0 to the mark on the horizontal line was measured in centimetres, with each animal scored from 0 to 100 for each adjective. Keepers were instructed not to discuss their ratings of individual giant pandas with cohort keepers before providing their scores. If questions arose during the survey, answers were provided immediately via the translator. Each animal was scored by one to six keepers. Surveys by three keepers were excluded from further analysis because of communication difficulties (all animals received the same score for each trait). Remaining data were analysed by binary logistics regression.

Historical data

Historical data were collected for all 61 giant pandas in the Biomedical Survey. However, behavioural trait data were collected for only 54 individuals (four were excluded because of juvenile status and three because of scoring errors (noted above). Table 5.2 indicates study subjects by location, sex and origin (captive born or wild born). Of the 34 females, 26 had been born in captivity and eight in the wild. Of the 20 males, 14 were captive born and six were wild born.

In addition to evaluating results on the basis of facility, sex and origin, the following historical variables also were analysed, including records for:

1. ability to successfully copulate;
2. number of litters born or sired over the course of the animal's lifetime;
3. number of litters surviving (one or more cubs per birthing event surviving at least until six months of age);
4. age (in years).

One-way analysis of variance for age distributions by facility revealed that the pandas at the Wolong Centre were younger ($p = 0.03$) than those in counterpart facilities (all of which did not differ from each other; $p > 0.05$). The other three variables were similar across institutions, with no substantial differences in copulatory ability, number of litters produced or number of litters surviving. These findings in the context of the categorisations used in the Biomedical Survey (Prime Breeder, Potential Breeder, Questionable Breeder, Poor Breeding Prospect; see Chapter 3) are shown in Table 5.3. Statistical comparisons on the basis of these categories were not possible due to inadequate sample sizes in some cells. Table 5.4 shows a breakdown of gender and origin of animals included in the behavioural survey. Wild-born giant pandas were 2.4 times more likely to be Prime Breeders than captive-born counterparts ($p < 0.01$ via a chi-squared contingency test). A

Table 5.2. *Giant panda distribution by location, gender and origin*

Institution	Total no. animals	Gender		Origin	
		Females	Males	Captive born	Wild born
Beijing Zoo	10	6	4	9	1
Chengdu Research Base	7	3	4	5	2
Chengdu Zoo	8	8	–	7	1
Chongqing Zoo	3	3	–	2	1
Wolong Centre	26	14	12	17	9
Total	54	34	20	40	14

comparison among animals of breeding age (>6 years old) revealed that there was no difference between older (>10.5 years) versus younger (≤10.5 years) pandas in terms of the proportion being Prime Breeders ($p > 0.05$ via chi square).

Behavioural survey data

The primary purpose of the behavioural survey was to determine if there were any relationships between one or more personality traits and reproductive success, i.e. the ability to naturally mate and produce offspring. A binary logistic (logit) regression was used to relate the behavioural scores (averaged across all keepers) for each animal to the response binary variable, which was the Prime Breeder (historical data) category described by Ellis *et al.* (see Chapter 3). This included animals that were in good-to-excellent weight and health, displaying past 'normal' behaviours and with a history of either successful reproduction or no reproductive deficiencies.

The first analytical step involved developing a correlation matrix of the 23 behaviours to determine those that were closely related. Table 5.5 demonstrates that of the 253 possible variable pairs, 41 were significant ($p < 0.05$). Twenty-five were positively inter-related, with correlation coefficients ranging from 0.35 to 0.80, and 16 variable pairs were negatively related, with correlation coefficients ranging from −0.35 to −0.80 (see Table 5.5). Not surprisingly, the strongest correlation ($r = 0.81$) occurred between 'oestrus strength' and 'interest in pandas during oestrus'. Positive and negative relationships were logical,

Table 5.3. *Relationship of Biomedical Survey categories to history of natural mating, litters born and litter survival*

Biomedical Survey category	History of natural mating					Litters born					Litter survival				
	No opportunity	Too young	Successful	Not successful	Total	Not assessed	None	1–4	>5	Total	Not assessed	None	1–4	>5	Total
Prime Breeder (1)	2	3	13	5	24	–	7	9	8	24	1	7	10	7	24
Potential Breeder (2)	1	16	–	–	17	1	16	–	–	17	1	16	–	–	17
Questionable Breeder (3)	2	2	2	–	6	1	5	–	–	6	1	5	–	–	6
Poor Breeding Prospect (4)	2	1	–	1	4	2	1	1	–	4	2	2	–	–	4
No assessment	–	1	2	–	3	–	3	–	–	3	–	3	–	–	3
Total	7	23	17	6	54	4	32	10	8	54	4	33	10	7	54

Table 5.4. *Gender and origin of animals included in the behavioural survey*

	Origin	
Sex	Captive born	Wild born
Female	26	8
Male	14	6
Total	40	14

validating the ability of the keepers to identify accurately behavioural characteristics of the animals that they evaluated. For example, in descending order, correlations were found between 'active' and 'spirited' ($r = 0.71$), 'aggressive' and 'fearful' of conspecifics ($r = 0.66$) and 'playful' and 'spirited' ($r = 0.55$).

The behavioural characteristic 'aggressive' was an important predictor of breeding: increased aggression in both males and females signified more likely reproductive success. However, when the data were analysed within sex, 'aggression' became a non-significant variable ($p > 0.05$). That is, the presence of only an assertive male or female (but not both) did not enhance reproductive success. This trait was also positively related to keepers' perceptions that these same animals were 'irritable' and 'eccentric', and inversely related to 'amiable' and 'friendly to people'.

Logit regression analysis, whereby the Prime Breeder binary (historical) variable was integrated with the behavioural traits in clusters, did not generate additional useful predictive information about breeding success. The trait 'aggressive' consistently produced significance values of $p < 0.01$ but was not enhanced further by evaluating in the context of other behaviours.

In summary, of the historical and behavioural traits examined, only origin (being wild born) and aggression to other pandas (when sexes were combined) were significant positive contributors to being an effective breeder. Interestingly, when sexes were separated, aggression was not a significant factor. This suggests that aggression in captive giant pandas may be a useful behavioural characteristic for both males and females even though male hyperaggression has been a main reason cited for breeding failure in captivity (Zheng *et al.*, 1997; see also Chapter 3). In nature, males are known to compete, sometimes

Table 5.5. *Statistically significant correlations between behaviours measured in the behavioural survey*

Correlations	Correlation coefficient (r)	p value
Positive		
Oestrus strength × interest in other pandas during oestrus	0.81	0.001
Active × spirited	0.71	0.001
Aggressive to conspecifics × fearful of conspecifics	0.66	0.001
Playful × spirited	0.55	0.001
Fearful of people × tense	0.54	0.001
Shy × solitary	0.54	0.001
Alert × anxious	0.53	0.001
Amiable × friendly to people	0.53	0.001
Anxious × eccentric	0.51	0.001
Active × playful	0.50	0.001
Active × curious	0.48	0.001
Curious × playful	0.48	0.001
Active × fearful of conspecifics	0.47	0.001
Irritable × eccentric	0.44	0.002
Aggressive to conspecifics × eccentric	0.43	0.003
Calm × secure	0.43	0.003
Fearful of conspecifics × shy	0.41	0.005
Curious × eccentric	0.39	0.008
Curious × spirited	0.38	0.009
Anxious × fearful of people	0.37	0.013
Friendly to conspecifics × secure	0.37	0.011
Alert × insecure	0.36	0.016
Anxious × spirited	0.36	0.016
Fearful of conspecifics × insecure	0.36	0.014
Insecure × shy	0.35	0.02
Negative		
Calm × insecure	−0.35	0.018
Fearful of conspecifics × spirited	−0.35	0.018
Playful × shy	−0.36	0.016
Aggressive to conspecifics × fearful of conspecifics	−0.37	0.012
Aggressive to conspecifics × friendly to people	−0.39	0.009
Amiable × eccentric	−0.39	0.008
Friendly to conspecifics × oblivious	−0.39	0.009
Amiable × interested in other giant pandas during oestrus	−0.40	0.019

Table 5.5. (cont.)

Correlations	Correlation coefficient (r)	p value
Friendly to conspecifics × insecure	−0.42	0.004
Fearful of conspecifics × playful	−0.44	0.003
Insecure × secure	−0.46	0.002
Oestrus strength × friendly to other pandas	−0.53	0.011
Friendly to conspecifics × shy	−0.54	0.001
Aggressive to conspecifics × amiable	−0.57	0.001
Friendly to people × irritable	−0.64	0.001
Amiable × irritable	−0.80	0.001

ferociously, for an oestrual female (Schaller *et al.*, 1985). It has been proposed that male hyperaggression in captivity may be an expression of a species-typical behaviour with no outlet and, thus, is misdirected toward females (see Chapter 12). The result is that in the *ex situ* environment, and in the absence of access to same-sex competitors, a male will often fight with a female, even one in oestrus. Thus it makes sense that a behaviourally aggressive female is more competent at handling an aggressive male, and perhaps breeding is less successful when only one of the animals is aggressive. In both cases there are many anecdotal reports of either shy males or shy females being so aggressively pursued by a cantankerous conspecific that mating is impossible. Being wild born also conferred a distinctive advantage, presumably appropriate aggressive behaviours are learned from the dam, and they better equip an individual for future agonistic encounters, including during mating.

PRIORITIES FOR THE FUTURE

The information in this chapter was important for three main reasons, the first demonstrating the need for Chinese panda institutions to begin systematically archiving fundamental historical information on every giant panda. Some data were recorded informally in individual keeper log books, and it was astonishing how much unrecorded information was in people's heads. Thus, one priority should be to establish a computerised record-keeping system, especially for the major breeding centres, which goes beyond the traditional Giant Panda Studbook (Xie & Gipps, 2003) and documents reproductive and health histories, including subjective assessments of important behaviours or personality traits.

Second, our analysis proved that no single behaviour or even cohort of traits was an absolute predictor of breeding success; this should not have been surprising. Just prior to the formal CBSG Biomedical Survey, Lindburg et al. (1997) examined a variety of factors potentially influencing reproductive success of male pandas housed in Chinese breeding centres and zoos. Lindburg and others (e.g. Zhang et al., 2004) determined that captive giant pandas often display inappropriate aggression when placed in mating situations. The present survey supported the importance of aggression as a determining variable, while additionally revealing a high correlation of mating success when both genders were characterised as 'aggressive'. It is apparent that there is some (as yet undefined) range of aggression that 'works' for this species – if an animal exceeds or shows less than the appropriate level, it probably will not successfully reproduce. Zhang and colleagues (2004) have suggested that a combination of temperament, stress and inadequate husbandry practices together contribute to reproductive failure. That study tracked 103 mating introductions involving a total of six males and 12 females where intromission did not occur. Of these copulatory failures, 47% were due to inability to mount properly (most likely due to sexual inexperience often related to age), and 33% were caused by lack of male interest. Zhang and colleagues (2004) suggested that timid females are less likely to breed, a result that the present study also supports, with an 'opposite' personality characteristic of aggression being one of the strongest predictors of breeding success. Of the six males in Zhang's study, the two that were the most successful breeders were also wild born, a finding also supported by the present study.

This is interesting given our third significant finding, that breeding success was highly related to origin: wild-born animals were more productive than their captive-born counterparts. Lindburg and colleagues (1997) have also addressed this issue briefly by stating that 'early rearing deficiencies could be one of several factors contributing to mating dysfunction in male adults'. Snyder and colleagues (2003, 2004; see also Chapter 14) have made progress in examining the potential effects of early rearing on later reproductive behaviour, particularly in males. In brief, behavioural development differs between dam- and hand-reared panda cubs, with the ultimate impact on later reproductive success still to be determined. Therefore, we agree that this factor merits intensive and immediate investigation, including identifying characteristics of the captive environment that perturb normal breeding behaviours. In fact, our finding of better overall reproduction in pandas of wild origin no doubt was linked to this same issue. It makes sense that offspring properly socialised by a mother in nature would have a better capacity to mate (even in captivity) compared to captive-born counterparts that receive less maternal care due to hand-rearing and/or early weaning. Mother-reared young ostensibly have also had the opportunity to learn appropriate expression of aggression or, conversely, to deal with aggression from conspecifics. Together, these findings re-affirm the assertions by Snyder and colleagues (2003; 2004; see also Chapter 14) that a priority for the future is not just sheer number of cubs but rather an emphasis on the 'quality' of offspring with the opportunity for normal behavioural development.

Nevertheless, a key question remains: what constitutes the behavioural essence of a 'quality' (i.e. successful) breeder – an individual that can consistently, naturally and successfully copulate and, for females, one that can always successfully rear cubs? While elucidating traits compatible with being a Prime Breeder, the present study (as part of the CBSG Biomedical Survey) has also generated even more questions. For example:

- What can be done to ensure that pandas are not only physically healthy but also behaviourally equipped to reproduce successfully and rear offspring?
- What are the enabling conditions necessary to achieve this goal in terms of enclosure design and husbandry practices associated with neonatal and juvenile management?
- Will this mean rejecting traditional approaches of maintaining giant panda pairs in favour of multigenerational groups to enhance both reproduction and socialisation (see Chapter 14)?
- Since aggression seems to be an inherent trait in wild giant pandas (especially males), how can we develop safe methods for encouraging breeding in such individuals?
- If an eventual goal of *ex situ* management is to contribute to reintroduction, then what are the behavioural competencies necessary for such animals to reproduce and survive in nature?
- How are such animals best 'created' *ex situ*?

All of these questions are of high priority as we seek to master the art of propagating giant pandas that are behaviourally, genetically and physiologically robust, manifesting the natural behaviours and behavioural flexibility that serve them so well in the wild.

ACKNOWLEDGEMENTS

We thank the giant panda keepers at Beijing Zoo, Chongqing Zoo, the Chengdu Research Base of Giant Panda Breeding, Chengdu Zoo and the China Conservation and Research Center for the Giant Panda for assistance in providing detailed information. Nadja Wielebnowski provided generous and substantive input in developing the keeper survey.

REFERENCES

Fagen, R. and Fagen, J. M. (1996). Individual distinctiveness in brown bears, *Ursus arctos. Ethology*, **102**, 212–26.

Feaver, J., Mendl, M. and Bateson, P. (1986). A method for rating individual distinctiveness of domestic cats. *Animal Behaviour*, **34**, 1016–25.

Gold, K. C. and Maple, T. L. (1994). Personality assessment in the gorilla and its utility as a management tool. *Zoo Biology*, **13**, 509–22.

Lewontin, R. C., Rose, S. and Kamin, L. J. (1984). *Not in Our Genes: Biology, Ideology and Human Nature*. New York, NY: Pantheon Books.

Lindburg, D. G., Huang, X. M. and Huang, S. Q. (1997). Reproductive performance of giant panda males in Chinese zoos. In *Proceedings of the International Symposium on the Protection of the Giant Panda (Ailuropoda melanoleuca)*, ed. A. Zhang and G. He. Chengdu: Sichuan Publishing House of Science and Technology, pp. 67–71.

Lorenz, K. (1965). *The Evolution and Modification of Behavior*. Chicago, IL: University of Chicago Press.

Schaller, G. B., Hu, J., Pan, W. and Zhu, J. (1985). *The Giant Pandas of Wolong*. Chicago, IL: University of Chicago Press.

Snyder, R. J., Zhang, A. J., Zhang, Z. H. *et al.* (2003). Behavioral and developmental consequences of early rearing experience for captive giant pandas. *Journal of Comparative Psychology*, **117**, 235–45.

Snyder, R. J., Lawson, D. P., Zhang, A. *et al.* (2004). Reproduction in giant pandas: hormones and behavior. In *Giant Pandas: Biology and Conservation*, ed. D. Lindburg and K. Baragona. Berkeley, CA: University of California Press, pp. 125–32.

Wielebnowski, N. C. (1999). Individual behavioral differences in captive cheetahs as predictors of breeding status. *Zoo Biology*, **18**, 335–49.

Wilson, E. O. (1975). *Sociobiology: The New Synthesis*. Cambridge, MA: Belknap Press of Harvard University Press.

Xie, Z. and Gipps, J. (2003). *The 2003 International Studbook for the Giant Panda (Ailuropoda melanoleuca)*. Beijing: Chinese Association of Zoological Gardens.

Zhang, G., Swaisgood R. R. and Zhang, H. (2004). An evaluation of the behavioural factors influencing reproductive success and failure in captive giant pandas. *Zoo Biology*, **23**, 15–31.

Zheng, S., Zhao, Q., Xie, Z., Wildt, D. E. and Seal, U. S. (1997). *Report of the Giant Panda Captive Management Planning Workshop*. Apple Valley, MN: IUCN–World Conservation Union/SSC Conservation Breeding Specialist Group.

6

Nutrition and dietary husbandry

MARK S. EDWARDS, GUIQUAN ZHANG, RONGPING WEI,
XUANZHEN LIU

INTRODUCTION

Nutrition involves a series of processes whereby an animal uses items in its external environment to support internal metabolism (Robbins, 1993). The nutrition and consequent nutritional status of an animal are basic to all aspects of health, including growth, reproduction and disease resistance. Thus, appropriate nutrition and feeding are essential to a comprehensive animal management and preventative medicine programme.

The giant panda's obligate dependence upon bamboo as a primary energy and nutrient source has been well described (Sheldon, 1937; Schaller et al., 1985). Many aspects of panda biology are directly related to its adaptations for utilisation of this highly fibrous, low energy density food, thus demonstrating the inseparable influence of nutrition on behaviour, reproduction and other physiological functions. There may be few other species that more effectively illustrate how an understanding of nutritional adaptations helps us interpret the species ecology.

This chapter describes insights into the nutritional adaptations of the giant panda while identifying priority research that will fill gaps in

Giant Pandas: Biology, Veterinary Medicine and Management, ed. David E. Wildt, Anju Zhang, Hemin Zhang, Donald L. Janssen and Susie Ellis. Published by Cambridge University Press. © Cambridge University Press 2006.

our understanding of these unique abilities. Historical and current strategies on feeding giant pandas in captivity are presented along with recommendations for improving nutrition and dietary husbandry to promote health and feeding behaviours.

ANATOMY, PHYSIOLOGY, GUIDELINES AND ASSESSMENT

Feeding ecology and anatomical adaptations to a herbivorous diet

More than 99% of the food consumed by the free-ranging giant panda consists of bamboo (Schaller *et al.*, 1985). Yet the giant panda is unique in that it has the relatively simple gastrointestinal tract of a carnivore. More specifically, it lacks the modifications found in most herbivores that promote increased digesta retention to facilitate microbial fermentation of ingested plant materials (Schaller *et al.*, 1985).

Anatomically, the giant panda exhibits several specialised adaptations for processing and utilising bamboo. Perhaps the most familiar is the panda's enlarged and elongated radial sesamoid bone, attached to the first metacarpal bone (Fig. 6.1) (Davis, 1964; Endo *et al.*, 1996, 1999). Using three-dimensional computed tomography, this adaptation can be visualised as a double pincer-like apparatus between the radial sesamoid and accessory carpal bones, allowing the panda to grasp and manipulate bamboo culm with remarkable dexterity, as though it had a thumb (Endo *et al.*, 2001).

The giant panda's large skull, with wide flaring zygomatic arches and a prominent sagittal crest (Fig. 6.2), supports heavy craniomandibular musculature, giving the species the crushing power required to masticate fibrous, highly lignified bamboo. Working in combination with these muscles are large, flat cheek teeth with elaborate crown patterns (Fig. 6.3), characteristic of herbivores. These dental characteristics are modifications of typical carnivore dentition (I3/3 C1/1 P4/4 M2/3 = 42, with P_1 degenerate in both jaws and sometimes absent from the upper jaw) (Chorn & Hoffman, 1978; Schaller *et al.*, 1985).

The giant panda's gastrointestinal tract, evolutionarily one of the most plastic organ systems in the animal world, is remarkably unspecialised for herbivory and is largely characteristic of omnivores. Once masticated, ingesta pass into an oesophagus with a tough and horny lining, followed by an uncompartmentalised stomach with a long, thick-walled pyloric section folded back on the cardia. The pylorus

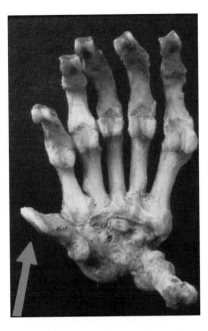

Figure 6.1. Paw anatomy of the giant panda with arrow depicting the enlarged and elongated radial sesamoid bone.

Figure 6.2. Giant panda skull illustrating the flaring zygomatic arches (arrows) and prominent sagittal crest.

Figure 6.3. Giant panda skull illustrating the large, flat teeth.

remotely resembles a gizzard in birds (see Chapter 18) and may knead and mix food with digestive juices.

The small intestine is a much-reduced segment of the gastro-intestinal tract, suggesting that limited digestion occurs in this region (Chorn & Hoffman, 1978). The caecum is absent. However, relative to other bear species, the surface area of the colon is enlarged (Chorn & Hoffman, 1978). Overall, intestinal length ranges from 4.1 to 5.5 times head and body length (Raven, 1937; Davis, 1964; Schaller *et al.*, 1985). The characteristics of the large intestine are consistent with being populated by microbial symbionts, although their contribution to digesta fermentation is probably limited due to rapid transit rates (see below) (Hirayama *et al.*, 1989).

Field biologists have noted that faecal boluses of the free-ranging giant panda are coated with a thin layer of mucus, which may lubricate the fibrous digesta, facilitating movement through the intestinal tract (Schaller *et al.*, 1985). Histological examination of the panda's gastrointestinal tract has revealed that the large intestine has a significant number of mucous cells (Wang *et al.*, 1982).

The consequence of consuming a highly fibrous diet and the rapid passage of ingesta through a relatively unspecialised gastrointestinal tract (i.e. without sacculations or compartments to retain digesta for microbial degradation) is reduced nutrient digestibility and absorption.

Food intake

Measurement of food intake is complicated by variability among pandas, food palatability and food items selected. Yet quantifying *ad libitum* intake as it relates to food quality is important in evaluating animal response to that food (Van Soest, 1994).

Ad libitum intake is typically measured in a controlled environment where food is offered at 15 to 20% in excess of the quantity usually consumed. Under these conditions, refused food may differ in composition from food offered, depending on the ability of the animal to select preferentially and ingest specific portions of that diet. This selective intake is of particular concern when evaluating nutrients supplied by bamboo because bamboo components (i.e. shoots, leaves, branches and culm) differ appreciably in composition.

Food intake quantified in giant pandas under field (free-ranging) and captive conditions is summarised in Table 6.1.

In conjunction with the CBSG Biomedical Survey of Giant Pandas (1998 to 2000; see Chapter 2), 24-hour food intake data were provided and/or measured for 34 pandas in captive facilities in China (Table 6.2).

Dry matter intake (as a percentage of body weight, BW) was comparable between the two groups of captive animals as summarised in Tables 6.1 and 6.2 (2.7 – 5.6% BW). However, dry matter intake (% BW) was lower across all captive specimens when compared to estimates of intake under field conditions. These differences were most likely due to divergence in digestible energy density of the dietary items consumed or to the dietary proportions of bamboo ingested (Van Soest, 1994). When increasing quantities of bamboo forage are consumed (as a percentage of total dry matter intake), digestible energy density of the total diet declines, requiring consumption of more food to maintain absolute caloric intake. Additionally, giant pandas in captivity would be expected to have lower dry matter intake and energy intake consistent with reduced activities (and, thus, lower energy requirements) compared to their wild counterparts.

A similar trend is seen among estimates of intake in animals under field conditions (see Table 6.1). Dry matter intake of bamboo

Table 6.1. *Daily food intake of giant pandas expressed as kilograms of fresh weight (FW), kilograms of dry matter (DM) or DM intake as a percentage of body weight (BW)*

Age class	Gender	Diet[a]	FW (kg)[b]	DM (kg)[b]	DM (%)BW[c]	Reference
Free-ranging						
Adult	Male	*Sinarundinaria*, winter	14.0	7.54	7.5	Schaller et al., 1985
Adult	Female	*Sinarundinaria*, winter	10.0	5.18	6.0	Schaller et al., 1985
Adult	Female	Bamboo, spring	12.8	7.20	8.4	Schaller et al., 1985
Adult	Female	Bamboo shoots	38.3	3.83	4.5	Schaller et al., 1985
Adult	–	Bamboo	17.0	9.46	9.5	Pan, 1988
Adult	–	*Bashania*, leaf and culm	12.9	7.06	7.1	Pan, 1988
Adult	–	*Fargesia*, culm	–	7.65	7.7	Pan, 1988
Adult	–	*Bashania*, shoots	–	4.76	4.8	Pan, 1988
Adult	–	*Fargesia*, shoots	–	4.70	4.7	Pan, 1988
Adult	–	Bamboo shoots	23.7	–	–	Pan, 1988

Adult	Male	Bamboo	11.3–13.6	6.3–7.6	6.3–7.3	Schaller, 1993
Captive						
Adult	Male	Mixed, 75% bamboo	–	6.9	5.6	Dierenfeld et al., 1982
	Female	Mixed, 54% bamboo	–	3.4	3.0	
Adult	Female	Mixed, 55% bamboo	11.3	2.98	2.7	Mainka et al., 1989
Subadult	Female	Mixed, 62% bamboo	14.4	2.28	3.7	
Adult	Male	Mixed, 77% bamboo	7.27	–	–	Edwards & Zhang, 1997
	Female	Mixed, 74% bamboo	7.28	–	–	Edwards & Zhang, 1997
Adult	Male	Mixed, 56% bamboo	–	2.86	2.73	Nickley et al., 1999
	Female	Mixed, 75% bamboo	–	3.84	3.59	Nickley et al., 1999
Subadult	–	Mixed	–	–	1.37–5.95	Liu et al., 2002

[a] Proportion of bamboo (leaves, branches and culm) consumed in mixed-ingredient diets expressed as percentage of dietary DM; [b] Values in italics calculated, assuming 45% leaf (48.4% DM) and 55% culm (61.6% DM) in FW of bamboo consumed; [c] Values in italics calculated, assuming 100 kg BW.

Table 6.2. *Food intake by giant pandas of three age classes over a 24-hour interval, expressed as kilograms dry matter (DM) or as percentage of body weight, (BW)*[a]

Age class[b]	Gender	Number of animals	Diet[c]	DM (kg)	DM (%BW)
Juvenile	Female	4	Mixed, 41% bamboo	1.35	2.9
Subadult	Male	3	Mixed, 62% bamboo	3.86	4.0
	Female	3	Mixed, 51% bamboo	2.85	3.3
Adult	Male	11	Mixed, 54% bamboo	3.40	3.0
	Female	13	Mixed, 59% bamboo	2.85	3.2

[a] CBSG Biomedical Survey (2001).
[b] Juvenile (12–18 months of age); subadult (19 months to 4.5 years); adult (>4.5 years).

leaves, branches and culm is higher (% BW) than when consuming only bamboo shoots, presumably because the latter is more digestible, providing more calories per unit consumed.

Digestive capabilities

The capacity of giant pandas to digest a diet comprised exclusively of bamboo or, more specifically, bamboo components (e.g. leaves, branches, culm) has yet to be determined in a controlled environment. However, several scientists have estimated the digestive efficiency of pandas through indirect methods in the field or direct methods when mixed-ingredient diets have been fed in captivity. These data are summarised in Table 6.3, while providing comparative data for the red panda (*Ailurus fulgens*).

Apparent digestibility is a measure of an item in the feed less that item in the faeces. In contrast to true digestibility, no corrections are made in this measure for dry matter, energy or nutrients in the faeces which originate (a) from the tissues or secretions of the study animal or (b) from microbial organisms within that animal's gastrointestinal tract which would appear in the faeces even when no food is consumed. Apparent digestibility (%) of dietary dry matter (AD_{DM}) is algebraically expressed as:

$$AD_{DM}(\%) = [(DM_{intake} - DM_{faeces})/DM_{intake}] \times 100$$

Table 6.3. Digestive capabilities of the red panda (Ailurus fulgens) and giant panda fed various diets[a]

Diet[b]	DMI (% BW)	Apparent digestibility (%)						Reference
		DM	NDF	ADF	HC	Cell.		
Red panda								
P. japonica, leaf	–	23.9	24.7	31.8	–	–		Warnell et al., 1989
Gruel	–	84.7	14.8	44.4	–	–		
B. spanostachya, leaf	11.9	29.6	–	–	28.2	3.5		Wei et al., 1999
B. spanostachya, shoots	7.8	46.0	–	–	44.4	–		
Giant panda								
S. fangiana, culm	7.6	12.5	–	–	21.5	–		Schaller et al., 1985
S. fangiana, leaf	5.0	23.3	–	–	26.0	–		
S. fangiana, mixed	6.4	18.7	–	–	18.2	–		
F. spathacea, shoots	3.8	40.0	–	–	–	–		
B. fargesii, leaf	–	17.2	–	–	17.5	0.5		Long et al., 2004
B. fargesii, culm	–	15.9	–	–	22.4	11.0		
F. spathacea, culm	–	12.7	–	–	15.2	7.2		
Mixed, 75% bamboo	5.6	38.3	19.1	–	28.2	3.6		Dierenfeld et al., 1982

Table 6.3. (cont.)

Diet[b]	DMI (% BW)	Apparent digestibility (%)						Reference
		DM	NDF	ADF	HC	Cell.		
Mixed, 54% bamboo	3.0	54.9	25.1	–	28.8	13.7		Kametaka et al., 1988
Mixed, 61% bamboo	2.6	62.2	52.0	48.0	56.0	47.0		
Mixed, ~45% bamboo	1.6	57.5	37.0	30.0	39.0	32.0		
Mixed, 62% bamboo	3.7	47.5	–	–	40.9	51.6		Mainka et al., 1989
Mixed, 55% bamboo	2.7	47.9	–	–	31.7	40.8		
Mixed	–	67.4	–	–	–	–		Zou et al., 1993
Mixed	–	61.6	–	–	–	–		

[a] DMI, DM intake; DM, dry matter; NDF, neutral detergent fibre; ADF, acid detergent fibre; HC, hemicellulose; Cell. cellulose; [b] Proportion of bamboo (leaves, branches and culm) consumed in mixed ingredient diets expressed as percentage of dietary DM.

where DM_{intake} is the mean dry matter intake (g day^{-1}), and DM_{faeces} is the average quantity of undigested dry matter (g day^{-1}) excreted in faeces (Van Soest, 1994).

Lignin, a polymerised, phenolic compound, is a highly indigestible component of the plant cell wall that has been used as an undigested marker to indirectly calculate food digestibility (Fahey & Jung, 1983; Van Soest, 1994). For example, when using lignin to indirectly estimate apparent digestibility (%) of dietary dry matter (AD_{DM}), the algebraic expression is:

$$AD_{DM}(\%) = [1 - (L_{feed}/L_{faeces})] \times 100$$

where L_{feed} is the concentration (%) of the indigestible marker (i.e. lignin) in the feed, and L_{faeces} is the concentration (%) of the indigestible marker in the faeces (Robbins, 1993).

Apparent digestibility (%) of a given nutrient (AD_{Nutr}) or component (e.g. protein or hemicellulose) can also be calculated using the indirect marker method:

$$AD_{Nutr}(\%) = [1 - (L_{feed} \times Nutr_{faeces})/(L_{faeces} \times Nutr_{feed})] \times 100$$

where $Nutr_{feed}$ is the concentration (%) of the nutrient or component measured in the feed, and $Nutr_{faeces}$ is the concentration (%) of the nutrient or component measured in the faeces. L_{feed} and L_{faeces} are the concentrations (%) of lignin in feed and faeces, respectively (NRC, 1993). Unfortunately, some of the lignin, although not digested, may not be recoverable from faeces. Thus, an error can be introduced into indirect estimates of apparent digestibility.

Long et al. (2004) estimated apparent digestibility of dry matter, hemicellulose and cellulose of bamboo by free-ranging giant pandas in the Qinling Mountains. Daily food intake was predicted from direct observations, an estimate of feeding rate (food mass per time unit) and the percentage of leaf material in foods consumed. Estimates of intake and lignin concentrations in foods consumed were used to estimate lignin intake. Lignin excretion was estimated by determining lignin concentrations in faeces produced by the focal animal. Based on these estimates and using the algebraic relationships described above, Long et al. (2004) estimated dry matter digestibilities to be 12.7 to 17.2% (see Table 6.3). Hemicellulose and cellulose digestibilities were estimated to be 15.2 to 22.4% and 0.5 to 11.0%, respectively.

Schaller et al. (1985) provided another estimate of bamboo dry matter and hemicellulose utilisation in free-ranging pandas in the

Wolong Nature Reserve. Using similar indirect methods of estimating total food intake, distribution of leaves, branches and culm consumed, and collection of faecal boluses, these authors calculated an average dry matter digestibility of 17% (range 12.5–23.3%; see Table 6.3). Calculated hemicellulose digestibility averaged 22% (range 18.2–26.0%; see Table 6.3).

Although a captive animal environment provides the level of control needed to conduct direct digestion trials, investigators have been reluctant or unable to offer such animals single ingredient diets (e.g. only bamboo) for evaluation. Assessment of bamboo digestibility extrapolated from animals fed mixed-ingredient diets is problematic at best. Such measurements assume that there is no interaction from the consumption of supplemental foods (e.g. vegetables, meat or dairy products), and that the digestibility of the basal component (i.e. bamboo portions consumed) is unaltered by the added supplement. Interactions or associative effects among feeds are common and often lead to under- or over-estimates of food digestibility (Van Soest, 1994).

Digestibility of dry matter by captive giant pandas fed mixed-ingredient diets with varying proportions of bamboo range from 38.3 to 67.4% (see Table 6.3), significantly higher than estimates made for animals under field conditions. Digestibility of hemicellulose shows a similar upward trend (28.2 to 56%; see Table 6.3). The differences observed between captive and wild populations are most likely due to variations in diets, levels of activity, methodology or a combination of these three factors. Although estimates made under field conditions include methodological limitations (particularly on intake), it appears that the giant panda has adapted to a diet primarily of bamboo. However, only 25% or less of the dry matter within that bamboo can be digested. Although the captive giant panda appears able to use more dietary dry matter when offered bamboo in the presence of other foods, the physiological impact of these supplements on gastrointestinal function and health has not been critically evaluated.

The ability of giant pandas to digest bamboo is limited by the rapid rate of digesta transit through the gastrointestinal tract. Several measures can be used to assess the time digesta spends in the tract, subject to the processes of mechanical mixing, digestion, microbial fermentation and absorption. When an inert digesta marker is fed, the time elapsing from ingestion to first appearance of the marker in the faeces is termed transit time (TT_1). Although this measure is useful, mean retention time (R_{git}) is more descriptive of the lag following initial marker excretion (TT_1) and total digesta turnover. Mean retention time is determined as follows:

$$R_{git} = [\Sigma(Y_i \times T_i)/\Sigma Y_i] - TT_1$$

where Y_i is the concentration of the marker at a given time and T_i is the time, since marker ingestion (Blaxter et al., 1956).

Transit time for a giant panda in captivity that consumes a diet of at least 50% bamboo ranges from 6 to 7 hours (Table 6.4) (Dierenfeld et al., 1982; Edwards, 2003). A similar result (mean 7.9 hours) was reported for another captive giant panda, although diet details were not provided (Schaller et al., 1985). Rapid digesta passage in pandas is not conducive to microbial fermentation of digesta, and cellulose digestibility is not significantly above zero (Van Soest, 1994). These observations are supported by low populations of obligate anaerobic bacteria in faeces (Hirayama et al., 1989).

When compared to other herbivores, the physical characteristics of individual faecal boluses produced by the captive giant panda suggest limited mixing of digesta in the gastrointestinal tract. We have measured the extent of mixing using two differently coloured acetate bead markers. Each marker was fed in a single pulse bolus, one at time 0 and the second three hours later. Two animals, whose bamboo intake was greater than 80% of diet dry matter, had no overlap of the two markers in the excreta. A single animal, whose bamboo intake was less than 50% of diet dry matter excreted 78% of the two markers in the same faecal bolus (Edwards, 2003). These results suggest that, unlike other herbivores, the giant panda passes digesta in a single pulse, or wave, through the gastrointestinal tract. This adaptation may have arisen from the consumption of a homogeneous diet, both in type and nutrient content. Implications for captive animal husbandry and feeding are addressed below.

Bamboo

Clearly, bamboo is the predominant food source for the giant panda (Sheldon, 1937; McClure, 1943; Wang & Lu, 1973; Schaller et al., 1985; Carter et al., 1999). Species of bamboo used by free-ranging giant pandas vary with habitat. Pandas in the Qinling Mountains, where nine bamboo species belonging to five genera have been described, consume Bashania fargesii and Fargesia spathacea preferentially (Pan, 1988; Long et al., 2004). These two genera are the dominant native bamboos in the

Table 6.4. *Digesta mean (± SEM) transit times (TT$_1$) and mean retention times (R$_{git}$) (hours) of diets fed to the red panda (Ailurus fulgens) versus giant panda*

Diet[a]	DMI (% BW)[b]	Marker	TT$_1$	R$_{git}$	Marker	TT$_1$	R$_{git}$	Reference
Red panda								
P. japonica, leaf	–	Wheat	5.4	–	–	–	–	Warnell et al., 1989
Gruel	–	Wheat	9.4	–	–	–	–	
B. spanostachya, leaf	11.9	Shoots	3.5	–	–	–	–	Wei et al., 1999
B. spanostachya, shoots	7.8	Leaves	2.5	–	–	–	–	
Giant panda								
Mixed	–	Shoots	7.9	–	–	–	–	Schaller et al., 1985
Mixed, 75% bamboo	5.6	Wheat	8.0 ± 3.0	–	PEG[c]	8.0 ± 3.0	–	Dierenfeld et al., 1982
Mixed, 54% bamboo	3.0	Wheat	8.0 ± 3.0	–	PEG[c]	8.0 ± 3.0	–	
Mixed, 62% bamboo	4.0	Wheat	6.5	–	–	–	–	Mainka et al., 1989
Mixed, 55% bamboo	2.7	Wheat	6.5	–	–	–	–	
Mixed	–	Bead	8.75	–	–	–	–	Edwards (unpublished data)
Mixed, 83% bamboo	4.9	Bead	6.0	0.65	–	–	–	Edwards, 2003
Mixed, 91% bamboo	6.6	Bead	6.0	0.65	–	–	–	
Mixed, 47% bamboo	2.2	Bead	11.0	4.87	–	–	–	

[a] Proportion of bamboo (leaves, branches and culm) consumed in mixed-ingredient diets expressed as percentage of dietary DM; [b] DMI (% BW), dry-matter intake as percentage of body weight; [c] Polyethylene glycol.

region. Giant pandas here also consume *Phyllostachys nigra*, a bamboo species introduced into the habitat and commonly used for feeding pandas *ex situ* (Edwards, 1996, 2003; Long *et al.*, 2004). Schaller *et al.* (1985) observed that giant pandas in the Wolong Nature Reserve fed mainly on *Sinarundinaria fangiana* and *Fargesia spathacea* with *Sinarundinaria nitida* also consumed, but to a much lesser extent. Giant pandas in the Min Mountains consume a single bamboo species, whereas animals in the Liang Mountains forage upon five bamboo species (Qing, 1977).

Bamboo taxonomy has been periodically revised based on new information, creating a long list of species synonyms. Because of earlier poor communication between Chinese and western scientists, different Latin binomials are often used for the same species (Reid & Hu, 1991; Carter *et al.*, 1999). Scientific names here were cross-referenced against the synonym lists compiled by the American Bamboo Society (Shor, 2001) and the European Bamboo Society (Masman, 1995). Table 6.5 lists those species reportedly consumed by free-ranging giant pandas.

A diversity of bamboo species has been offered to giant pandas in captivity, primarily due to logistical and climatic limitations in cultivating some of the bamboo species consumed in the wild. Giant pandas living *ex situ* have been noted to demonstrate particular preferences for specific bamboo species and portions of species, as well as selecting non-bamboo foods preferentially over bamboo when given that opportunity. However, studies quantifying these preferences and the factors that influence them have yet to be published. Table 6.6 lists species commonly offered to, and consumed by, giant pandas in captivity.

Selected proximate, fibre, mineral and energy concentrations in bamboos consumed by pandas in the field and in captivity are provided in Tables 6.A.1 and 6.A.2 in Appendix 6.A.

Consumption of non-bamboo foods

Although bamboo constitutes the primary source of nutrition for wild giant pandas, other food items are eaten, typically in less than 1% of total food intake (Schaller *et al.*, 1985; Pan, 1988). Giant pandas have been reported to consume more than 25 wild plant species (Schaller *et al.*, 1985) and occasionally animal-based foods (e.g. eggs, small or infant animals and carrion). Consumption of these non-bamboo items is generally opportunistic, and their contribution to the free-living panda's overall nutrient intake is probably insignificant.

Geophagia (deliberate consumption of soils not necessarily associated with a mineral lick) has been observed in at least one giant panda in

Table 6.5. *Bamboo species consumed by free-ranging giant pandas, current taxonomic synonyms for those species and regions where found (modified from Carter et al., 1999)*

Species name	Synonym	Region	References
Bashania fargesii		Min, Qinling, Qionglai	Pan, 1988; Wang, 1989; Li, 1997
Chimonobambusa pachystachys		Min, Qionglai, Xiangling, Liang	Wang, 1989; Li, 1997
Fargesia denudata		Min	Wang, 1989; Li, 1997
F. ferax		Xiangling, Liang	Wang, 1989; Li, 1997
F. rufa		Min	Wang, 1989; Li, 1997
F. scrabrida		Min, Qionglai	Wang, 1989; Li; 1997
F. robusta	*F. spathacea*	Min, Qinling, Qionglai	Schaller *et al.*, 1985; Pan, 1988; Wang, 1989; Li, 1997
Qiongzhuea opienensis		Xiangling, Liang	Wang, 1989; Li, 1997
Phyllostachys nigra		Qinling	Pan, 1988
Yushania chungii	*Sinarundinaria chungii*	Min, Qinling, Qionglai	Wang, 1989
Bashania fangiana	*Sinarundinaria fangiana*	Min, Xiangling, Liang	Schaller *et al.* 1985; Wang, 1989; Li, 1997
F. nitida	*Sinarundinaria nitida*	Min, Qinling, Qionglai	Schaller *et al.* 1985; Wang, 1989; Li, 1997
Y. confusa		Xiangling, Liang	Wang, 1989

Table 6.6. *Bamboo species commonly offered to giant pandas in captivity (modified from Nickley, 2001)*

Scientific name	Reference
Bambusa beecheyana	Edwards, 2003
B. blumeana	Liu *et al.*, 2002
B. glaucescens	Edwards, 2003
B. multiplex	Edwards, 2003
B. oldhamii	Edwards, 2003
B. textiles	Edwards, 2003
B. tuldoides	Edwards, 2003
B. ventricosa	Edwards, 2003
B. vulgaris	Edwards, 2003
Bashania fargesii	Edwards, 2003
B. fargiana	Liu *et al.*, 2002
Chimnobambusa quadrangularis	Edwards, 2003
Fargesia fungosa	Edwards, 2003
F. nitida	Edwards, 2003
F. robusta	Liu *et al.*, 2002
F. spathacea	Mainka *et al.*, 1989
Phyllostachys aurea	Mainka *et al.*, 1989; Edwards, 2003
Ph. aureosulcata	Dierenfeld *et al.*, 1982; Edwards, 2003; Tabet *et al.*, 2004
Ph. bambusoides	Edwards, 2003
Ph. bisetti	Tabet *et al.*, 2004
Ph. mitis	Kametaka *et al.*, 1988
Ph. nigra	Crouzet & Frädrich, 1985; Edwards, 2003; Tabet *et al.*, 2004
Pseudosasa japonica	Tabet *et al.*, 2004

the field and in another in captivity (Schaller *et al.*, 1985). Soil consumption is believed to have several possible functions, including satisfying trace mineral requirements, chelating metal ions and binding secondary plant metabolites, thus inhibiting their absorption (Robbins, 1993).

Estimated nutrient requirements and availability

Information on nutrient requirements of giant pandas is lacking, yet managers of captive specimens could benefit from reference standards of adequate dietary nutrient concentrations. Such estimates are possible if one considers the species' natural feeding ecology and its gastrointestinal anatomy, as well as the defined nutrient requirements of similar species (e.g. the domestic dog). Using actual annual food consumption by three adult giant pandas and chemical analysis of consumed foods, we predicted mean intake of selected nutrients. As part of these estimates, intake of bamboo species was based on equal distribution of five commonly offered species across total bamboo intake. Consumption of the individual bamboo components (leaf, branch, culm, each being unique in nutrient composition) was calculated using prediction equations based on the relationship between total bamboo fresh mass and bamboo component dry mass (Edwards *et al.*, unpublished data). Estimated adequate nutrient concentrations in offered diets and calculated nutrient intake for the giant panda in captivity are summarized in Table 6.7. For comparative purposes, recommended adult dog nutrient allowances for maintenance and dog food nutrient profiles also are provided.

Users of this information should consider the physiological differences between the giant panda and other non-ruminants (e.g. dog, rat and pig) from which these nutrient guidelines are extrapolated. In particular, although not known, it is possible that the rapid passage of fibrous digesta through the panda gastrointestinal tract could reduce digestion and/or absorption of certain nutrients, including major minerals, trace elements and fat-soluble vitamins. However, the comparatively higher level of food intake by the giant panda may well compensate for reduced nutrient availability. A research priority is determining if nutrient availabilities implicit in recommendations for better-studied models (e.g. the dog) are appropriate for the giant panda, or if higher nutrient levels are warranted as a precaution against low accessibility.

Nutrient guidelines

Water

Water is an essential and often overlooked nutrient. Water, which comprises 99% of all molecules within the animal's body, functions as a solvent, is involved in hydrolytic reactions, temperature control, transport of metabolic products, excretion, lubrication of skeletal joints and sound and light transport within the ear and eye (MacFarlane & Howard, 1972; Robbins, 1993).

Table 6.7. *Estimated adequate nutrient concentrations in diets offered to, and calculated nutrient intake (on a dry matter basis) by, giant pandas in captivity, recommended adult dog maintenance nutrient allowances and dog food nutrient profiles (on a dry-matter basis)*

Nutrient	Giant panda Estimate[a]	Giant panda Intake[b]	Dog NRC[c]	Dog AAFCO[d]
Crude protein (%)	16–18	9.3	10	18
Arginine (%)	0.7	nd	0.35	0.51
Histidine (%)	–	nd	0.19	0.18
Isoleucine (%)	0.7	nd	0.38	0.37
Leucine (%)	–	nd	0.68	0.59
Lysine (%)	0.8	nd	0.35	0.63
Methionine + cystine (%)	0.6	nd	0.65	0.43
Phenylalanine + tyrosine (%)	–	nd	0.74	0.73
Threonine (%)	0.58	nd	0.43	0.48
Tryptophan (%)	0.2	nd	0.14	0.16
Crude fat (%)[g]	3.0	2.3	5.5	5.0
Essential n-6 fatty acids (%)	1.0	nd	1.1	1.0
Essential n-3 fatty acids (%)	–	nd	0.05	–
Neutral detergent fibre (%)	50	70.9	–	–
Acid detergent fibre (%)	25	46.3	–	–
Calcium (%)	1.0	0.4	0.3	0.6
Total phosphorus (%)	0.8	0.2	0.3	0.5

Nutrient	Giant panda Estimate[a]	Giant panda Intake[b]	Dog NRC[c]	Dog AAFCO[d]
Iron (mg kg^{-1})[e]	100	127	30	80
Copper (mg kg^{-1})[f]	8	4	6	7.3
Manganese (mg kg^{-1})	40	40	5.0	5.0
Zinc (mg kg^{-1})	120	36	60	120
Iodine (mg kg^{-1})	1.5	nd	0.9	1.5
Selenium (mg kg^{-1})	0.3	0.07	0.35	0.11
Chromium (mg kg^{-1})	–	nd	0.7	–
Vitamin A (IU kg^{-1})	5000	nd	4545	5000
β carotene (mg kg^{-1})	50	nd	–	–
Vitamin D (IU kg^{-1})	1200	nd	550	500
Vitamin E (IU kg^{-1})	200	nd	30	50
Vitamin K (mg kg^{-1})	–	nd	1	–
Thiamin (mg kg^{-1})[h]	3	nd	2.25	1.0
Riboflavin (mg kg^{-1})	5	nd	5.25	2.2
Pantothenic acid (mg kg^{-1})	15	nd	15	10
Available niacin (mg kg^{-1})	50	nd	15	11.4
Vitamin B6 (mg kg^{-1})	3	nd	1.5	1.0

Table 6.7. (cont.)

Nutrient	Giant panda Estimate[a]	Intake[b]	Dog NRC[c]	AAFCO[d]
Magnesium (%)	0.2	0.11	0.06	0.04
Potassium (%)	0.85	0.95	0.4	0.6
Sodium (%)	0.2	0.10	0.04	0.06
Chloride (%)	0.3	nd	0.06	0.09

Nutrient	Giant panda Estimate[a]	Intake[b]	Dog NRC[c]	AAFCO[d]
Biotin (μg kg^{-1})[i]	200	nd	–	–
Folate (μg kg^{-1})	500	nd	270	180
Vitamin B$_{12}$ (μg kg^{-1})	30	nd	35	22
Vitamin C (mg kg^{-1})	100	nd	–	–
Choline (mg kg^{-1})	1500	nd	1700	1200

[a] Estimated adequate nutrient concentrations of diets offered to captive animals are based on species' natural diets, gastrointestinal morphology and established nutrient requirements of similar species; [b] Nutrient intake calculated based on actual annual food consumption of three adult giant pandas and chemical analysis of those foods; intake of bamboo species based on equal distribution of actual bamboo intake over five commonly offered species; intake of bamboo components (leaf, branch, culm) based on prediction equations of the relationship between total bamboo fresh mass and bamboo component dry mass (Edwards et al., unpublished data); nd, nutrient concentrations not determined by chemical analysis; [c] Recommended nutrient allowances (4.0 kcal ME g^{-1}) to support maintenance in adult dogs (NRC, 2004); [d] Minimum nutrient profiles for dog food (3.5 kcal ME g^{-1}) to support maintenance in adult dogs (AAFCO, 2004); [e] Due to poor bioavailability, iron from carbonate or oxide sources that are added to the diet should not be considered in determining minimum nutrient concentration; [f] Due to poor bioavailability, copper from oxide sources that are added to the diet should not be considered in determining minimum nutrient concentration; [g] Although a true requirement for crude fat has not been established, the minimum concentration was based on the recognition of crude fat as a source of essential fatty acids and as a carrier of fat-soluble vitamins; [h] As feed manufacturing may destroy up to 90% of the thiamin in the diet, allowances in the formation of supplemental foods should be made to ensure that the minimum nutrient level is met after processing; [i] Normal diets not containing raw egg white provide adequate biotin; administration of oral antibiotics may require supplementation.

Schaller *et al.* (1985) reported that the giant panda excretes more water in its faeces than it obtains from food, except during seasons when high-moisture *Fargesia* shoots are consumed. Pandas compensate for this water deficit through consumption of wet bamboo during rainy periods and by drinking free water from depressions and pools at least once daily (Yong, 1981; Schaller *et al.*, 1985).

Approximations of free-water intake underestimate total water requirements because of omission of water in food and metabolic water (Robbins, 1993). An adult female giant panda (100 kg BW) consuming a daily average of 6.2 kg of bamboo (on a fresh-weight basis) would ingest 3.8 kg of preformed water associated with bamboo tissues (*Phyllostachys aurea* leaves contain 62% moisture). It is recommended that potable water be available *ad libitum* for captive animals.

Carbohydrates

Excluding water, carbohydrates are the most abundant compounds in plants. As a storage form of photosynthetic energy, carbohydrates are quantitatively a significant source of food energy for herbivores (Van Soest, 1994). A diverse group, carbohydrates are classified according to chemical structure, which, in turn, influences how they are used by animals and how they are quantified analytically (NRC, 2003).

Some carbohydrates (e.g. starch) are readily degraded by endogenous mammalian enzymes, such as salivary and pancreatic α-amylases. However, endogenous enzymes cannot degrade others, such as the structural carbohydrates cellulose and hemicellulose, where use depends on microbial fermentation by symbiotic gastrointestinal anaerobes yielding beneficial volatile fatty acids. Gastrointestinal adaptations which promote an environment conducive to this symbiotic relationship are pronounced in herbivores. However, as we have already indicated, these gastrointestinal adaptations for the giant panda appear to be limited only to an enlarged surface area of the colon with an absent fermentation compartment.

Dietary fibre collectively includes those indigestible components that comprise the plant cell wall (NRC, 2003). Neutral detergent fibre (NDF) includes the total insoluble fibre in the plant cell wall, primarily cellulose, hemicelluloses and lignin. Acid detergent fibre (ADF) is primarily cellulose and lignin. Schaller (1993) described the limited digestive capacity of the giant panda by suggesting that much of what is eaten is useless bulk. Yet that fraction of the diet that is not digested, absorbed or metabolised may play an important role in maintaining gastrointestinal tract health. Although fibre is not generally considered a dietary essential for simple-stomached mammals, its benefits have been demonstrated for many species (NRC, 1985, 2003), including for herbivores (Edwards, 1995). Conversely, those species that are mostly herbivorous experience the most negative health and behavioural effects when dietary fibre is reduced or eliminated.

Fibre concentrations in bamboos vary among species, plant parts (e.g. leaves, branches, culm), within species and across seasons (see Table 6.A.1 in Appendix 6.A). The ability of the giant panda to utilise various plant cell wall fractions is summarized in Table 6.3.

Protein

Animals require protein and about ten constituent amino acids to support growth, maintenance of body tissues and other essential metabolic functions. Dietary protein requirements are not only influenced by life stage, metabolic disorders and environmental factors, but also by the digestibility and quality of the protein source. Protein quality is determined by the relationship between amino acid composition and the animal's amino acid requirements.

Little research has been conducted to quantify protein or amino acid requirements for the giant panda at different life stages. Crude protein consumption by free-ranging giant pandas (based on observations of feeding behaviour) has been estimated to be 246.1 to 645.6 g per day by Schaller *et al.* (1985) and 145.3 to 664.5 g per day by Pan (1988). These observers concluded that protein concentrations consumed in the free-ranging panda diet were from 1.9 to 9.4% on a dry-matter basis. Based on this information, Schaller *et al.* (1985) and Pan (1988) calculated that the crude protein requirement of giant pandas is 100 g per day and 75.3 to 93.0 g per day for an 80-kg individual, respectively. This estimated requirement could also be expressed as 1.3 to 1.6% crude protein on a dry-matter basis.

It is important to note here that one cannot assume that the intake of protein (or any other nutrient) by a free-ranging animal is representative of 'what is required'. Such measures cannot be done accurately in the field, based solely on behavioural observations. Furthermore, a free-living animal may be consuming diets far above (or below) actual requirements, particularly in seasonal environments. It is likely that animals in the wild attempt to compensate over long time-frames to periods of shortage versus abundance by changing protein intake and lean body mass. And, when unable to do so during intervals of drought, harsh winters or even serious human disturbances, the animal may experience reproductive failure, increased disease susceptibility and mortality. Also, it may well be possible that extant wild panda populations are indeed feeding on diets that differ substantially from their ancestors. Human-induced habitat restrictions and alterations now prevent giant pandas from food resources normally found only at lower elevations.

It is interesting that estimates of crude protein intake and protein required for animals under field conditions and for maintenance and growth of subadults are lower than anticipated based on known needs for other species. For comparative purposes, crude protein require-ments for the dog using purified ingredient diets (lactalbumin for protein) with 4.2 kcal of metabolic energy per gram are from 9.3 g per day $BWKg^{0.75}$ (11.7% crude protein in an air-dry diet) to 14.0 g per day $BWKg^{0.75}$ (15.0% crude protein in an air-dry diet). It has been demon-strated that an air-dry diet (3.5 to 4.0 kcal of metabolic energy per gram) providing 6.0% crude protein from casein will maintain nitrogen bal-ance in the adult dog for a short time (Burns et al., 1982; NRC, 1985). It is emphasised that the protein sources used to develop these require-ments for the dog are highly digestible and have a high biological value. Thus, they represent minimum requirements. Animals fed natural in-gredient diets, with a lower digestibility and biological value, require higher protein intake levels to support metabolic needs.

Dierenfeld (1997) has proposed that protein requirements of the giant panda are readily met by bamboo (average 9% crude protein on a dry-matter basis) based on intakes of 6 to 15% BW (fresh-weight basis).

Fat

Having the greatest energy density among dietary fractions, fats are a concentrated source of calories. Fats also contain fatty acids, several of which cannot be synthesised from precursors by mammalian tissues (dietary essential fatty acids). Additionally, fat facilitates the absorption of fat-soluble vitamins (vitamins A, D, E and K). Bamboo, like other grasses, is typically low in fat. Dierenfeld (1997) proposed that dietary crude fat concentrations should be about 5% (dry-matter basis) for the captive giant panda. Additionally, this investigator proposed a min-imum dietary concentration of 1% linoleic acid (C18:2 n-6; dry-matter basis).

Minerals and vitamins

Studies of mineral and vitamin metabolism and utilisation in the giant panda are lacking. There are limited references identifying mineral concentrations in bamboo species consumed by giant pandas under field conditions (See Table 6.A.2 in Appendix 6.A). Again, the rapid passage of food through the giant panda gut could impact mineral and fat-soluble vitamin requirements. Further, it is unknown if the rapid transit phenomenon is compensated for by increased food intake. Nonetheless, the concentrations offered in Table 6.A.1 in Appendix 6.A (in conjunction with established requirements for the domestic dog) might serve as dietary guidelines until controlled studies are conducted.

Management and dietary husbandry

Despite the nearly obligatory relationship between the wild giant panda and the bamboo upon which it feeds, pandas in captivity have been offered, and consume, a diversity of foods. The rationale behind offering non-bamboo foods is unclear but may be related to reports that giant pandas in the wild opportunistically consume various items (Dierenfeld *et al.*, 1982, 1995, Schaller *et al.*, 1985). The exclusive use of food habits in the field for developing captive management schemes is problematic, largely because nutritional interactions within a natural ecosystem have not yet been characterised. It is clear, however, that consuming a novel food in a captive environment does not in itself imply nutritional adequacy of that food (Ullrey, 1989; Robbins, 1993). Furthermore, when formulating a dietary husbandry plan for a species it is essential to understand that the nutrients required in the diet can be derived from a broad range of ingredients. The giant panda adheres to this principle, although bamboo continues to be the single most important component of the diet in captivity.

The most limiting factors to dietary husbandry of the giant panda are the logistical efforts required in cultivating, harvesting, transporting and presenting bamboo, and selecting the most appropriate species. Of more specific concern is the large volume of bamboo needed to support a single animal (upwards of 14 kg of edible material daily on a fresh-weight basis). This amount is required because not all that is offered is selected for consumption. A review of diets offered to giant pandas at five Chinese facilities indicated that bamboo contributed

17 to 82% of the offered diet (on an as-fed basis) (Dierenfeld et al., 1995). In the CBSG Biomedical Survey of giant panda diets at five Chinese facilities (from 1998 to 2000), the average proportion of bamboo in the offered diet was 72.1% of dry matter (46.3 to 91.6%; n = 22 animals). Bamboo contributed 53.3% of dietary dry matter (19 to 88%; n = 36) in these diets based on consumed amounts.

Harvested bamboo should be presented to giant pandas in a manner that promotes acceptance, using storage and handling methods that minimise changes (e.g. desiccation) post harvest. These methods include multiple presentations of fresh bamboo, at least three times daily, with more frequent feedings in environments with elevated ambient temperatures. This strategy addresses the rapid desiccation of some bamboo species and also encourages extended bouts of feeding that are more typical of panda feeding behaviour in nature.

Selection of bamboo species to be cultivated for an *ex situ* management programme may be influenced by availability and local growing conditions. Differences in nutrient composition among bamboo species have been described (Tables 6.A.1 and 6.A.2 in Appendix 6.A; Nickley, 2001; Tabet et al., 2004), and a range of species has been offered and accepted by giant pandas (see Table 6.6). However, too little research has focused on the nutritional benefits of utilising one species over another. Without these data, it seems intuitive to begin a programme by selecting bamboo species that are consumed by free-ranging animals. Additional species might be selected on the basis of similarities in nutrient content, taxonomic relationship or a geographic range that are compatible with the panda. For example, it would be more appropriate to utilise bamboos that grow in temperate rather than tropical regions.

If non-bamboo foods are used as supplements, their effects on gastrointestinal function must be considered. Anecdotally (based on faecal characteristics), the giant panda in captivity has limited ability to use certain foods that would be considered quite digestible by other omnivores (e.g. bears and pigs). This constraint, combined with the lack of apparent digesta mixing in the intestinal tract, may contribute to gastrointestinal disease in captive giant pandas fed foods physically and nutritionally dissimilar to bamboo. Mainka et al. (1989) reported mucous stools and constipation in animals consuming diets containing less than 60% bamboo (dry-matter basis). Others have also described negative effects of limiting offered bamboo on faecal consistency and gastrointestinal health in this species (Goss, 1940; Nickley, 2001).

Traditionally, diets fed giant pandas in captivity in China have included a high-starch bread or gruel. These foods serve as vehicles supplementing a variety of proteins, minerals and vitamins. Their composition, which sometimes constitutes the primary source of dietary calories, is widely variable amongst institutions (Dierenfeld *et al.*, 1995). Ingredients used include a range of grains, grain by-products, legumes, milk, milk products, meat and eggs. Because of elevated concentrations of starch, sugars and other nonstructural carbohydrates, these items are significantly different in composition from bamboo.

A high-fibre supplemental biscuit has been formulated to replace the traditional 'bread' offered to captive giant pandas in China (Edwards & Zhang, 1997). This supplemental food delivers amino acids, minerals and vitamins within a fibre matrix that mimics the structural carbohydrate content of bamboo leaves (21% NDF and 16% ADF on a dry-matter basis). Thus, these nutrients enter the gastrointestinal tract in a form that is less likely to result in excessively rapid fermentation, digestive upset, unsatisfactory faecal consistency and gastrointestinal disease.

Evaluation and assessment

Body mass and body condition are important criteria for evaluating the nutritional and overall health status of an animal. A standardised body condition scoring system for objectively evaluating subcutaneous adipose tissue distribution and lean body mass has been developed for the giant panda (see Chapter 4). Interested readers should refer to this chapter for details about assessing body mass and condition, as well as understanding these findings associated with the CBSG Biomedical Survey.

The appearance and 'quality' of giant panda faeces is influenced by differences in nutrient content, physical form of diet and the diversity of foods offered. Faecal consistency is important as it is often used as an indicator of food utilisation and gastrointestinal health. A standardised faecal grading system customised for the giant pandas has been used routinely to monitor and document faecal consistency, facilitate communication among animal care professionals and evaluate the relationship between dietary factors and faecal quality (Fig. 6.4; Plate VII) (Edwards & Nickley, 2000).

Crouzet and Frädrich (1985) noted that giant panda faecal boluses are normal appearing when the animals consume long fibre in the form of bamboo. A significant inter-relationship between bamboo intake (as a

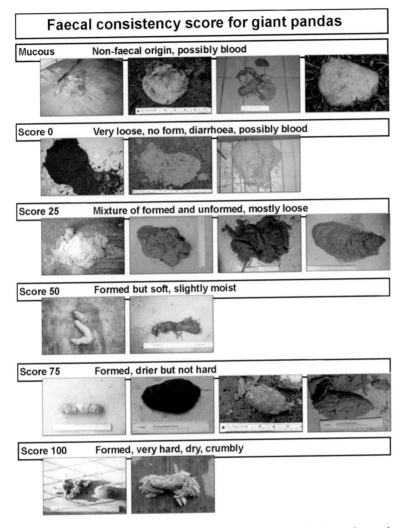

Figure 6.4. A standardised faecal grading system to routinely monitor and document faecal consistency (reproduced with permission from Edwards & Nickley, 2000). (See also Plate VII.)

percentage of total dry-matter intake) and faecal consistency has also been described in two giant pandas (Nickley, 2001). An animal consuming 53% bamboo (dry-matter basis) produced more faecal boluses (12.5% of total) of undesirable grade (0 to 25; see Fig. 6.4; Plate VII) than another animal consuming 70% bamboo (dry-matter basis) that produced only 0.6% faecal boluses of undesirable grade.

Although free-ranging giant pandas excrete faecal boluses coated with a thin, protective layer of mucus, captive counterparts excrete variable quantities of mucus that are not associated with faeces. Panda caregivers commonly refer to these free boluses as 'mucous stools'. The production of mucous stools, unrelated to faecal material, has not been observed in the free-ranging giant panda (W. Pan, pers. comm.). The excretion frequency and volume of mucous stools in captive specimens vary across individuals, locations and seasons. A significant inverse relationship has been reported between the amount of bamboo consumed by a captive giant panda and the frequency of mucous stool excretion (Goss, 1940; Mainka *et al.*, 1989). For example, a male giant panda that consumed 53% bamboo (dry-matter basis) produced 39 mucous stools over a 636-day period compared to a female at the same location who consumed 70% bamboo and produced six mucous stools over the same interval (Nickley, 2001).

Life stage considerations

Giant panda life stages are generally categorised into four groups: neonates (0–365 days old); juveniles (12–18 months); subadults (19 months to 4.5 years); and adults (>4.5 years). Nutritional support of the neonatal giant panda has been addressed in detail by Edwards *et al.* (see Chapter 13). The teeth of neonates begin to erupt by the third month (Chorn & Hoffman, 1978), and young pandas in the wild may sample bamboo and other solid foods as early as five to six months of age (Schaller *et al.*, 1985). Thereafter, there is a gradual transition from highly digestible mother's milk to the highly fibrous, poorly digestible diet of bamboo. Nonetheless, the juvenile giant panda is not nutritionally independent from its dam until about 18 to 24 months of age (Pan, 1988; Schaller, 1993; Zhu *et al.*, 2001).

When applying this information to captive giant panda management, it seems appropriate to allow cubs to remain with their mothers to as late as 24 months of age, even though the female may re-breed a few months before weaning. During this two-year interval, the cub becomes familiar with and starts to use the diet offered to its dam. Quantities of these foods can be increased as the cub's solids intake increases. If a management protocol requires removing an early stage cub from the dam, hand-rearing protocols and human-made milk formulae are available (see Chapter 13).

Subadult age is a particularly vulnerable time for captive giant pandas (Liu *et al.*, 2002). In one review of captive births over 30 years, only 18.2% of cubs (26 of 143 births) survived to three years of age and, of these, 79.8% of mortalities were associated with gastroenteritis and malnutrition (Feng *et al.*, 1993). Inadequate nutritional support of cubs weaned from their mothers as early as four months of age no doubt contributed to this high death rate. Interestingly, little information is available on the growth and development of giant pandas that were exclusively mother reared. This is due, in part, to the common practices of early weaning as well as giving cubs that remain with their mother supplemental foods to promote rapid weight gains. Some investigators (Liu *et al.*, 2002) have proposed that subadults that are 19 months to two years of age should gain 3 kg per month, whereas subadults of 2.1 to 3.9 years should gain 1 kg per month. It also was proposed that giant pandas four years of age or older should have no net weight gain.

Given the advances in nutrition and health care, certainly the number of giant pandas reaching geriatric age will increase. Proper nutritional support for these individuals is an essential feature of responsible palliative care. Of particular concern is the impact of declining dental health on the ability to consume adequate long fibre in the form of bamboo (see Chapter 15), which, in turn, will result in abnormal faecal consistency. One potential solution to support normal gastrointestinal health in older animals may be the mechanical chopping of bamboo followed by mixing with other suitable foods. An example would be the use of a nutritionally balanced, commercially available, high-fibre extruded diet which would provide a stable source of nutrients and energy along with a pre-processed source of plant fibre.

PRIORITIES FOR THE FUTURE

Studies to date have confirmed that bamboo must be the primary food in all captive giant panda diets. To ensure that the most suitable material is provided for feeding:

1. Consistent sources of several bamboo species should be identified in amounts sufficient to permit free-choice consumption.
2. Supplies should be readily available and easily accessible.
3. Routine nutrient analyses and heavy-metal screening should be performed to monitor seasonal variability and to guarantee high quality.

Although there are no quantitative data on the impact of toxicants (including heavy metals) on giant panda health, some breeding facilities are located in, or adjacent to, densely populated urbanised areas where pollution is common. Thus, the issue of influence of environmental contaminants may well be worthy of attention.

In terms of research priorities, there remains a serious lack of scholarly information on the adaptations that permit a giant panda to use and thrive on bamboo as the primary dietary component. The implications of such a diet have been well described (Schaller *et al.*, 1985; Pan, 1988) but the mechanisms by which this species is able to accomplish this lifestyle have yet to be defined. Additionally, much of the knowledge regarding the digestive ontogeny of giant pandas is derived from studies of captive specimens. Such information is biased by the production-style management techniques that are generally used. Feeding diets that better mimic the physical forms and nutrient and energy concentrations of foods consumed by wild counterparts would yield new insights into normal growth, development and maturation of mother-reared animals in captivity. Furthermore, it appears that the phenomenon of mucous stools and irregular stool output may be related to consumption of low-fibre/high-starch diets and/or the effect of sporadic or irregular feeding of foods that vary widely in nutrient content (e.g. bamboo versus nutrient- or energy-dense items). More details on what provokes a mucous stool and how to avoid it would be worthwhile.

There is also a need for continued standardisation of data collection, including the routine use of body and faecal scoring systems, which could further our understanding about the physiology and health of the collective *ex situ* population. Gathering parallel data from individuals living in nature would also allow improving the development, evaluation and routine application of better dietary management strategies *ex situ*.

Finally, an essential priority is to develop sufficient bamboo supplies to support the rapidly growing captive population. Selecting appropriate bamboo species and then producing adequate and sustainable quantities of high-quality plants are critical to the long-term viability of captive giant pandas in China and in the west.

ACKNOWLEDGEMENTS

The authors thank Drs. Duane E. Ullrey and Olav T. Oftedal for important comments and contributions to this chapter.

Table 6.A.1. Selected proximate, energy and fibre composition of foods consumed by free-ranging and captive giant pandas (Ailuropoda melanoleuca)

Description	Dry matter (%)	Gross energy (kJ g^{-1})	Crude protein (%)	Ether extract (%)	Ash (%)	CHO (%)	NDF (%)	ADF (%)	Cellulose (%)	Hemicellulose (%)	AD lignin (%)	Reference
Cereal products												
Baby cereal, Fei Er Fen brand	89.9	16.10	5.1	1.5	1.0	81.8	2.1	–	–	–	–	CBSG, 2001
	100.0	17.91	5.7	1.7	1.1	91.0	2.3	–	–	–	–	
Baby cereal, high protein, Heinz®	93.9	–	36.0	5.9	5.3	46.7	–	–	–	–	–	CBSG, 2001
	100.0	–	38.3	6.3	5.6	49.7	–	–	–	–	–	
Peanut cracker	97.4	–	6.0	26.0	1.2	63.5	4.0	3.1	3.1	0.9	0.0	CBSG, 2001
	100.0	–	6.2	26.7	1.2	65.2	4.1	3.2	3.2	0.9	0.0	
Animal products												
Ailuropoda melanoleuca milk, 77 days post-partum	32.0	7.60	8.3	18.4	–	2.7	–	–	–	–	–	Pan et al., 1998
	100.0	23.75	26.1	57.5	–	8.4	–	–	–	–	–	
Bos taurus milk, whole, powder Nespray®	97.5	–	26.3	26.7	6.0	38.4	–	–	–	–	–	CBSG, 2001
	100.0	–	27.0	27.4	6.2	39.4	–	–	–	–	–	

Table 6.A.1. (cont.)

Description	Dry matter (%)	Gross energy (kJ g^{-1})	Crude protein (%)	Ether extract (%)	Ash (%)	CHO (%)	NDF (%)	ADF (%)	Cellulose (%)	Hemicellulose (%)	AD lignin (%)	Reference
Egg, chicken whole	24.7	–	12.5	10.0	0.9	1.2	–	–	–	–	–	CBSG, 2001
	100.0	–	50.6	40.6	3.8	5.0	–	–	–	–	–	
Bamboo												
Bashania fargesii												
leaf, young		–	–	–	–	–	–	–	–	–	–	Pan, 1988;
	100.0	–	9.6	–	3.52	–	63.7	33.7	20.2	30.0	10.0	Long et al., 2004
leaf, old		–	–	–	–	–	–	–	–	–	–	Pan, 1988
	100.0	–	10.2	–	–	–	–	–	–	–	–	
stem, young		–	–	–	–	–	–	–	–	–	–	Pan, 1988;
	100.0	–	1.6	–	–	–	81.3	45.9	30.8	36.3	14.7	Long et al., 2004
stem, old		–	–	–	–	–	–	–	–	–	–	Pan, 1988
	100.0	–	1.0	–	–	–	–	–	–	–	–	
Bambusa glaucescens												
leaf, 6 months	37.8	7.13	7.6	–	3.6	–	26.2	12.2	10.4	13.9	1.8	Nickley, 2001
	100.0	18.87	20.1	–	9.4	–	69.2	32.4	27.6	36.8	4.8	

Species / sample												Reference
leaf, 12 months	45.3	8.00	7.4	—	7.2	—	31.2	16.0	13.1	15.2	2.9	Nickley, 2001
	100.0	17.66	16.3	—	15.9	—	68.9	35.4	28.9	33.5	6.5	
Bambusa oldhamii												
leaf, 6 months	37.5	6.78	8.6	—	3.6	—	25.5	12.0	10.1	13.5	2.0	Nickley, 2001
	100.0	18.07	22.8	—	9.5	—	68.1	32.1	26.8	36.0	5.3	
leaf, 12 months	42.0	7.43	8.2	—	5.8	—	28.9	14.1	11.6	14.7	2.5	Nickley, 2001
	100.0	17.70	19.6	—	13.8	—	68.7	33.5	27.6	35.1	6.0	
culm, 6 months	38.3	6.86	1.1	—	1.2	—	34.4	26.1	20.2	8.3	5.9	Nickley, 2001
	100.0	17.91	2.9	—	3.2	—	89.7	68.1	52.7	21.6	15.4	
culm, 12 months	51.6	9.28	1.2	—	1.2	—	46.0	35.0	27.2	11.0	7.7	Nickley, 2001
	100.0	17.99	2.3	—	2.4	—	89.1	67.8	52.8	21.3	15.0	
Chimonobambusa quadragularis												
leaf, dried	89.5	17.53	3.6	1.5	3.3	—	72.0	54.6	41.5	17.4	13.1	Edwards,
	100.0	19.58	4.0	1.7	3.7	—	80.4	61.0	46.4	19.4	14.6	unpublished data
Fargesia robusta												
leaf, dried	89.1	16.59	4.5	2.0	2.7	35.0	71.9	50.5	37.6	21.4	12.9	Edwards,
	100.0	18.62	5.0	2.2	3.0	39.3	80.7	56.7	42.2	24.0	14.5	unpublished data
Fargesia spathacea												
shoot, new	—	—	—	—	—	—	—	—	—	—	—	Pan, 1988
	100.0	—	4.0	—	—	—	—	—	—	—	—	
leaf	—	—	—	—	—	—	—	—	—	—	—	Pan, 1988
	100.0	—	11.3	—	—	—	—	—	—	—	—	

Table 6.A1. (cont.)

Description	Dry matter (%)	Gross energy (kJ g⁻¹)	Crude protein (%)	Ether extract (%)	Ash (%)	CHO (%)	NDF (%)	ADF (%)	Cellulose (%)	Hemicellulose (%)	AD lignin (%)	Reference
leaf	33.5	–	9.2	–	–	–	23.0	12.2	11.4	10.9	0.7	Mainka et al., 1989
	100.0	–	27.4	–	–	–	68.8	36.3	34.1	32.5	2.2	
leaf, all ages	48.4	–	6.8	–	4.3	–	34.5	18.5	13.8	16.0	4.7	Schaller et al., 1985
	100.0	–	14.1	–	8.8	–	71.2	38.2	28.5	33.0	9.7	
branch, all ages	50.5	–	2.2	–	2.3	–	40.9	26.2	18.9	14.7	7.3	Schaller et al., 1985
	100.0	–	4.3	–	4.5	–	81.0	51.9	37.5	29.1	14.4	
branch	40.3	–	7.4	–	–	–	25.5	13.5	13.0	12.0	0.4	Mainka et al., 1989
	100.0	–	18.4	–	–	–	63.3	33.4	32.3	29.9	1.1	
stem, young	–	–	–	–	–	–	–	–	–	–	–	Pan, 1988;
	100.0	–	2.1	–	–	–	91.8	46.2	32.6	45.6	13.6	Long et al., 2004
stem, old	–	–	–	–	–	–	–	–	–	–	–	Pan, 1988
	100.0	–	1.5	–	–	–	–	–	–	–	–	
stem, old shoot	45.9	–	1.1	–	1.1	–	39.4	28.9	21.8	10.5	7.0	Schaller et al., 1985
	100.0	–	2.5	–	2.3	–	85.8	62.9	47.6	22.9	15.3	
stem, two-year and old	61.6	–	0.7	–	0.7	–	53.3	39.5	29.4	13.7	10.1	Schaller et al., 1985
	100.0	–	1.2	–	1.2	–	86.5	64.2	47.8	22.3	16.4	

Phyllostachys aurea

leaf	45.4	—	9.0	—	—	—	32.1	14.9	13.7	17.2	1.2	Mainka et al., 1989
	100.0	—	19.8	—	—	—	70.6	32.8	30.1	37.8	2.7	
leaf, 6 months	39.6	7.21	7.0	—	4.1	—	28.2	13.1	11.2	15.1	1.9	Nickley, 2001
	100.0	18.20	17.7	—	10.3	—	71.1	33.0	28.2	38.1	4.8	
leaf, 12 months	45.3	8.28	7.0	—	5.6	—	31.5	14.9	12.0	16.7	2.9	Nickley, 2001
	100.0	18.28	15.4	—	12.3	—	69.6	32.8	26.5	36.8	6.3	
culm	46.1	—	2.3	—	—	—	36.2	26.3	24.6	9.9	1.8	Mainka et al., 1989
	100.0	—	5.0	—	—	—	78.6	57.1	53.3	21.5	3.8	
culm, 6 months	46.8	8.52	1.1	—	0.9	—	43.1	31.0	23.9	12.1	7.2	Nickley, 2001
	100.0	18.20	2.3	—	2.0	—	92.1	66.3	51.0	25.8	15.3	
culm, 12 months	54.0	9.99	1.0	—	0.9	—	49.6	35.7	27.2	13.9	8.5	Nickley, 2001
	100.0	18.49	1.8	—	1.6	—	91.8	66.1	50.4	25.7	15.7	
branch	45.6	—	3.7	—	—	—	37.1	23.3	22.5	13.8	0.8	Mainka et al., 1989
	100.0	—	8.1	—	—	—	81.3	51.1	49.4	30.2	1.7	
shoot	23.9	—	5.1	—	—	—	16.5	7.7	7.0	8.8	0.6	Mainka et al., 1989
	100.0	—	21.3	—	—	—	68.9	32.1	29.4	36.8	2.7	
shoot, ≤0.9-cm base diameter	15.6	—	2.3	0.3	0.3	2.3	10.1	5.0	4.8	5.1	0.3	Edwards et al., unpublished data
	100.0	—	14.8	2.0	2.0	14.7	65.0	32.2	30.5	32.8	1.7	
shoot, 1.0–1.2-cm base diameter	14.6	—	2.1	0.2	5.7	2.2	9.6	4.6	4.2	5.0	0.4	Edwards et al., unpublished data
	100.0	—	14.7	1.7	5.9	14.8	65.6	31.5	28.9	34.1	2.6	

Table 6.A1. (cont.)

Description	Dry matter (%)	Gross energy (kJ g⁻¹)	Crude protein (%)	Ether extract (%)	Ash (%)	CHO (%)	NDF (%)	ADF (%)	Cellulose (%)	Hemicellulose (%)	AD lignin (%)	Reference
shoot, ≥1.3-cm base diameter	17.4	—	2.3	0.3	1.0	1.8	12.7	5.9	5.4	6.8	0.5	Edwards et al., unpublished data
	100.0	—	13.5	1.5	5.7	10.5	73.2	33.9	30.8	39.3	3.1	
Phyllostachys aureosulcata												
leaf	52.0	10.44	7.0	—	4.1	—	34.1	14.4	11.2	18.6	3.2	Dierenfeld et al., 1982
	100.0	20.08	13.4	—	7.8	—	65.6	27.7	21.5	35.7	6.2	
leaf, early spring	42.7	—	7.0	—	2.1	—	27.2	14.5	12.1	12.7	2.4	Tabet et al., 2004
	100.0	—	16.3	—	4.9	—	63.7	34.0	28.4	29.7	5.5	
leaf, late spring	49.4	—	6.6	—	1.9	—	30.4	17.4	13.9	13.0	3.6	Tabet et al., 2004
	100.0	—	13.3	—	3.8	—	61.6	35.3	28.1	26.3	7.2	
leaf, early summer	46.6	—	5.6	—	1.7	—	31.0	18.1	14.0	13.0	4.1	Tabet et al., 2004
	100.0	—	12.0	—	3.6	—	66.6	38.8	30.0	27.8	8.8	
leaf, late summer	56.2	—	8.7	—	2.1	—	37.3	20.9	16.5	16.3	4.4	Tabet et al., 2004
	100.0	—	15.4	—	3.7	—	66.3	37.2	29.4	29.1	7.8	
leaf, autumn	54.1	—	8.9	—	1.3	—	37.4	20.3	15.6	17.2	4.7	Tabet et al., 2004
	100.0	—	16.4	—	2.3	—	69.2	37.5	28.8	31.8	8.6	
leaf, winter	58.8	—	9.1	—	2.7	—	37.1	20.3	16.7	16.9	3.6	Tabet et al., 2004
	100.0	—	15.4	—	4.6	—	63.2	34.5	28.4	28.7	6.1	

culm, <5 mm diameter	61.7	11.88	2.3	—	1.4	49.1	27.3	21.3	15.5	6.1	Dierenfeld et al., 1982
	100.0	19.25	3.8	—	2.3	79.6	44.2	34.5	25.2	9.9	
culm, 5–15 mm diameter	57.1	10.99	1.3	—	0.7	50.5	27.5	23.5	13.0	4.1	Dierenfeld et al., 1982
	100.0	19.25	2.2	—	1.2	88.4	48.2	41.1	22.8	7.1	
culm, 15–25 mm diameter	67.2	13.21	2.4	—	0.8	54.9	30.1	25.1	15.7	5.0	Dierenfeld et al., 1982
	100.0	19.66	3.5	—	1.2	81.7	44.8	37.4	23.4	7.4	
culm, >25 mm diameter	59.4	—	1.4	—	0.7	50.2	27.3	23.8	13.7	3.5	Dierenfeld et al., 1982
	100.0	—	2.3	—	1.1	84.5	45.9	40.0	23.0	5.9	
pith, early spring	56.2	—	2.3	—	0.0	44.4	32.9	23.2	11.5	9.8	Tabet et al., 2004
	100.0	—	4.0	—	0.1	79.1	58.6	41.2	20.5	17.4	
pith, late spring	67.0	—	1.0	—	0.0	54.3	40.4	29.6	13.8	10.8	Tabet et al., 2004
	100.0	—	1.5	—	0.1	81.1	60.4	44.3	20.7	16.1	
pith, early summer	51.5	—	0.6	—	0.3	42.5	32.4	23.6	10.1	8.8	Tabet et al., 2004
	100.0	—	1.1	—	0.6	82.5	62.8	45.8	19.6	17.1	
pith, late summer	46.3	—	1.1	—	0.0	39.1	29.1	21.3	10.1	7.8	Tabet et al., 2004
	100.0	—	2.3	—	0.1	84.5	62.8	46.0	21.8	16.8	
pith, autumn	50.6	—	2.3	—	—	40.4	31.8	24.0	8.6	7.8	Tabet et al., 2004
	100.0	—	4.5	—	—	79.8	62.8	47.5	17.0	15.3	
pith, winter	59.4	—	0.7	—	0.2	46.6	32.8	23.7	13.8	9.1	Tabet et al., 2004
	100.0	—	1.2	—	0.3	78.4	55.2	39.9	23.2	15.3	
shoot	8.9	—	2.1	—	0.0	3.9	2.0	1.9	1.9	0.1	Tabet et al., 2004
	100.0	—	23.4	—	0.3	44.1	22.5	21.3	21.6	1.2	

Table 6.A.1. (cont.)

Description	Dry matter (%)	Gross energy (kJ g⁻¹)	Crude protein (%)	Ether extract (%)	Ash (%)	CHO (%)	NDF (%)	ADF (%)	Cellulose (%)	Hemicellulose (%)	AD lignin (%)	Reference
Phyllostachys bisetti												
leaf, early spring	55.3	—	7.2	—	3.9	—	31.7	19.2	16.0	12.5	3.2	Tabet *et al.*, 2004
	100.0	—	12.9	—	7.1	—	57.3	34.8	29.0	22.5	5.8	
leaf, late spring	30.4	—	3.8	—	0.8	—	19.3	10.6	9.1	8.7	1.5	Tabet *et al.*, 2004
	100.0	—	12.6	—	2.5	—	63.5	35.0	30.1	28.5	4.9	
leaf, early summer	43.2	—	4.5	—	2.2	—	28.7	17.2	15.0	11.5	2.2	Tabet *et al.*, 2004
	100.0	—	10.5	—	5.1	—	66.5	39.8	34.7	26.7	5.1	
leaf, late summer	46.7	—	5.8	—	2.0	—	28.6	16.7	14.4	11.9	2.3	Tabet *et al.*, 2004
	100.0	—	12.4	—	4.3	—	61.3	35.8	30.8	25.5	5.0	
leaf, autumn	51.2	—	9.1	—	4.1	—	32.6	19.3	16.0	13.3	3.3	Tabet *et al.*, 2004
	100.0	—	17.7	—	8.0	—	63.6	37.6	31.2	26.0	6.5	
leaf, winter	54.5	—	7.8	—	2.8	—	35.6	19.9	16.3	15.7	3.6	Tabet *et al.*, 2004
	100.0	—	14.3	—	5.2	—	65.4	36.6	30.0	28.8	6.6	
pith, early spring	69.0	—	0.6	—	0.0	—	53.3	34.1	24.7	19.3	9.3	Tabet *et al.*, 2004
	100.0	—	0.8	—	0.1	—	77.3	49.4	35.8	27.9	13.5	
pith, late spring	27.5	—	0.6	—	0.0	—	21.5	15.3	12.4	6.3	2.9	Tabet *et al.*, 2004
	100.0	—	2.0	—	0.2	—	78.4	55.6	45.1	22.8	10.5	

pith, early summer	40.1	—	0.4	0.1	—	34.6	26.3	18.9	8.4	7.3	Tabet et al., 2004
	100.0	—	0.9	0.2	—	86.4	65.6	47.3	20.9	18.3	
pith, late summer	47.0	—	0.4	0.1	—	39.6	30.5	22.1	9.1	8.4	Tabet et al., 2004
	100.0	—	0.8	0.2	—	84.3	65.0	47.1	19.3	17.9	
pith, autumn	45.9	—	0.7	0.2	—	39.3	29.8	21.5	9.5	8.2	Tabet et al., 2004
	100.0	—	1.6	0.4	—	85.7	64.9	47.0	20.8	17.9	
pith, winter	49.1	—	0.7	0.1	—	35.9	28.5	21.3	7.4	7.2	Tabet et al., 2004
	100.0	—	1.3	0.2	—	73.1	58.0	43.4	15.1	14.6	
shoot	13.6	—	2.7	0.1	—	7.1	3.6	3.4	3.5	0.2	Tabet et al., 2004
	100.0	—	19.6	0.8	—	52.2	26.6	24.8	25.6	1.8	
Phyllostachys nigra											
leaf, 6 months	37.4	6.95	7.0	3.8	—	27.4	12.5	10.2	14.9	2.3	Nickley, 2001
	100.0	18.58	18.6	10.2	—	73.2	33.4	27.2	39.8	6.2	
leaf, 12 months	45.3	8.06	7.5	6.9	—	31.3	16.4	13.0	14.9	3.4	Nickley, 2001
	100.0	17.78	16.6	15.2	—	69.2	36.3	28.8	32.9	7.5	
leaf, early spring	60.4	—	8.5	3.5	—	34.0	19.2	15.7	14.8	3.5	Tabet et al., 2004
	100.0	—	14.1	5.8	—	56.3	31.9	26.1	24.5	5.8	
leaf, late spring	51.9	—	7.4	3.1	—	29.5	18.3	14.9	11.1	3.5	Tabet et al., 2004
	100.0	—	14.4	6.0	—	56.8	35.4	28.6	21.4	6.7	
leaf, early summer	45.2	—	7.7	1.3	—	30.0	15.6	12.0	14.5	3.6	Tabet et al., 2004
	100.0	—	17.0	3.0	—	66.5	34.5	26.5	32.0	8.0	
leaf, late summer	48.8	—	6.2	2.7	—	31.5	18.0	14.3	13.5	3.7	Tabet et al., 2004
	100.0	—	12.8	5.6	—	64.6	36.9	29.3	27.7	7.6	

Table 6.A.1. (cont.)

Description	Dry matter (%)	Gross energy (kJ g⁻¹)	Crude protein (%)	Ether extract (%)	Ash (%)	CHO (%)	NDF (%)	ADF (%)	Cellulose (%)	Hemicellulose (%)	AD lignin (%)	Reference
leaf, autumn	54.0	—	6.0	—	4.7	—	34.2	21.8	17.2	12.5	4.5	Tabet et al., 2004
	100.0	—	11.2	—	8.6	—	63.4	40.3	31.9	23.1	8.4	
leaf, winter	59.1	—	8.4	—	4.5	—	34.9	22.1	18.1	12.8	4.1	Tabet et al., 2004
	100.0	—	14.1	—	7.6	—	59.0	37.4	30.5	21.6	6.9	
pith, early spring	61.6	—	0.8	—	0.0	—	51.1	37.3	27.6	13.8	9.7	Tabet et al., 2004
	100.0	—	1.4	—	0.0	—	82.9	60.5	44.8	22.4	15.7	
pith, late spring	59.3	—	0.7	—	0.0	—	49.3	35.5	26.6	13.8	8.9	Tabet et al., 2004
	100.0	—	1.2	—	0.0	—	83.1	59.8	44.9	23.3	14.9	
pith, early summer	59.9	—	0.7	—	0.0	—	51.1	39.3	28.6	11.9	10.7	Tabet et al., 2004
	100.0	—	1.2	—	0.1	—	85.4	65.6	47.8	19.8	17.8	
pith, late summer	56.2	—	0.8	—	0.0	—	45.7	33.4	24.7	12.3	8.7	Tabet et al., 2004
	100.0	—	1.4	—	0.1	—	81.3	59.4	43.9	22.0	15.5	
pith, autumn	51.9	—	0.4	—	0.0	—	43.4	32.9	24.5	10.5	8.4	Tabet et al., 2004
	100.0	—	0.9	—	0.0	—	83.5	63.3	47.2	20.2	16.2	
pith, winter	53.1	—	0.5	—	0.0	—	42.5	31.3	23.8	11.1	7.5	Tabet et al., 2004
	100.0	—	0.9	—	0.0	—	79.9	59.0	44.9	20.9	14.1	
shoot	12.1	—	1.9	—	0.0	—	6.5	3.4	3.2	3.0	0.2	Tabet et al., 2004
	100.0	—	15.8	—	0.4	—	53.3	28.3	26.7	25.0	1.5	

Pseudosasa japonica

												Reference
leaf	24.1	4.39	3.1	0.7	2.2	—	18.5	11.8	9.8	6.7	2.0	Warnell, 1988
	100.0	18.20	12.7	2.7	9.1	—	76.8	48.9	40.5	27.9	8.4	
leaf, early spring	56.0	—	7.0	—	4.0	—	32.2	21.4	16.0	10.8	5.4	Tabet et al., 2004
	100.0	—	12.6	—	7.2	—	57.6	38.3	28.6	19.3	9.7	
leaf, late spring	49.4	—	5.7	—	3.6	—	30.1	20.4	17.3	9.7	3.0	Tabet et al., 2004
	100.0	—	11.5	—	7.2	—	60.9	41.3	35.1	19.6	6.2	
leaf, early summer	47.0	—	5.2	—	2.9	—	30.0	20.4	16.6	9.6	3.8	Tabet et al., 2004
	100.0	—	11.1	—	6.1	—	63.8	43.3	35.2	20.5	8.1	
leaf, late summer	53.0	—	6.1	—	2.2	—	33.0	21.0	16.2	12.1	4.8	Tabet et al., 2004
	100.0	—	11.4	—	4.2	—	62.3	39.5	30.6	22.8	9.0	
leaf, autumn	45.8	—	6.9	—	1.7	—	32.4	19.8	16.1	12.6	3.6	Tabet et al., 2004
	100.0	—	15.2	—	3.7	—	70.7	43.2	35.2	27.6	7.9	
leaf, winter	56.7	—	7.9	—	2.9	—	35.6	22.4	18.2	13.2	4.2	Tabet et al., 2004
	100.0	—	14.0	—	5.1	—	62.8	39.6	32.1	23.3	7.5	
pith, early spring	57.5	—	1.1	—	0.0	—	43.1	26.9	20.6	16.2	6.3	Tabet et al., 2004
	100.0	—	1.9	—	0.1	—	75.0	46.7	35.8	28.3	11.0	
pith, late spring	59.5	—	1.1	—	0.0	—	45.3	28.3	21.4	17.0	6.9	Tabet et al., 2004
	100.0	—	1.9	—	0.1	—	76.1	47.6	36.0	28.5	11.6	
pith, early summer	48.2	—	0.9	—	0.1	—	39.1	28.3	21.4	10.8	6.9	Tabet et al., 2004
	100.0	—	1.9	—	0.1	—	81.1	58.7	44.4	22.3	14.4	
pith, late summer	47.3	—	1.0	—	0.0	—	38.9	28.8	21.7	10.0	7.2	Tabet et al., 2004
	100.0	—	2.1	—	0.1	—	82.2	60.9	45.8	21.2	15.1	
pith, autumn	31.5	—	1.5	—	0.0	—	25.7	19.8	15.6	5.8	4.2	Tabet et al., 2004

Table 6.A.1. (cont.)

Description	Dry matter (%)	Gross energy (kJ g^{-1})	Crude protein (%)	Ether extract (%)	Ash (%)	CHO (%)	NDF (%)	ADF (%)	Cellulose (%)	Hemicellulose (%)	AD lignin (%)	Reference
	100.0	–	4.8	–	0.1	–	81.6	63.0	49.6	18.6	13.4	
pith, winter	58.7	–	1.4	–	0.0	–	45.5	34.0	26.2	11.5	7.8	Tabet et al., 2004
	100.0	–	2.4	–	0.0	–	77.5	57.9	44.7	19.6	13.2	
shoot	16.4	–	1.9	–	0.3	–	10.4	5.7	5.0	4.7	0.6	Tabet et al., 2004
	100.0	–	11.5	–	1.9	–	63.3	34.6	30.8	28.7	3.8	
Sinarundinaria fangiana												
leaf, all ages	48.3	–	7.5	–	4.1	–	34.7	17.6	13.4	17.1	4.2	Schaller et al., 1985
	100.0	–	15.5	–	8.4	–	71.9	36.4	27.8	35.5	8.6	
branch, all ages	49.5	–	3.2	–	3.2	–	38.8	23.9	17.5	14.9	6.4	Schaller et al., 1985
	100.0	–	6.5	–	6.5	–	78.3	48.3	35.3	30.0	13.0	
stem, old shoot	43.8	–	1.9	–	1.2	–	36.7	26.4	20.5	10.3	6.0	Schaller et al., 1985
	100.0	–	4.4	–	2.7	–	83.9	60.3	46.7	23.6	13.6	
stem, 2 year and old	58.8	–	1.4	–	1.2	–	50.0	36.4	26.9	13.6	9.5	Schaller et al., 1985
	100.0	–	2.4	–	2.0	–	85.0	61.9	45.7	23.1	16.2	

Figures in italics indicate these values calculated in the appropriate references.

Table 6.A.2. *Mineral composition of foods consumed by free-ranging and captive giant pandas (Ailuropoda melanoleuca)*

Description	Dry matter (%)	Ca (%)	P (%)	Na (%)	K (%)	Mg (%)	Cu (ppm)	Fe (ppm)	Mn (ppm)	Se (ppm)	Zn (ppm)	Reference
Cereal products												
Baby cereal Fei Er Fen brand	89.9	0.02	0.15	–	–	0.04	2.6	42.1	–	–	14.8	CBSG, 2001
	100.0	0.02	0.17	–	–	0.05	2.9	46.8	–	–	16.5	
Baby cereal high protein, Heinz®	93.9	0.72	0.61	–	–	0.23	12.8	475.0	–	–	44.2	CBSG, 2001
	100.0	0.77	0.65	–	–	0.24	13.6	505.9	–	–	47.1	
Peanut cracker	97.4	0.03	0.15	0.29	0.15	0.05	1.9	31.5	7.5	<0.1	10.3	CBSG, 2001
	100.0	0.03	0.15	0.30	0.15	0.05	2.0	32.3	7.7	<0.1	10.6	
Animal products												
Ailuropoda melanoleuca milk, 77 days post-partum	32.0	–	–	–	–	–	–	–	–	–	–	Pan et al., 1998
	100.0	–	–	–	–	–	–	–	–	–	–	
Bos taurus milk, whole, powder Nespray®	97.5	0.92	0.78	–	–	0.09	0.8	4.7	–	–	33.3	CBSG, 2001
	100.0	0.94	0.80	–	–	0.09	0.8	4.8	–	–	34.2	
Egg, chicken whole	24.7	0.05	0.18	–	–	0.01	0.1	14.4	–	–	11.0	CBSG, 2001
	100.0	0.20	0.72	–	–	0.04	0.6	58.4	–	–	44.6	

Table 6.A.2. (cont.)

Description	Dry matter (%)	Ca (%)	P (%)	Na (%)	K (%)	Mg (%)	Cu (ppm)	Fe (ppm)	Mn (ppm)	Se (ppm)	Zn (ppm)	Reference
Bamboo												
Bashania fargesii												
leaf, young	–	–	–	–	–	–	–	–	–	–	–	Pan, 1988;
	100.0	–	–	–	–	–	–	–	–	–	–	Long et al., 2004
leaf, old	–	–	–	–	–	.	–	–	–	–	–	Pan, 1988
	100.0	–	–	–	–	–	–	–	–	–	–	
stem, young	–	–	–	–	–	–	–	–	–	–	–	Pan, 1988;
	100.0	–	–	–	–	–	–	–	–	–	–	Long et al., 2004
stem, old	–	–	–	–	–	–	–	–	–	–	–	Pan, 1988
	100.0	–	–	–	–	–	–	–	–	–	–	
Bambusa glaucescens												
leaf, 6 months	37.8	–	–	–	–	–	–	–	–	–	–	Nickley, 2001
	100.0	–	–	–	–	–	–	–	–	–	–	
leaf, 12 months	45.3	–	–	–	–	–	–	–	–	–	–	Nickley, 2001
	100.0	–	–	–	–	–	–	–	–	–	–	

Bambusa oldhamii

leaf, 6 months	37.5	–	–	–	–	–	–	–	–	–	–	Nickley, 2001
	100.0	–	–	–	–	–	–	–	–	–	–	
leaf, 12 months	42.0	–	–	–	–	–	–	–	–	–	–	Nickley, 2001
	100.0	–	–	–	–	–	–	–	–	–	–	
culm, 6 months	38.3	–	–	–	–	–	–	–	–	–	–	Nickley, 2001
	100.0	–	–	–	–	–	–	–	–	–	–	
culm, 12 months	51.6	–	–	–	–	–	–	–	–	–	–	Nickley, 2001
	100.0	–	–	–	–	–	–	–	–	–	–	
Chimonobambusa quadragularis												
leaf, dried	89.5	0.12	0.06	–	–	2.0	220.2	–	< 0.1	2.0		Edwards, unpublished data
	100.0	0.13	0.07	–	–	2.2	246.0	–	<0.1	2.2		
Fargesia robusta												
leaf, dried	89.1	0.10	0.06	<0.02	0.50	0.04	4.0	112.3	71.3	<2.0	12.0	Edwards, unpublished data
	100.0	0.11	0.07	<0.02	0.56	0.05	4.5	126.0	80.0	<2.0	13.5	
Fargesia spathacea												
shoot, new	–	–	–	–	–	–	–	–	–	–	–	Pan, 1988
	100.0	–	–	–	–	–	–	–	–	–	–	
leaf	–	–	–	–	–	–	–	–	–	–	–	Pan, 1988
	100.0	–	–	–	–	–	–	–	–	–	–	
leaf	33.5	–	–	–	–	–	–	–	–	–	–	Mainka *et al.*, 1989
	100.0	–	–	–	–	–	–	–	–	–	–	
leaf, all ages	48.4	–	–	–	–	–	–	–	–	–	–	Schaller *et al.*, 1985
	100.0	–	–	–	–	–	–	–	–	–	–	

Table 6.A.2. (cont.)

Description	Dry matter (%)	Ca (%)	P (%)	Na (%)	K (%)	Mg (%)	Cu (ppm)	Fe (ppm)	Mn (ppm)	Se (ppm)	Zn (ppm)	Reference
branch, all ages	50.5	–	–	–	–	–	–	–	–	–	–	Schaller et al., 1985
	100.0	–	–	–	–	–	–	–	–	–	–	Mainka et al., 1989
branch	40.3	–	–	–	–	–	–	–	–	–	–	Mainka et al., 1989
	100.0	–	–	–	–	–	–	–	–	–	–	
stem, young	–	–	–	–	–	–	–	–	–	–	–	Pan, 1988;
	100.0	–	–	–	–	–	–	–	–	–	–	Long et al., 2004
stem, old	–	–	–	–	–	–	–	–	–	–	–	Pan, 1988
	100.0	–	–	–	–	–	–	–	–	–	–	
stem, old shoot	45.9	–	–	–	–	–	–	–	–	–	–	Schaller et al., 1985
	100.0	–	–	–	–	–	–	–	–	–	–	
stem, two-year and old	61.6	–	–	–	–	–	–	–	–	–	–	Schaller et al., 1985
	100.0	–	–	–	–	–	–	–	–	–	–	
Phyllostachys aurea												
leaf	45.4	–	–	–	–	–	–	–	–	–	–	Mainka et al., 1989
	100.0	–	–	–	–	–	–	–	–	–	–	
leaf, 6 months	39.6	–	–	–	–	–	–	–	–	–	–	Nickley, 2001
	100.0	–	–	–	–	–	–	–	–	–	–	
leaf, 12 months	45.3	–	–	–	–	–	–	–	–	–	–	Nickley, 2001
	100.0	–	–	–	–	–	–	–	–	–	–	
culm	46.1	–	–	–	–	–	–	–	–	–	–	Mainka et al., 1989

Item												Reference
culm, 6 months	46.8	–	–	–	–	–	–	–	–	–	–	Nickley, 2001
	100.0	–	–	–	–	–	–	–	–	–	–	
culm, 12 months	54.0	–	–	–	–	–	–	–	–	–	–	Nickley, 2001
	100.0	–	–	–	–	–	–	–	–	–	–	
branch	45.6	–	–	–	–	–	–	–	–	–	–	Mainka et al., 1989
	100.0	–	–	–	–	–	–	–	–	–	–	
shoot	23.9	–	–	–	–	–	–	–	–	–	–	Mainka et al., 1989
	100.0	–	–	–	–	–	–	–	–	–	–	
shoot, ≤0.9-cm base diameter	15.6	–	–	–	–	–	–	–	–	–	–	Edwards et al., unpublished
	100.0	–	–	–	–	–	–	–	–	–	–	
shoot, 1.0–1.2-cm base diameter	14.6	–	–	–	–	–	–	–	–	–	–	Edwards et al., unpublished
	100.0	–	–	–	–	–	–	–	–	–	–	
shoot, ≥1.3-cm base diameter	17.4	–	–	–	–	–	–	–	–	–	–	Edwards et al., unpublished
	100.0	–	–	–	–	–	–	–	–	–	–	
Phyllostachys aureosulcata												
leaf	52.0	–	–	–	–	–	–	–	–	–	–	Dierenfeld et al., 1982
	100.0	–	–	–	–	–	–	–	–	–	–	
leaf, early spring	42.7	0.21	0.07	–	0.38	0.09	–	–	153.5	–	–	Tabet et al., 2004
	100.0	0.49	0.16	–	0.89	0.21	–	–	359.6	–	–	
leaf, late spring	49.4	0.35	0.06	–	0.33	0.11	–	–	151.7	–	–	Tabet et al., 2004
	100.0	0.70	0.13	–	0.67	0.22	–	–	307.2	–	–	
leaf, early summer	46.6	0.20	0.07	–	0.67	0.07	–	–	95.2	–	–	Tabet et al., 2004
	100.0	0.43	0.15	–	1.44	0.15	–	–	204.4	–	–	

Table 6.A.2. (cont.)

Description	Dry matter (%)	Ca (%)	P (%)	Na (%)	K (%)	Mg (%)	Cu (ppm)	Fe (ppm)	Mn (ppm)	Se (ppm)	Zn (ppm)	Reference
leaf, late summer	56.2	0.31	0.07	–	0.58	0.12	–	–	421.4	–	–	Tabet et al., 2004
	100.0	0.56	0.12	–	1.04	0.21	–	–	750.0	–	–	
leaf, autumn	54.1	0.20	0.06	–	0.57	0.10	–	–	214.9	–	–	Tabet et al., 2004
	100.0	0.38	0.12	–	1.06	0.18	–	–	397.5	–	–	
leaf, winter	58.8	0.33	0.08	–	0.27	0.17	–	–	292.9	–	–	Tabet et al., 2004
	100.0	0.57	0.14	–	0.46	0.28	–	–	497.9	–	–	
culm, <5 mm diameter	61.7	–	–	–	–	–	–	–	–	–	–	Dierenfeld et al., 1982
	100.0	–	–	–	–	–	–	–	–	–	–	
culm, 5–15 mm diameter	57.1	–	–	–	–	–	–	–	–	–	–	Dierenfeld et al., 1982
	100.0	–	–	–	–	–	–	–	–	–	–	
culm, 15–25 mm diameter	67.2	–	–	–	–	–	–	–	–	–	–	Dierenfeld et al., 1982
	100.0	–	–	–	–	–	–	–	–	–	–	
culm, >25 mm diameter	59.4	–	–	–	–	–	–	–	–	–	–	Dierenfeld et al., 1982
	100.0	–	–	–	–	–	–	–	–	–	–	

pith, early spring	56.2	0.02	0.09	–	0.21	0.03	–	25.3	–	–	Tabet *et al.*, 2004
	100.0	0.03	0.15	–	0.37	0.05	–	45.1	–	–	
pith, late spring	67.0	0.02	0.07	–	0.10	0.02	–	21.2	–	–	Tabet *et al.*, 2004
	100.0	0.03	0.11	–	0.16	0.03	–	31.6	–	–	
pith, early summer	51.5	0.01	0.04	–	0.21	0.02	–	16.2	–	–	Tabet *et al.*, 2004
	100.0	0.03	0.07	–	0.40	0.04	–	31.5	–	–	
pith, late summer	46.3	0.01	0.03	–	0.38	0.01	–	19.4	–	–	Tabet *et al.*, 2004
	100.0	0.02	0.06	–	0.82	0.02	–	42.0	–	–	
pith, autumn	50.6	0.02	0.02	–	0.28	0.01	–	49.9	–	–	Tabet *et al.*, 2004
	100.0	0.04	0.03	–	0.56	0.02	–	98.6	–	–	
pith, winter	59.4	0.02	0.03	–	0.24	0.03	–	61.2	–	–	Tabet *et al.*, 2004
	100.0	0.03	0.05	–	0.41	0.04	–	102.9	–	–	
shoot	8.9	0.01	0.05	–	0.32	0.01	–	6.7	–	–	Tabet *et al.*, 2004
	100.0	0.07	0.53	–	3.62	0.15	–	74.7	–	–	
Phyllostachys bisetti											
leaf, early spring	55.3	0.41	0.05	–	0.51	0.05	–	46.4	–	–	Tabet *et al.*, 2004
	100.0	0.74	0.10	–	0.92	0.08	–	83.8	–	–	
leaf, late spring	30.4	0.08	0.09	–	0.62	0.05	–	12.3	–	–	Tabet *et al.*, 2004
	100.0	0.27	0.29	–	2.04	0.17	–	40.6	–	–	
leaf, early summer	43.2	0.13	0.07	–	0.96	0.07	–	12.8	–	–	Tabet *et al.*, 2004
	100.0	0.30	0.17	–	2.21	0.16	–	29.6	–	–	
leaf, late summer	46.7	0.19	0.07	–	0.98	0.07	–	26.3	–	–	Tabet *et al.*, 2004
	100.0	0.40	0.14	–	2.10	0.16	–	56.2	–	–	
leaf, autumn	51.2	0.33	0.08	–	0.44	0.07	–	27.5	–	–	Tabet *et al.*, 2004

Table 6.A.2. (cont.)

Description	Dry matter (%)	Ca (%)	P (%)	Na (%)	K (%)	Mg (%)	Cu (ppm)	Fe (ppm)	Mn (ppm)	Se (ppm)	Zn (ppm)	Reference
leaf, winter	100.0	0.64	0.15	–	0.85	0.14	–	–	53.7	–	–	Tabet et al., 2004
	54.5	0.28	0.06	–	0.54	0.06	–	–	30.9	–	–	
	100.0	0.51	0.12	–	0.99	0.11	–	–	56.8	–	–	Tabet et al., 2004
pith, early spring	69.0	0.03	0.03	–	0.21	0.02	–	–	8.8	–	–	Tabet et al., 2004
	100.0	0.04	0.04	–	0.30	0.03	–	–	12.7	–	–	
pith, late spring	27.5	0.00	0.02	–	0.34	0.01	–	–	1.1	–	–	Tabet et al., 2004
	100.0	0.02	0.07	–	1.25	0.03	–	–	4.1	–	–	
pith, early summer	40.1	0.01	0.02	–	0.47	0.01	–	–	0.4	–	–	Tabet et al., 2004
	100.0	0.01	0.06	–	1.17	0.02	–	–	1.0	–	–	
pith, late summer	47.0	0.01	0.02	–	0.48	0.01	–	–	0.9	–	–	Tabet et al., 2004
	100.0	0.01	0.05	–	1.02	0.02	–	–	2.0	–	–	
pith, autumn	45.9	0.01	0.05	–	0.48	0.02	–	–	1.5	–	–	Tabet et al., 2004
	100.0	0.02	0.10	–	1.04	0.05	–	–	3.3	–	–	
pith, winter	49.1	0.01	0.05	–	0.62	0.01	–	–	2.8	–	–	Tabet et al., 2004
	100.0	0.02	0.10	–	1.26	0.03	–	–	5.8	–	–	
shoot	13.6	0.02	0.06	–	0.46	0.02	–	–	2.6	–	–	Tabet et al., 2004
	100.0	0.13	0.45	–	3.40	0.12	–	–	19.0	–	–	

Phyllostachys nigra

													Reference
leaf, 6 months	37.4	–	–	–	–	–	–	–	–	–	–	–	Nickley, 2001
	100.0	–	–	–	–	–	–	–	–	–	–	–	
leaf, 12 months	45.3	–	–	–	–	–	–	–	–	–	–	–	Nickley, 2001
	100.0	–	–	–	–	–	–	–	–	–	–	–	
leaf, early spring	60.4	0.27	0.10	–	0.45	0.05	–	–	14.1	–	–	–	Tabet et al., 2004
	100.0	0.46	0.17	–	0.75	0.08	–	–	23.4	–	–	–	
leaf, late spring	51.9	0.30	0.08	–	0.45	0.06	–	–	17.4	–	–	–	Tabet et al., 2004
	100.0	0.57	0.15	–	0.87	0.12	–	–	33.5	–	–	–	
leaf, early summer	45.2	0.17	0.09	–	0.87	0.09	–	–	57.8	–	–	–	Tabet et al., 2004
	100.0	0.37	0.20	–	1.93	0.19	–	–	127.9	–	–	–	
leaf, late summer	48.8	0.23	0.10	–	0.60	0.08	–	–	44.5	–	–	–	Tabet et al., 2004
	100.0	0.48	0.21	–	1.23	0.17	–	–	91.2	–	–	–	
leaf, autumn	54.0	0.33	0.12	–	0.47	0.06	–	–	36.2	–	–	–	Tabet et al., 2004
	100.0	0.61	0.21	–	0.88	0.12	–	–	66.9	–	–	–	
leaf, winter	59.1	0.30	0.15	–	0.58	0.07	–	–	21.0	–	–	–	Tabet et al., 2004
	100.0	0.50	0.25	–	0.99	0.11	–	–	35.5	–	–	–	
pith, early spring	61.6	0.01	0.07	–	0.25	0.01	–	–	6.0	–	–	–	Tabet et al., 2004
	100.0	0.02	0.11	–	0.40	0.02	–	–	9.8	–	–	–	
pith, late spring	59.3	0.01	0.10	–	0.20	0.01	–	–	5.9	–	–	–	Tabet et al., 2004
	100.0	0.01	0.17	–	0.33	0.02	–	–	9.9	–	–	–	
pith, early summer	59.9	0.01	0.05	–	0.16	0.01	–	–	10.3	–	–	–	Tabet et al., 2004
	100.0	0.02	0.08	–	0.26	0.02	–	–	17.2	–	–	–	

Table 6.A.2. (cont.)

Description	Dry matter (%)	Ca (%)	P (%)	Na (%)	K (%)	Mg (%)	Cu (ppm)	Fe (ppm)	Mn (ppm)	Se (ppm)	Zn (ppm)	Reference
pith, late summer	56.2	0.01	0.08	–	0.21	0.01	–	–	8.8	–	–	Tabet et al., 2004
	100.0	0.02	0.14	–	0.38	0.02	–	–	15.7	–	–	
pith, autumn	51.9	0.01	0.09	–	0.22	0.01	–	–	5.8	–	–	Tabet et al., 2004
	100.0	0.02	0.17	–	0.42	0.03	–	–	11.2	–	–	
pith, winter	53.1	0.01	0.10	–	0.26	0.01	–	–	3.5	–	–	Tabet et al., 2004
	100.0	0.01	0.19	–	0.49	0.02	–	–	6.5	–	–	
shoot	12.1	0.01	0.05	–	0.42	0.01	–	–	2.0	–	–	Tabet et al., 2004
	100.0	0.05	0.40	–	3.43	0.10	–	–	16.7	–	–	
Pseudosasa japonica												
leaf	24.1	–	–	–	–	–	–	–	–	–	–	Warnell, 1988
	100.0	–	–	–	–	–	–	–	–	–	–	
leaf, early spring	56.0	0.30	0.06	–	0.44	0.09	–	–	321.7	–	–	Tabet et al., 2004
	100.0	0.53	0.11	–	0.78	0.17	–	–	574.9	–	–	
leaf, late spring	49.4	0.30	0.05	–	0.43	0.09	–	–	274.3	–	–	Tabet et al., 2004
	100.0	0.62	0.09	–	0.88	0.19	–	–	555.8	–	–	
leaf, early summer	47.0	0.24	0.05	–	0.51	0.10	–	–	193.4	–	–	Tabet et al., 2004
	100.0	0.51	0.11	–	1.08	0.21	–	–	411.3	–	–	

Part											Reference
leaf, late summer	53.0	0.29	0.06	–	0.52	0.10	–	213.2	–	–	Tabet et al., 2004
	100.0	0.54	0.10	–	0.97	0.20	–	402.3	–	–	
leaf, autumn	45.8	0.19	0.07	–	0.50	0.10	–	149.2	–	–	Tabet et al., 2004
	100.0	0.41	0.14	–	1.08	0.21	–	325.7	–	–	
leaf, winter	56.7	0.31	0.06	–	0.40	0.11	–	329.1	–	–	Tabet et al., 2004
	100.0	0.55	0.11	–	0.71	0.19	–	580.3	–	–	
pith, early spring	57.5	0.01	0.03	–	0.39	0.02	–	14.7	–	–	Tabet et al., 2004
	100.0	0.02	0.05	–	0.67	0.03	–	25.5	–	–	
pith, late spring	59.5	0.01	0.02	–	0.30	0.02	–	14.9	–	–	Tabet et al., 2004
	100.0	0.02	0.03	–	0.51	0.03	–	25.1	–	–	
pith, early summer	48.2	0.01	0.02	–	0.39	0.01	–	9.5	–	–	Tabet et al., 2004
	100.0	0.02	0.03	–	0.80	0.03	–	19.7	–	–	
pith, late summer	47.3	0.01	0.02	–	0.10	0.01	–	5.0	–	–	Tabet et al., 2004
	100.0	0.02	0.04	–	0.20	0.03	–	10.5	–	–	
pith, autumn	31.5	0.01	0.02	–	0.63	0.01	–	3.2	–	–	Tabet et al., 2004
	100.0	0.02	0.08	–	2.00	0.04	–	10.1	–	–	
pith, winter	58.7	0.01	0.03	–	0.25	0.02	–	21.7	–	–	Tabet et al., 2004
	100.0	0.02	0.05	–	0.42	0.03	–	37.0	–	–	
shoot	16.4	0.02	0.04	–	0.44	0.02	–	12.6	–	–	Tabet et al., 2004
	100.0	0.10	0.24	–	2.66	0.14	–	77.1	–	–	
Sinarundinaria fangiana											
leaf, all ages	48.3	–	–	–	–	–	–	–	–	–	Schaller et al., 1985
	100.0	–	–	–	–	–	–	–	–	–	

Table 6.A.2. (cont.)

Description	Dry matter (%)	Ca (%)	P (%)	Na (%)	K (%)	Mg (%)	Cu (ppm)	Fe (ppm)	Mn (ppm)	Se (ppm)	Zn (ppm)	Reference
branch, all ages	49.5	–	–	–	–	–	–	–	–	–	–	Schaller et al., 1985
	100.0	–	–	–	–	–	–	–	–	–	–	
stem, old shoot	43.8	–	–	–	–	–	–	–	–	–	–	Schaller et al., 1985
	100.0	–	–	–	–	–	–	–	–	–	–	
stem, two-year and old	58.8	–	–	–	–	–	–	–	–	–	–	Schaller et al., 1985
	100.0	–	–	–	–	–	–	–	–	–	–	

REFERENCES

AAFCO (Association of American Feed Control Officials) (2004). *2004 Official Publication*. Oxford, In: Association of American Feed Control Officials, Inc.

Blaxter, K. L., McGraham, N. and Wainman, F. W. (1956). Some observations on the digestibility of food by sheep and on related problems. *British Journal of Nutrition*, **10**, 69–91.

Burns, R. A., LeFaivre, M. H. and Milner, J. A. (1982). Effects of dietary protein quantity and quality on the growth of dogs and rats. *Journal of Nutrition*, **112**, 1843–53.

Carter J., Ackleh, A. S., Leonard, B. P. and Wang, H. (1999). Giant panda (*Ailuropoda melanoleuca*) population dynamics and bamboo (subfamily Bambusoideae) life history: a structured population approach to examining carrying capacity when the prey are semelparous. *Ecological Modeling*, **123**, 207–23.

CBSG (Conservation Breeding Specialist Group) (2001). *Report on 1999–2000 CBSG Biomedical Survey of Giant Pandas in Captivity in China*. Apple Valley, MN: IUCN–World Conservation Union/SSC Conservation Breeding Specialist Group.

Chorn, J. and Hoffman, R. S. (1978). *Ailuropoda melanoleuca. Mammalian Species*, **110**, 1–6.

Crouzet, Y. and Frädrich, H. (1985). Bamboo as a panda diet. *Bongo*, **10**, 21–6.

Davis, D. D. (1964). *The Giant Panda. A Morphological Study of Evolutionary Mechanisms. Fieldiana: Zoological Memoirs. Volume 3*, Chicago, IL: Chicago Natural History Museum Press.

Dierenfeld, E. S. (1997). Chemical composition of bamboo in relation to giant panda nutrition. In *The Bamboos*, ed. G. P. Chapman. London: Linnean Society of London, pp. 205–11.

Dierenfeld, E. S., Hintz, H. F., Robertson, J. B., Van Soest, P. J. and Oftedal, O. T. (1982). Utilization of bamboo by the giant panda. *Journal of Nutrition*, **112**, 636–41.

Dierenfeld, E. S., Qiu, X., Mainka, S. A. and Liu, W. (1995). Giant panda diets fed in five Chinese facilities: an assessment. *Zoo Biology*, **14**, 211–22.

Edwards, M. S. (1995). *Comparative Adaptations to Folivory in Primates*. Doctoral dissertation. East Lansing, MI: Michigan State University.

Edwards, M. S. (1996). *Nutritional management of giant pandas* (Ailuropoda melanoleuca) *at the Zoological Society of San Diego*. Technical Conference of Giant Panda Breeding (Chengdu) Beijing: Chinese Association of Zoological Gardens.

Edwards, M. S. (2003). Nutrition of zoo animals. *Recent Advances in Animal Nutrition in Australia*, **14**, 1–9.

Edwards, M. S. and Nickley, J. K. (2000). Development of a fecal consistency scoring system for giant pandas (*Ailuropoda melanoleuca*). In *Panda 2000: Priorities for the New Millennium*, ed. D. Lindburg and K. Baragona. San Diego, CA: Zoological Society of San Diego, p. 367.

Edwards, M. S. and Zhang, G. (1997). Preliminary observations on the use of a higher fiber biscuit as a supplemental food item for giant pandas (*Ailuropoda melanoleuca*). In *International Symposium on the Protection of the Giant Panda*. Chengdu: Sichuan Publishing House of Science and Technology, pp. 50–2.

Endo, H., Sasaki, M., Yamagiwa, D. *et al.* (1996). Functional anatomy of the radial sesamoid bone in the giant panda (*Ailuropoda melanoleuca*). *Journal of Anatomy*, **189**, 587–92.

Endo, H., Makita, T., Sasaki, M. *et al.* (1999). Comparative anatomy of the radial sesamoid bone in the polar bear (*Ursus maritimus*), the brown bear (*Ursus arctos*) and the giant panda (*Ailuropoda melanoleuca*). *Journal of Veterinary Medical Science*, **61**, 903–7.

Endo, H., Sasaki, M., Hayashi, Y. *et al.* (2001). Carpal bone movements in gripping action of the giant panda (*Ailuropoda melanoleuca*). *Journal of Anatomy*, **198**, 243–6.

Fahey, G. C., Jr. and Jung, H. G. (1983). Lignin as a marker in digestion studies: a review. *Journal of Animal Science*, **57**, 220–6.

Feng, W. H., Zhang, F. X., Huang, X. M. *et al.* (1993). Study on the numeral change and artificial breeding effect of giant panda. In *Minutes of the International Symposium on the Protection of the Giant Panda*, ed. A. Zhang and G. He. Chengdu: Sichuan Publishing House of Science and Technology, pp. 226–30.

Goss, L. J. (1940). Acute hemorrhagic gastro-enteritis in a giant panda. *Zoological Scientific Contributions of the New York Zoological Society*, **25**, 261–2.

Hirayama, K., Kawamura, S., Mitsuoka, T. and Tashiro, K. (1989). The faecal flora of the giant panda (*Ailuropoda melanoleuca*). *Journal of Applied Bacteriology*, **67**, 411–15.

Kametaka, M., Takahashi, M., Nakazato, R. *et al.* (1988). Digestibility of dietary fiber in giant pandas and its importance for their nutrition. In *Proceedings of the Second International Symposium on Giant Panda*. Tokyo: Tokyo Zoological Park Society, pp. 143–52.

Li, C. G. (ed.) (1997). *A Study of Staple Food Bamboo for the Giant Panda*. Guiyang, Guizhou: Guizhou Scientific Publisher. (In Chinese with English summaries.)

Liu, X., Yu, J., Li, M., Yang, Z. and Li, G. (2002). Study of crude protein intake and growth response in captive subadult giant pandas (*Ailuropoda melanoleuca*). *Zoo Biology*, **21**, 223–32.

Long, Y., Lu, Z., Wang, D. *et al.* (2004). The nutritional strategy of giant pandas in the Qinling Mountains of China. In *Giant Pandas: Biology and Conservation,* ed. D. Lindburg and K. Baragona. Berkeley, CA: University of California Press, pp. 90–100.

MacFarlane, W. V. and Howard, B. (1972). Comparative water and energy economy of wild and domestic mammals. *Symposia of the Zoological Society of London*, **31**, 261–96.

Mainka, S. A., Guanlu, Z. and Mao, L. (1989). Utilization of a bamboo, sugar cane and gruel diet by two juvenile giant pandas (*Ailuropoda melanoleuca*). *Journal of Zoo and Wildlife Medicine*, **20**, 39–44.

Masman, W. (1995). Bamboo names and synonyms. http://home.iae.nl/users/pms/wmas_dbase/synonyms.html (accessed May 2005).

McClure, F. (1943). Bamboo as panda food. *Journal of Mammology*, **24**, 267–8.

Nickley, J. K. (2001). *Giant Pandas: Bamboo Intake and Fecal Analysis*. Master of Science Thesis. Pomona, CA: California State Polytechnic University.

Nickley, J. K., Edwards, M. S. and Bray, R. E. (1999). The effect of bamboo intake on fecal consistency in giant pandas (*Ailuropoda melanoleuca*). In *Proceedings of the Third Conference of the American Zoo and Aquarium Association Nutrition Advisory Group*. Columbus, OH, pp. 46–50.

NRC (National Research Council) (1985). *Nutrient Requirements of Dogs*. Washington, DC: National Academy Press.

NRC (National Research Council) (1993). *Nutrient Requirements of Fish*. Washington, DC: National Academy Press.

NRC (National Research Council) (2003). *Nutrient Requirements of Nonhuman Primates*, 2nd rev. edn. Washington, DC: National Academy Press.

NRC (National Research Council) (2004). *Nutrient Requirements of Dogs and Cats*. Washington, DC: National Academy Press.

Pan, W. (1988). The panda's food and nutritional value. In *The Giant Panda's Natural Refuge in the Qinling Mountains*, ed. W. Pan, Z. Gao, Z. Lu *et al.* pp. 129–46. Beijing: Peking University Publisher. (In Chinese with an English summary.)

Pan, W., Oftedal, O. T., Zhu, X. *et al.* (1998). Milk composition and nursing in a giant panda (*Ailuropoda melanoleuca*). *Acta Scientiarum Naturalium*, **34**, 350–1.

Qing, S. (1977). Bamboos and the giant panda in the Min Mountains. *Botanical Journal*, **3**, 38–9. (In Chinese.)

Raven, H. C. (1937). Notes on the anatomy and viscera of the giant panda. *American Museum Novitiates*, **877**, 1–23.

Reid, D. G. and Hu, J. (1991). Giant panda selection between *Bashania fangiana* bamboo habitats in Wolong Reserve, Sichuan, China. *Journal of Applied Ecology*, **28**, 228–43.

Robbins, C. T. (1993). *Wildlife Feeding and Nutrition*, 2nd edn. San Diego, CA: Academic Press.

Schaller, G. B. (1993). *The Last Panda*. Chicago, IL: University of Chicago Press.

Schaller, G. B., Hu, J., Pan, W. and Zhu, J. (1985). Feeding strategy. In *The Giant Pandas of Wolong*, ed. G. B. Schaller, J. Hu, W. Pan and J. Zhu. Chicago, IL: University of Chicago Press, pp. 37–107.

Sheldon, W. (1937). Notes on the giant panda. *Journal of Mammalogy*, **18**, 13–19.

Shor, G. (ed.) (2001). *Bamboo Source List, Number 21*. Albany, NY: American Bamboo Society.

Tabet, R. B., Oftedal, O. T. and Allen, M. E. (2004). Seasonal differences in composition of bamboo fed to giant pandas (*Ailuropoda melanoleuca*) at the National Zoo. In *Proceedings of the Fifth Comparative Nutrition Society Symposium*. Hickory Corners, MI, pp. 176–83.

Ullrey, D. E. (1989). Nutritional wisdom (editorial). *Journal of Zoo and Wildlife Medicine*, **20**, 1–2.

Van Soest, P. J. (1994). *Nutritional Ecology of the Ruminant*, 2nd edn. Ithaca, NY: Cornell University Press.

Wang, M. (ed.) (1989). *A Comprehensive Survey Report On China's Giant Panda and Its Habitat*. Ministry of Forestry, Beijing, and the World Wildlife Fund for Nature. (In Chinese.)

Wang, S. and Lu, C. (1973). Giant pandas in the wild. *Natural History*, **82**, 70–1.

Wang, P., Cao, C. & Chen, M. (1982). Histological survey of the alimentary tract of the giant panda. *Zoological Research*, **3** (suppl.), 27–8. (In Chinese.)

Warnell, K. J. (1988). Feed intake, digestibility, digesta passage and fecal microbial ecology of the red panda (*Ailurus fulgens*). Master of Science Thesis. East Lansing, MI: Michigan State University.

Warnell, K. J., Crissey, S. D. and Oftedal, O. T. (1989). Utilization of bamboo and other fiber sources in red panda diets. In *Red Panda Biology*, ed. A. R. Glatson. The Hague: SPB Academic Publishing, pp. 51–6.

Wei, F., Feng, Z., Wang, Z., Zhou, A. and Hu, J. (1999). Use of the nutrients in bamboo by the red panda (*Ailurus fulgens*). *Journal of Zoology (London)*, **248**, 535–41.

Yong, Y. (1981). The preliminary observations on the giant panda in Foping Natural Reserve. *Wildlife*, **4**, 10–16. (In Chinese.)

Zhu, X., Lindburg, D. G., Pan, W., Forney, K. A. and Wang, D. (2001). The reproductive strategy of giant pandas (*Ailuropoda melanoleuca*): infant growth and development and mother-infant relationships. *Journal of Zoology (London)*, **253**, 141–55.

Zou, X., Wang, A., Zou, Q. *et al.* (1993). The experiment on giant pandas' digestion and metabolism. In *International Symposium on the Protection of the Giant Panda*, ed. A. Zhang and G. He. Chengdu: Chengdu Foundation of Giant Panda Breeding, pp. 284–289. (Chinese manuscript with English abstract.)

7

Male reproductive biology in giant pandas in breeding programmes in China

JOGAYLE HOWARD, ZHIHE ZHANG, DESHENG LI, YAN HUANG, RONG HOU,
GUANGHAN LI, MEIJIA ZHANG, ZHIYONG YE, JINGUO ZHANG, SHIQIANG
HUANG, REBECCA E. SPINDLER, HEMIN ZHANG, DAVID E. WILDT

INTRODUCTION

The goal of the giant panda *ex situ* breeding programme is to produce healthy, genetically diverse and reproductively sound offspring. However, reproduction in this species has been poor, in part, due to lack of male libido or aggressive behaviours towards conspecific females. Although giant panda breeding facilities have made progress in producing more surviving young, only about 29% of captive male giant pandas have ever sired offspring (Lindburg *et al.*, 1998), and most of these males were wild born. Of the 104 giant pandas in the *ex situ* population in China in 1996 (at the time of the first masterplanning meeting in China; Zheng *et al.*, 1997; see also Chapter 2), there were 33 adult males of reproductive age (6–26 years old). Only five (15.2%) had ever mated naturally and sired young. This was the main reason for 'male reproduction' being a primary target of the Biomedical Survey conducted under the umbrella of the Conservation Breeding Specialist Group (CBSG) (see Chapter 2).

We had three goals, the first being to measure the presence or absence of any obvious physiological or anatomical abnormalities. The second was to learn more about species reproductive biology, specifically comparing males of different ages, successful versus unsuccessful

Giant Pandas: Biology, Veterinary Medicine and Management, ed. David E. Wildt, Anju Zhang, Hemin Zhang, Donald L. Janssen and Susie Ellis. Published by Cambridge University Press. © Cambridge University Press 2006.

breeders and wild-born versus captive born. Our approach also allowed a third opportunity: studies that would enhance our understanding on how better to use male gametes (sperm) to advance genetic management (see Chapter 21). In this case, our focus was on:

1.   sperm morphology and acrosomal integrity;
2.   testes development during the breeding season;
3.   sperm processing that would allow consistently successful artificial insemination (AI) with fresh or thawed spermatozoa.

The latter objective involved a comparative examination of media, cryodiluents, freezing methods, freezing rates and sperm function (via assessments of biological phenomena known as capacitation, the acrosome reaction, decondensation and zona pellucida penetration).

METHODS

Animals

Adult male giant pandas were evaluated in February and March (during the early breeding season) in 1998, 1999 and 2000. Of the 24 males in the Biomedical Survey, 17 were of an age that allowed a detailed fertility assessment. They were distributed among four facilities: Chengdu Zoo (n = 1 male); the Chengdu Research Base of Giant Panda Breeding (n = 3); Beijing Zoo (n = 4); and the China Conservation and Research Centre for the Giant Panda (n = 9) (within the Wolong Nature Reserve). Animals were 5.5 to 16.5 years of age: 5.5 years (n = 4 males), 6.5 (n = 2), 7.5 (n = 1), 8.5 (n = 1), 10.5 (n = 3), 12.5 (n = 1), 13.5 (n = 2), 14.5 (n = 2) and 16.5 (n = 1). All males were maintained in individual enclosures with indoor/outdoor access. Diets consisted of freshly cut bamboo and a concentrate in the form of gruel or steamed bread, all of which was fed several times a day. Historical data were collected on each male that included origin (wild-born or captive-born), date of birth and reproductive history, including past opportunities to breed (see Chapter 5).

Physical examination, morphometric traits and testicular development

Each male was induced into a surgical plane of anaesthesia using injectable ketamine HCl (Ketaset, Fort Dodge Laboratories, Inc., Fort Dodge, IA) with maintenance often sustained with isoflurane gas (for details, see Chapter 4). Body weight and numerous morphometric measurements, including neck girth, chest girth and body length, were obtained (see Chapter 4). Testes were examined for location (i.e. scrotal sac versus inguinal) and palpated for tone and consistency. Each testis was measured for length and width, which permitted calculating testis volume using the formula for a prolate spheroid: volume $= (4/3)\pi ab^2$ where a is 1/2 length and b is 1/2 width) (Beyer, 1987). For quick reference, the formula was recalculated (volume $= 0.524 \times L \times W^2$ where L is length and W is width). Total testicular volume per male was determined by adding the right and left testis volumes.

Semen collection and evaluation

Semen was collected by an already described electroejaculation technique for the giant panda (Platz *et al.*, 1983; Howard, 1993, 1999) which relied on a 2.6-cm diameter rectal probe with three longitudinal, stainless-steel electrodes and a 60-Hz sine wave stimulator (P. T. Electronics, Boring, OR) (Fig. 7.1). Each male was placed in a dorsal recumbent position for electroejaculation. The probe was inserted into the rectum with the middle electrode positioned ventrally. A standardised set of low-voltage stimulations (2–8 V) over three series of 30 stimuli each was adequate to elicit ejaculation. The electrical stimuli consisted of a slow steady rise from 0 V to initial peak voltage over a three second time period. The current was held at this peak for two to three seconds and then returned rapidly to zero. Voltage was increased after administering ten stimuli at the same voltage. The entire semen collection interval generally required less than 20 minutes. A predetermined amount of semen (0.5 or 1.0 ml) from each male was allocated to these characterisation assessments and comparative sperm cryopreservation studies. The remaining semen was used for AI or cryopreserved for long-term storage.

Semen assessments were conducted on 16 of the 17 males. The remaining male (8.5 years) was in poor health with Stunted Development Syndrome (see Chapter 4) and bilaterally retained inguinal testes, so only morphometric traits were measured. Freshly obtained semen from the other males was evaluated immediately for ejaculate volume, pH (via test strips; EM Science, Gibbstown, NJ) and sperm concentration (via a standard haemocytometer method; Howard, 1993). Subjective estimates of sperm motility and forward progression (i.e. the type of forward movement of sperm from a rapid, straight direction to

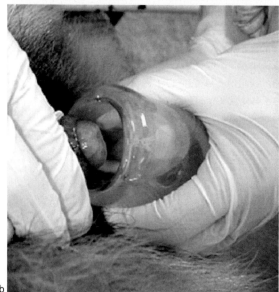

Figure 7.1. (a) Electroejaculator and rectal probe used for semen collection in the giant panda, and (b) close-up of the glans penis and semen collection vial.

quivering with no progression on a scale of 0 to 5; 5 being best) was determined at ×400 using a microscope with stage warmer (Howard, 1993). Spermatozoa were maintained in a plastic vial (Vangard International, Inc., Neptune, NJ) at 37°C in a water bath or dry-bath incubator.

Sperm morphology and acrosomal membranes

Sperm morphology was determined from fixed aliquots of raw semen as previously described (Howard, 1993) (Fig. 7.2). A 5 μl aliquot of the fresh electroejaculate was added to 100 μl of fixative (0.3% glutaraldehyde in saline) for morphological examination of spermatozoa by phase-contrast microscopy ($\times$1000) (Howard, 1993). Sperm were categorised as normal or as having one of the following anomalies: an abnormal head

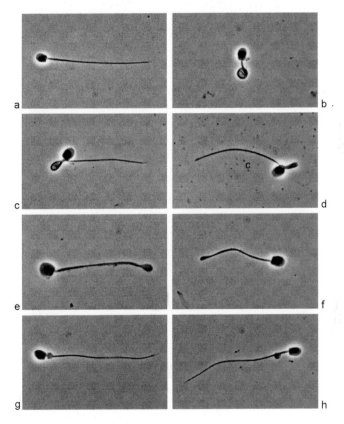

Figure 7.2. Sperm morphology in the giant panda including (a) a normal or (b) abnormal spermatozoon with a coiled flagellum, (c) bent midpiece with cytoplasmic droplet, (d) bent midpiece without cytoplasmic droplet, (e) bent flagellum with cytoplasmic droplet, (f) bent flagellum without cytoplasmic droplet, (g) proximal cytoplasmic droplet or (h) distal cytoplasmic droplet.

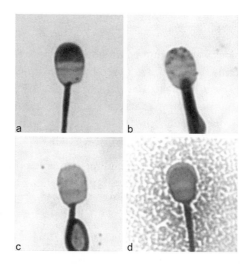

Figure 7.3. Acrosomal morphology of the giant panda spermatozoon with a (a) normal apical ridge, (b) damaged apical ridge, (c) missing apical ridge or (d) loose acrosomal cap. (See also Plate VIII.)

(i.e. macrocephaly, microcephaly or bicephaly); abnormal acrosome; coiled flagellum; bent midpiece with cytoplasmic droplet; bent midpiece without cytoplasmic droplet; bent flagellum with cytoplasmic droplet; bent flagellum without cytoplasmic droplet; proximal cytoplasmic droplet; and distal cytoplasmic droplet (Howard, 1993) (see Fig. 7.2). Sperm with an abaxial attachment of the neck were not considered abnormal. This unusual offset of the head with the neck region (most easily seen in Fig. 7.2a, f, h) is a common feature of the giant panda spermatozoon (Moore *et al.*, 1984).

The acrosome is the outer membrane of the spermatozoon, which is critical to successful fertilisation. Damage to this structure as a result of inherent physiological anomalies, physical mishandling or suboptimal cryopreservation techniques of semen is a common cause of infertility. Acrosomal integrity was evaluated in the giant panda using the rose bengal/fast green stain (Pope *et al.*, 1991) (Fig. 7.3; Plate VIII). Raw semen was diluted in Ham's F10 tissue culture medium (Irvine Scientific, Santa Ana, CA) supplemented with 25-mM HEPES buffer and 5% heat-treated fetal calf serum (one part semen to 10 parts Ham's). For staining, 1 μl of diluted semen was added to 9 μl of rose bengal/fast green stain, incubated for 90 seconds and then smeared on a glass slide. Using bright-field microscopy (×1000), a minimum of 100 sperm acrosomes per sample was assessed as having a:

1.   normal intact apical ridge (uniform staining of the acrosome over the anterior half of sperm head);
2.   damaged apical ridge (non-uniform staining with ruffled or folded acrosome);
3.   missing apical ridge (lack of staining due to acrosome absence);
4.   loose acrosomal cap (loose membrane protruding above the level of the sperm head) (Howard, 1993) (see Fig. 7.3; Plate VIII).

Semen processing and sperm dilution

Successful sperm cryopreservation requires using a cryoprotectant – an antifreeze to protect the sperm from lethal ice damage during freezing and thawing. For the giant panda, the cryoprotectant of choice has been glycerol, which in this case was added to two cryodiluents used historically by the USA and Chinese colleagues. A cryodiluent called TEST was prepared in the USA using a commercially available 'Freezing Medium–TEST Yolk Buffer (TYB) with Glycerol', which is marketed by Irvine Scientific for freezing human sperm. This is a 20% egg-yolk solution containing 12% glycerol combined with the commercially available 'Refrigeration Medium-TEST Yolk Buffer (TYB) without Glycerol' to yield a 5% glycerol concentration. The medium was prepared in the USA, then re-frozen for shipment to China. The second cryodiluent, called SFS (for Chinese 'Sperm Freezing Solution'), was prepared fresh each day in China and consisted of 20% egg yolk, 12% sucrose and 5% glycerol. This SFS 5% glycerol cryodiluent was maintained at 4°C and not frozen. For comparative studies, both diluents were also prepared with no glycerol (i.e. TEST 0% and SFS 0% glycerol) for initial dilution of specific treatments.

Following collection, giant panda semen was diluted (1:3; semen: diluent) immediately with TEST or SFS diluent containing 0% or 5% glycerol at 37°C. Diluted aliquots were evaluated for sperm motility and acrosomal traits at 37°C, then placed in a 400-ml water-jacket for slow cooling over 4 hours to 4°C. Aliquots were either maintained at 4°C to assess duration of motility for 48 hours or further diluted in glycerolated diluents and prepared for cryopreservation.

Sperm cryopreservation and thawing

Three freezing methods (pellets vs. straws vs. cryovials) were evaluated, with cooling rates monitored by a thermocouple device (Omega Engineering, Inc., Stanford, CT). After cooling, additional aliquots of glycerolated TEST and SFS were added to the sperm suspensions to achieve a final 4% glycerol concentration before freezing. Two pellet-freezing techniques were tested that involved a plastic tray within a stainless-steel box or a metal wire-mesh screen in a Polystyrene box (Fig. 7.4). To compare the impact of pellet size, semen was also frozen in 40-µl (~100°C per minute) or 80-µl (~50°C per minute) droplets, then packaged in plastic cryovials (Vangard International, Inc.), placed on a metal cane and maintained immersed in liquid nitrogen. For straws, cooled semen was pipetted into 0.25-ml sterile plastic straws (Veterinary Concepts, Spring Valley, WI) and sealed at the open end (Fig. 7.5). A two-step

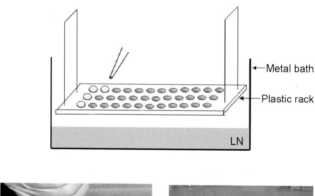

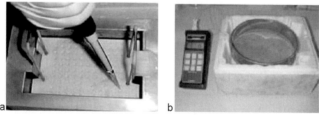

Figure 7.4. Semen cryopreservation using the pellet-freezing method on a plastic tray within (a) a stainless-steel box or (b) a wire mesh-screen in a Styrofoam box positioned at a consistent level above liquid nitrogen to achieve a constant temperature of −96°C. Pellets remained on the tray or wire mesh for three minutes before plunging into liquid nitrogen.

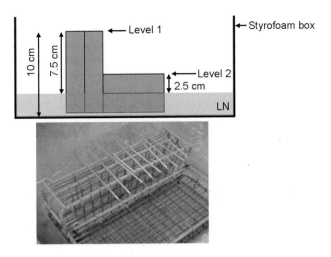

Figure 7.5. Semen cryopreservation using a two-step straw-freezing
method in liquid nitrogen (LN) vapour. Straws (0.25-ml) containing cooled
semen are placed at 7.5 cm (Level 1; $-30°C$; for one minute) and 2.5 cm
(Level 2; $-130°C$; for one minute) above liquid nitrogen in a Styrofoam
box, then plunged into liquid nitrogen.

method was used whereby each straw was placed 7.5 cm over liquid
nitrogen for 1 minute (Level 1; $-30°C$) and then 2.5 cm above liquid nitrogen for 1 minute (Level 2; $-130°C$) to achieve rapid cryopreservation
(about $-70°C$ per minute) before plunging into liquid nitrogen (see
Fig. 7.5). Finally, 0.5-ml aliquots of semen cooled in 1.8-ml cryovials
(Vangard International, Inc.) were frozen on a thin Polystyrene platform
floating on 7.5 cm of liquid nitrogen for 15 minutes before storing
immersed in liquid nitrogen (Fig. 7.6). The cryovial method resulted in
the slowest freezing rate ($\sim20°C$ per minute; Fig. 7.7).

Two tissue culture media were evaluated for thawing and post-
thaw sperm dilution, Ham's F10 medium and Tyrode's 199, each containing 25-mM HEPES buffer to maintain pH. The vial containing pellets
was maintained in liquid nitrogen, while one pellet was removed for
thawing. The pellet was held in air for 10 seconds, then placed into a
sterile 12 × 75-mm glass tube containing 0.5 ml of either thaw medium
(for 45 seconds in a 37°C water bath). The glass tube was agitated and
the thawed suspension transferred into a plastic Eppendorf tube (Brink-
man Instruments, Inc., Westbury, NY) and maintained for 90 minutes
(at 37°C). For the straw method, each straw was held in air for 10

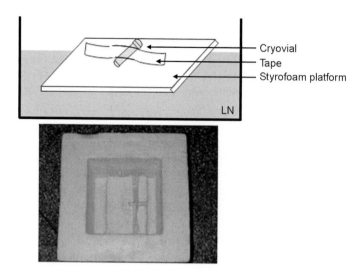

Figure 7.6. Semen cryopreservation using a cryovial method of freezing. Cooled semen (0.5-ml) in a 1.8-ml cryovial is placed and taped onto a thin Styrofoam platform floating on 7.5 cm of liquid nitrogen for 15 minutes before plunging into liquid nitrogen.

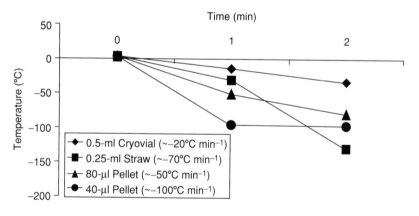

Figure 7.7. Comparison of freezing rates using the pellet, straw and cryovial method of sperm cryopreservation.

seconds, then plunged into the 37°C water bath for 45 seconds. The sealed end of the straw was cut off, then thawed semen was expelled and diluted in either thaw medium at 37°C (half in 1.25-ml Ham's; half in 1.25-ml Tyrode's 199). For the cryovial method, each vial was simply thawed in the 37°C water bath. All samples were maintained at 37°C while assessed for sperm motility, progression and acrosomal integrity at 0, 30, 60 and 90 minutes after thawing.

Sperm capacitation, acrosome reaction, zona penetration and decondensation

Most traditional sperm assessments have relied largely on simple evaluations of motility (cell movement). Modern fertility evaluations also test the *functionality* of spermatozoa by examining how effectively they undergo the biochemical and cellular processes associated with preparation for, or the actual, fertilisation event (Yanagimachi, 1994). In our studies, we chose several such phenomena, including the ability of sperm to undergo:

1. capacitation, or a series of physiological changes (alterations or removal of substances) on the sperm plasma membrane that renders the spermatozoon capable of undergoing the acrosome reaction;

2. the acrosome reaction, or multiple fusions between the outer acrosomal membrane and overlaying plasma membrane, that enable the contents of the acrosome to escape through fenestrated membranes.

For these two assessments, fresh or thawed semen was diluted in Ham's F10 (37°C) containing 5% fetal calf serum and 25-mM HEPES, centrifuged (200g, 8 minutes) and the pellet resuspended in fresh Ham's F10 (37°C). At 0, 3, 6 and 9 hours after resuspension, aliquots were removed and assessed for sperm motility and acrosomal integrity. Additional aliquots were removed and diluted in Ham's F10 only (control) or Ham's F10 containing heterologous (domestic cat) solubilized zonae pellucidae for 30 minutes. Following incubation, acrosomes were evaluated as above. Percentage of capacitated sperm was defined as the proportion of sperm with intact acrosomes after exposure to solubilized cat zonae subtracted from the proportion of sperm with intact acrosomes after exposure only to Ham's F10 medium with no zonae (control).

The impact of cryopreservation on the functional ability of the sperm nucleus to undergo decondensation, or the release of disulphide bonds within the spermatozoon, also was evaluated. Nuclear

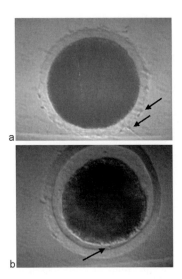

Figure 7.8. Sperm penetration of the zona pellucida of a (a) homologous giant panda and (b) heterologous cat salt-stored oocyte used to assess sperm function before and after cryopreservation. (See also Plate IX.)

decondensation allows the sperm chromosomes to become accessible to the oocyte (Mahi & Yanagimachi, 1975). Fresh and frozen–thawed giant panda sperm were pre-incubated in Ham's F10 for 6 hours to induce capacitation, followed by a cat solubilized zonae pellucidae solution for 30 minutes to induce the acrosome reaction (as described above). To assess decondensation, aliquots were then removed at 0, 2 and 4 hours and incubated in Ham's F10 only (control) or cat oocyte cytoplasmic emulsion at 37°C. Sperm were categorised as normal (condensed cytoplasm) or having partial or complete decondensation (based on sperm head transparency and enlargement).

The sperm penetration assay using zona-intact, salt-stored oocytes (Fig. 7.8; Plate IX) has been used to assess sperm–oocyte interaction in numerous species (Drobnis *et al.*, 1988). Spermatozoa must undergo capacitation and the acrosome reaction for subsequent zona penetration; thus this assay evaluates several stages of fertilisation. Also, heterologous oocytes are easily obtained and provide an attractive approach for evaluating the ability of sperm to bind and penetrate the zona pellucida, an often limiting factor for fertilisation. In a preliminary study, the ability of giant panda sperm to bind and penetrate the zonae pellucidae of salt-stored, domestic cat oocytes (which were readily available) was investigated.

RESULTS AND DISCUSSION

Male breeding history and assessment of morphometric and testicular traits

Of the 17 evaluated males, 10 were captive-born within China and seven were wild-born. Twelve of the 17 (70.6%) had never mated naturally, despite most having the opportunity with an oestrual female. Of these, four (33.3%) were wild-born and the remaining eight (66.7%) were captive-born. Of the five males that had naturally mated, three were wild-born and two captive-born.

The mean (and range) measures for body weight, chest girth, neck girth and body length for 16 of the 17 sampled males are provided in Table 7.1. These same body metrics for SB 356 (weight 67 kg, chest girth 84 cm, neck girth 52 cm and body length 150 cm) were excluded from these average values because of his Stunted Development Syndrome (see Chapter 4). Similarly, no attempt was made to collect semen from this male due to his extraordinarily small inguinal testes (total testicular volume, 4.2 cm^3). Of the other 16 giant panda males, 15 had large testes with a total testicular volume averaging 300 cm^3 (range 100–455 cm^3) (see Table 7.1). On average, each testis was more than 8 cm long and 5 cm wide with no size difference ($p > 0.05$) between the right and left gonads.

The remaining male (SB 400) was 5.5 years old with bilaterally small testes held closely adjacent to the body and a total testicular volume of only 99.8 cm^3 (Table 7.2). This was probably due to his youth or a delayed pubertal onset, because his other age-matched counterparts (three 5.5-year-old males) all had larger testes and total testicular volumes (see Table 7.2). Interestingly, SB 400 also had the lowest body weight (87 kg), chest girth (97 cm) and neck girth (61 cm) despite having an overall good conformation and above-average body length (184 cm) (see Table 7.2).

Seminal quality, sperm morphology and acrosomal integrity

Most reports on male reproduction in giant pandas have focused on semen collection and processing techniques for AI with fresh spermatozoa

Table 7.1. *Morphometric, testicular and ejaculate traits of giant pandas evaluated during the 1998 to 2000 Giant Panda Biomedical Survey (16 males, 5.5 to 16.5 years of age)*

Traits	Mean (± SEM)	Range
Morphometry		
Body weight (kg)	111.7 ± 3.1	86.8–130.5
Chest girth (cm)	107.9 ± 1.6	97.2–124.0
Neck girth (cm)	73.4 ± 1.3	61.3–83.0
Body length (tip of nose to base of tail) (cm)	184.1 ± 1.8	166.5–199.8
Testicular		
Right testis length (cm)	8.9 ± 0.2	5.8–10.1
Right testis width (cm)	5.6 ± 0.2	2.8–6.5
Right testis volume (cm^3)	148.2 ± 10.4	23.8–214.5
Left testis length (cm)	9.1 ± 0.2	6.0–10.1
Left testis width (cm)	5.8 ± 0.2	3.6–6.9
Left testis volume (cm^3)	164.3 ± 13.1	41.1–259.9
Total testes volume (cm^3)	300.0 ± 22.2	99.8–455.2
Ejaculate		
Ejaculate volume (ml)	3.9 ± 0.6	0.5–9.7
Ejaculate pH	8.1 ± 0.1	6.8–8.7
Sperm concentration/ml ($\times 10^6$)	1182.4 ± 220.6	3–3664
Sperm concentration/ejaculate ($\times 10^6$)	4299.3 ± 878.7	1.6–15 022.04
Sperm motility (%)	68.8 ± 5.2	0–95
Forward progression (0 to 5; 5 = best)	2.8 ± 0.2	0–4
Abaxial sperm head attachment (%)	89.5 ± 0.9	82–95
Normal sperm (%)	66.3 ± 6.4	4–96
Abnormal sperm (%)		
acrosomal defects	8.2 ± 4.2	1–74
coiled flagellum	5.3 ± 1.4	0–18
bent midpiece with droplet	6.3 ± 1.4	1–21
bent midpiece without droplet	2.5 ± 0.8	0–11
bent flagellum with droplet	2.6 ± 1.2	0–20
bent flagellum without droplet	3.5 ± 2.3	0–40
proximal cytoplasmic droplet	1.3 ± 0.4	0–5
distal cytoplasmic droplet	4.0 ± 1.3	0–15
Acrosomal integrity (%)		
normal apical ridge	91.8 ± 4.2	26–99
damaged apical ridge	1.9 ± 0.4	0–6
missing apical ridge	4.8 ± 3.3	0–57
loose acrosomal cap	1.5 ± 1.0	0–17

Table 7.2. *Ejaculate and morphometric traits of individual giant pandas (age 5.5 years) compared to older (6.5 to 16.5 years) males*

	SB 392 5.5 years	SB 400 5.5 years	SB 399 5.5 years	SB 413 5.5 years	Older males (n = 12)[a] 6.5 to 16.5 years
Ejaculate volume (ml)	4.1	0.5	2.5	3.8	4.2 ± 0.7
Ejaculate pH	8.7	6.8	7.9	8.2	8.1 ± 0.1
Sperm concentration/ml ($\times 10^6$)	3664.0	3.0	793.0	531.0	1162.3 ± 180.9
Sperm concentration/ejaculate ($\times 10^6$)	15022.4	1.6	1982.5	2017.8	4161.0 ± 647.8
Sperm motility (%)	75	0	50	95	73.1 ± 2.8
Forward progression (0 to 5; 5 = best)	3.5	0	2.5	3.0	2.9 ± 0.1
Normal sperm (%)	42	8	4	68	77.3 ± 3.2
Abnormal sperm (%)					
acrosomal defects	5	74	7	5	3.7 ± 0.7
coiled flagellum	12	12	18	1	3.6 ± 1.1
bent midpiece defects	21	4	9	14	7.9 ± 1.0
bent flagellum defects	1	2	60	7	2.6 ± 0.6
proximal cytoplasmic droplet	4	0	1	1	1.2 ± 0.4
distal cytoplasmic droplet	15	0	1	4	3.8 ± 1.5
Acrosomal integrity (%)					
normal apical ridge	95	26	93	95	96.3 ± 0.7
damaged apical ridge	1	0	1	4	2.0 ± 0.5
missing apical ridge	2	57	6	1	1.2 ± 0.5
loose acrosomal cap	2	17	0	0	0.5 ± 0.2

Table 7.2. (cont.)

	SB 392 5.5 years	SB 400 5.5 years	SB 399 5.5 years	SB 413 5.5 years	Older males (n = 12)[a] 6.5 to 16.5 years
Body weight (kg)	125.0	86.8	113.1	110.1	112.7 ± 3.5
Chest girth (cm)	103.0	97.2	109.5	107.0	109.0 ± 1.8
Neck girth (cm)	77.0	61.3	75.5	78.1	73.5 ± 1.3
Body length (tip of nose to base of tail) (cm)	184.0	184.3	195.3	183.8	183.3 ± 2.2
Right testis					
length (cm)	8.4	6.6	8.0	9.4	9.2 ± 0.2
width (cm)	6.2	4.1	5.0	5.4	5.7 ± 0.2
testis volume (cm^3)	166.0	58.6	105.8	144.7	158.0 ± 10.6
Left testis					
length (cm)	9.7	6.0	8.8	8.8	9.3 ± 0.2
width (cm)	6.7	3.6	5.2	5.7	5.9 ± 0.2
testis volume (cm^3)	226.1	41.1	124.1	151.1	173.8 ± 12.4
Total testes volume (cm^3)	392.1	99.8	229.9	295.7	314.0 ± 22.4

[a] Values are means ± SEM.

(Liu, 1981; Platz et al., 1983; Moore et al., 1984; Masui et al., 1989). Semen collection by electroejaculation is used widely in China and is considered safe when conducted by trained staff using appropriate equipment. The major breeding facilities (i.e. in Chengdu, Beijing and Wolong) commonly use this approach to evaluate male fertility and to collect sperm for AI. Because electroejaculation is used extensively, there are substantial data on basic semen and sperm characteristics, especially on semen volume, sperm motility and number of motile sperm per ejaculate (Platz et al., 1983; Moore et al., 1984; Howard, 1993). Individual sperm appearance and size have also been described. The giant panda spermatozoon has a rounded head (mean ± SEM, i.e. standard error of the mean; length 5.0 ± 0.1 μm, width 4.2 ± 0.1 μm), narrow midpiece (length 7.2 ± 0.1 μm, width 0.8 ± 0.04 μm), flagellum (length 39.0 ± 1.3 μm) and an average total length of 51.2 ± 0.4 μm (Platz et al., 1983). Overall spermatozoon shape is generally similar to other bears although the total length is shorter than that reported for the sloth bear (72.5 μm), Kodiak bear (69.4 μm) and spectacled bear (73.1 μm) (Platz et al., 1983). Other than a few reports discussing ultrastructure (Chen et al., 1984; Moore et al., 1984; Sun et al., 1996), there is little published information on giant panda sperm morphology and membrane integrity.

Of the 16 electroejaculated males in our survey, all produced spermic ejaculates (see Table 7.1). The semen of the giant panda (average volume, 3.9 ± 0.6 ml) contained prodigious numbers of sperm (mean 1182.4 ± 220.6 × 10^6 sperm ml^{-1}) that were highly motile (~69%; see Table 7.1). Thus, an average ejaculate can contain almost three billion motile sperm, which is slightly higher than other bear species (Ishikawa et al., 1998) but much greater than what has been reported for other carnivores such as felids (Howard, 1993; Wildt & Swanson, 1998) and canids (Amann, 1986).

Most giant panda ejaculates contained high proportions of structurally normal sperm (>65%), including a high proportion (>85%) of sperm with an abaxial attachment of the neck. The observed defects included acrosomal anomalies, a coiled flagellum, bent midpiece, bent flagellum or cytoplasmic droplet (see Table 7.1 and Fig. 7.2). Similarly, other bear species produce high numbers of morphologically normal sperm (Ishikawa et al., 1998). Most giant panda spermatozoa had an intact acrosome with normal apical ridge (>90%) with a low incidence of a damaged apical ridge (<2%), a missing apical ridge (<5%) or loose acrosomal cap (<2%) (see Table 7.1 and Fig. 7.3).

Influence of age on reproductive traits

Seminal traits, including sperm morphology, were influenced by age (see Table 7.2) but only near the onset of puberty. Ejaculate quality in adult males 6.5 to 16.5 years old was consistently similar (regardless of age) and superior to the four 5.5-year-old pubertal counterparts (see Table 7.2). Semen from the latter group had the greatest variability in sperm concentration (range 3–3664 $\times 10^6$ sperm ml^{-1}), motility (range 0–95%) and forward progression scores (range 0–3.5). The 5.5-year-old males also produced fewer structurally normal sperm (mean 30.5 $\pm$ 15.2%; range 4–68%) than older giant pandas (see Table 7.2). Even so, in general, acrosomal integrity in these younger males was similar to the older group, with the exception of male SB 400, which ejaculated sperm with only 26% normal acrosomes (see Table 7.2). This individual had a subadult-sized penis and produced an extremely low sperm concentration (3×10^6 sperm ml^{-1}) compared to all other males (see Table 7.2). All spermatozoa in SB 400's ejaculate lacked motility with the ejaculate containing many spermatids (immature cells). Combined with his smaller testes and many body metrics, we concluded that SB 400 probably was experiencing delayed development and/or puberty. And, although having a history of chronic gastrointestinal disease, SB 400 was not classified as having Stunted Development Syndrome, largely because of his overall normal body length.

Reproductive traits in wild- versus captive-born, and breeding versus non-breeding males

One of the most important findings from our contribution to the Biomedical Survey was that male physiological function was not compromised. Of the 12 giant pandas of more than six years of age, reproductive traits and body measures were the same for the seven wild-born versus the five captive-born males (Table 7.3). Despite many captive-born males failing to reproduce, there were no differences ($p > 0.05$) in ejaculate traits, sperm morphology, acrosomal integrity, body measurements and testicular size compared to pandas of a wild origin. However, there was a trend ($p = 0.06$) for the wild-born males to have larger testes (see Table 7.3). Ejaculate and morphometric traits of the five males that had sired offspring were also the same as the seven non-breeding counterparts (see Table 7.3). Both groups produced similarly ($p > 0.05$)

Table 7.3. *Ejaculate and morphometric traits in wild-born versus captive-born, and breeding versus nonbreeding male giant pandas*[a]

	Wild-born males (n = 7)	Captive-born males (n = 5)	Breeding males (n = 5)	Non-breeding males (n = 7)
Ejaculate volume (ml)	4.8 ± 1.1	3.4 ± 0.7	5.5 ± 1.4	3.1 ± 0.4
Ejaculate pH	8.1 ± 0.1	8.2 ± 0.1	8.0 ± 0.1	8.2 ± 0.1
Sperm concentration/ml ($\times 10^6$)	1079.3 ± 236.3	1295.2 ± 247.8	919.7 ± 289.9	1370.3 ± 196.4
Sperm concentration/ ejaculate ($\times 10^6$)	4144.6 ± 665.0	4187.1 ± 1100.5	3983.5 ± 864.4	4313.1 ± 923.3
Sperm motility (%)	77.5 ± 3.5	66.0 ± 2.2	74.2 ± 2.7	72.1 ± 4.3
Forward progression (0 to 5; 5 = best)	2.9 ± 0.2	3.0 ± 0.1	3.0 ± 0.2	2.9 ± 0.2
Normal acrosomes (%)	97.3 ± 0.7	94.8 ± 1.0	95.7 ± 1.5	96.9 ± 0.4
Normal sperm (%)	77.8 ± 4.9	76.6 ± 3.3	81.3 ± 4.3	73.9 ± 4.0
Abnormal sperm (%)				
acrosomal defects	2.8 ± 0.7	5.2 ± 1.0	4.3 ± 1.5	3.1 ± 0.4
coiled flagellum	3.1 ± 1.8	4.4 ± 1.0	3.2 ± 1.4	4.0 ± 1.6
bent midpiece defects	9.4 ± 1.4	5.4 ± 0.9	5.3 ± 1.1	10.0 ± 1.4
bent flagellum defects	1.9 ± 0.5	3.8 ± 1.1	2.1 ± 0.8	3.0 ± 0.4
proximal cytoplasmic droplet	1.5 ± 0.7	0.8 ± 0.3	1.0 ± 0.5	1.4 ± 0.6
distal cytoplasmic droplet	3.8 ± 2.1	3.8 ± 1.8	2.8 ± 2.5	4.6 ± 1.6
Body weight (kg)	107.5 ± 4.8	121.0 ± 2.3	117.6 ± 4.9	108.4 ± 4.2
Chest girth (cm)	109.4 ± 3.1	108.3 ± 1.1	113.2 ± 3.0	105.5 ± 1.1

Table 7.3. (cont.)

	Wild-born males (n = 7)	Captive-born males (n = 5)	Breeding males (n = 5)	Non-breeding males (n = 7)
Body length (tip of nose to base of tail) (cm)	179.2 ± 2.4	189.8 ± 2.1	185.2 ± 3.3	181.7 ± 2.8
Right testis length (cm)	9.4 ± 0.2	8.7 ± 0.2	9.6 ± 0.1	8.8 ± 0.2
Right testis width (cm)	5.7 ± 0.3	5.6 ± 0.1	6.0 ± 0.1	5.3 ± 0.2
Right testis volume (cm^3)	163.9 ± 16.7	146.2 ± 7.6	184.3 ± 9.0	131.7 ± 11.0
Left testis length (cm)	9.5 ± 0.2	9.0 ± 0.1	9.4 ± 0.3	9.3 ± 0.1
Left testis width (cm)	6.0 ± 0.3	5.7 ± 0.2	6.3 ± 0.1	5.6 ± 0.2
Left testis volume (cm^3)	183.5 ± 17.0	154.6 ± 14.3	199.7 ± 14.9	155.4 ± 13.4
Total testes volume (cm^3)	347.4 ± 30.8	260.6 ± 15.3	363.4 ± 32.2	271.7 ± 18.8

[a] Values are means ± SEM in male giant pandas >6 years of age. The four young 5.5-year-old males were not included in analyses due to influence of age on seminal quality (as shown in Table 7.2).

high numbers of structurally normal, motile sperm which contained an intact acrosome with normal apical ridge. The most predominant sperm abnormality for both groups was a bent midpiece defect. Although sperm production was similar, there was an interesting trend ($p = 0.06$) for the breeding males to have larger testes than non-breeding males.

Testicular hypoplasia or atrophy

Testicular abnormalities and undescended testes were discovered in three males (SB 356, SB 323 and SB 345) and a closely related male SB 181 (already deceased; see below) (Table 7.4). SB 356 was a small (67 kg), emaciated, 8.5-year-old, captive-born male with a history of severe chronic illness. He was the progeny of a wild-born sire (SB 201) and captive-born dam (SB 278). This male, which met our criteria for Stunted Development Syndrome (see Chapter 4), also had a small penis and bilaterally retained inguinal testes with severe testicular hypoplasia. The testes were adequately positioned to allow a crude measurement but due to their small size no attempt was made to collect semen.

SB 323 was a robust 12.5-year-old male with one retained testis (near the inguinal canal; see Table 7.4) which adhered to the body wall. He was the offspring of a wild-born sire (SB 186) and captive-born dam (SB 148) who had wild-born parents. SB 323 was an active proven breeder with multiple living offspring. Similar to SB 356, the left retained testis adjacent to the inguinal canal could be measured. Zoo records revealed that this gonad had never developed normally or descended into the scrotum, indicating unilateral testicular hypoplasia. Even with only one normal-sized testis, excellent semen quality was measured (see Table 7.4). We speculated that testicular hypoplasia may be inherited because a related male (SB 181 and a sibling to SB 148, the dam of SB 323) also had only one testis. SB 181 was deceased at the time of the survey but his testicular anomaly was clearly evident in zoo records and historical photographs. In terms of the normality of SB 323's offspring, at least one son (SB 392) has two descended testes, but another (SB 469, born in 1998) has only one descended testis, suggesting that testicular hypoplasia may have some paternal mode of inheritance.

In contrast to hypoplasia, SB 345 appeared to have experienced testicular atrophy. Historical records indicated that both testes had developed and increased in size during puberty, although the right

Table 7.4. *Testicular abnormalities, morphometrics and seminal quality in giant pandas with a unilaterally or bilaterally retained testis*

	SB 356	SB 323[a]	SB 345
	8.5 years	12.5 years	10.5 years
Right testis	Retained – inguinal	Descended – scrotal	Retained – inguinal
length (cm)	4.0	9.6	5.8
width (cm)	1.0	5.5	2.8
testis volume (cm^3)	2.1	152.2	23.8
Left testis	Retained – inguinal	Retained – inguinal	Descended – scrotal
length (cm)	4.0	7.8	9.5
width (cm)	1.0	4.3	6.5
testis volume (cm^3)	2.1	75.6	210.3
Total testes volume (cm^3)	4.2	227.8	234.1
Body weight (kg)	67.0	129.5	123.5
Chest girth (cm)	84.0	107.0	106.0
Neck girth (cm)	52.0	70.5	69.8
Body length (tip of nose to base of tail) (cm)	150.0	199.8	187.8
Ejaculate volume (ml)	No collection due to poor health, stunted growth and retained testes	3.3	0.5
Sperm concentration/ ml ($\times 10^6$)	–	809.0	1154.0
Sperm concentration/ ejaculate ($\times 10^6$)	–	2669.7	577.0
Sperm motility (%)	–	65.0	60.0
Forward progression (0 to 5; 5 = best)	–	3.5	3.0
Normal acrosomes (%)	–	92.0	98.0
Normal sperm (%)	–	70.0	94.0
Abnormal sperm (%)			
acrosomal defects	–	8	2
coiled flagellum	–	7	1
bent midpiece defects	–	13	3
bent flagellum defects	–	1	0
cytoplasmic droplets	–	1	0

[a] Male SB 323 was related to SB 181 (deceased), which also had only one descended testis and similar testicular hypoplasia.

testis was consistently smaller than the left. After puberty, at about 7.5 years of age, the right testis decreased in size and atrophied with no evidence of trauma during the animal's lifetime. At the time of the survey, this captive-born male (produced from wild-born parents, male SB 183 and female SB 162) was 10.5 years old, and the right testis was small and adhered to the body wall near the inguinal canal (see Table 7.4). Similar to SB 323, SB 345 produced a high-quality ejaculate, suggesting that a single, normally descended testis probably compensates for compromised function in the contralateral gonad.

Delayed testicular development and cryptorchidism can be a manifestation of reduced genetic variation and inbreeding. For example, in a small population of about 50 wild Florida panthers (*Puma concolor coryi*) in the southern USA, delayed testicular descent was documented in 23% of juveniles (Mansfield & Land, 2002). Cryptorchidism (one or more retained testes in the inguinal canal) was associated with incestuous matings and occurred in 56% of males (Roelke *et al.*, 1993), compared to only 3.9% in other puma populations (Barone *et al.*, 1994). Following genetic 'enhancement' of Florida panthers with Texas puma (*Puma concolor stanleyana*) introductions, there was a complete absence of cryptorchidism in all resulting progeny (Mansfield & Land, 2002), supporting a link between genetics and cryptorchidism in a carnivore. A low incidence (circa 2%) of unilateral cryptorchidism has been observed in wild American black bear populations (Mark Cunningham, pers. comm.). Unilateral cryptorchidism has also been reported in maned wolves and black-footed ferrets produced in captivity where the populations have originated from four and seven founders, respectively. Finally, cryptorchidism is known to be highly heritable in the dog and pig, with all affected males excluded from breeding programmes. Until this condition is better understood in the giant panda, we would not recommend breeding any male with testicular hypoplasia or atrophy.

USE OF FRESH AND COLD-STORED SEMEN FOR ASSISTED REPRODUCTION

Artificial insemination with fresh semen is a valuable management tool for the captive breeding of giant pandas due to the high incidence of sexual incompatibility. It is common practice in China to combine natural breeding and AI. Generally, the one available 'breeding' male is used to copulate with the female after which she is immediately inseminated with semen from one or more males in an attempt to ensure pregnancy and promote genetic diversity. Paternity analysis has revealed that siring success is heavily skewed to the initial mating male rather than to AI (see Chapter 10). This finding is not an indictment of AI efficiency, but more likely a reflection of optimal timing for sperm deposition – the copulating male benefits from peak oestrus and the earlier deposit of large sperm numbers compared to the later AI. Furthermore, AI when used alone has been shown to be as effective at achieving pregnancy as the usual natural mating plus AI approach (Liu, 1981; Feng & Zhang, 1988; Huang *et al.*, 2002; see also Chapter 20). Pregnancies strictly through AI also have been achieved in Spain, Japan and the USA (Moore *et al.*, 1984; Masui *et al.*, 1989; Durrant *et al.*, 2003).

The Biomedical Survey, which produced multiple semen samples, allowed us to conduct further studies which will eventually improve AI efficiency. For example, most current Chinese protocols that rely on fresh sperm mandate that a male be anaesthetised daily for semen collection. So a logical question is – would it be possible to cold-store giant panda sperm over the course of several days without losing functionality? In fact, others have already proven that cooled giant panda sperm from a single ejaculate can be used for AI over two to three days to produce offspring (in Spain: Moore *et al.*, 1984; China, Wolong Nature Reserve: Huang *et al.*, 2002; USA: Durrant *et al.*, 2003). The ramifications are profound because this means that a single semen sample can be collected and used over the two- to three-day oestrus avoiding repeated re-anaesthetisation of the male. An additional benefit is steering clear of sperm cryodamage that occurs at lower temperatures.

Using semen collected during the Biomedical Survey, we further examined the sensitivity of giant panda sperm to cold storage by comparing the influence of two egg-yolk diluents (TEST vs. SFS) and glycerol (0 vs. 5%) on sperm motility and acrosomal integrity over 48 hours at 4°C (Olson *et al.*, 2003; Table 7.5). Percentage sperm motility and normal acrosomes declined over time ($p < 0.05$). The type of egg-yolk diluent affected sperm motility ($p < 0.05$) but not acrosomes. After 48 hours of storage at 4°C, motility was lower ($P < 0.05$) in the SFS (mean range 33–36%) compared to the TEST (52–65%) diluents (see Table 7.5). Glycerol had no influence ($p > 0.05$) on acrosomal integrity over time in either diluent. However, adding glycerol to TEST reduced ($p < 0.05$) sperm motility at 48 hours (52%) compared to TEST without glycerol (65%), and both were superior to SFS with or without glycerol. Overall, TEST with 0% glycerol was the most effective diluent for maintaining the

Table 7.5. *Influence of refrigeration (4°C) of giant panda semen diluted in TEST or SFS diluent containing 0% or 5% glycerol (n = 6 males) on sperm motility and acrosomal integrity*[a]

	TEST 0%	TEST 5%	SFS 0%	SFS 5%
Sperm motility (%)				
0 hour	77.5 ± 4.8	77.5 ± 5.4	77.5 ± 4.8	77.5 ± 4.8
4 hour	65.2 ± 2.2	66.7 ± 3.0	66.2 ± 4.0	59.3 ± 8.1
8 hour	61.2 ± 3.2	64.7 ± 2.8	57.7 ± 6.5	50.0 ± 8.7
24 hour	64.0 ± 3.2[b]	54.3 ± 6.1[b,c]	42.8 ± 7.0[c]	40.2 ± 8.2[c]
48 hour	65.2 ± 4.4[b]	51.7 ± 8.2[c]	33.3 ± 6.2[d]	36.0 ± 8.3[d]
Normal acrosomes (%)				
0 hour	92.5 ± 1.0	92.5 ± 1.0	92.5 ± 1.0	92.5 ± 1.0
4 hour	92.8 ± 1.3	91.2 ± 1.6	91.7 ± 1.5	91.7 ± 1.8
8 hour	86.2 ± 7.2	90.3 ± 1.5	92.0 ± 1.4	92.2 ± 1.6
24 hour	87.3 ± 2.7	80.3 ± 3.5	87.2 ± 2.0	85.8 ± 2.0
48 hour	64.8 ± 5.1	66.7 ± 5.9	70.5 ± 8.6	72.7 ± 5.9

[a] Values are means ± SEM. Semen was diluted (1:3; semen:diluent) in TEST or SFS diluent, cooled slowly and maintained at 4°C for 48 hours; [b,c,d] Within a time period, means with different superscripts are different ($p < 0.05$).

highest sperm motility (>65%) over 48 hours (see Table 7.5). These data confirm that short-term, cold storage at 4°C is effective for maintaining excellent sperm motility and intact acrosomes for at least 48 hours in the giant panda.

Benefits of semen cryopreservation and a genome resource bank

Having access to an organised Giant Panda Genome Resource Bank (GRB; a repository of sperm as well as tissue, blood and DNA) would be a valuable resource for helping to maintain genetic diversity in the *ex-situ* giant panda population, both inside and outside China (Wildt *et al.*, 1997; Howard, 1999; see Chapter 21). Frozen semen could be used to move genes among geographically disparate breeding centres to avoid inbreeding. Although most AI procedures in China rely on fresh semen, AI with frozen–thawed sperm has been used with modest success, the first birth occurring in 1980 at the Chengdu Zoo (Ye *et al.*, 1991; Zhang *et al.*, 1991). A more recent evaluation of available data is presented in this book (see Chapter 20).

To survive 'cryo-stress', sperm require an appropriate:

1. seminal cryodiluent and cryoprotectant;
2. cooling rate (to a temperature just above freezing);
3. storage package;
4. freezing method/rate;
5. thawing method/rate;
6. post-thaw dilution to remove cryoprotectant.

The Biomedical Survey allowed the systematic examination of the following:

Impact of freezing method and cryodiluent

The traditional method for freezing giant panda sperm is pelleting, a technique originally developed for bull, boar, dog and cat semen (Platz *et al.*, 1983; Howard *et al.*, 1986; Chen *et al.*, 1992; Howard, 1993). This cryotechnique was first used in 1979 for the giant panda Hsing Hsing (SB 121) at the Smithsonian's National Zoological Park and involved a cryodiluent of 20% egg yolk, 11% lactose and 4% glycerol (Platz *et al.*, 1983). The method itself consists of pipetting cooled, liquid seminal drops into indentations on a block of dry ice (i.e. solid carbon dioxide at $-96°C$) and then leaving them in place for three minutes before plunging into liquid nitrogen. In China, this pelleting method had been modified to avoid the need for dry ice. Rather a plastic tray or wire-mesh screen was suspended over liquid nitrogen vapour (see Fig. 7.4).

Initially during the survey, the Chinese pelleting technique and SFS cryodiluent were tested, but resulted in wide-ranging, unreliable post-thaw sperm motility and acrosomal ratings. We speculated that discrepancies were related to inconsistent:

1. heights above the nitrogen vapour causing varied temperatures on the tray or mesh during pelleting;
2. pellet sizes; and/or
3. length of time pellets remained on the mesh or tray before plunging into liquid nitrogen.

Each of these factors are well known to influence freezing rate and, thus, the incidence of intracellular ice formation that can cause membrane damage and cell death (Mazur, 1974; Howard *et al.*, 1986; Hammerstedt *et al.*, 1990).

To promote consistency in technique, we developed a means to stabilize the plastic tray or wire-mesh screen above the liquid nitrogen to achieve a constant temperature of $-96°C$ during pelleting (to mimic dry ice temperature). A high-quality pipetting device was used to ensure a standardised pellet size of 40 μl. We also decided to test another cryomethod, specifically, a manual straw container technique. This simple two-step straw method of placing the straws on a test-tube rack at two different levels in liquid nitrogen vapour proved to be an excellent and highly portable method for cryopreservation of panda spermatozoa (see Fig. 7.5; Table 7.6). Two tissue culture solutions (Ham's F10 and Tyrode's 199) also were evaluated as thawing media for pellets and dilution media for straws. For the former method, one pellet was thawed in 0.5-ml of each medium at 37°C. For the straw method, one 0.25-ml straw was thawed in a 37°C water-bath for 45 seconds, then the thawed semen split into two aliquots and diluted in Ham's F10 or Tyrodes 199 culture medium at 37°C.

Neither cryodiluent nor freezing method influenced ($p > 0.05$) sperm viability (see Table 7.6). There was no difference ($p > 0.05$) in pre-freeze sperm motility or forward progression after dilution in the TEST egg-yolk diluent (commercially available) or SFS egg-yolk diluent (made fresh daily), each containing 5% glycerol (see Table 7.6). Although sperm quality was not affected, motility traits were visualised more easily microscopically in the TEST compared to the SFS diluent. This was no doubt due to the sophisticated processing of TEST during preparation by the manufacturer. In contrast, SFS was not filtered, thus making it difficult to view sperm easily because of particulate matter in the egg yolk. Overall, diluting giant panda semen in TEST 5% and SFS 5% cryodiluents at 37°C provided similar ($p > 0.05$) sperm protection during freezing and thawing (see Table 7.6). Likewise, post-thaw sperm motility, forward progression, longevity of sperm motility *in vitro* (data not shown) and acrosomal integrity (Table 7.6) were similar after cryopreservation in TEST or SFS using the pellet-versus-straw method. Tissue culture medium used for thawing and dilution also had no impact ($p > 0.05$) on sperm viability (including acrosomal integrity) after thawing (data not shown).

Impact of freezing rate and cryomethod

The challenge during sperm freezing is not the cell's endurance to an ultra-low temperature (-80 to $-196°C$) (Mazur, 1974). Rather, the hazard is the intermediate temperature zone (-15 to $-60°C$) that the

Table 7.6. *Influence of cryodiluents (TEST vs. SFS) and freezing methods (straws vs. pellets) on cryopreservation of giant panda semen (n = 14 males)*[a]

	Sperm motility (%)	Forward progression[b]	Acrosomal integrity (%)			
			Normal apical ridge	Damaged apical ridge	Missing apical ridge	Loose acrosomal cap
Pre-freeze						
TEST 5% glycerol	68.0 ± 4.0	3.1 ± 0.1	95.7 ± 0.6	2.1 ± 04	1.7 ± 0.5	0.5 ± 0.2
SFS 5% glycerol	67.9 ± 4.2	2.9 ± 0.1	95.7 ± 0.6	2.1 ± 04	1.7 ± 0.5	0.5 ± 0.2
Post-thaw						
TEST 5% glycerol/straws	59.3 ± 3.9	3.5 ± 0.2	75.5 ± 3.9	11.8 ± 1.7	7.5 ± 1.9	5.2 ± 1.3
TEST 5% glycerol/pellets	56.7 ± 5.6	3.5 ± 0.2	67.0 ± 6.6	13.3 ± 2.7	13.0 ± 2.6	6.8 ± 2.3
SFS 5% glycerol/straws	59.6 ± 4.2	3.4 ± 0.1	72.4 ± 5.7	12.8 ± 2.7	6.2 ± 0.9	2.5 ± 0.5
SFS 5% glycerol/pellets	55.0 ± 7.5	3.4 ± 0.2	62.6 ± 8.4	9.3 ± 2.8	17.9 ± 4.9	10.2 ± 4.9

[a] Values are means ± SEM. Fresh semen was diluted (1:3; semen:cryodiluent) in TEST 5% glycerol or SFS 5% glycerol at 37°C, cooled to 4°C over 4 hours, then frozen in 0.25-ml straws or 40-μl pellets. Straws were thawed at 37°C, then diluted in Ham's F10/HEPES medium at 37°C. Pellets were thawed in Ham's F10/HEPES medium at 37°C. Sperm acrosomes were evaluated immediately post thaw. Data presented for percentage sperm motility and forward progression were at 60 minutes post-thaw; [b] Sperm forward progression was based on a scale of 0 to 5; 5 = best.

cell must traverse twice, once during freezing and once during thawing. Different approaches for packaging semen affect the freezing rate through this critical zone by providing different surface-to-volume ratios (Mazur, 1974; Hammerstedt et al., 1990). The 40-μl pellet and straw freezing techniques have very rapid freezing rates (~100°C per minute and ~70°C per minute, respectively; see Fig. 7.7), largely because of the significant surface area exposed to liquid nitrogen. And, as we saw above, when these rates were applied to giant panda sperm, the result was high post-thaw sperm viability and acrosomal integrity. Because larger volumes freeze more slowly, the impact of a slow freezing rate can be assessed by cryopreserving semen in large volumes such as in a bulk vial. Therefore, we took the opportunity to evaluate a cryovial method for giant panda sperm by placing 0.5-ml of semen in a 1.8-ml cryovial, cooling slowly to 4°C, then transferring the container onto a Styrofoam platform floating on 7.5 cm of liquid nitrogen for 15 minutes before plunging into liquid nitrogen. As anticipated, the content of the cryovial froze slowly (at about −20°C per minute) due to the larger volume of semen and low surface area of the vial (see Fig. 7.7). Therefore, to further assess cryomethods, this study examined the efficacy of the cryovial versus the 0.25-ml straw versus two sizes of pellets. The first was the standard 40-ml pellet that freezes at about 100°C per minute versus an 80-ml drop that freezes at about 50°C per minute. All semen samples were diluted in TEST, cooled slowly to 4°C over 4 hours and then frozen using one of these four methods.

Freezing rate had a profound influence on post-thaw sperm survival in the giant panda (Table 7.7). The faster freezing rates associated with either pelleting method or with the 0.25-ml straw consistently produced excellent post-thaw sperm motility (>70%) and acrosomal integrity (>85%). Slower freezing via a cryovial constantly resulted in almost a 20% decrease in sperm motility and a more than 30% reduction in the number of sperm with normal acrosomes ($p < 0.05$). One result was many more sperm ($p < 0.05$) with a damaged (35%) or missing (11%) apical ridge. Overall, results confirm that giant panda sperm prefer a rapid cryopreservation rate, whereas slower freezing damages both motility and the acrosomal apparatus.

Impact of glycerol temperature and duration of exposure

The temperature of glycerol when added to the semen and the duration of glycerol exposure can influence sperm viability. In livestock species,

Table 7.7. *Impact of freezing rate using pellets, straws and cryovials on cryopreservation of giant panda sperm (n = 6 males)*[a]

	Sperm motility (%)	Forward progression[b]	Acrosomal integrity (%)			
			Normal apical ridge	Damaged apical ridge	Missing apical ridge	Loose acrosomal cap
Pre-freeze						
TEST cryodiluent	76.7 ± 6.0[c]	3.1 ± 0.2	96.2 ± 1.3[c]	2.3 ± 0.6[c]	1.5 ± 0.9[c]	0 ± 0
Post-thaw						
40 µl pellets (~100°C/min)	70.8 ± 5.2[c]	4.0 ± 0.1	86.3 ± 1.7[d]	5.8 ± 1.2[d]	6.7 ± 0.8[d]	1.2 ± 0.3
80 µl pellets (~50°C/min)	73.3 ± 6.0[c]	4.0 ± 0.1	87.2 ± 1.6[d]	5.8 ± 1.1[d]	5.5 ± 0.8[d]	1.5 ± 0.6
0.25 ml straws (~70°C/min)	70.8 ± 4.4[c]	4.1 ± 0.2	87.5 ± 1.9[d]	5.5 ± 0.8[d]	5.7 ± 1.8[d]	1.3 ± 0.4
0.5 ml cryovial (~20°C/min)	52.8 ± 4.2[d]	3.4 ± 0.2	54.0 ± 2.9[e]	34.8 ± 3.2[e]	10.7 ± 4.5[d]	0.5 ± 0.3

[a] Values are means ± SEM. Semen was diluted (1:2; semen:cryodiluent) with TEST 0% or 5% cryodiluent at 37°C, cooled to 4°C over 4 hours and frozen in pellets, straws or cryovials at a 4% final glycerol concentration. Sperm acrosomes were evaluated immediately post-thaw. Data presented for percentage sperm motility and forward progression were at 60 minutes post-thaw; [b] Sperm forward progression was based on a scale of 0 to 5; 5 = best; [c,d,e] Within columns, values with different superscripts differ (p < 0.05).

glycerol must often be added to the semen at 4°C to minimise osmotic injury to sperm (Hammerstedt *et al.*, 1990). As discussed above, we initially discovered that exposing giant panda sperm to TEST or SFS with glycerol at 37°C, followed by slow cooling to 4°C before freezing, produced a high incidence of intact acrosomes (see Table 7.6). Here, we further assessed the sensitivity of giant panda sperm to glycerol by examining addition at 37°C versus 4°C and a duration of glycerol exposure of 4 hours versus 1 hour. A treatment involving short-term glycerol exposure at a lower temperature (4°C for only 1 hour following a 3-hour cooling interval) allowed evaluating the protective or toxic effects of glycerol. Thus, for this study, semen was diluted in TEST or SFS with 0% or 5% glycerol at 37°C, cooled slowly to 4°C over 3 hours and then diluted in TEST or SFS with 8% or 5% glycerol for an additional 1 hour of cooling before freezing in 0.25-ml straws. Adding glycerol at 4°C for only 1 hour failed to enhance or compromise ($p > 0.05$) sperm quality post-thawing (Table 7.8). Sperm motility traits and the incidence of normal acrosomes were similar ($p > 0.05$) after cryopreservation regardless of the temperature or duration of glycerol exposure.

Sperm capacitation, the acrosome reaction and decondensation

For optimal post-thaw functionality, a spermatozoon must have progressive motility, an intact acrosome and the ability to undergo capacitation, the acrosome reaction, zona pellucida penetration and decondensation in the oocyte's cytoplasm (Yanagimachi, 1994). Although giant panda sperm capacitation (Sun *et al.*, 1996), the acrosome reaction (Chen *et al.*, 1989a) and oocyte penetration *in vitro* (Moore *et al.*, 1984; Chen *et al.*, 1989a,b) have been studied, these functional events have not been evaluated with frozen–thawed sperm.

During the Biomedical Survey, giant panda sperm capacitation and the acrosome reaction were evaluated by exposing fresh and thawed sperm to heterologous (cat) solubilised zonae pellucidae (Spindler *et al.*, 2004). Results revealed that giant panda sperm were capable of capacitating *in vitro* over six hours (Table 7.9). Because the acrosome reaction was elicited using heterologous (felid) zonae emulsions, our findings supported earlier data indicating that the triggers to this phenomenon are not species specific (Yanagimachi, 1994). Most importantly, freeze-thawing had no detrimental influence on these functional

Table 7.8. *Influence of cryodiluent (TEST vs. SFS) and temperature of glycerol addition (37°C vs. 4°C) using the straw method on sperm cryopreservation in the giant panda (n = 6 males)*[a]

	Pre-freeze			Post-thaw		
	Sperm motility (%)	Forward progression[b]	Normal apical ridge	Sperm motility (%)	Forward progression[b]	Normal apical ridge
TEST/glycerol at 37°C	76.7 ± 6.0	3.1 ± 0.2	95.3 ± 1.2	70.8 ± 4.4	4.1 ± 0.2	87.5 ± 1.9
TEST/glycerol at 4°C	75.0 ± 5.6	3.3 ± 0.2	95.3 ± 1.2	69.2 ± 4.5	3.9 ± 0.2	85.5 ± 2.8
SFS/glycerol at 37°C	75.0 ± 6.1	2.8 ± 0.1	95.3 ± 1.2	70.8 ± 4.0	3.8 ± 0.2	86.3 ± 2.3
SFS/glycerol at 4°C	76.6 ± 6.0	2.9 ± 0.2	95.3 ± 1.2	68.3 ± 4.2	3.8 ± 0.2	85.5 ± 2.2

[a] Values are means ± SEM. Fresh semen was diluted (1:2) in TEST 0% or 5% glycerol or SFS 0% or 5% glycerol at 37°C and cooled to 4°C over 3 hours. Aliquots were then diluted in TEST 8% or 5% (1:1) or SFS 8% or 5% (1:1) for a final 4% glycerol concentration. Samples were cooled for an additional hour, then frozen in 0.25 ml straws. For thawing, straws were held at 37°C for 45 seconds, then diluted in Ham's F10/HEPES medium at 37°C. Sperm acrosomes were evaluated immediately post-thaw. Data presented for percentage sperm motility and forward progression were at 60 minutes post-thaw; [b] Sperm forward progression was based on a scale of 0 to 5; 5 = best.

Table 7.9. *Percentage (mean ± SEM) of fresh and frozen–thawed giant panda spermatozoa that demonstrated capacitation after exposure to heterologous (cat) solubilised zonae pellucidae (n = 9 males)*[a]

Time	0 hour	3 hours	6 hours	9 hours
Fresh sperm – capacitated	3.0 ± 0.2	28.0 ± 0.7	43.6 ± 1.1	41.3 ± 1.1
Frozen–thawed sperm – capacitated	5.3 ± 0.5	27.7 ± 1.2	43.3 ± 1.1	49.6 ± 1.1[b]

[a] Capacitation was defined as the proportion of sperm with intact acrosomes after exposure to solubilised cat zonae pellucidae subtracted from the proportion of sperm with intact acrosomes after exposure to only Ham's F10 medium with no zonae (control). For cryopreservation, semen was diluted in TEST egg-yolk diluent with 5% glycerol, cooled slowly and frozen in 0.25-ml straws; [b] Different from fresh counterparts ($p < 0.05$).

events and did not compromise the ability of giant panda sperm to undergo capacitation and the acrosome reaction.

Once zona penetration has been achieved, decondensation is essential to fertilisation, whereby disulphide bonds in the spermatozoon are released so that the chromosomes become accessible to the oocyte (Mahi & Yanagimachi, 1975). Since cryopreservation exposes the cell to expansion, shrinkage, dehydration and significant temperature fluctuations which can cause lethal damage to chromatin structure and bond stability (Gao *et al.*, 1997), we also examined the impact of cryopreservation on the sperm nucleus and chromatin decondensation in the giant panda. Following pre-incubation and exposure to a solubilized cat zonae solution (to induce capacitation and the acrosome reaction), fresh and frozen (TEST/straw)–thawed giant panda sperm were exposed to heterologous (cat) oocyte cytoplasmic emulsion for assessment of decondensation. There was no effect of cryopreservation on the subsequent ability of giant panda sperm to decondense (Table 7.10). More than half of the fresh sperm underwent decondensation after two hours of incubation in the cytoplasmic emulsion with nearly three-quarters after four hours. The total number of thawed sperm undergoing decondensation did not differ ($p > 0.05$) from fresh counterparts at any time point (see Table 7.10). Thus, we conclude that cryopreserved as well as fresh giant panda sperm have equivalent capacity for maintaining chromosomal and nuclear stability.

Table 7.10. *Percentage (mean ± SEM) of fresh and frozen–thawed giant panda spermatozoa that demonstrated decondensation during incubation in cat oocyte cytoplasmic emulsion (n = 8 males)*[a]

Time	0 hour	2 hours	4 hours
Fresh sperm – decondensed	2.8 ± 0.4	51.4 ± 3.7	69.8 ± 5.9
Frozen-thawed sperm – decondensed	3.8 ± 0.4	58.1 ± 6.2	71.5 ± 4.9

[a] Giant panda spermatozoa were pre-incubated in Ham's F10 medium for 6 hours to induce sperm capacitation, then exposed to a cat solubilised zonae pellucidae solution for 30 minutes to induce the acrosome reaction prior to the assessment of decondensation during incubation in cat oocyte cytoplasmic emulsion. For cryopreservation, semen was diluted in TEST egg-yolk diluent with 5% glycerol, cooled slowly and frozen in 0.25-ml straws.

Sperm–ovum interaction and zona penetration

The ultimate assay for assessing sperm function after freezing is zona penetration and fertilisation of oocytes *in vivo* or *in vitro* (Drobnis *et al.*, 1988). But, of course, testing functionality of giant panda sperm on giant panda oocytes is limited because of a lack of eggs from the latter. One alternative is to use the oocytes of another species. A salt-stored zona penetration assay, developed in the hamster (Yanagimachi *et al.*, 1979; Boatman *et al.*, 1988) and adapted for cat oocytes (Andrews *et al.*, 1992), may be one future approach for testing the ability of panda sperm to bind and penetrate the zona pellucida, the oocyte's primary barrier to fertilisation. This could, for example, be valuable in testing new sperm cryopreservation protocols. Salt-stored zonae also retain the ability to distinguish between capacitated and non-capacitated sperm (Boatman *et al.*, 1988; Andrews *et al.*, 1992).

Our preliminary studies have demonstrated that zona-intact, salt-stored cat oocytes can be penetrated by fresh and thawed giant panda spermatozoa (see Fig. 7.8). Giant panda sperm binding and penetration of cat oocytes appeared similar to what has been measured with salt-stored giant panda oocytes recovered from ovaries post-mortem (see Fig. 7.8). This interesting finding implies that the zona receptor on the giant panda spermatozoon and the ligand on the cat zona pellucida may be the same, that is, conserved across these quite different species. Due to this non-specificity, we predict that the ready availability of cat eggs, and thus their zonae, could be a useful tool for studying sperm function and gamete interaction in the giant panda.

The Biomedical Survey gave us access to significant numbers of giant pandas that, in turn, allowed a substantial increase in our understanding of male reproductive biology. We now know that:

1. giant pandas held *ex situ* generally produce high numbers of motile, structurally normal spermatozoa during the breeding season;
2. sperm quality is similar between individuals that were born in nature versus captivity as well in males of proven fertility versus nonbreeders;
3. males as young as 5.5 years old produce sperm although the quality is not as good as that of older counterparts;
4. there is a hypoplastic testicular anomaly of unknown etiology in a giant panda subpopulation;
5. rapid freezing rates are optimum for sperm cryopreservation;
6. giant panda sperm subjected to freeze-thawing exhibit good motility, morphology and functionality *in vitro*.

Although much has been learned, there are high research priorities for the future, especially linking new knowledge in male and female physiology with behaviour and management to enhance natural or assisted-breeding success. The highest priority is to optimise the practical use of frozen–thawed sperm to meet the genetic management goals identified by the managers of the *ex situ* population (see Chapter 21). It is evident that physiological infertility in male giant pandas is uncommon, and that sperm are robust and comparatively resistant *in vitro* to the stresses of cryopreservation. So, even though preliminary positive data are available (see Chapter 20), there is a need to conduct systematic studies of AI with thawed sperm, especially to identify the optimal numbers of sperm and timing of deposition in the oestrual female. Once it is determined that cryopreserved sperm can consistently result in pregnancies, many of the political and logistical concerns associated with moving animals between breeding centres become moot. Rather, it will be possible simply to transport semen from the most desirable male to the facility holding the most genetically compatible mate. Associated with this priority will be an eventual need to prove the ability to collect and freeze sperm from free-living males, another strategy for infusing new genes into the *ex situ* population while leaving wild pandas in nature. All of these priorities also are related to an ability to develop genome resource banks (see Chapter 20), organised stores of biomaterials, especially sperm, that will allow meeting genetic management goals. Developing these repositories and sharing the best technologies and the banked biomaterials will require high levels of cooperation.

There is also a need for more fundamental studies of male reproductive biology. One high priority issue is male seasonality: does the adult giant panda produce viable sperm of similar quality throughout the year, or is it highly seasonal like its female conspecific? A seasonal pattern in spermatogenesis has practical implications, as it will limit the times of the year when semen can be collected for cryobanking. In contrast, if males produce sperm throughout the year, then it will be possible to collect and freeze sperm during the nonbreeding season, potentially increasing the total amounts of genetic material that can be stored and shared. Secondly, although we now have some information on the age-related onset of sperm production, there is a need to identify the earliest onset of male fertility. For example, can 5- or 6-year-old males be used for breeding or AI? Finally, there is a need to fully understand the testicular hypoplasia/atrophy anomaly identified during the Biomedical Survey. If indeed this malformation has genetic origins, then it will be necessary to eliminate affected or carrier animals from the breeding population.

ACKNOWLEDGEMENTS

We thank the staff of the Chengdu Research Base of Giant Panda Breeding, Beijing Zoo and the China Conservation and Research Centre for the Giant Panda for their hospitality. We thank Lena May Bush for technical assistance in acquiring donated equipment (dry shippers from Air Gas, Inc.). Financial support, in part, for the National Zoo scientists was provided by Friends of the National Zoo.

REFERENCES

Amann, R. P. (1986). Reproductive physiology and endocrinology of the dog. In *Current Therapy in Theriogenology II*, ed. D. A. Morrow. Philadelphia, PA: W. B. Saunders Co., pp. 532–8.

Andrews, J. C., Howard, J. G., Bavister, B. D. and Wildt, D. E. (1992). Sperm capacitation in the domestic cat (*Felis catus*) and leopard cat (*Felis bengalensis*) as studied with a salt-stored zona pellucida penetration assay. *Molecular Reproduction and Development*, 31, 200–7.

Barone, M. A., Roelke, M. E., Howard, J. G., Anderson, A. E. and Wildt, D. E. (1994). Reproductive characterization of male Florida panthers: comparative studies from Florida, Texas, Colorado, Chile and North American zoos. *Journal of Mammology*, 75, 150–62.

Beyer, W. H. (1987). *CRC Standard Mathematical Tables*, 28th Edn. Boca Raton, FL CRC Press, p. 131.

Boatman, D. E., Andrews, J. C. and Bavister, B. D. (1988). A quantitative assay for capacitation: evaluation of multiple sperm penetration through the zona pellucida of salt-stored hamster eggs. *Gamete Research*, 19, 19–29.

Chen, D., Ye, Z. and Zhang, Z. (1984). Ultrastructure of giant panda spermatozoa. *Acta Zoologica Sinica*, 30, 301–4.

Chen, D., Shi, Q., Zhao, X. *et al.* (1989a). Studies of heterofertilization *in vitro* between giant panda and golden hamster. *Acta Zoologica Sinica*, 35, 276–380.

Chen, D., Song, X., Zhou, X. and Duan, C. (1989b). Study on *in vitro* sperm capacitation and egg penetration of giant panda. *Science in China*, 32, 435–41.

Chen, D., Song, X., Feng, W. *et al.* (1992). Study on the thawing solutions of cryopreserved semen pellets in the giant panda. *Chinese Journal of Zoology*, 27–63.

Drobnis, E., Yudin, A., Cherr, G. and Katz, D. (1988). Hamster sperm penetration of the zona pellucida: kinematic analysis and mechanical implications. *Developmental Biology*, 130, 311–23.

Durrant, B. S., Olson, M. A., Amodeo, D. *et al.* (2003). Vaginal cytology and vulvar swelling as indicators of impending estrus and ovulation in the giant panda (*Ailuropoda melanoleuca*). *Zoo Biology*, 22, 313–21.

Feng, W. and Zhang, A. (1988). *Giant Panda Theriogenology and Assisted Reproduction*. Chengdu: Publishing Company of Sichuan University, pp. 32–8.

Gao, G., Mazur, P. and Critser, J. (1997). Fundamental cryobiology of mammalian spermatozoa. In *Reproductive Tissue Banking: Scientific Principles*, ed. A. Karow and J. Critser. San Diego, CA: Academic Press, pp. 263–328.

Hammerstedt, R. H., Graham, J. K. and Nolan, J. P. (1990). Cryopreservation of mammalian sperm: what we ask them to survive. *Journal of Andrology*, 11, 73–88.

Howard, J. G. (1993). Semen collection and analysis in nondomestic carnivores. In *Zoo and Wild Animal Medicine: Current Therapy III*, ed. M. E. Fowler. Philadelphia, PA: W. B. Saunders Co., pp. 390–9.

Howard, J. G. (1999). Assisted reproductive techniques in nondomestic carnivores. In *Zoo and Wild Animal Medicine: Current Therapy IV*, ed. M. Fowler and E. Miller. Philadelphia, PA: W. B. Saunders Co., pp. 449–57.

Howard, J. G., Bush, M. and Wildt, D. E. (1986). Semen collection, analysis and cryopreservation in nondomestic mammals. In *Current Therapy in Theriogenology II*, ed. D. Morrow. Philadelphia, PA: W. B. Saunders Co., pp. 1047–53.

Huang, Y., Wang, P., Zhang, G. *et al.* (2002). Use of artificial insemination to enhance propagation of giant pandas at the Wolong Breeding Center. In *Proceedings of the Second International Symposium on Assisted Reproductive Technology for the Conservation and Genetic Management of Wildlife*, ed. N. Loskutoff, Omaha's Henry Doorly Zoo, Omaha, NE, pp. 172–9.

Ishikawa, A., Matsui, M., Tsuruga, H. *et al.* (1998). Electroejaculation and semen characteristics of the captive Hokkaido brown bear (*Ursus arctos yesoensis*). *Journal of Veterinary Medicine Research*, **60**, 965–8.

Lindburg, D. G., Huang, X. M. and Huang, S. Q. (1998). Reproductive performance of male giant panda in Chinese zoos. In *Proceedings of the International Symposium on the Protection of the Giant Panda*, Chengdu: Sichuan Publishing House of Science and Technology, pp. 65–71.

Liu, W. (1981). A note on the artificial insemination of the giant panda. *Acta Zoologica Sinica*, **12**, 73–6.

Mahi, C. A. and Yanagimachi, R. (1975). Induction of nuclear decondensation of mammalian spermatozoa *in vitro*. *Journal of Reproduction and Fertility* **44**, 293–6.

Mansfield, K. G. and Land, E. D. (2002). Cryptorchidism in Florida panthers: prevalence, features and influence of genetic restoration. *Journal of Wildlife Diseases*, **38**, 693–8.

Masui, M., Hiramatsu, H., Nose, N. *et al.* (1989). Successful artificial insemination in the giant panda (*Ailuropoda melanoleuca*) at Ueno Zoo. *Zoo Biology*, **8**, 17–26.

Mazur, P. (1974). Freezing of living cells: mechanisms and implications. *American Journal of Physiology*, **247**, C125–C142.

Moore, H., Bush, M., Celma, M. *et al.* (1984). Artificial insemination in the giant panda (*Ailuropoda melanoleuca*). *Journal of Zoology (London)*, **203**, 269–78.

Olson, M. A., Huang, Y., Li, D. *et al.* (2003). Assessment of motility, acrosomal integrity and viability of giant panda (*Ailuropoda melanoleuca*) sperm following short-term storage at 4°C. *Zoo Biology*, **22**, 529–44.

Platz, C., Wildt, D. E., Howard, J. G. and Bush, M. (1983). Electroejaculation and semen analysis and freezing in the giant panda (*Ailuropoda melanoleuca*). *Journal of Reproduction and Fertility*, **67**, 9–12.

Pope, C., Zhang, Y. and Dresser, B. (1991). A simple staining method for evaluating acrosomal status of cat spermatozoa. *Journal of Zoo and Wildlife Medicine*, **22**, 87–95.

Roelke, M. E., Martenson, J. S. and O'Brien, S. J. (1993). The consequences of demographic reduction and genetic depletion in the endangered Florida panther. *Current Biology*, **3**, 340–50.

Spindler, R. E., Huang, Y., Howard, J. G. *et al.* (2004). Acrosomal integrity and capacitation are not influenced by sperm cryopreservation in the giant panda. *Reproduction*, **127**, 547–56.

Sun, Q., Liu, H., Li, X. *et al.* (1996). The role of calcium and protein kinase C in the acrosome reaction of giant panda spermatozoa. *Theriogenology*, **46**, 359–67.

Wildt, D. E. and Swanson, W. F. (1998). Reproduction in cats. In *Encyclopedia of Reproduction*, ed. E. Knobil and J. Neil. New York, NY: Academic Press, Inc., pp. 497–510.

Wildt, D. E., Rall, W. F., Critser, J. K., Monfort, S. L. and Seal, U. S. (1997). Genome resource banks: 'living collections' for biodiversity conservation. *BioScience*, **47**, 689–98.

Yanagimachi, R. (1994). Mammalian fertilization. In *The Physiology of Reproduction*, ed. E. Knobil and J. Neil. New York, NY: Raven Press, pp. 189–317.

Yanagimachi, R., Lopata, A., Odom, C. B. *et al.* (1979). Retention of biologic characteristics of zona pellucida in highly concentrated salt solution: the use of salt-stored eggs for assessing the fertilizing ability of spermatozoa. *Fertility and Sterility*, **31**, 562–74.

Ye, Z. Y., He, G. X., Zhang, A. J. *et al.* (1991). Studies on the artificial pollination method of giant panda. *Journal of Sichuan University (Natural Science)*, **28**, 50–3.

Zhang, A. J., Ye, Z. Y., He, G. X. *et al.* (1991). Studies on conception effect of frozen semen in the giant panda. *Journal of Sichuan University (Natural Science)*, **28**, 54–9.

Zheng, S., Zhao, Q., Xie, Z., Wildt, D. E. and Seal, U. S. (1997). *Giant Panda Captive Planning Management Workshop Report*. Apple Valley, MN: Conservation Breeding Specialist Group.

8

Endocrinology of the giant panda and application of hormone technology to species management

KAREN J. STEINMAN, STEVEN L. MONFORT, LAURA MCGEEHAN, DAVID C. KERSEY, FERNANDO GUAL-SIL, REBECCA J. SNYDER, PENGYAN WANG, TATSUKO NAKAO, NANCY M. CZEKALA

INTRODUCTION

Increasing breeding success in the giant panda requires a better understanding of its complex reproductive biology. We know that the female is typically mono-oestrus during a breeding season which occurs from February to May (within and outside China). Behavioural and physiological changes associated with pro-oestrus and oestrus last one to two weeks, during which the female exhibits proceptive behaviours, such as scent marking, to advertise her sexual receptivity (Lindburg et al., 2001). During the peri-ovulatory interval, receptive behaviours (e.g. tail-up lordotic posture) climax with copulation generally occurring over a one- to three-day interval. Birthing occurs from June to October with a gestation of 85 to 185 days (Zhu et al., 2001). This unusually wide gestation span is due to the phenomenon of delayed implantation, a varied interval before the conceptus implants in the uterus and begins foetal development. The driving force behind implantation in this species is unknown. The giant panda also experiences pseudopregnancy, whereby the female exhibits behavioural, physiological and hormonal changes similar to pregnancy.

Giant Pandas: Biology, Veterinary Medicine and Management, ed. David E. Wildt, Anju Zhang, Hemin Zhang, Donald L. Janssen and Susie Ellis. Published by Cambridge University Press. © Cambridge University Press 2006.

Behavioural and physiological cues associated with both pregnancy and pseudopregnancy include decreased appetite, nest-building and cradling behaviours, vulvar swelling and colouration, mammary gland enlargement and lethargy. Additionally, temporal and quantitative progesterone patterns (tracked by assessing urinary hormone by-products and progestins) are indistinguishable between pregnancy and pseudopregnancy. Therefore, no definitive test currently exists for identifying pregnant from pseudopregnant giant pandas. However, endocrine and behavioural traits are useful for broadly estimating impending parturition and, in the absence of a birth, hormonal measures provide retrospective evidence of a presumptive pseudopregnancy. Although there is no doubt that the giant panda experiences pseudopregnancy, some presumed pseudopregnancies almost certainly represent failed pregnancies accompanied by undetected embryonic or foetal loss.

Endocrinology, particularly when combined with behavioural observations, has provided valuable insights into giant panda reproduction. Especially important has been the ability to measure hormones in animal excreta, historically urine and, more recently, faeces. These noninvasive measures provide animal care staff and researchers with a simple means of securing important physiological data without the stress associated with physical restraint or anaesthesia (generally required for collecting blood samples). In fact, the physiological stress resulting from such manipulations may obscure the ability to use blood samples to discern the normal underlying hormonal milieu. Additionally, evaluating steroid metabolites in urine or faeces represents a snapshot of hormonal activity while permitting long-term studies of reproductive patterns in individuals, populations or species, all without disturbing the animal (Monfort, 2003).

Urinary endocrine monitoring has been used for nearly 20 years in the giant panda. However, fewer than 25 females have been studied, and almost nothing is known about male reproductive endocrinology. For females, urinary oestrogens increase during the one- to two-week peri-oestrual interval when proceptive behaviours are prevalent and then decrease during oestrus; in fact, excreted oestrogen concentrations are rapidly declining at the time of mating (Bonney *et al.*, 1982; Hodges *et al.*, 1984; Chaudhuri *et al.*, 1988; Monfort *et al.*, 1989; McGeehan *et al.*, 2002; Czekala *et al.*, 2003). Endocrine databases for both sexes are now being rapidly expanded, largely because of international collaborations between Chinese breeding facilities and western zoos. The net effect has been more animals for study.

To provide a contemporary perspective of the unique reproductive endocrinology of the giant panda, this chapter reviews previous work as well as recent and novel research activities. Our focus is on the value of noninvasive endocrine monitoring (urinary and faecal), including radioimmunoassay (RIA) and enzyme immunoassay (EIA) techniques. Case studies of normal and abnormal endocrine cycles are presented, as well as endocrine responses to exogenous gonadotrophin treatments to stimulate ovarian activity. Data on androgen profiles in the male panda are offered for the first time, and preliminary evidence is presented on the potential for evaluating adrenal status (i.e. 'stress') in this species. We conclude by illustrating how this information can be used in an applied way to improve giant panda management.

HORMONAL ANALYSIS AND INTERPRETATION METHODOLOGIES

Collaborators and animals

Zoological Society of San Diego

The following information is summarised in Table 8.1. Male studbook (SB) 381 at the San Diego Zoo was originally wild caught but never bred while in China or when given opportunities in San Diego after pairing with female SB 371. Urinary endocrine analyses were conducted in-house year round beginning in 1997.

Smithsonian's National Zoological Park

From 1987 to 1991, urinary progestins were measured in SB 112 from oestrual onset to the end of pregnancy, failed pregnancy or pseudopregnancy. Upon arrival of a new pair of subadult giant pandas (male SB 458 and female SB 473) in 2001, annual monitoring of urinary and faecal hormones commenced for both individuals.

Zoo Atlanta

Urinary hormone monitoring for female SB 461 began in 2001. Beginning in 2002, concurrent faecal hormone measurements were initiated.

Table 8.1. *Husbandry and life history of study animals by studbook (SB) number and name*

SB no. (name)	DOB[a] or WC[b] Sex (M/F)	Location	± M/F access[c]	Oe[d]/M[e]/AI[f]/EEJ[g]	Offspring[c]
371 (Bai Yun)	1991/F	SDZ[h]	+M	Oe, AI (1999)	1 (1999)
381 (Shi Shi)	WC 1978/M	SDZ	+F	EEJ (1999)	1 (1999)
112 (Ling Ling)	WC 1970/F	SNZP[i]	+M	Oe, M (1987–91)	5 (1987) Foetal loss (1989)
473 (Mei Xiang)	1998/F	SNZP	+M	Oe (2002)	0
461 (Lun Lun)	1997/F	ZA[j]	+M	Oe (2001, 2002)	0
452 (Yang Yang)	1997/M	ZA	+F	No M or EEJ	0
390 (Yong Ming)	1992/M	AW[k]	−F (2000) +F (2001)	M, EEJ (2001)	2 (1999)
414 (No. 20)	WC 1990/F	CCRCGP[l]	+M	Oe (1998, 1999, 2000) M (1999, 2000) AI (1999, 2000)	1 (2000)
374 (Lei Lei)	WC 1989/F	CCRCGP	+M	Oe (1998, 1999, 2000) M (1998, 1999, 2000) AI (1998, 1999, 2000)	1 (2000)
332 (Shuan Shuan)	1987/F	ZC[m]	+ M (1999, 2001)	Oe (1998, 1999, 2001) AI (1999, 2001) No Oe (2000)	0

[a] DOB = date of birth; [b] WC = wild caught; [c] male/female access, oestrus/mating and offspring data are confined to the study period; [d] Oe, oestrus; [e] M, mating; [f] AI, artificial insemination; [g] EEJ, Electroejaculation; [h] San Diego Zoo, San Diego, CA; [i] Smithsonian's National Zoological Park, Washington, DC; [j] Zoo Atlanta, Atlanta, GA; [k] Adventure World, Wakayama, Shirahama, Japan; [l] China Conservation and Research Centre for the Giant Panda (Wolong Nature Reserve), China; [m] Zoológico de Chapultepec, Mexico City, Mexico.

Adventure World (Wakayama, Japan)

Urinary androgens were measured in male SB 390 under two different conditions – when housed singly (2000) and when he was paired with a female conspecific (2001).

Zoológico de Chapultepec (Mexico City, Mexico)

Female SB 332 was housed in visual contact with two other females (a sibling and niece). Assessments of urinary ovarian and adrenal steroids commenced in 1998.

China Conservation and Research Centre for the Giant Panda (Wolong Nature Reserve, China)

Two case studies at this location are presented illustrating the usefulness of urinary hormone monitoring for evaluating hormonal therapy. Because SB 414 had never exhibited strong behavioural oestrus or mated, twice daily intramuscular (i.m.) injections of follicle stimulating hormone (FSH, 75 international units, IU, Metrodin HP, Serono, Rockland, MA) were administered for seven days starting on 17 March 1999. The dose was doubled on 24 March (to 150 IU), and twice-daily injections continued up to 29 March. Although mating occurred on 30 March, human chorionic gonadotrophin (hCG; 10 000 IU, Profasi 500, Serono) was administered (i.m.) on this day, and artificial insemination (AI) was performed 48 hours later (1 April); no pregnancy resulted. In 2000, SB 414 exhibited strong oestrous behaviours, mated and later gave birth.

The other case involved SB 374 who consistently displayed strong behavioural oestrus, naturally mated and also was artificially inseminated, but never produced offspring. This animal was given hCG (10 000 IU) on 25 March 1999 during her usual robust oestrus. She mated the next day (26 March), was artificially inseminated on 28 and 29 March, was given additional hCG (10 000 IU) on 30 March, but failed to give birth. In 2000, SB 374 exhibited strong behavioural oestrus again, mated and gave birth.

Urine and faeces collection

At each institution, urine and/or faeces were collected four to seven times per week. Typically, specimen collection was year-round, with the exception of female SB 414 and 374 for which collections occurred from March to May (1998 to 2000). Each urine sample (2–5 ml) was aspirated from the enclosure floor with a plastic syringe and stored in a 12 × 75-mm plastic tube or 2-ml plastic cryovials (−20°C); any sample containing faecal matter was discarded. Generally, each urine sample was collected in the morning (0700 to 0900 hours) and within one to two hours post-urination. If freshly excreted urine was unavailable, samples voided during the night were collected and marked as 'overnight' specimens. Freshly voided faeces (10 to 20 g) were collected by one to two hours post-defaecation and each stored in a sealed plastic bag (−20°C). All specimen containers were marked with the animal's identification number and date, and samples were either processed immediately or stored frozen for up to one year before analysis.

The value of urinary creatinine

Creatinine (Cr; a by-product of muscle degradation, which occurs normally in all mammals) is excreted in urine at a constant rate under homeostatic conditions. Thus Cr concentration, which can be determined colourimetrically (Taussky, 1954), serves as an indirect index of an animal's day-to-day fluid balance. If a urine specimen is 'dilute', because an animal is over-hydrated or because the urine is water contaminated (e.g. with water used to clean an enclosure floor), Cr concentration is low. Conversely, animal dehydration elevates Cr. The concentration of other excreted by-products, including hormones, rise and fall in parallel with Cr. Thus, dividing urinary hormone content (mass hormone per ml of urine excreted) by Cr concentration (mg Cr per ml urine excreted) 'indexes' the hormone concentration, which then is expressed as mass of hormone excreted per mg of Cr excreted (e.g. x ng hormone per mg Cr).

For Cr determinations in the giant panda, 0.1 ml urine (diluted 1:100 in distilled water) was added in duplicate to 96-well, flat-bottomed microtitre plates (Costar, Cambridge, MA), combined with 0.05 ml of 0.4 N picric acid and 0.05 ml of 0.75 N NaOH and incubated at room temperature (25°C) for 15 minutes. Optical density was measured at 490 nm (reference 620 nm) using a microplate reader (Benchmark, Biorad, Hercules, CA), and urine samples were compared to a single reference Cr standard (0.001 mg ml^{-1}, Sigma Chemical Co., St Louis, MO) assayed in quadruplicate. Urine samples with Cr concentrations

<0.1 mg Cr per ml were considered too dilute (probably from water contamination). This criterion generally excluded about 7% of all samples.

Faecal extraction

Steroids must be transferred or 'extracted' from the solid faeces into a supernatant (or liquid medium) to allow measuring the hormone concentration. Numerous methods are available for processing, extracting and accurately quantifying faecal steroid metabolites (Wasser *et al.*, 2000) but typically employ an aqueous-based, alcohol-containing (10–100%) solvent, with variation, depending on whether samples are extracted wet or dry, if procedural losses are documented and if heat is used during processing (Monfort, 2003). For the giant panda, a fixed mass of dried (lyophilised; 0.09–0.1 g) or wet (0.45–0.55 g) faeces was placed in a 16 × 125-mm glass tube, combined with 80% ethanol (5 ml) in water and boiled for 20 minutes (90°C). During and after boiling, ethanol (80%) was added to maintain the pre-boil volume (5 ml). Extraction tubes were centrifuged (1500g, 20 minutes), and the supernatant was decanted into a clean glass tube. Faeces were resuspended in 80% ethanol (5 ml) and recentrifuged; both extractants were then combined and evaporated to dryness under a stream of air in a 37°C water-bath. To concentrate dried extractant and rinse particulates adhering to the vessel wall, tubes were successively rinsed with 4, 3 and 1 ml of ethanol, allowing evaporation under a stream of air after each alcohol addition. Finally, extractants were resuspended in 1 ml of dilution buffer (195 ml of 0.2 M NaH_2PO_4; 305 ml of 0.2M Na_2HPO_4; 500 ml distilled water; 8.7 g NaCl, pH 7.0), vortexed and sonicated until dissolved. Hormonal data were expressed as mass of steroid per gram of faeces excreted.

Immunoassays

Minute hormone concentrations can be quantified in biological fluids using antibody-based immunoassays that employ radioactive (i.e. radio-immunoassay or RIA) or non-radioactive (i.e. enzyme immunoassay or EIA) reagents. RIA requires expensive equipment capable of quantifying radioisotopes whereas EIA is colourimetric, relying on less expensive, portable instruments. While both provide equivalent results, each has its advantages and disadvantages (Table 8.2).

Table 8.2. *Comparison of radioimmunoassay (RIA) versus enzyme immunoassay (EIA)*

	RIA	EIA
Advantages	Ease of use	Non-hazardous, disposable reagents
	Assay many samples at once	Transportable
	Gold standard for immunoassays	Less expensive
Disadvantages	Requires radioisotopes	Limited number of samples per assay
	Safety, environmental concerns	More difficult to use
	Large, expensive instrumentation	
	Not mobile	

Radioimmunoassays

OESTROGEN CONJUGATE RIA

The predominant urinary oestrogen in giant panda urine is oestrone conjugated (or attached) to a sulphate molecule (oestrone-3-sulphate; Czekala *et al.*, 2003). Urinary oestrogen conjugates were quantified using a single antibody, charcoal separation RIA previously validated for this species (Monfort *et al.*, 1989; Czekala *et al.*, 2003). Oestrone-3-glucuronide antiserum (0.1 ml, 1:5000; D. Collins, University of Kentucky, Lexington, KY) and tritiated oestrone-3-sulphate (0.1 ml, 10 000 counts per minute; New England Nuclear, Dupont, Boston, MA) were added to unprocessed panda urine (0.05–0.002 ml depending on reproductive state) or oestrone-3-sulphate standards (39–5000 pg per tube) in duplicate. After a one-hour incubation (4°C), bound and free fractions were separated by adding a charcoal dextran solution (0.25 ml), incubating for 30 minutes and centrifuging for 15 minutes (1500 g). Supernatant was counted for radioactivity in a beta radiation counter (Wallac 1409, Perkin Elmer, Gaithersburg, MD). Intra- and interassay coefficients of variation were 8% (n = 10) and 14.8% (n = 39), respectively. Interassay variation for female SB 374 and SB 14 was 7.0% and 14.5% (n = 10), respectively.

PREGNANEDIOL-3-GLUCURONIDE (PDG) RIA

Urinary progestins were quantified by RIA (Hodges *et al.*, 1984; Chaudhuri *et al.*, 1988; Monfort *et al.*, 1989; Mainka *et al.*, 1990). Briefly,

0.1 ml of urine was added to 0.9 ml of pure ethanol, vortexed and centrifuged for 10 minutes at 1500 g. Supernatants were poured off, dried down and reconstituted in 0.5 ml of phosphate buffer saline (PBS). Urine samples (0.01 ml) and standards (19.5–5000 pg per tube), in duplicate, were added to 0.1 ml of PdG antiserum (Courtauld Institute of Biochemistry, London) and tritiated PdG (7000 counts per minute) and incubated overnight at 4°C. Bound and free fractions were separated by adding charcoal dextran (0.3 ml) and incubating for 45 minutes. After centrifugation, supernatant was poured off and counted. Intra- and interassay variation was <5% (n = 10) and 6.2% and 7.3%, respectively.

ANDROGEN RIA

Urinary androgens were measured using a single antibody testosterone RIA. Testosterone standards (7.8–1000 pg per tube; ICN Biomedicals, Costa Mesa, CA) and 0.01 ml of unprocessed panda urine were combined, in duplicate, with testosterone antiserum (0.1 ml, 1:16 000; ICN Biomedicals) and tritiated testosterone (10 000 counts per minute per 0.1 ml diluted in PBS; New England Nuclear). Assays were incubated overnight (4°C), and bound and free ligands were separated by adding a charcoal dextran solution (0.25 ml) for 30 minutes (4°C). Samples were centrifuged (1500 g, 15 minutes), and the supernatant was counted for radioactivity. Intra- and interassay variation was <10% and 20%, respectively (n = 30).

CORTICOSTEROID RIA

Urinary corticosteroids were measured using a single antibody RIA. Cortisol standards (7.8–1000 pg per tube; ICN Biomedicals) and 0.01–0.002 ml of unprocessed panda urine (depending on appropriate binding) were combined in duplicate with cortisol antiserum (0.1 ml, 1:3000; ICN Biomedicals) and tritiated hydrocortisone (10 000 counts per minute per 0.1 ml; New England Nuclear). Assays were incubated overnight (4°C), and bound and free ligands were separated by adding charcoal dextran (0.25 ml) for 30 minutes (4°C). Samples were centrifuged (1500 g, 15 minutes), and the supernatant was counted for radioactivity. Intra-assay variation was 8% (n = 20), and interassay variation was 20% (n = 5); for females SB 374 and SB 414, interassay variation was 8% and 5% (n = 11), respectively.

Enzyme immunoassays

Oestrogen conjugate EIA

Urinary and faecal oestrogen conjugates were measured by single anti-body EIA (Stabenfeldt *et al.*, 1991) with minor modifications. Unprocessed urine (0.4–5.0 µl) or reconstituted faecal extract (1.6–5.0 µl) or oestrone-3-sulphate standards (1.95–500 pg per well) were added in duplicate to 96-well microtitre plates (Maxisorp, Nalge Nunc International, Rochester, NY) coated with the R583 oestrogen metabolite antibody (C. Munro, University of California, Davis, CA). An enzyme conjugate (0.05 ml) then was added to all wells and, after a two-hour incubation at room temperature (25°C), 0.1 ml of substrate solution (ABTS, in citrate buffer; Sigma Chemical Co.) was added. After a 30- to 60-minute incubation (25°C), depending on how fast the reaction/colour change occurred, optical density readings (405 nm, reference 540 nm) were taken using a microplate reader (MRX, Dynex Technologies, Chantilly, VA). Intra- and interassay variation was <10% and 22% (n = 15), respectively.

Progestin EIA

Urinary progestins were measured using an established EIA (McGeehan *et al.*, 2002). Unprocessed urine (0.01–0.001 ml) and PdG standards (range 1.98–625 pg per well) were added, in duplicate, to 96-well micro-titre plates (Immulon 1, Dynex Technologies) coated with a pregnan-ediol metabolite specific antibody (C. Munro). An enzyme conjugate (0.05-ml) was added to all wells and, after a two-hour incubation (25°C), a 0.1-ml substrate solution (tetramethylbenzidine in phosphate citrate buffer; Sigma Chemical Co.) was added to all wells. After a 30-minute incubation (25°C), stop solution (0.4 N H_2SO_4, 0.05 ml) was added to all wells, and optical density readings (450 nm, reference 620 nm) were taken (as above). Intra- and interassay variation was 5% (n = 16) and 18% (n = 49), respectively. Interassay variation for female SB 374 and SB 414 was 8% (n = 16).

Faecal progestins were measured by a different broad spectrum progestin EIA (Cl 425; C. Munro). Extracted faeces (2.5–0.31 µl) or pro-gesterone standards (0.78–200 pg per well; Sigma Chemical Co.) were added to 96-well microtitre plates coated with the Cl 425 progestin antibody followed by adding an enzyme conjugate (0.05 ml) and incubation at 25°C for two hours. Then an ABTS substrate solution was added to all wells (0.1 ml) and incubated (25°C) for 30 to 60 minutes; and plates were read (as above). Inter- and intra-assay variation was 9% and 5%, respectively.

Data analysis

Weekly mean hormone concentrations were expressed as the mean ± SEM (standard error of the mean) of urine samples grouped in one-week increments. Values were aligned to the Julian calendar by month, week or day from oestrogen peak or by week from natural mating. Statistical differences were measured using a Student's *t*-test with equal variance. For females SB 374 and SB 414, all data were aligned to the day of peak oestrogen excretion (Day 0). Baseline oestrogen concentrations were defined for each individual female as the mean oestrogen concentration outside breeding and pregnancy/pseudopregnancy intervals. Beginning in February of each year, oestrogen concentrations were determined weekly for each female. When hormone concentrations increased two-fold above an individual's own baseline, frequency of endocrine monitoring was increased to one to three times daily, depending on if hormonal data were being used to schedule a timed mating or AI. A greater than six-fold increase in urinary oestrogen above basal, followed by a rapid decline to baseline concentrations within one to four days was considered presumptive evidence of ovulation. During the post-ovulatory interval, progestins were monitored bi-weekly until a female's mean progestin concentrations increased two-fold above concentrations observed during the previous week. This inflection point generally occurred four to five months post-ovulation (40 to 50 days pre-partum), at which time daily progestin assays were conducted to identify a sharp and steady decline in hormone concentrations that could alert animal care and veterinary staff that parturition was imminent.

REPRODUCTIVE HORMONE PATTERNS

Average peri-oestrual endocrine profile

Mean urinary oestrogen in a representative female (SB 112; 1987 to 1991), with superimposed reproductive behavioural data during the peri-oestrus interval are depicted in Figure 8.1. Oestrogen concentrations increased slightly by Day -12 (7 ± 2 ng mg^{-1} Cr), were eight-fold above baseline (19 ± 3 ng mg^{-1} Cr) by Day -6, remained at that level for four days and then peaked on Day 0 (63 ± 5 ng mg^{-1} Cr). Oestrogen excretion then declined, returning to baseline by Day $+4$. Scent mark behaviours began on Day -6, peaked on Day $+1$ and ceased by Day $+3$. Bleating was observed on Days -4 to $+5$ with a dramatic increase in number of bleats beginning on Day 0 (75 bleats), which were sustained through Day $+4$.

Typical endocrine profile during a normal pregnancy

Urinary oestrogen and progestin excretion for a female (SB 371) that gave birth to a healthy cub is depicted in Figure 8.2a. By Week -1,

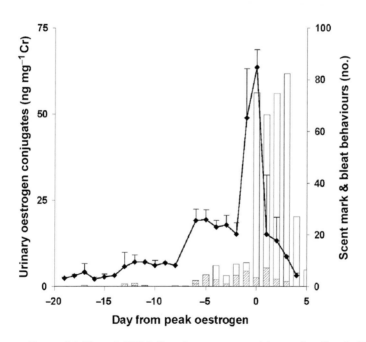

Figure 8.1. Mean $\pm$ SEM daily urinary oestrogen ($\blacklozenge$) excretion (female SB 112, n = 5 oestrous cycles over 5 years) from 20 days before to five days after the oestrogen peak (Day 0). Frequency of scent marking (▨▨▨) and bleating (▭▭▭) also are shown.

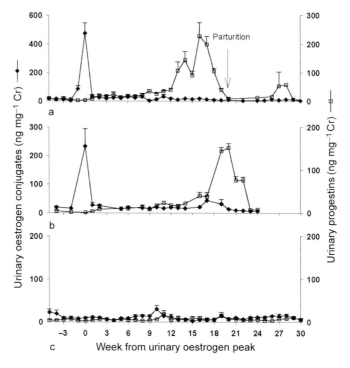

Figure 8.2. Mean ± SEM weekly urinary oestrogens (♦) and progestins (□) during a single representative (a) pregnancy, (b) pseudopregnancy (non-mated oestrual female) and (c) anovulation. Values were aligned to week from urinary oestrogen peak (Week 0; a, b) or an arbitrary midpoint of the breeding season (c). Arrow denotes parturition.

urinary oestrogen concentrations had increased (86 ± 21 ng mg^{-1} Cr) more than 20-fold above baseline (13 ± 4 ng mg^{-1} Cr, Weeks -13 to -2), peaked the next week (474 ± 72 ng mg^{-1} Cr, Week 0) and then declined to 25 ± 4 ng mg^{-1} Cr (from Week $+2$ to $+8$). Urinary progestins were basal from Weeks -13 to $+1$ (7 ± 3 ng mg^{-1} Cr), but increased three-fold ($p < 0.01$) after the pre-ovulatory oestrogen surge (24 ± 8 ng mg^{-1} Cr, Week $+2$ to $+12$). By Week $+13$, progestins exhibited a secondary three-fold increase (107 ± 31 ng mg^{-1} Cr; $p < 0.05$), whereas progestin excretion peaked during Week $+16$ (227 ± 47 ng mg^{-1} Cr) and declined thereafter, returning to baseline (8 ± 4 ng mg^{-1} Cr, Week $+20$) coincident with parturition.

Typical endocrine profile during a pubertal oestrus

Urinary hormones in a 4.5-year-old pubertal female (SB 461) are depicted in Figure 8.2b. She was paired with a male but no mating occurred. Although inconsistent urine collection prevented a more precise characterisation of the temporal oestrogen rise and fall during peri-oestrus, mean oestrogen concentrations increased ~20-fold to peak during Week 0 (230 ± 150 ng mg^{-1} Cr) and declined to basal by Week +1. Proceptive behaviours, including bleating and chirping, commenced on Day −5, peaked on Day +2 and returned to zero by Day +10. Spontaneous tail-up and anogenital presentation behaviours commenced on Day +1, peaked on Day +2 and ended by Day +7. Mean urinary progestins remained basal (2 ± 1 ng mg^{-1} Cr) from Week −4 to +1, but increased subtly but significantly during the post-ovulatory interval (Week +2, 9 ± 4 ng mg^{-1} Cr; $p < 0.05$) through Week +16 (17 ± 3 ng mg^{-1} Cr). The secondary rise in progestin excretion occurred during Week +17 (38 ± 1 ng mg^{-1} Cr; $p < 0.05$), was sustained through Week +20 (152 ± 24 ng mg^{-1} Cr) and then declined steadily to baseline by Week +23.

Typical endocrine profile during an anovulatory breeding season

In 1999, an 11.5-year-old female (SB 332) exhibited oestrus and was artificially inseminated but failed to give birth. During the following (2000) breeding season, SB 332 failed to exhibit oestrus, and oestrogen and progestin excretion profiles (Fig. 8.2c) were indicative of anovulation. Because this female exhibited oestrus in 2001, the anovulatory cycle of 2000 was not due to reproductive senescence. It appears that fewer than 7% of adult, non-lactating giant pandas are likely to be anovulatory in a given year (Zhang *et al.*, 2004).

Urinary progestin excretion over successive years

Female SB 112 mated each year from 1987 to 1991. Her sequential progestin profiles are shown in Figure 8.3, with hormone values aligned to week of copulation (Week 0). Birth occurred in 1987, foetal loss in 1989 and no parturition or foetal loss was observed in other years. In 1987 (see Fig. 8.3a), urinary progestins increased from Week

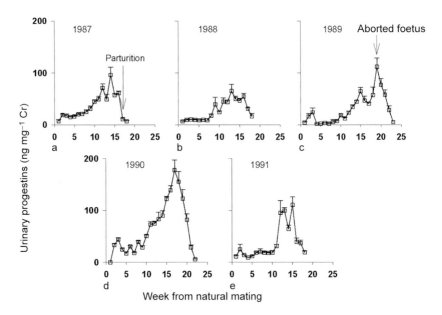

Figure 8.3. Mean ± SEM weekly urinary progestins in a single female, SB 112: (a) 1987, pregnancy; (b) 1988, mating, no birth; (c) 1989, aborted foetus; (d) 1990, mating, no birth; and (e) 1991, mating, no birth. Values were aligned to week from natural mating (Week 0). Arrows indicate dates of parturition or abortion.

+9 to +14 before decreasing from Week +15 to +16 leading to parturition during Week +17. In 1989 (see Fig. 8.3c), an initial post-ovulatory progestin increase (Weeks +2 to +3) was followed by basal concentrations (Weeks +4 to +11) and then the secondary rise (Weeks +12 to +19), which lasted eight weeks. In the face of peak progestin excretion (112 ± 17 ng mg^{-1} Cr), abortion of a single foetus occurred during Week +19 followed by hormone decline to nadir over four weeks. For all other years, progestins were elevated for seven-, eight- and five-week intervals during 1988 (see Fig. 8.3b; Weeks +9 to +15), 1990 (see Fig. 8.3d; Weeks +10 to +17) and 1991 (Fig. 8.3e; Weeks +11 to +15), respectively. Progestin concentrations then declined to baseline over two- (1988), four- (1990) and two- (1991) week intervals, respectively. Although no births occurred in these years, it is unknown if the intervals of elevated luteal activity represented pseudopregnancy or failed pregnancy.

Faecal hormone analysis

The validity of faecal analysis was demonstrated, in part, by comparing urinary and faecal oestrogen excretion in matched samples collected from a pubertal female (SB 473) across her inaugural peri-oestrual interval (Fig. 8.4; Day 0 = day of peak urinary oestrogen excretion). Basal faecal oestrogen concentrations were ~50 ng g^{-1} faeces while urinary oestrogen concentrations were <10 ng mg^{-1} Cr. Faecal oestrogens were increased above basal on Day −5, rose four-fold higher to a pre-ovulatory peak of 214 ng g^{-1} faeces (Day −2) and then declined to basal by Day +3. Although correlation between the faecal and urinary oestrogen measures (Day −25 to +16) was high ($r = 0.87$), faecal oestrogen peaked two days before urinary oestrogen.

Matched urine and faecal samples were collected from SB 473 the following breeding season (2003), during which copulation occurred, but no birth (Fig. 8.5). Again, correlation between urinary and faecal oestrogen measures was high ($r = 0.75$; see Fig. 8.5a) but faecal oestrogen peaked one day before urinary oestrogen; both urinary and faecal oestrogens remained low thereafter. Urinary and faecal progestins were both elevated above basal on Day +2 (see Fig. 8.5b). A secondary rise in urinary progestins began on Day +82, peaked on Day +118 (188 ng mg^{-1} Cr) and returned to basal on Day +133 (3 ng mg^{-1} Cr). The secondary rise in faecal progestins was delayed until 25 days later (Day +109), peaked on Day +125 (24 µg g^{-1} Cr) and declined to basal by Day +131. Overall, the duration of the secondary progestin rise was considerably shorter in faeces (23 days) compared to urine (52 days).

Urinary profiles in untreated versus gonadotrophin-treated females

SB 414

Urinary ovarian steroid excretion patterns aligned to the day of peak oestrogen excretion (Day 0) are depicted for this female from 1998 to 2000 (Fig. 8.6). In all years, spontaneous oestrogen excretion increased above baseline by Day −10. In 1998 (the 'control' year when no gonadotrophin treatment was administered), oestrogen concentrations during the 30-day interval preceding ovulation (Days −30 to 0) ranged from 5 to 283 ng mg^{-1} Cr, with oestrogen returning to baseline by Day +5. In the

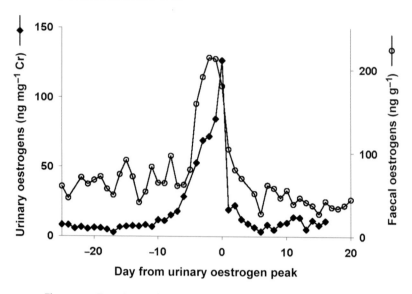

Figure 8.4. Faecal (○) and urinary (♦) oestrogen excretion in female SB 473. Values, aligned to day of urinary oestrogen peak (Day 0), were highly correlated ($r = 0.87$).

absence of mating or AI, this female nevertheless displayed typical oestrual behaviours, including bleating and chirping. A lordotic tail-up response was observed as urinary oestrogens declined. In this same year, post-ovulatory progestins increased by Day +3 with values from Days 0 through 40 exceeding ($p < 0.05$) pre-ovulatory concentrations.

In 1999, this same female was treated with FSH and hCG (see Methods) in an attempt to induce oestrus and increase the chance of mating. Despite seven days of FSH treatment (Days −13 to −8), oestrogens did not increase appreciably (see Fig. 8.6b). Therefore, the FSH dose was increased on Day −6, which induced a rapid increase in oestrogen excretion over the next six days (~150 ng mg^{-1} Cr). On the day of mating (Day 0), hCG was administered, and AI was performed two days later (Day +2). Within five days, a precipitous and sustained 24-day increase in oestrogen production ensued (peak concentrations on Day +19, 5713 ng mg^{-1} Cr), before oestrogen concentrations declined to pre-treatment levels (Day +33). Urinary progestins increased after mating and remained elevated in a profile that was indistinguishable from that observed in 1998 (see Fig. 8.6a,b). For the 1999 season, this female chirped on Day −15 (before gonadotrophins were given), scent-marked

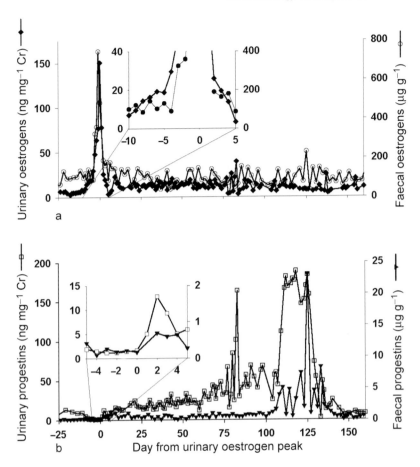

Figure 8.5. Faecal and urinary steroid measures in female SB 473:
(a) faecal (○) and urinary (♦) oestrogen excretion; and (b) faecal (▾) and
urinary (□) progestins. Values were aligned to day of urinary oestrogen
peak (Day 0).

on Days −9 to −7 and exhibited tail-up behaviour on the last day of FSH
treatment. There was a brief mating the next day followed by three days
of receptive behaviours (tail-up and lordosis); however, the female
showed little interest in any male, and no birth resulted.

During 2000, oestrogens increased over a ten-day interval to a pre-
ovulatory peak of 671 ng mg^{-1} Cr (see Fig. 8.6c, Day 0) followed by a
steep two-day decline to baseline. Urinary progestins increased ($p <$
0.05) above pre-ovulatory concentrations by Day +7 and were elevated

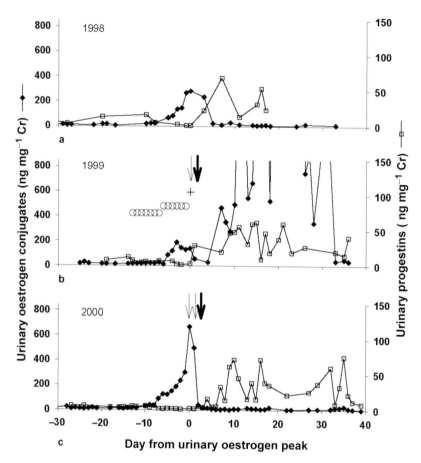

Figure 8.6. Daily urinary oestrogens (♦) and progestins (□) in female SB 414: (a) 1998, a 'control' natural oestrus; (b) 1999, exogenous gonadotrophins, mating and AI; and (c) 2000, natural oestrus, mating and AI. Values were aligned to day of peak oestrogen (Day 0), except for 1999, when values were aligned to the day of mating (Day 0). →, mating; → (bold), AI; ○, FSH; +, hCG.

(~22 ng mg^{-1} Cr) during the remainder of the sampling period. The female bleated on Day −8 and exhibited tail-up and lordotic postures on Day −1. After naturally mating twice (Day 0, Day +1), AI was performed on Day +4, progestins increased by Day +8, and a healthy cub was delivered 121 days post-first copulation.

SB 374

During 1998, when no hormone treatments were administered, urinary oestrogens increased over an 11-day interval to peak pre-ovulatory concentrations of 321 ng mg^{-1} Cr (Day 0), followed by a return to basal by Day +5 (Fig. 8.7a). Progestins increased by Day +14 and remained elevated (~19 ng mg^{-1} Cr) throughout the luteal phase. Proceptive behaviours were minimal, although bleating and chirping were observed on Day 0, and mating occurred three and four days later followed by AI on Day +5; no birth occurred.

In 1999, oestrogens increased (see Fig. 8.7b) above basal on Day −7 and peaked (185 ng mg^{-1} Cr) on Day 0. In an attempt to promote ovulation, hCG was administered (Day +2), the female mated one day later (Day +3), and AI was performed on Days +5 and +6. Oestrogens declined on Day +5 (36 ng mg^{-1} Cr), but a second hCG injection (Day +7) markedly boosted oestrogen excretion by Day +12 (167 ng mg^{-1} Cr), which remained elevated for a 15-day interval (peak on Day +26, 259 ng mg^{-1} Cr). Although urine specimens were not collected for the next ten days, oestrogens were basal by Day +37. Urinary progestins increased by Day +6 (28 ng mg^{-1} Cr) and remained elevated within the normal range (~24 ng mg^{-1} Cr) throughout the sampling interval. In addition to mating on Day +3, this female chirped on Days −3 through −1 and exhibited tail-up behaviours on Days +5 and +6; no birth occurred during 1999.

During 2000, oestrogens increased over an eight-day interval to peak pre-ovulatory concentrations of 256 ng mg^{-1} Cr (Day 0), followed by a return to basal concentrations by Day +3 (see Fig. 8.7c). Urinary progestins increased by Day +9 (26 ng mg^{-1} Cr) and remained elevated in the normal range (~12 ng mg^{-1} Cr). This female exhibited water play and scent-marking on Day −3, tail-up activity on Day −2 and lordosis/ tail-up behaviours on Day +2 and +5. Mating occurred on Day +3 followed by AI on Day +4. Twin cubs were produced (with only one surviving) 142 days after the day of mating.

Urinary corticosteroid measures in response to anaesthesia

Immunoreactive corticosteroids were assessed before, during and after female SB 371 was anaesthetised on three occasions from 1998 to 1999 to conduct AI and/or physical examination (Fig. 8.8).

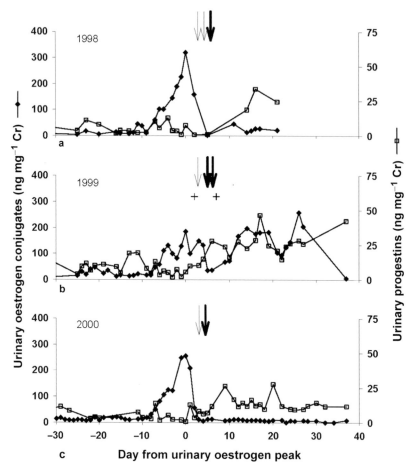

Figure 8.7. Daily urinary oestrogens (♦) and progestins (□) in female SB 374: (a) 1998, control natural oestrus, mating and AI; (b) 1999, exogenous gonadotrophins, mating and AI; and (c) 2000, mating and AI. Values were aligned to day of peak oestrogen (Day 0), except for 1999 when values were aligned to the day of the first oestrogen peak (Day 0). →, mating; → (bold), AI; +, hCG.

Small sample size precluded statistical analysis, but excretion patterns were strikingly similar across years for all anaesthetic procedures. In general, corticosteroids were low (38–82 ng mg^{-1} Cr) pre-anaesthesia (from −10 to 0 hours, with Time 0 being the time of anaesthetic administration), increased eight- to ten-fold (to 543–648 ng mg^{-1} Cr)

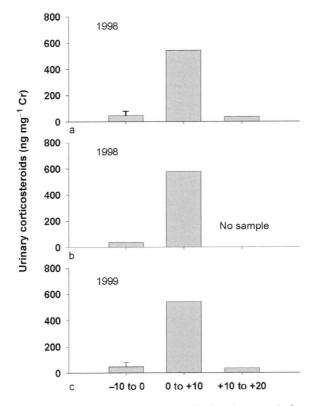

Figure 8.8. Urinary corticosteroid values in female SB 371 before, during and after anaesthetic episodes: two for 1998 (a, b) and one for 1999 (c). Values were aligned to time (in hours) from anaesthetic administration.

during the ten-hour post-induction interval and then declined to basal (39–116 ng mg^{-1} Cr) within 10 to 20 hours post-anaesthetic induction.

Urinary androgen profiles in males

Weekly urinary androgen excretion is shown in Figure 8.9a for a male (SB 381) that sired an offspring the same year (1999). Average androgen concentrations were highest ($p < 0.05$) from January to April (58 ± 4 ng mg^{-1} Cr) compared to May to December (29 ± 2 ng mg^{-1} Cr). Androgen excretion in a male (SB 390) maintained in isolation from females (2000) is depicted in Figure 8.9b. The following year (2001; see Fig. 8.9c), this

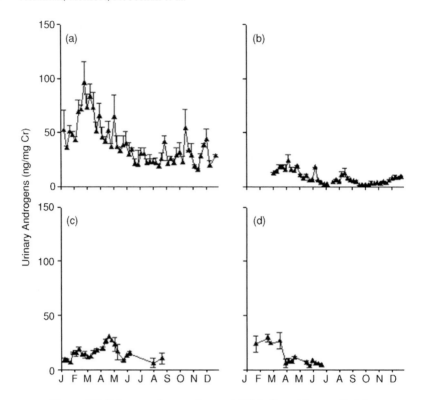

Figure 8.9. Urinary androgens (mean ± SEM) aligned by month: (a) an adult proven sire (SB 381); (b) an adult (SB 390) maintained in isolation; (c) the same previously isolated male (SB 390) with access to adult females; and (d) a subadult male (SB 452) with female access but no mating.

same male mated, sired an offspring and served as a semen donor for AI. While isolated, mean urinary androgen was typically <25 ng mg^{-1} Cr (see Fig. 8.9b), although average androgen from March to June (14 ± 1 ng mg^{-1} Cr) was higher ($p < 0.01$) than from July to December (5 ± 1 ng mg^{-1} Cr). In 2001 (see Fig. 8.9c), androgen excretion was higher ($p < 0.01$) from April to May (20 ± 3 ng mg^{-1} Cr) compared to January to March (13 ± 1 ng mg^{-1} Cr). Interestingly, this male participated in a rare autumnal mating in 2001, but no urine samples were collected during this interval. No among-year differences ($p > 0.05$) in androgen concentrations were evident despite the lack of mating opportunities in 2000.

Androgen excretion patterns in a 5-year-old subadult male (SB 452) are presented in Figure 8.9d. This individual was exposed to an oestrual female but showed little interest and failed to mate. Although sample collections were inconsistent, his profile revealed higher ($p < 0.01$) androgen concentrations from January to March (26 ± 1 ng mg^{-1} Cr) compared to April to June (7 ± 1 ng mg^{-1} Cr).

DISCUSSION AND PRIORITIES FOR THE FUTURE

Hormones are the keys to reproductive success as well as the regulators of animal well-being. In this chapter, we described the unique endocrinology of the giant panda, from the peri-oestrus interval, pregnancy, pseudopregnancy and seasonality to adrenal responsiveness to a physiological 'stressor' (anaesthesia). Data revealed the extraordinary potential that non-invasive endocrine methods provide for biologists to understand the fundamental mechanisms associated with endocrine control of general health and reproductive success.

Urinary oestrogens and associated phenomena

During a giant panda ovarian cycle, it is now known that urinary oestrogens increase gradually over one to two weeks, rapidly accelerate to pre-ovulatory concentrations, decline precipitously coincident with ovulation and return to baseline two to five days later (Bonney et al., 1982 ; Hodges et al., 1984; Chaudhuri et al., 1988; Monfort et al., 1989; Mainka et al., 1990; Lindburg et al., 2001; McGeehan et al., 2002; Czekala et al., 2003). Thus, peak oestrous behaviours, including sexual receptivity (e.g. tail-up and lordosis) and mating, usually occur when circulating and excreted oestrogens are declining (Bonney et al., 1982; Kleiman, 1985; Monfort et al., 1989). Although mating has been observed when oestrogens are increasing (or are at peak concentrations), those incidences generally have been associated with the first of successive copulatory bouts (Bonney et al., 1982; Monfort et al., 1989). In 1999, the San Diego Zoo female (SB 371) became pregnant following AI conducted on three successive days; AI was initiated only after oestrogen concentrations were declining. Likewise, the cub produced at the Smithsonian's National Zoo in 2005 was born to SB 473 who was artificially inseminated one time after urinary oestrogen levels were clearly already declining. Thus, optimal fertility (to mating or AI) occurs in the late peri-ovulatory interval in an endocrine atmosphere characterised by declining oestrogen influence (Lindburg *et al.*, 2001), a phenomenon also observed in the gray wolf (Seal *et al.*, 1979), domestic dog (Wildt *et al.*, 1979) and mink (Pilbeam *et al.*, 1979).

Urinary oestrogen patterns, in combination with reproductive behaviours and vaginal cytology (see Chapter 9), provide the best means of accurately evaluating reproductive status of the female giant panda. For example, when only behaviours were monitored, 80% of captive giant pandas were classified as experiencing a 'weak' or 'silent' oestrus (Zheng *et al.*, 1997). However, subsequent evaluations of oestrogen excretion indicated that most of these females actually ovulated (McGeehan *et al.*, 2002). Nonetheless, more work is warranted to compare 'weak' versus 'strong' oestrual females, and to determine if variations in behavioural intensity at oestrus are associated with fertility and fecundity.

Intermittent oestrous acyclicity or anoestrus has been correlated to lactational suppression of ovulation in the giant panda (Schaller *et al.*, 1985). However, for the first time, we report that non-lactating females with a history of regular ovarian cyclicity can experience idiopathic acyclicity (SB 332; see Fig. 8.2c). Skipping an annual ovarian cycle does not appear to be rare in giant pandas, and we have not detected any substantive changes in management procedures that would suggest that psychosocial or physiological stress contribute to this phenomenon. Periodic absence of ovarian activity indicates that the giant panda may not be evolutionarily programmed to sustain pregnancy and lactation on an annual cycle due to the long-term maternal investment required to nurture young until weaning (18–24 months; Schaller *et al.*, 1985).

Exogenous gonadotrophins have been used to stimulate follicular development and/or induce ovulation in many wildlife species in captivity (Howard, 1999), including the giant panda (Chaudhuri *et al.*, 1988). Our findings revealed that the sequential injection of FSH and/or hCG could cause serious hyper- and protracted oestrogen secretion. Even hCG alone caused this effect, probably by stimulating too many ovarian follicles and/or interfering with normal steroidogenic pathways in luteinised follicles or the corpus luteum (CL). More work is needed to develop appropriate ovulation induction regimens for the giant panda. Promising new research to clone and express giant panda FSH and LH (L. Mingjuan, pers. comm.) is in progress. Regardless, non-invasive hormone monitoring will continue to be valuable for tracking the effectiveness and safety of new endocrine therapies.

Our data also revealed that faecal steroid analysis was effective for assessing ovarian status via both oestrogens and progestins. The strong correlation between urinary and faecal steroids, both of which coincided with appropriate behaviours associated with oestrus and ovulation, provided strong incentive for further applying this technique. However, because faecal analysis requires a time-consuming extraction procedure, urinary analysis provides a quicker evaluation of oestrual activity for captive management. Faecal steroid measures are better suited for longitudinal assessments, perhaps eventually including studying free-living giant pandas.

Urinary progestins and associated phenomena

For the giant panda, an initial rise in excreted progestins occurs immediately after the peri-ovulatory oestrogen peak (Hodges et al., 1984; Monfort et al., 1989; Mainka et al., 1990; McGeehan et al., 2002; Czekala et al., 2003). This initial post-ovulatory rise in urinary progestins, although not as robust as the later secondary rise, signals the occurrence of ovulation (Mainka et al., 1990; McGeehan et al., 2002). Due to temporal variation in speed and amplitude of this initial rise, however, the urinary progestin profile is most effective as an ovulation marker when used in concert with the dramatic pre-ovulatory oestrogen surge.

A secondary progestin excretion rise (74–122 days post-ovulation) occurs 40 to 50 days before parturition, suggesting that the giant panda experiences delayed implantation (Hodges et al., 1984; Chaudhuri et al., 1988; Monfort et al., 1989; Mainka et al., 1990; McGeehan et al., 2002). At the onset of the secondary progestin rise, it is believed that the blastocyst becomes implanted in the uterine wall with foetal development further supported by continued progestin production. Delayed implantation occurs in all bears (Sandell, 1990), the mink (Cochrane & Shackelford, 1962), badger (Bonnin et al., 1978) and western spotted skunk (Foresman & Mead, 1974), presumably to ensure that birth occurs at a time most conducive to offspring survival. This 'delay' is believed to be due to an insufficiency of the CL which, when exposed to the pituitary hormone prolactin, becomes reactivated to produce significant amounts of CL-derived progesterone (Mead, 1993; Sato et al., 2001). Under progesterone influence, the conceptus implants to resume development.

Delayed CL reactivation in bears is obligate once ovulation occurs, whether or not a female conceives. This is why some nonpregnant females are classified as pseudopregnant (Mainka et al., 1990; Sato et al., 2001). Females have no way of assessing their own pregnancy status, so pseudopregnancy may function as a hormonal 'insurance policy' protecting pregnancy should a female be carrying a diapausing embryo awaiting implantation. Qualitative and temporal progestin (and oestrogen) excretion patterns are similar between the pregnant and non-pregnant giant panda (Hodges *et al.*, 1984; Chaudhuri *et al.*, 1988). There also appears to be no difference in hormonal patterns among a pubertal pseudopregnancy, pregnancy and adult pseudo-pregnancy. It is likely that some females classified as pseudopregnant actually have been pregnant but suffered undetected pre-implantation embryonic or post-implantation foetal loss. Therefore, we suspect that the incidence of pseudopregnancy may be overestimated. We recommend that, until improved pregnancy detection methods are developed, this term should be limited to describing unmated/ovulatory females that exhibit concurrent endocrine and/or behavioural signs of pseudopregnancy. Comparisons between known pregnant and non-mated, ovulatory females – a rare group given the emphasis on breeding all adult giant pandas – could help to establish endocrine differences between these two reproductive states.

More research is also needed regarding specific factors for identifying pregnancy in the giant panda. For example, embryo-derived platelet activity factor (EDPAF) has recently been shown to be useful for predicting pregnancy in this species (R. Hou, pers. comm.). However, this approach requires a minimum of two blood samples collected during the peri-ovulatory interval (immediately before and after presumed conception); thus, it may be considered too invasive, especially if anaesthesia or physical restraint is required for venipuncture. Despite the lack of a definitive pregnancy diagnostic test, a decline in the secondary urinary progestin surge (occurring at the end of pregnancy or pseudopregnancy) is a valuable sentinel measure. The approaching return to baseline either signals an impending birth (allowing managers to mobilise resources to monitor the female) or, in the case of pseudopregnancy, indicates the end of a neccessary pregnancy watch (Chaudhuri *et al.*, 1988; Monfort *et al.*, 1989; Czekala *et al.*, 2003).

Urinary corticosteroids and stress

Physiological or psychosocial stress increases corticosteroid secretion, long a marker for 'stress' in diverse mammals (Morton *et al.*, 1995). The blood collection procedure itself (restraint and/or anaesthesia plus venipuncture) can be a stressor. Therefore, serum corticoid values may not accurately represent undisturbed, baseline adrenal activity (Monfort, 2003). For the first time, we have demonstrated the ability to assess adrenal activity in the giant panda non-invasively by assessing urine. A cause and effect between administering anaesthesia (a physiological 'stressor') and subsequent increased corticosteroid excretion was observed. The brief elevation (~10 hours) in this stress hormone suggested that more frequent sampling (e.g. multiple urines per day) would better document the impact of short-term, acute stressors. Infrequent sampling could incorrectly lead to the conclusion that a physiological stressor had no impact on corticosteroids. For example, if samples in our study had only been collected after ten hours, anaesthesia would not have appeared to increase corticosteroid excretion. As the giant panda defaecates more frequently than it urinates, faeces may be more useful for these type of studies; the potential of faecal corticosteroid measures is now being examined.

In a preliminary study (data not shown), an injection of adreno-corticotrophic hormone (ACTH) to a single giant panda male markedly increased both faecal and urinary corticoids in <24 hours (Kersey, unpublished data). Interestingly, the post-ACTH corticosteroid increase occurred earlier in faeces (nine hours) than urine (19 hours). Now that faecal monitoring has been shown to be a physiological reflection of true adrenal activity, there is a need to determine if the giant panda exhibits seasonal fluctuations in corticosteroid activity, as reported for other bears (Palumbo et al., 1983; Harlow et al., 1990). In preliminary studies (data not shown), there were no differences ($p > 0.05$) in urinary corticosteroids (1) within females over the four-month breeding season (February to May), (2) in spontaneously ovulating versus gonadotrophin-treated females, or (3) among mated, non-mated, anovulatory, pregnant or pseudopregnant females. The latter findings are important because they suggest that anovulation or 'weak oestrus' in the giant panda is probably unrelated to stress.

Urinary androgens and associated phenomena

In the only previous such study in a male giant panda, urinary androgens increased during the reproductive season when the mate was sexually receptive, but not during the next year when the female failed to exhibit sexual behaviours (Bonney et al., 1982). Anticipation of mating or mating itself can induce androgen production in the rat (Taleisnik *et al.*, 1966), rabbit (Saginor & Horton, 1968) and bull (Katongole *et al.*, 1971). Our studies of multiple male giant pandas revealed that urinary androgens were elevated consistently during the normal mating season (February to April) regardless of whether the male was sexually active. The extent to which the male giant panda is responding to photoperiodic cues is unknown, but occasional out-of-season matings (SB 390 mated with a female in autumn 2001) and sperm production (see Chapter 7) suggest that testicular function in this species is not strictly seasonal. More research is needed to understand the role of androgens in modulating spermatogenesis and the expression of reproductive behaviours in the giant panda.

Hormonal monitoring for applied management now and in the future

It is well accepted that endocrine monitoring is an essential tool in the routine management of a breeding pair (or population) of giant pandas. The technology is critical for establishing the receptive period of the female to optimise timed mating or AI. This is important because paired animals can be seriously antagonistic outside peak sexual receptivity. Knowing the precise status of ovarian function (combined with behavioural and vaginal cytology data) can be a comfort to managers who are responsible for the tricky introduction of such individuals at the best time to achieve successful mating. Additionally, the technology is now sufficiently advanced to provide informative endocrine data within three to four hours of recovering a urine sample. The same advantage holds for using urinary progestins at the end of gestation to estimate day of parturition accurately or, in the case of pseudopregnancy, end an intensive pregnancy watch so that staff can return to normal duties. With the advent of EIA systems, endocrine laboratories can (and have been) established on-site at giant panda facilities with minimal cost and instrumentation.

Presently, studies are in place or planned to use non-invasive endocrine monitoring in captive giant pandas to evaluate:

1. more reproductive cycles for more individuals of diverse ages (e.g. pubertal onset or reproductive senescence) or in the breeding versus non-breeding season (e.g. what triggers seasonal ovarian activity);

2.   behavioural cues predictive of reproduction;
3.   optimal timing for AI using fresh versus thawed sperm;
4.   pregnancy and expected parturition;
5.   the influence of social and husbandry conditions (e.g. single vs. multiple animals) or medical procedures (e.g. restraint or anaesthesia) on animal well-being and reproductive fitness.

Particularly exciting is the possibility of rapidly monitoring physiological indices of stress, information that will assist managers of captive giant pandas in optimising husbandry protocols to improve both quality of life and reproductive success. For example, enclosure space could be modified (or enriched) according to corticosteroid patterns until baseline 'stress' levels are reached. Stress could actually be measured between introduced male and female giant pandas in an attempt to understand the causes of behavioural sexual incompatibility and perhaps to develop remediation approaches.

This technology is also highly relevant for future field-related studies, including examining the influence of the potentially controversial procedure of radiocollaring wild pandas or of human disturbance or environmental disrupters (i.e. forestry practices, agriculture, pollutants and toxicants) on animal well-being. Faecal endocrine monitoring, in particular, holds great promise for integrating endocrinology with giant panda field ecology studies. The possibility of measuring adrenal activity to assess levels of human disturbance and encroachment in natural habitats is intriguing. For example, non-invasive corticosteroid monitoring has been used to assess stress in free-living spotted owls (Wasser et al., 1997), and wolves and elk (Creel et al., 2001) subjected to logging and recreational snowmobile activity, respectively. This type of information can inform policies to minimise disruption of wild breeding areas to increase chances for species survival. Such technology may well be applicable to the giant panda, especially in the future when reintroduction might be used for repopulating reserves. Knowing how individual giant pandas react to specific reintroduction techniques or habitats could substantially enhance the emerging field of reintroduction science.

ACKNOWLEDGEMENTS

The authors thank the animal care staff of San Diego Zoo, the Smithsonian's National Zoological Park, Zoo Atlanta, Adventure World,

Zoológico de Chapultepec and the China Conservation and Research Centre for the Giant Panda for assisting in sample collection and record-keeping. We are especially grateful to Desheng Li for technical assistance. We thank Lesley Northrop, Edy MacDonald, Angeles Pintada, Alan Fetter, Jean-Pierre Montagne, Kendall Mashburn and Nicole Presley. Funding was provided by the Alice C. Tyler Perpetual Trust, Friends of the National Zoo, the Zoological Society of San Diego and the Smithsonian's National Zoological Park.

REFERENCES

Bonney, R. C., Wood, D. J. and Kleiman, D. G. (1982). Endocrine correlates of behavioural oestrus in the female giant panda (*Ailuropoda melanoleuca*) and associated hormonal changes in the male. *Journal of Reproduction and Fertility*, **64**, 209–15.

Bonnin, M., Canivenc, R. M. and Ribes, C. I. (1978). Plasma progesterone levels during delayed implantation in the European badger (*Meles meles*). *Journal of Reproduction and Fertility*, **52**, 55–8.

Chaudhuri, M., Kleiman, D. G., Wildt, D. E. *et al.* (1988). Urinary steroid concentrations during natural and gonadotrophin-induced estrus and pregnancy in the giant panda (*Ailuropoda melanoleuca*). *Journal of Reproduction and Fertility*, **84**, 23–8.

Cochrane, R. L. and Shackelford, R. M. (1962). Effects of exogenous estrogen alone and in combination with progesterone on pregnancy in the intact mink. *Journal of Endocrinology*, **25**, 101–6.

Creel, S., Fox, J. E., Hardy, A. *et al.* (2001). Snowmobile activity and glucocorticoid stress responses in wolves and elk. *Conservation Biology*, **16**, 809–14.

Czekala, N. M., McGeehan, L., Steinman, K. J., Li, X. and Gual-Sil, F. (2003). Endocrine monitoring and its application to the management of the giant panda. *Zoo Biology*, **22**, 389–400.

Foresman, K. R. and Mead, R. A. (1974). Pattern of luteinizing hormone secretion during delayed implantation in the spotted skunk (*Spilogale putorius latifrons*). *Biology of Reproduction*, **11**, 475–80.

Harlow, H. J., Beck, T. D. I., Walters, L. M. and Greenhouse, L. W. (1990). Seasonal serum glucose, progesterone and cortisol levels of black bears (*Ursus americanus*). *Canadian Journal of Zoology*, **68**, 183–7.

Hodges, J. K., Bevan, D. J., Celma, M. *et al.* (1984). Aspects of the reproductive endocrinology of the female giant panda (*Ailuropoda melanoleuca*) in captivity with special reference to the detection of ovulation and pregnancy. *Journal of Zoology London*, **203**, 253–67.

Howard, J. G. (1999). Assisted reproductive technologies in nondomestic carnivores. In *Zoo and Wildlife Medicine*, ed. M. E. Fowler and R. E. Miller. W. B. Saunders Co., Philadelphia, PA: pp. 449–57.

Katongole, C. B., Naftolin, F. and Short, R. V. (1971). Relationship between blood levels of luteinizing hormone and testosterone in bulls and the effects of sexual stimulation. *Journal of Endocrinology*, **50**, 457–66.

Kleiman, D. G. (1985). Social and reproductive behavior of the giant panda (*Ailuropoda melanoleuca*). In *Proceedings of the International Symposium of the Giant Panda*, ed. H. G. Klös and H. Frädrich. Berlin Zoo, **10**, 45–58.

Lindburg, D. G., Czekala, N. M. and Swaisgood, R. R. (2001). Hormonal and behavioral relationships during estrus in the giant panda. *Zoo Biology*, **20**, 537–43.

Mainka, S. A., Cooper, R. M., Mao, L. and Guanlu, Z. (1990). Urinary hormones in two juvenile female giant pandas (*Ailuropoda melanoleuca*). *Journal of Zoo and Wildlife Medicine*, **21**, 334–41.

McGeehan, L., Li, X., Jackintell, L. *et al.* (2002). Hormonal and behavioral correlates of estrus in captive giant pandas. *Zoo Biology*, **21**, 449–66.

Mead, R. A. (1993). Embryonic diapause in vertebrates. *Journal of Experimental Zoology*, **266**, 629–41.

Monfort, S. L., Dahl, K. D., Czekala, N. M. *et al.* (1989). Monitoring ovarian function and pregnancy in the giant panda (*Ailuropoda melanoleuca*) by evaluating urinary bioactive FSH and steroid metabolites. *Journal of Reproduction and Fertility*, **85**, 203–12.

Monfort, S. L. (2003). Non-invasive endocrine measures of reproduction and stress in wild populations. In *Reproduction and Integrated Conservation Science*, ed. W. V. Holt, A. Pickard, J. Rodger and D. E. Wildt. Cambridge: Cambridge University Press, pp. 147–165.

Morton, D. J., Anderson, E., Foggin, C. M., Kock, M. D. and Tiran, E. P. (1995). Plasma cortisol as an indicator of stress due to capture and translocation in wildlife species. *Veterinary Research*, **136**, 60–3.

Palumbo, P. J., Wellik, D. L., Bagley, N. A. and Nelson, R. A. (1983). Insulin and glucagon responses in the hibernating black bear. *International Conference of Bear Research and Management*, **5**, 291–6.

Pilbeam, T. E., Concannon, P. W. and Travis, H. F. (1979). The annual reproductive cycle of the mink (*Mustela vison*). *Journal of Animal Science*, **48**, 578–84.

Saginor, M. and Horton, R. (1968). Reflex release of gonadotrophin and increased plasma testosterone concentration in male rabbits during copulation. *Endocrinology*, **82**, 627–30.

Sandell, M. (1990). The evolution of seasonal delayed implantation. *Quarterly Review of Biology*, **65**, 23–42.

Sato, M., Tsubota, T., Komatsu, T. *et al.* (2001). Changes in sex steroids, gonadotrophins, prolactin and inhibin in pregnant and nonpregnant Japanese black bears (*Ursus thibetanus japonicus*). *Biology of Reproduction*, **65**, 1006–13.

Schaller, G. B., Hu, J., Pan, W. and Zhu, J. (1985). *The Giant Pandas of Wolong*. Chicago, IL, University of Chicago Press.

Seal, U. S., Plotka, E. D., Packard, J. M. and Mech, L. D. (1979). Endocrine correlates of reproduction in the wolf: serum progesterone, estradiol and LH during the estrous cycle. *Biology of Reproduction*, **21**, 1057–66.

Stabenfeldt, G. H., Daels, P. F., Munro, C. J. *et al.* (1991). An oestrogen conjugate enzyme immunoassay for monitoring pregnancy in the mare: limitations of the assay between days 40 and 70 of gestation. *Journal of Reproduction and Fertility*, (Suppl.), **44**, 37–44.

Taleisnik, S., Caligaris, L. and Astrada, J. J. (1966). Effect of copulation on the release of gonadotrophins in male and female rats. *Endocrinology*, **79**, 49–54.

Taussky, H. H. (1954). A micrometric determination of creatine concentration by Jaffe reaction. *Journal of Biological Chemistry*, **208**, 853–61.

Wasser, S. K., Bevis, K., King, G. and Hanson, E. (1997). Non-invasive physiological measures of disturbance in the northern spotted owl. *Conservation Biology*, **11**, 1019–22.

Wasser, S. K., Hunt, K. E., Brown, J. L. *et al.* (2000). A generalized fecal glucocorticoid assay for use in a diverse array of nondomestic mammalian and avian species. *General and Comparative Endocrinology*, **120**, 260–75.

Wildt, D. E., Panko, W. B., Chakraborty, P. K. and Seager, S. W. J. (1979). Relationship of serum estrone, estradiol-17β and progesterone to LH, sexual behavior and time of ovulation in the bitch. *Biology of Reproduction*, **20**, 648–58.

Zhang, G., Swaisgood, R. R. and Zhang, H. (2004). Evaluation of behavioral factors influencing reproductive success and failure in captive giant pandas. *Zoo Biology*, **23**, 15–31.

Zheng, S., Zhao, Q., Xie, Z., Wildt, D. E. and Seal, U. S. (1997). *Report of the Giant Panda Captive Management Planning Workshop*. Apple Valley, MN: IUCN–World Conservation Union/SSC Conservation Breeding Specialist Group, p. 266.

Zhu, X., Lindburg, D. G., Pan, W., Forney, K. A. and Wang, D. (2001). The reproductive strategy of giant pandas (*Ailuropoda melanoleuca*): infant growth and development and mother–infant relationships. *Journal of Zoology (London)*, **253**, 141–55.

9

The value and significance of vaginal cytology

BARBARA S. DURRANT, MARY ANN OLSON, AUTUMN ANDERSON, FERNANDO
GUAL-SIL, DESHENG LI, YAN HUANG

INTRODUCTION

The giant panda is seasonally monoestrus, experiencing a single oestrus with spontaneous ovulation in the spring (Schaller *et al.*, 1985). Although natural breeding produces the majority of cubs in captivity (Xie & Gipps, 2001), the number of sexually competent breeding males is insufficient to create or maintain a genetically diverse population (Hu, 1990; Xie & Gipps, 2001). Inclusion of males that are behaviourally incapable of mating, but that are genetically valuable, is possible through artificial insemination (AI) (see Chapter 20). Accurate monitoring of the oestrous cycle to pinpoint the time of ovulation is critical for timed matings and, especially, AI success.

The vaginal epithelium of many mammalian species is responsive to changes in circulating oestrogen concentrations. The value of vaginal cytology in monitoring the oestrous cycle of rodents (Zylicz *et al.*, 1967; Parakkal, 1974) and domestic carnivores (Shutte, 1967; Mills *et al.*, 1979) is widely recognised. In routine practice, evaluating vaginal cytology in these taxa involves quantifying proportions of mature exfoliated epithelial cells, also known as superficial, cornified or keratinised cells. Increasing proportions of mature cells are correlated with the pre-oestrual rise in oestrogen as well as oestrous behaviours.

Giant Pandas: Biology, Veterinary Medicine and Management, ed. David E. Wildt, Anju Zhang, Hemin Zhang, Donald L. Janssen and Susie Ellis. Published by Cambridge University Press. © Cambridge University Press 2006.

Despite the logistical difficulty of obtaining vaginal cells from most wildlife species, the oestrous cycles of several small carnivores (raccoon dog: Valtonen *et al.*, 1977; river otter: Stenson, 1988; tayra: Poglayen-Neuwall *et al.*, 1989; multiple ferret species: Mead *et al.*, 1990; Williams *et al.*, 1992; mink: Klotchkov *et al.*, 1998; fox: Boue *et al.*, 2000) have been described by analysing vaginal cytology. The few large carnivores studied to date include the sun bear (Onuma *et al.*, 2002), American black bear (Reynolds & Beecham, 1980) and giant panda (Moore *et al.*, 1984). These investigations have been based on only a single or a very few vaginal swabs per individual, which has yielded only modest information about cytological changes throughout the oestrous cycle. These studies also have relied on monochrome staining, which is effective for distinguishing nucleated cells from superficial (anucleated) cells but fails to identify changes in staining patterns over time. The application of a trichrome stain significantly enhances the ability to characterise vaginal cells that, in turn, provides more useful information on female reproductive status.

Among ursids, the sequential collection of vaginal cells from multiple individuals has been described only in the giant panda (Durrant *et al.*, 2002). With behavioural conditioning, the captive giant panda can be trained to allow vaginal swabbing without chemical restraint or associated stress. This chapter reviews the value of this technology for monitoring female ovarian status and illustrates the positive correlation of oestrogen patterns with vaginal cell morphology. Therefore, this rapid, noninvasive technique is a useful tool for monitoring the dynamic events associated with oestrus and ovulation, thereby facilitating male introductions and/or AI.

MATERIALS AND METHODS

Four captive-born, adult female giant pandas comprised the study group (Table 9.1). Each was trained to enter a squeeze chute, where modest restraint was applied. Using operant conditioning and food as positive reinforcement, each giant panda was trained to accept vaginal swabbing. Training involved focusing the animal's attention on the keeper while desensitising her to other stimuli, including manipulating the vulva and vaginal insertion of a cotton swab. In most cases, one to four weeks of training were required before the female tolerated the insertion of the swab, usually while she was in a sitting position facing the collector (Fig. 9.1). The female housed at the San Diego Zoo

Table 9.1. *Study animals*

Studbook number	Institution animal name	Year of birth	Year entered study (age in years)	Location during study	Number of oestrous cycles evaluated
291	Xiu Hua	1985	1999 (14)	Chapultepec Zoo	4
332	Shuan Shuan	1987	1999 (12)	Chapultepec Zoo	4
360	Xin Xin	1990	1999 (9)	Chapultepec Zoo	4
371	Bai Yun	1991	1997 (6)	San Diego Zoo	5

Figure 9.1. Female SB 371 moderately restrained in a squeeze cage. The upright sitting position was maintained with positive food reward during the collection of vaginal cells.

(Studbook, SB, 371) was also trained to allow swabbing while in dorsal recumbency (Fig. 9.2).

A phosphate buffer saline-moistened sterile cotton swab (Fisher Scientific, Tustin, CA) was inserted 2 to 3 cm into the vagina, rotated

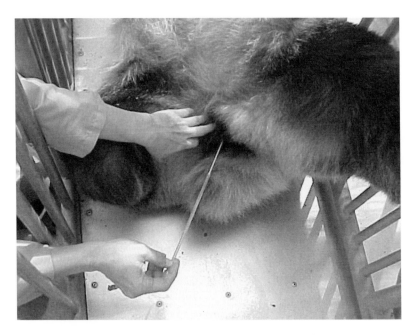

Figure 9.2. Female SB 371 in the 'down' position for vaginal swabbing.
Dorsal recumbency was maintained with positive food reward.

and removed. Care was taken not to scrape cells from the vaginal
epithelium. Each swab was rolled gently onto a clean microscope slide,
which was immediately sprayed with cell fixative (1:1, methanol:ether)
and allowed to air dry. Cells were stained with modified trichrome
Papanicolaou (Fisher Scientific) (Papanicolaou, 1942) as described by
Durrant *et al.* (2002). In brief, this involved increasing the haematoxylin
immersion from two to five minutes and reducing subsequent rehydra-
tion and clearing times. Coverslipped slides were evaluated immediately, then cured for three to four days in 37°C dry heat before
archiving. A minimum of 200 cells was evaluated per slide (usually at
×40 magnification) and classified as basophilic (stained blue), acidophilic (stained pink), keratinised (stained yellow to orange), in addition
to the traditional categories of basophil, intermediate and superficial
(Fig. 9.3; Plate X) (Shutte, 1967; Feldman & Nelson, 1987). A predomin-
ance of basophils in the vaginal smear is correlated with low circulating
oestrogen. Intermediate cells reflect increasing oestrogen concentrations
as the female approaches oestrus, and superficial cells are indicative of

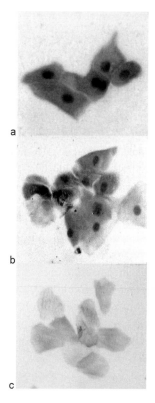

Figure 9.3. Exfoliated vaginal epithelial cells of the giant panda stained
with modified trichrome Papanicolaou. (a) Nucleated basophilic cells
($\times$60); (b) nucleated acidophilic cells ($\times$40); and (c) anucleated, keratinised
cells ($\times$40). (See also Plate X.)

peak oestrogen associated with oestrus. Cells were examined twice, the
first time to record colour only and the second to record morphology
regardless of colour.

Swabs were obtained daily during the breeding season beginning at
the first behavioural or hormonal signs of pro-oestrus and continuing until
behavioural oestrus was complete, or the last AI had been conducted.
Behavioural indications of impending oestrus included increased vocalisation, water play and/or scent marking (see Chapter 11). Urinary
oestrone sulphate conjugate concentrations (E_1C) were assayed daily by
enzyme immunoassay or radioimmunoassay procedures (see Chapter 8).
An increase in E_1C above 25 ng mg^{-1} creatinine (Cr) was considered

evidence of approaching oestrus. The day oestrogen metabolites fell significantly from peak value was designated the day of ovulation (Day 0).

RESULTS

Morphology and staining traits

The morphology and staining characteristics of vaginal cells are illustrated in Figure 9.3. Basophilic cells (see Fig. 9.3a) stained blue, regardless of morphology. This classification included cells traditionally classified as basophils (small, round cells) as well as intermediates (larger, more angular cells). Acidophilic cells (see Fig. 9.3b) included all pink-staining cells regardless of morphology. The cells pictured in Figures 9.3a,b are not morphologically distinct; if a monochrome stain had been used, the cells would have been incorrectly classified together as the same type. We have categorised the sudden shift from blue to pink cells (without morphological transformation) as the 'first chromic shift'. A 'second chromic shift' also occurs when cells abruptly change from pink to yellow. A keratinised cell (see Fig. 9.3c) stains orange, yellow or tan and only occasionally contains a visible nucleus. This distinction often results in higher numbers of keratinised cells than superficial cells.

Relationship of vaginal cytology patterns to urinary oestrogen profiles

Figure 9.4 aligns vaginal cytology with a typical urinary oestrone conjugate (E_1C) profile. Vaginal cytology data (see Fig. 9.4a) represent average values for eight cycles from the four evaluated giant pandas. Endocrine data (see Fig. 9.4b) represent five E_1C cyclic patterns of two of the females (SB 371 and 291) and are a subset of the vaginal cytology cycles. Superficial cells began to rise sharply eight days (Day −8) prior to ovulation (Day 0, as defined above as the first day of significant E_1C decline post-peak concentration). At Day −8, the proportion of acidophilic cells surpassed the proportion of basophilic cells in the first chromic shift (arrow 1 in Fig. 9.4a). The increase in acidophilic cells was not closely aligned with the increase in superficial cells. In most cycles, the percentages of acidophilic cells exceeded the proportion of superficial cells until Day −5. Superficial cells then climbed steadily from Day −4 until Day 0. On Day −2, the second chromic shift occurred (arrow 2 in Fig. 9.4a), which was characterised as a decline in acidophilic

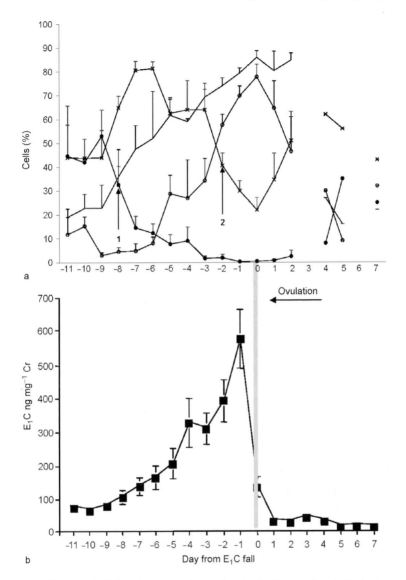

Figure 9.4. (a) Relative proportions of vaginal cell types for eight giant panda cycles (mean ± SEM; four females; •, basophilic; ×, acidophilic; ○, kerantinised; —, superficial. (b) correlated with urinary oestrone sulphate conjugate (E_1C) patterns (mean ± SEM; for a subset of two females and five cycles). Data are standardised to Day 0, defined as ovulation on the basis of the first significant E_1C decline post-peak concentration. Arrows 1 and 2 in panel 'a' indicate days of the first and second chromic shifts, respectively. Vaginal swabs were not obtained on Days +3 and +6.

cells with a concomitant rise in keratinised cells such that the latter predominated. While superficial cells remained elevated, the relative percentages of acidophilic and keratinised cells reversed again on Day +2 as keratinised cells declined rapidly. By Day +5, superficial and keratinised cells had returned to baseline, acidophilic cells were decreasing and basophilic cells were rising.

Concentrations of E_1C in urine rose above 25 ng mg^{-1} Cr on Day −12, gradually climbing until Day −3 when the rate of increase accelerated (see Fig. 9.4b). Peak levels were recorded on Day −1, and the rapid decline on Day 0 was considered the day of ovulation.

Case studies and representative profiles

Figure 9.5 illustrates a representative vaginal cytology profile from an individual giant panda, SB 371, in 1998. The first chromic shift occurred eight days before the second shift. The significance of the chromic shifts had not yet been recognised. However, in this case, the female was artificially inseminated with fresh semen on the day before, and the day of, the second shift, and no pregnancy resulted. Hormonal data (not shown) indicated that E_1C concentrations peaked two days after the second chromic shift. In retrospect, conception failure may be attributed to premature insemination (at least two days before ovulation).

The following year, 1999, AI of this same female with fresh semen resulted in the birth of a single female offspring 135 days later. Figure 9.6 depicts a false (unsustained) second chromic shift on 2 April, seven days after the first shift. On this date, urinary E_1C concentrations were steadily rising to ~150 ng mg^{-1} Cr (data not shown). Another false second chromic shift occurred two days later on 4 April. Urinary E_1C dipped for a single day on 5 April, corresponding to the decrease in keratinised cells, which indicated the false nature of this second chromic shift. The authentic, sustained second shift occurred two days later on 6 April and initiated a four-day peak in keratinised cells, at which time superficial cells peaked. On 8 April, two days after the authentic second chromic shift, urinary E_1C concentrations both peaked at >800 ng mg^{-1} Cr and then declined precipitously to <100 ng mg^{-1} Cr. Artificial insemination was performed for three consecutive days beginning on 9 April. False second shifts were not observed in vaginal cytology profiles for other individual pandas or cycles.

In 2002, SB 291 experienced a first chromic shift on 20 February, but no second shift occurred (Fig. 9.7). Keratinised cells never exceeded

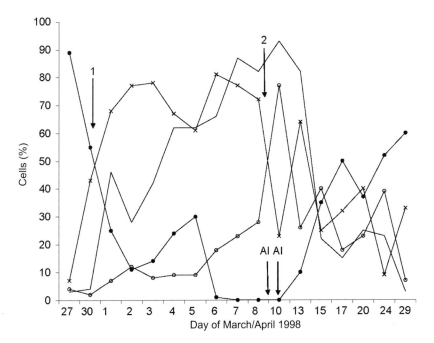

Figure 9.5. Relative proportions of vaginal cell types for SB 371 during a normal oestrus in 1998. Two artificial inseminations were performed (arrows labelled AI) but no cub was born. Days of the two chromic shifts are also indicated with arrows labelled 1 and 2, respectively. Based on the second chromic shift, ovulation was predicted to occur on 10 April. Note that after 8 April vaginal swabs were not obtained daily. •, basophilic; ×, acidophilic; ○, kerantinised; ——, superficial.

42%, and superficial cell counts did not rise above 63%. Urinary E_1C concentrations rose slightly, but failed to peak (data not shown), mimicking the vaginal cytology profile. This anomalous vaginal cytology and E_1C profile may have reflected an irregular or anovulatory oestrus, perhaps related to this nulliparous female's advanced age (17 years old in 2002).

PRIORITIES FOR THE FUTURE

Because peak urinary E_1C values range from ~100 to more than 800 ng mg^{-1} Cr before ovulation (Czekala et al., 2003; see also Chapter 8) and vary significantly among and within individual giant pandas, it is

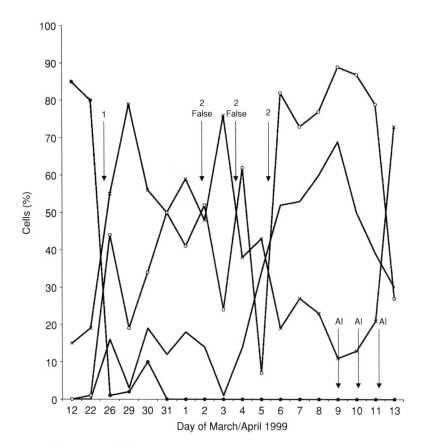

Figure 9.6. Relative proportions of vaginal cell types for SB 371 during a
normal oestrus in 1999 that included three artificial inseminations
(arrows labelled AI) followed by a pregnancy and birth of a living young.
Chromic shifts are also indicated by arrows (1 and 2), including two false
second shifts (labelled 2 false) before the actual shift. Based on the actual
second chromic shift, ovulation was predicted for 9 April. •, basophilic; ×,
acidophilic; ○, kerantinised; ——, superficial.

impossible to predict the onset of the E_1C fall that is indicative of
ovulation. Chromic shifts in vaginal cells represent a rapid, inexpensive
and dependable reflection of oestrogen and, thus, ovarian status. Furthermore, the consistent timing of the second chromic shift is predictive of the E_1C fall. As emphasised repeatedly throughout this book, the
successful breeding and management of the giant panda depend on

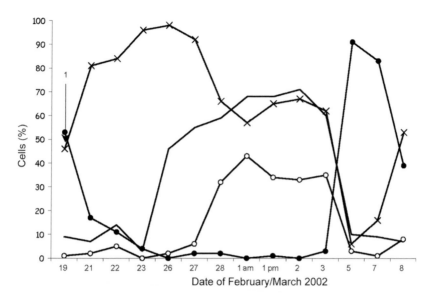

Figure 9.7. Relative proportions of vaginal cell types for SB 291 during an abnormal oestrus in 2002. The first chromic shift is indicated by arrow 1, but the second shift failed to occur. This female was not artificially inseminated. •, basophilic; ×, acidophilic; ○, kerantinised; ——, superficial.

integrating disciplines and techniques. Vaginal cytology is now a proven tool to be added to intensive management strategies and is especially useful in the peri-oestrual interval.

Despite its effectiveness, vaginal cytology has not become a routine technique in most holding facilities, probably because of the need for behavioural conditioning to permit easy, nonstressful collection of vaginal cells. Thus, one priority is to apply this tool to more pandas in more facilities, especially as behavioural conditioning is becoming more common (see Chapter 11). It would be especially interesting to evaluate cytological profiles in females with a history of infertility (particularly in parallel with urinary hormone monitoring and behavioural assessments). This may provide new insights into the aetiology of suboptimal oestrus and the occasional phenomenon of failed ovulation (see Chapter 8).

Due to the value of this technique as a management tool, it would also be useful to explore the precise mechanism that induces nucleated intermediate cells to shift from basophilic to acidophilic in the vaginal smear. No change in the rate or concentration of E_1C appears to correlate with the shift. However, this first chromic shift (easily apparent with modified trichrome Papanicolaou staining) is a definitive precursor to peak oestrus and ovulation in this species. Occurring eight or nine days before ovulation, it is a reliable indicator of impending oestrous events, thereby allowing staff to mobilise resources for a host of events ranging from intensive behavioural monitoring to the need to collect more frequent urine samples for measuring the E_1C profile. No giant panda oestrous cycle examined to date has failed to exhibit a first chromic shift.

When it is impossible to collect daily vaginal smears, the observance of a preponderance of acidophilic cells is a clear signal that the first shift has occurred. Occasionally, acidophilic cells outnumber basophilic cells for one to two days before falling again below a 50% threshold. These false first chromic shifts can be misleading when pandas are not swabbed daily, thus emphasising the need to develop animal handling methods and facilities that allow swabbing as frequently as possible. Swabbing less frequently than once a day can result in misinterpretation of transient changes in vaginal cells. However, daily sampling can virtually always identify an authentic chromic shift that is irreversible and reflective of a forthcoming oestrus and ovulation.

As circulating oestrogens increase near the time of ovulation, a second chromic shift is associated with keratinisation of acidophilic cells one day before the urinary oestrogen peak and two days before the fall in this hormone. Again, false shifts can occur, the best example in our experience being the single, conceptive oestrous cycle in SB 371, a female that exhibited two false, second chromic shifts (see Fig. 9.6). Although this uneven rise in keratinised cells altered the normal interval between the two chromic shifts, the temporal relationship between the second chromic shift and ovulation remained unchanged. Similar to the first chromic shift, there was no clear rate change or consistent threshold of E_1C concentrations that correlated to the second chromic shift. To fully understand the value of vaginal cytology as an index of the physiological dynamics of ovarian activity in the giant panda, further basic studies would be worthwhile. For example, investigating other forms of oestrogens or seeking out alternative hormone ratios might be useful to understand the driving forces behind altered vaginal cytology, including the chromic shifts.

ACKNOWLEDGEMENTS

The authors are grateful for the expert assistance of the San Diego Zoo's giant panda keepers, Kathy Hawk, Dallas Ripsky, Kim Bacon and Susan Euing and the keeper and veterinary staff of Chapultepec Zoo.

REFERENCES

Boue, F., Delhomme, A. and Chaffaux, S. (2000). Reproductive management of silver foxes (*Vulpes vulpes*) in captivity. *Theriogenology*, **53**, 1717–28.

Czekala, N., McGeehan, L., Steinman, K., Li, X. and Gual-Sil, F. (2003). Endocrine monitoring and its application to the management of the giant panda. *Zoo Biology*, **22**, 389–400.

Durrant, B., Czekala, N., Olson, M. *et al.* (2002). Papanicolaou staining of exfoliated vaginal epithelial cells facilitates the prediction of ovulation in the giant panda. *Theriogenology*, **57**, 1855–64.

Feldman, E. C. and Nelson, R. W. (1987). *Canine and Feline Endocrinology and Reproduction*. Philadelphia, PA: W. B. Saunders Co., pp. 399–480.

Hu, J. (1990). Studies on reproductive biology of the giant panda. In *Proceedings of the Second International Symposium on Giant Panda*, ed. S. Asakura and S. Nakagawa. Tokyo: Tokyo Zoological Park Society, pp. 73–6.

Klotchkov, D. V., Trapezov, O. V. and Kharlamova, A. V. (1998). Folliculogenesis, onset of puberty and fecundity of mink (*Mustela vision* Schreb) selectively bred for docility or aggressiveness. *Theriogenology*, **49**, 1545–53.

Mead, R. A., Neirinckx, S. and Czekala, N. M. (1990). Reproductive cycle of the steppe polecat (*Mustela eversmanni*). *Journal of Reproduction and Fertility*, **88**, 353–60.

Mills, J. N., Valli, V. E. and Lumsden, J. H. (1979). Cyclical changes of vaginal cytology in the cat. *Canadian Veterinary Journal*, **20**, 95–101.

Moore, H. D. M., Bush, M., Celma, M. *et al.* (1984). Artificial insemination in the giant panda (*Ailuropoda melanoleuca*). *Journal of Zoology London*, **203**, 269–78.

Onuma, M., Suzuki, M., Uchida, E., Niiyama, M. and Ohtaishi, N. (2002). Annual changes in fecal estradiol-17β concentrations of the sun bear (*Helarctos malayanus*) in Sarawak, Malaysia. *Journal of Veterinary Medical Science*, **64**, 309–13.

Papanicolaou, G. N. (1942). A new procedure for staining vaginal smears. *Science*, **95**, 438–9.

Parakkal, P. F. (1974). Cyclical changes in the vaginal epithelium of the rat seen by scanning electron microscopy. *Anatomical Record*, **178**, 529–37.

Poglayen-Neuwall, I., Durrant, B. S., Swansen, M. L., Williams, R. C. and Barnes, R. A. (1989). Estrous cycle of the tayra, *Eira barbara*. *Zoo Biology*, **8**, 171–7.

Reynolds, D. G. and Beecham, J. J. (1980). Home range activities and reproduction of black bears in west-central Idaho. In *Bears: Their Biology and Management*, ed. C. J. Martinka and K. L. McArthur. Knoxville, TN: International Association for Bear Research and Management, pp. 181–90.

Schaller, G. B., Hu, J., Pan, W. and Zu, J. (1985). *The Giant Pandas of Wolong*. Chicago, IL: University of Chicago Press.

Shutte, A. P. (1967). Canine vaginal cytology. I. Technique and cytological morphology. *Journal of Small Animal Practice*, **8**, 301–6.

Stenson, G. B. (1988). Oestrus and vaginal smear cycle of the river otter, *Lutra canadensis*. *Journal of Reproduction and Fertility*, **83**, 605–10.

Valtonen, M. H., Rajakowski, E. J. and Makela, J. I. (1977). Reproductive features in the female raccoon dog (*Nyctereutes procyonoides*). *Journal of Reproduction and Fertility*, **51**, 17–18.

Williams, E. S., Thorne, E. T., Kwiatkowski, D. R., Lutz, K. and Anderson, S. L. (1992). Comparative vaginal cytology of the estrous cycle of black-footed ferrets (*Mustela nigripes*), Siberian polecats (*M. eversmanni*) and domestic ferrets (*M. putorius furo*). *Journal of Veterinary Diagnostic Investigation*, **4**, 38–44.

Xie, Z. and Gipps, J. (2001). *The 2001 International Studbook for Giant Panda (Ailuropoda melanoleuca)*. Beijing: Chinese Association of Zoological Gardens.

Zylicz, E., Sips, D., Levy, E. and Peters, H. (1967). The vaginal smear in mice: a correlation of smear type and oocyte number. *Acta Cytologica*, **11**, 483–5.

10

Parentage assessment among captive giant pandas in China

VICTOR A. DAVID*, SHAN SUN*, ZHIHE ZHANG, FUJUN SHEN, GUIQUAN ZHANG, HEMIN ZHANG, ZHONG XIE, YA-PING ZHANG, OLIVER A. RYDER, SUSIE ELLIS, DAVID E. WILDT, ANJU ZHANG, STEPHEN J. O'BRIEN

INTRODUCTION

While many recent advances have been made in the breeding of giant pandas *ex situ*, historically this species has never reproduced well in captivity. Sexual incompatibility, health problems, low fecundity and a juvenile mortality rate in excess of 70% have contributed to low reproductive success (O'Brien & Knight, 1987; O'Brien *et al.*, 1994; Peng *et al.*, 2001a,b). Wild- and captive-born giant pandas, particularly those captured at a young age, traditionally had difficulty producing off-spring in captivity upon becoming adults (Lu & Kemf, 2001). As a result, the *ex-situ* giant panda population has not been self-sustaining and, until recently, its growth has relied on introducing animals captured from nature. In some cases, this included individuals that appeared ill (rescues) or cubs that were believed to be neglected or abandoned by their mothers. Later field studies, however, revealed that females often leave cubs alone for four to eight hours while foraging, and in one documented case for 52 hours (Lu *et al.*, 1994). Recently, China has

* These authors contributed equivalently to this work.

The content of this publication does not necessarily reflect the views or policies of the Department of Health and Human Services, nor does mention of trade names, commercial products or organisations imply endorsement by the US Government.

Giant Pandas: Biology, Veterinary Medicine and Management, ed. David E. Wildt, Anju Zhang, Hemin Zhang, Donald L. Janssen and Susie Ellis. Published by Cambridge University Press. © Cambridge University Press 2006.

placed a general moratorium on capturing wild giant pandas for captive breeding (Lu & Kemf, 2001), a move that forces the breeding community to develop a self-sustaining population.

The goal, however, is not only ensuring demographic self-sustainability but also the maintenance of genetic diversity. The deleterious effects of inbreeding are well recognised (O'Brien, 1994a; Frankham, 1995; Hedrick & Kalinowski, 2001; Frankham *et al.*, 2002). For an outstanding example, one need look no further than the Florida panther, perhaps the flagship species for illustrating the dramatic biological consequences of severe inbreeding (Roelke *et al.*, 1993; O'Brien, 1994b). In the case of this felid subspecies, a small remnant population of pumas underwent a population crash followed by incestuous matings that resulted in a high incidence of malformed sperm, missing or undescended testicles, congenital cardiac abnormalities and high microbial parasite disease loads. There is no doubt that inbreeding depression occurs in poorly managed wildlife populations and must be avoided.

For the giant panda, there are multiple challenges to genetic management (see Chapter 21). For example, there are bureaucratic obstacles to transferring animals between institutions to allow the easier mixing of diverse genes. Additionally, the breeding process itself presents some unusual complications because sexual incompatibility among designated pairs can be high. Historically, a given Chinese giant panda breeding facility has been likely to maintain only one to three unrelated males that are capable of naturally mating. Other males in these collections fail to mate, but generally are available as semen donors for artificial insemination (AI). To maximise both reproductive success and genetic representation, a common management practice is to arrange copulation with one or more competent males during the short two- to three-day oestrus. This then is followed immediately by AI with semen from a male (or males) incapable of natural mating. The result has been that some females have been naturally mated with up to three different males and artificially inseminated with semen from up to another three different males. Thus, paternity has been unknown for the majority of giant panda cubs born in captivity. To allow accurate genetic management and to avoid future inbreeding, it is imperative to verify the paternity of all animals born in captivity.

The first case of giant panda paternity was resolved 20 years ago. At that time, Ling Ling, Studbook 112 (SB 112), the female panda at the Smithsonian's National Zoo in Washington, was the subject of much media attention. For years, Ling Ling had been sexually incompatible with her resident mate Hsing Hsing (SB 121), much to the dismay of the American public. After years of futile courtship, there was a successful mating in 1983. Nonetheless, to maximise the chances of a pregnancy, Ling Ling was also artificially inseminated with semen from Chia Chia (SB 141), the male panda living at London Zoo. A cub was produced which unfortunately died soon after birth. By typing electrophoretic protein polymorphisms, O'Brien and colleagues (1984) determined that Hsing Hsing (and not Chia Chia) indeed sired the cub.

More recently, paternity issues have been addressed using DNA fingerprinting probes (Fang *et al.*, 1997b) and a panel of seven panda-specific microsatellites (Zhang *et al.*, 1994; Ding *et al.*, 2000). Microsatellites represent a class of abundant and highly polymorphic genetic markers, randomly dispersed among the genomes of vertebrate species that are easily isolated and assayed. They are incorporated in genetic maps as tools for locating genes associated with heritable genetic disorders, for human forensic analysis, for paternity assessment and for conservation genetics (Bruford & Wayne, 1993; Goldstein & Schlotterer, 1999; Lu *et al.*, 2001; Driscoll *et al.*, 2002). In this report, we apply multilocus microsatellite assessment to 50 captive giant pandas in the two largest giant panda breeding centres: the Chengdu Research Base of Giant Panda Breeding and the China Conservation and Research Centre for the Giant Panda, Wolong Nature Reserve. We demonstrate parentage of 50 cubs and present a validated pedigree of giant pandas produced in these facilities from 1990 to 2000.

MATERIALS AND METHODS

This project was based on significant collaboration amongst research laboratories, zoos, breeding centres and conservation organisations in China and the USA. As pointed out earlier in this volume (see Chapter 2), early decisions were made to avoid the analysis of raw materials outside of China. This was important both to build in-country capacity as well as to avoid increasingly difficult export/import issues (both CITES – Convention on International Trade in Endangered Species of Wild Flora and Fauna – and governmental) associated with the international transfer of biomaterials. Therefore, our first step for paternity analysis was to develop a genotyping facility in the Laboratory of Genetics and Reproductive Studies at the Chengdu Research Base of

Giant Panda Breeding (Fig. 10.1). Since a modern genetics laboratory already existed at the Chengdu Research Base, creating the genotyping centre only required transferring the necessary additional tools and technology. A team from the Laboratory of Genomic Diversity (National Institutes of Health, USA) travelled to Chengdu. An Applied Biosystems Model 373 DNA sequencer (generously donated by the manufacturer) was installed and tested. A training seminar (focused on instrument operation and data collection/analysis) was conducted for scientists and technicians from the Chengdu Research Base as well as representatives from Beijing Zoo and the Key Laboratory of Cellular and Molecular Evolution (Kunming Institute of Zoology) (Fig. 10.2). Laboratory set-up and technology transfer were sufficient to allow this new facility to be used long into the future for resolving paternity issues relative to giant pandas living *ex situ* as well as *in situ*.

DNA was extracted from 71 individual giant panda samples. In some cases, samples were collected during the CBSG Biomedical Survey (see Chapter 3); in others, samples were collected specifically to be included here. DNA was isolated as previously described for 40 blood samples (Lu *et al.*, 2001), 30 buccal swabs (Fig. 10.3) and one faecal

Figure 10.1. A modern state-of-the-art giant panda microsatellite genotyping centre was established in the Laboratory of Genetics and Reproductive Studies at the Chengdu Research Base of Giant Panda Breeding.

Figure 10.2. A training session for performing fluorescence-based microsatellite genotyping was held at the Laboratory of Genetics and Reproductive Studies at the Chengdu Research Base of Giant Panda Breeding. Shan Sun and Victor A. David led the discussions attended by scientists from Chengdu, Beijing Zoo and the Kunming Institute of Zoology.

sample using a commercially available kit (Qiagen, Germantown, MD). Animal names and studbook numbers were obtained from the *Giant Panda International Studbook* (Xie & Gipps, 2001). Mating information for each animal was obtained from the breeding records of the respective institutions.

Polymerase chain reaction (PCR) amplifications were performed as previously described (Lu *et al.*, 2001) with the following changes. To reduce 'plus A' addition problems (Brownstein *et al.*, 1996), the DNA sequence GTGTCTT was added to the 5′ end of the unlabelled primer for all primers except AME-µ28A and −µ28B. This addition resulted in PCR products that were eight base pairs larger than previously described. The PCR conditions were modified to a touchdown PCR procedure (95°C, 10 minutes; two cycles of 95°C for 15 seconds, annealing temperature 30 seconds, 72°C for one minute, at annealing temperatures of 60, 58, 56, 54, 52 and 30 cycles at annealing temperature of 50°C, 72°C for five minutes), and Taq Gold DNA polymerase was used (Applied Biosystems, Foster City, CA). The PCR products were electrophoresed on an Applied

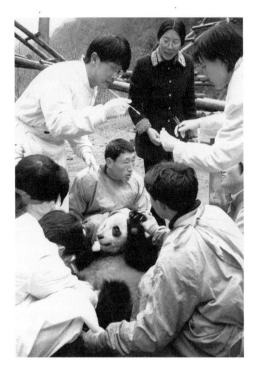

Figure 10.3. Buccal cells were collected from giant pandas at the China Conservation and Research Centre for the Giant Panda (Wolong Nature Reserve) by brushing the inside of the animal's cheek with a soft brush. DNA was extracted from the samples in the laboratory at the Chengdu Research Base. In the photograph, the giant panda is being distracted with a piece of apple so the sample could be collected.

Biosystems Model 373 automated DNA sequencer, and the data were analysed using Genescan Version 1.2.2-1 and Genotyper Version 2.5 software (Applied Biosystems).

The computer program Cervus (available at http://helios.bto.ed.ac. uk/evolgen/) was used to evaluate potential paternity based upon a maximum likelihood assessment of the various potential sires' composite microsatellite genotypes (Marshall *et al.*, 1998). The paternity questions at the two breeding centres were evaluated independently. In each case, all animals for which genotype data were available at that site were used to generate a reference population for allelic frequencies. These values, in turn, were used by the Cervus program to determine match statistics. Offspring were checked against all available sires and dams for possible parentage. The calculations presumed a maximum of 1% error in genotyping using 10 000 simulation cycles. If one parent was assigned as known, the program inferred the other parent's alleles and then calculated probability of non-exclusion for the unknown parent based on the allelic frequencies of the microsatellites in the reference population. This value indicated the probability that an animal that was not the true parent would fail to be excluded. The exclusion probability for the sire was calculated after assigning the known dam as the mother. The value for the dam was calculated by fixing the putative sire as the father. In some cases, the non-exclusion value for the dam was absent because there was not a DNA sample for the known dam. Pedigrees were constructed using PedDraw version 5 (available at http://www.sfbr.org/sfbr/public/software/pedraw/peddrw. html).

Genetic relatedness was estimated from the genotype data using the program Relatedness version 5.0.8 (available at http://gsoft.smu.edu/ GSoft.html; Queller & Goodnight, 1989). A subset of relatedness estimates employed 44 animals (25 from Wolong and 19 from Chengdu) for which there were no missing data at ten microsatellite loci (Ame-μ10, 11, 13, 14, 15, 22, 25, 26, 27 and 28B).

PARENTAGE RESULTS

An example of a paternity assessment

We addressed parentage among 50 giant pandas in Chinese breeding centres using a panel of 17 previously described polymorphic microsatellite loci (Lu *et al.*, 2001). A heterozygosity table was generated based on the genotypes obtained for 15 wild-born giant pandas in this study (Table 10.1). The number of alleles ranged from two to ten with an average of 5.4, and the heterozygosity ranged from 0.07 to 0.84 with an average of 0.63. In this study, Ame-μ19 was inferred to reside on the X chromosome based upon the presence of two Ame-μ19 microsatellite alleles in females and one allele in all males due to their carrying a hemizygous X chromosome. Ame-μ70 has an unusually large size range with alleles ranging from 190 to 202 base pairs and 274 to 296 base pairs with a 72 base pair gap, probably due to an insertion/deletion event in the flanking sequence of the locus.

To illustrate the interpretation process for microsatellite genotypes, Table 10.2 presents a summary of paternity assessment for Tian

Table 10.1. *Allelic diversity of 17 giant panda microsatellites*

Microsatellite[a]	No. of alleles	Estimated heterozygosity		Allele range (base pairs)
Ame-μ5	4	0.42[b]	0.57[c]	149–157[d]
Ame-μ10	6	0.61	0.78	145–161
Ame-μ11	5	0.44	0.73	238–246
Ame-μ13	7	0.51	0.69	151–179
Ame-μ14	5	0.54	0.79	146–154
Ame-μ15	4	0.09	0.07	129–139
Ame-μ16	5	0.38	0.56	134–147
Ame-μ19 (X-linked)	4	0.33	0.74	160–168
Ame-μ21	8	0.66	0.84	164–178
Ame-μ22	2	0.27	0.30	134–136
Ame-μ24	5	0.35	0.59	258–268
Ame-μ25	7	0.49	0.76	228–246
Ame-μ26	3	0.46	0.67	122–128
Ame-μ27	6	0.29	0.76	140–156
Ame-μ28A	3	0.32	0.67	118–132
Ame-μ28B	7	0.36	0.43	169–190
Ame-μ70	10	0.66	0.79	190–202, 274–296
Average	5.4	0.42	0.63	

[a] Microsatellite locus name and primer sequence after Lu *et al.*, 2001; [b] From Lu *et al.*, 2001; [c] This study; [d] These data are based on results from 15 wild-born giant pandas. The unlabelled primers for all microsatellites except AME-μ28A and AME-μ 28B have the sequence GTGTCTT attached to the 5′ end; thus, the PCR products are larger than previously reported (Lu *et al.*, 2001).

Tian (SB 458), the male giant panda on loan to the Smithsonian's National Zoological Park in Washington, DC. Tian Tian's dam is Yong Ba (SB 397). His sire is uncertain because Yong Ba was naturally mated to Pan Pan (SB 308) and artificially inseminated with semen from Lin Nan (SB 298) and Lo Lo (SB 305). Microsatellite markers are inherited in a Mendelian fashion so, barring a mutation, Tian Tian's genotype must be derived from a combination of alleles present in Yong Ba and his sire. Lin Nan is excluded as the sire because he does not share any allele with Tian Tian at Ame-μ13, Ame-μ27 and Ame-μ70. (This failure of allele sharing between an offspring and a possible sire is termed an Offspring [Off]-Sire loci mismatch in Table 10.3, column 12; see below.) Lin Nan

Table 10.2. Microsatellite genotype data for Tian Tian (SB 458), his dam and potential sires. Numbers represent allele size in base pairs of PCR amplification products for 15 microsatellites[a]

Name	Sire check	AME-μ05	AME-μ10	AME-μ11	AME-μ13	AME-μ14	AME-μ15	AME-μ19	AME-μ21	AME-μ22	AME-μ24	AME-μ25	AME-μ26	AME-μ27	AME-μ28b	AME-μ70
Pan Pan	NM[b] SB 308	149 149	147 147	238 242	173 171	148 148	129 129	160 0	172 172	134 134	260 260	242 242	124 128	154 154	171 171	194 194
Lin Nan	AI[c] SB 298	155 155	147 153	242 244	151 **151**[d]	**152**[e] 154	129 129	162 0	172 178	134 134	260 264	242 244	124 128	**140 140**	171 171	**190 198**
Lo Lo	AI SB 305	149 153	145 147	242 242	173 173	148 150	129 129	162 0	168 172	134 134	260 260	236 242	122 128	**140** 152	171 171	**198 280**
Dam																
Yong Ba	SB 397	0[f] 0	147 151	240 246	167 171	152 152	129 129	160 160	170 172	134 134	0 0	0 0	0 0	140 156	171 171	0 0
Offspring																
Tian Tian	SB 458	149 155	147 147	242 246	171 173	148 152	129 129	160 0	170 172	0 0	260 260	0 0	128 128	154 156	171 171	194 282

[a] Yong Ba (SB 397) is the known dam of Tian Tian (SB 458). Yong Ba was naturally mated to SB 308 and artificially inseminated with semen from SB 298 and SB 305 so the paternity of Tian Tian is uncertain. Both SB 298 and SB 305 are excluded as potential sires at multiple loci. Tian Tian resulted from the natural mating with Pan Pan (SB 308). [b] NM, natural mating. [c] AI, artificial insemination. [d] Bold, Excluded as sire. [e] ■, Excluded with selected dam Yong Ba (SB 458). [f] O, no product detected.

Table 10.3. Microsatellite loci exclusions and probabilities for dams and potential sires[a]

Studbook number	Off loci typed	Dam	Dam loci typed	OffDam loci compared	OffDam loci mismatch	Sire non-exclusion probability	Dam non-exclusion probability	Sire	Sire loci typed	OffSire loci compared	OffSire loci mismatch	OffSire-Dam loci compared	OffSire-Dam loci mismatch	Consult all sires: next most likely sire — OffSire loci compared	OffSire loci mismatch	OffSire-Dam loci mismatch	Based upon breeding records: closest true possible sire — OffSire loci compared	OffSire loci mismatch	OffSire-Dam loci mismatch	Date of Birth
Chengdu																				
362	17	278	16	16	0	1.50×10^{-4}	5.11×10^{-6}	202	17	17	0	16	0	17	6	5	(343)13	8	8	8/1990
363	17	278	16	16	0	1.89×10^{-7}	4.77×10^{-5}	202	17	16	0	15	0	12	6	6	(343)12	6	6	8/1990
386	17	312	17	17	0	2.27×10^{-9}	9.78×10^{-5}	343	13	13	0	13	0	17	7	10	(298)17	8	11	7/1992
387	17	278	16	16	0	9.67×10^{-6}	2.95×10^{-5}	202	17	17	0	16	0	6	2	2	(34)36	2	2	9/1992
400	7	278	16	7	0	1.01×10^{-2}	2.17×10^{-2}	342	17	7	0	7	0	7	0	0(202)[b]	(298)7	2	2	9/1993
401	17	278	16	16	0	5.55×10^{-8}	4.14×10^{-5}	342	17	17	0	16	0	6	5	5	(298)17	6	13	9/1993
407	16	312	17	16	0	1.88×10^{-3}	3.42×10^{-6}	342	17	16	0	16	0	16	1	1	(342)16	4	9	8/1994
408	17	312	16	16	0	2.71×10^{-4}	6.38×10^{-6}	298	17	17	0	16	0	6	1	3	(34)36	1	3	8/1994
425	17	314	17	17	0	1.34×10^{-3}	6.84×10^{-5}	298	17	17	0	17	0	6	1	2	(34)36	1	2	8/1995
428	17	278	16	16	0	7.65×10^{-7}	4.60×10^{-4}	342	17	17	0	16	0	17	2	5	(298)17	2	10	8/1995
453	10	278	16	9	0	3.29×10^{-2}	1.40×10^{-2}	287	17	10	0	9	0	10	1	1	(298)10	1	1	9/1997
454	13	278	16	13	0	1.75×10^{-3}	6.53×10^{-4}	287	17	13	0	13	0	13	1	2	(298)13	1	2	9/1997

Sample	Date																						
490	9/1999	5	3	(386)15	2	2	6	0	15	0	15	2.21×10^{-2}	3.87×10^{-4}	287	15	17	0	15	0	15	17	408	15
491	9/1999	7	3	(386)15	2	2	6	0	15	0	15	4.17×10^{-3}	1.21×10^{-4}	287	15	17	0	15	0	15	17	408	15
493	9/1999	8	8	(386)15	3	2	15	0	15	0	15	1.60×10^{-3}	3.21×10^{-4}	287	15	17	0	15	0	15	17	362	15
494	9/1999	6	6	(386)14	3	0	14	0	14	0	14	3.84×10^{-2}	2.94×10^{-3}	287	14	17	0	14	0	14	17	362	14
515	8/2000	5	6	(386)14	2	1	14	0	13	0	14	9.34×10^{-2}	1.43×10^{-2}	287	14	17	0	13	0	13	16	278	14
519	9/2000	9	5	(386)15	5	2	6	0	15	0	15	1.28×10^{-3}	2.04×10^{-4}	287	15	17	0	15	0	15	17	314	15
520	9/2000	10	4	(386)15	6	4	15	0	15	0	15	6.64×10^{-4}	7.12×10^{-6}	287	15	17	0	15	0	15	17	314	15
522	9/2000				2	1	6	0	15	0	15	1.42×10^{-6}	2.24×10^{-3}	386	15	17	0	15	0	15	17	297	15
523	9/2000				1	1	6	0	15	0	15	2.51×10^{-5}	3.02×10^{-5}	386	15	17	0	15	0	15	17	297	15
Wolong																							
371	9/1991	4	5	(298)15	1	0	12	0	13	0	15	7.91×10^{-3}	1.04×10^{-5}	308	15	15	0	13	0	13	13	358	15
394	9/1992	3	3	(298)11	0(399)	0	9	0	11	0	12	3.33×10^{-2}	6.37×10^{-4}	308	12	15	0	11	0	11	13	358	11
399	9/1993	2	1	(329)12	0(394)	0	9	0	11	0	12	2.71×10^{-2}	1.43×10^{-3}	308	12	15	0	11	0	11	13	358	12
404	11/1993		1	(308)10		0(305)	10	0	11	0	10	ND	4.00×10^{-5}	329	10	15	0	0	0	0	0	230	10
413	10/1993					0(298)	12	0	0	0	12	ND	1.16×10^{-4}	308	12	15	0	0	0	0	0	230	12
432	8/1995				0(399)	0	9	0	8	0	10	7.03×10^{-2}	6.79×10^{-5}	308	10	15	0	8	0	8	10	397	10
458	8/1997	1	2	(305)13	0(433)	0	6	0	9	0	13	1.99×10^{-2}	4.76×10^{-4}	308	13	15	0	9	0	9	10	397	13
460	10/1997	3	3	(298)9	1	1	8	0	9	0	9	5.69×10^{-2}	1.40×10^{-4}	308	9	15	0	9	0	9	13	358	9
473	7/1998				0	0	0	0	4	0	4	2.70×10^{-2}	4.21×10^{-2}	298	4	15	0	4	0	4	14	444	4
474	8/1998				2	2	11	0	10	0	11	1.67×10^{-4}	3.74×10^{-4}	298	11	15	0	10	0	10	12	382	11
476	8/1998	4	3	(298)10	1	1	8	0	8	0	10	1.05×10^{-2}	2.56×10^{-3}	308	10	15	0	8	0	8	10	397	10
477	9/1998	3	2	(298)7	2	2	7	0	6	0	6	1.34×10^{-2}	2.77×10^{-4}	394	11	11	0	7	0	7	13	446	7

Table 10.3. (cont.)

														Consult all sires: next most likely sire			Based upon breeding records: closest true possible sire			
Studbook number	Off loci typed	Dam	Dam loci typed	Off-Dam loci compared	Off-Dam loci mismatch	Sire non-exclusion probability	Dam non-exclusion probability	Sire	Sire loci typed	Off-Sire loci compared	Off-Sire loci mismatch	Off-Sire-Dam loci compared	Off-Sire-Dam loci mismatch	Off-Sire loci compared	Off-Sire loci mismatch	Off-Sire-Dam loci mismatch	Off-Sire loci compared	Off-Sire loci mismatch	Off-Sire-Dam loci mismatch	Date of Birth
478	15	418		0	0	ND	3.10×10^{-5}	394	11	11	0			12	2		(298)15	4		8/1999
479	14	418		0	0	ND	7.82×10^{-4}	394	11	11	0			11	1		(298)14	4		8/1999
485	15	397	10	10	0	1.64×10^{-3}	4.15×10^{-6}	308	15	15	0	10	0	14	3	2	(394)11	2	2	8/1999
492	14	444	14	14	0	1.03×10^{-2}	3.22×10^{-4}	308	15	14	0	14	0	13	1	1	(329)14	3	2	9/1999
495	14	446	13	13	0	5.44×10^{-4}	2.22×10^{-5}	329	15	14	1	13	1	15	1	2	(298)14	4	5	9/1999
496	15	446	13	13	0	6.07×10^{-3}	2.68×10^{-7}	308	15	15	0	13	1	14	1	1	(329)15	2	3	9/2000
509	14	418		0	0	ND	1.98×10^{-7}	329	15	14	1			11	2		(357)13	4		8/2000
510	13	418		0	0	ND	1.06×10^{-7}	329	15	13	1			13	2		(357)12	5		8/2000
511	15	414		0	0	ND	5.91×10^{-5}	394	11	11	0			12	2		(399)12	2		8/2000
512	15	432	10	10	0	1.50×10^{-4}	2.79×10^{-4}	305	15	15	0	10	0	15	0	2				8/2000
513	14	382	12	11	0	2.31×10^{-3}	7.81×10^{-6}	394	11	11	0	10	0	11	1	1	(305)14	4	3	8/2000
514	12	382	12	9	0	4.97×10^{-4}	4.54×10^{-4}	394	9	9	0	8	0	9	1	1	(305)12	3	3	8/2000
516	15	374		0	0	ND	2.81×10^{-6}	308	15	15	0			12	0(413)					8/2000

518	15	404	10	10	0	5.28×10^{-4}	1.78×10^{-4}	394	11	11	0	10	0	12	0	0(399)	(357)14	3	8/2000
524	15	397	10	10	0	2.70×10^{-3}	4.67×10^{-5}	394	11	11	0	9	0	12	0	1	(305)15	2	9/2000
525	15	397	10	10	0	6.95×10^{-3}	2.31×10^{-5}	394	11	11	0	9	0	12	0	1	(305)15	2	9/2000
526	15	446	13	13	0	2.53×10^{-4}	5.04×10^{-5}	329	15	15	0	13	0	15	1	2	(357)14	6	9/2000

[a] In all cases the known dam was confirmed by the microsatellite data. The putative sire was always one of the males used in the natural mating or AI. Giant pandas SB 495 and SB 509 have one mismatch with the putative sire, and this is probably due to the presence of a null allele for locus Ame-µ25 (see text). Exclusion data are shown for all male pandas at the breeding station and for only those sires deemed possible from the breeding records; [b] Next most likely sires shown in parenthesis; [c] ND, not determined, no DNA available for dam.

(SB 298) is further excluded at Ame-µ14, because his two alleles cannot be combined with the alleles of Yong Ba to generate the Tian Tian genotype. (This exclusion is labelled in Table 10.3, column 14 under 'Off-Sire-Dam mismatch') Similarly, Lo Lo is excluded as a sire at loci Ame-µ27 and Ame-µ70 due to Off-Sire loci mismatches. Pan Pan matches as the sire and fits with the dam at all loci tested; he is implicated as Tian Tian's sire. We conclude therefore that the natural mating by Pan Pan (SB 308) to Yong Ba (SB 397) produced the offspring Tian Tian (SB 458).

Validation of the genotyping results with blinded samples

To validate parentage analysis, 17 cubs born at the China Conservation and Research Centre for Giant Pandas (Wolong) in 1999 and 2000 were tested in a blinded fashion. Buccal swabs were obtained from each animal and given a code number. The scientists performing the DNA isolation and parentage testing were not given any information concerning the animals' possible parents. All breeding age males and females at the Wolong facility were checked as possible parents, although DNA samples were not available for some dams. Correct maternal assignments were made for every cub when the dam's DNA was available. For cubs in which the mother's DNA was absent, all available and tested females were excluded as potential dams. Four of the cubs were correctly inferred to be females since they were heterozygous for microsatellite Ame-µ19. Five cubs were correctly inferred to be males based on the fact that they did not express an X chromosome allele from their sire at locus Ame-µ19. In the case of Long Shen's (SB 518) paternity, two full brothers (SB 394 and SB 399) were both possible sires, and in the case of Long Xin (SB 516), a father/son pair were found to be possible sires (SB 308/413). Inspection of the breeding records made it possible to exclude all but a single sire in both of these instances (see Table 10.3). In all 17 cases, the implicated paternity from the blind analysis was consistent with one of the named possible sires based upon natural mating and AI.

Addressing unknown paternities

Tables 10.3 and 10.4 list the maternity/paternity assignments plus the parameters of microsatellite genotype allele matching for 50 cubs born at the Chengdu and Wolong facilities. For each paternity queried, the

Table 10.4. *Microsatellite confirmed parentage at the Chengdu and Wolong breeding facilities*

Birthplace	Studbook number	Name	Sex	Current location	Sire	Dam	Date of birth	Potential sires in order of mating/ insemination (confirmed sire in bold)		
Chengdu	362	Ya Ya	F	Chengdu	202	278	1990	**NM 202**[a]	AI 201[b]	AI 343
	363	Jun Jun	M	deceased	202	278	1990	**NM 202**	AI 201	AI 343
	386	Kebi	M	Chengdu	343	312	1992	**AI 343**	AI298	
	387	Li Li	F	Chengdu	202	278	1992	AI 343	**NM 202**	
	400	Jian Jian	M	Wuhan	342	278	1993	**AI 342**	AI 298	
	401	Eryatou	F	Chengdu	342	278	1993	**AI 342**	AI 298	
	407	Zhiqiangs	F	Chengdu	298	312	1994	AI 342	**AI 298**	
	408	Mei Mei	F	Wakayama	298	312	1994	AI 342	**AI 298**	
	425	Jiao Zi	F	Chengdu	298	314	1995	**AI 298**		
	428	Xiaoyatou	F	Chengdu	342	278	1995	**AI342**	AI 298	
	453	Da Shuang	F	Chengdu	287	278	1997	**NM 287**	AI 298	AI 231
	454	Xiao Shuan	M	Chengdu	287	278	1997	**NM 287**	AI 298	AI 231
	490	Qi Zhen	F	Chengdu	287	408	1999	**AI 287**	AI 306	NM 287
	491	Yuan Yuan	F	Chengdu	287	408	1999	**AI 287**	AI 306	NM 287
	493	Yalaoda	F	Chongqing	287	362	1999	AI 386	**NM 287**	
	494	Yalaoer	F	Chengdu	287	362	1999	AI 386	**NM 287**	
	515	Liang Liang	M	Chengdu	287	278	2000	**NM 287**	AI 386	
	519	Bingxin	M	Chengdu	287	314	2000	**NM 287**	AI 287	AI 386

Table 10.4. (cont.)

Birthplace	Studbook number	Name	Sex	Current location	Sire	Dam	Date of birth	Potential sires in order of mating/ insemination (confirmed sire in bold)		
	520	Bingdian	M	Chengdu	287	314	2000	**NM 287**	AI 287	AI 386
	522	Chenggong	F	Chengdu	386	297	2000	**AI 386**		
	523	Chengji	F	Chengdu	386	297	2000	**AI 386**		
Wolong	371	Bai Yun	F	San Diego Zoo	308	358	1991	**NM 308**	AI 298	
	394	Dadi	M	Wolong	308	358	1992	**NM 308**	AI 298	
	399	Xi Meng	M	Wolong	308	358	1993	**NM 308**	NM329	
	404	Yue Yue	F	Wolong	329	230	1993	NM 308	**NM 329**	
	413	Di Di	M	Wolong	308	230	1994	**NM 308**		
	432	Fei Fei	F	Wolong	308	397	1995	**NM 308**	AI 298	
	458	Tian Tian	M	National Zoo	308	397	1997	**NM 308**	AI 305	AI 298
	460	Ding Ding	M	Louguanta	308	358	1997	**NM 308**	AI 298	
	473	Mei Xiang	F	National Zoo	298	444	1998	**AI 298**		
	474	You You	F	Wolong	298	382	1998	**AI 298**		
	476	Xi Xi	F	Wolong	308	397	1998	**NM 308**	AI 298	
	477	Gongzhu	F	Wolong	394	446	1998	AI 298	**AI 394**	
	478	Xiu Xiu	F	deceased	394	418	1999	**NM 394**	AI 298	
	479	Qing Qing	M	Wolong	394	418	1999	**NM 394**	AI 298	
	485	Guo Qing	M	Wolong	308	397	1999	**NM 308**	NM 394	NM 329 AI 298

492	Peng Peng	M	Wolong	308	444	1999	NM 308	AI 329		
495	Ye Ye	F	Wolong	329	446	1999	NM 329	NM 308	AI 298	
496	Gu Gu	M	Wolong	308	446	1999	**NM 329**	**NM 308**	AI 298	
509	Zhu Zhu	F	Wolong	329	418	2000	NM 329	NM 308	AI 357	
510	Chuangchuang	M	Wolong	329	418	2000	NM 329	NM 308	AI 357	
511	Xi Mei	F	Wolong	394	414	2000	NM 394	NM 399	AI 357	
512	Le Sheng	F	Wolong	305	432	2000	AI 305			
513	Liang Liang	M	Wolong	394	382	2000	NM 394	AI 305		
514	Yang Yang	F	Wolong	394	382	2000	NM 394	AI 305		
516	Long Xin	F	Wolong	308	374	2000	NM 308	AI 308		
518	Long Shen	M	Wolong	394	404	2000	NM 394	AI 357		
524	Long Ten	M	Wolong	394	397	2000	NM 394	NM 308	NM 329	AI 305
525	Long Fei	M	Wolong	394	397	2000	NM 394	NM 308	NM 329	AI 305
526	Long Hui	M	Wolong	329	446	2000	NM 329	AI 357		

[a] NM, natural mating; [b] AI, Artificial insemination.

known dam was shown to have zero Off-Dam mismatches and, thus, was concordant with maternity. The known dam was then assigned and all resident males at the centre (regardless of age) were queried for possible paternity. Table 10.3 lists the mismatch results for the putative sire, for the next most likely male at the breeding centre and for the next most likely sire based on inspection of only those animals that were naturally mated or supplied semen for AI of the dam. In every case, the implicated dam was naturally mated to (or artificially inseminated with semen from) the putative sire.

Of 50 parentage assessments, there were only three cases with a single locus exclusion for the putative sire. For SB 495 and SB 509, both the putative sire (SB 329) and the offspring have a different single (homozygous) allele for Ame-μ25, resulting in an exclusion (i.e. they inherited an allele from the dam but did not show an allele from the sire). The absence of a paternal allele can be explained by a polymorphic nucleotide site within the panda's homologous Ame-μ25 primer sequence that abrogates PCR amplification due to a primer sequence mismatch. The presence of such 'null' alleles has been previously recognised (Callen *et al.*, 1993), and we have observed them in the genotyping of a large multigenerational pedigree during construction of a microsatellite-based genetic linkage map of the domestic cat (Menotti-Raymond *et al.*, 1999). Finally, SB 495 and SB 496 are twins resulting from the natural mating of sires SB 329 and SB 308 to the dam SB 446, who also was artificially inseminated with semen from SB 298. The putative sire of SB 495, as discussed above, is SB 329, and the putative sire of her twin SB 496 is SB 308. There is, however, one Off-Sire-Dam mismatch for SB 496 with SB 308 at locus Ame-μ25. This is not consistent with the presence of a null allele because SB 496 has a unique allele not present in either parent. It is possible that the offspring allele is a new mutation or a genotyping error.

There were nine cases where more than a single male was listed as a possible sire in the blinded analysis of all males at the centre. In three cases, no genotype data existed for the dam, thereby reducing the exclusion power. In five cases, two included sires were close relatives, and the last situation involved microsatellite amplification failure. In all cases of uncertain paternity in the blind analysis, it was possible to exclude all but a single male when checking against the potential sires based on breeding records.

Insufficient genotypes to assign paternity were obtained for animals SB 388, SB 433, SB 434, SB437, SB439 and SB455 and, thus, were excluded from Tables 10.3 and 10.4. Four offspring were typed that were included in a previous microsatellite paternity study (Zhang *et al.*, 1994). The results between the two independent laboratories agreed for off-spring SB 394 and SB 399. Zhang *et al.* (1994) excluded both potential sires for twins SB 387 and SB 388 and suggested that one of the sire samples (SB 202) was misidentified. In contrast, our results implicated SB 202 as the sire of SB 387. Also, Zhang and colleagues proposed SB 201 as the sire of SB 386 based on exclusion of SB 343. However, breeding records at the Chengdu Research Base listed SB 343 and SB 298 as potential sires, and our data confirmed that SB 343 was the sire of SB 386.

Accessing paternity: natural mating versus artificial insemination

Fifty paternity questions were addressed with results presented in Table 10.4 and Figures 10.4 and 10.5. Of the 50 paternities, seven were not in question since offspring resulted from either AI with semen from a single sire (SB 425, SB 522, SB 523, SB 473, SB 474 and SB 512) or natural mating plus AI using a single sire (SB 516). In these confirmation cases, the presumptive sire was sustained in every case.

There were 13 cases where offspring resulted from only using AI (see Table 10.4). Six offspring were born as a result of AI with semen from a single sire and seven from multiple sires. Clearly, AI is a viable procedure, and recent evidence indicates nearly identical pregnancy/birth success rates when using natural mating versus AI with fresh semen (Huang *et al.*, 2001; see Chapter 20). Four offspring were pro-duced after a single male was used for both natural mating and AI, and then a second male was used for AI (SB 490, SB 491, SB 519, SB 520; see Table 10.4). In these four cases, the male used for natural mating was the true sire but it is impossible to determine if fertilisation occurred with semen that was deposited naturally or by AI.

In 29 cases, females were mated naturally to one or more males and then artificially inseminated by one or more different males. In all 29 instances, the offspring resulted from natural mating. The predom-inance of natural mating success was probably due to the advantage of early semen deposition because females are virtually always allowed to mate naturally before being artificially inseminated. Thus the copu-lating male benefits by having his sperm at the site of fertilisation

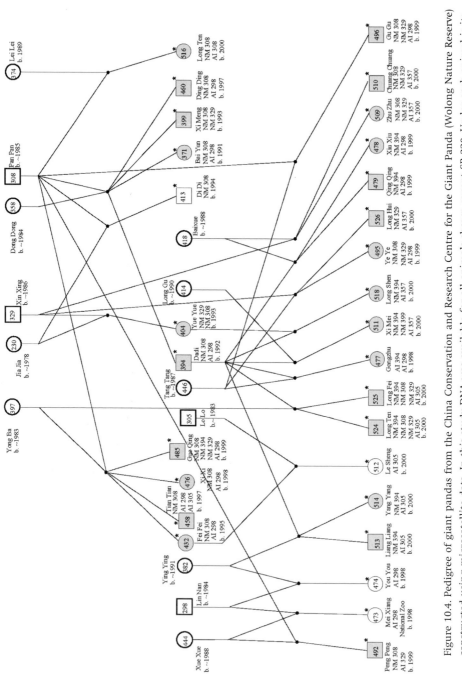

Figure 10.4. Pedigree of giant pandas from the China Conservation and Research Centre for the Giant Panda (Wolong Nature Reserve) constructed using microsatellite data. In this study DNA was available for all animals shown except SB 230. Under each animal is its name, potential sire by natural mating (NM) or AI, and date of birth. □ male, ○ female, ◯ bold - wild born, ▨ grey shading - multiple potential sires, * - paternity determined or confirmed with microsatellites, \ - deceased.

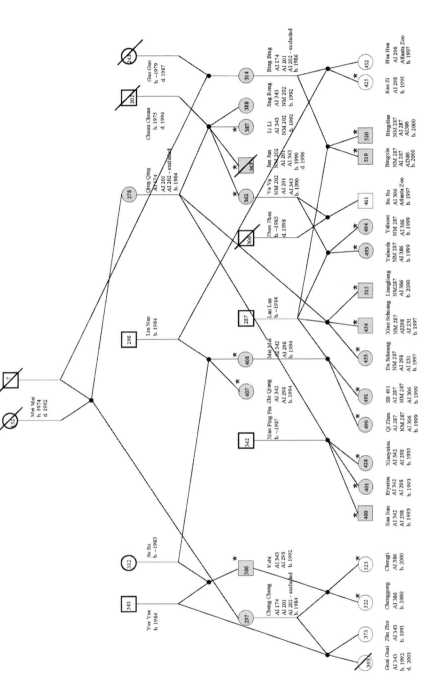

Figure 10.5. Pedigree of giant pandas from the Chengdu Research Base facility constructed using microsatellite data. DNA was available from all displayed pandas except SB 152, SB 243, SB 373, SB 393, SB 452 and SB 461. Under each animal is its name, potential sire by natural mating (NM) or AI, and date of birth and death. □ male, ○ female, ○ bold - wild born, ■ grey shading - multiple potential sires, * - paternity determined or confirmed with microsatellites, \ - deceased.

(oviduct) during early oestrus and at least a few hours (if not a day or more) before the competing AI donor.

Data analysis for twins and the issue of genetic over-representation

Paternity information for 15 sets of twins is presented in Table 10.5. Of 13 cases where multiple paternity was possible, we observed only a single case (SB 495/496) where offspring were sired by different males, through natural mating. All other pairs were sired by singleton males via copulation or AI. There were four sets of twins born by AI only. Of the 15 pairs, all were dizygotic, ten of mixed sex, suggesting that monozygotic twins are rare if they occur at all (Fang *et al.*, 1997c).

At the Wolong facility, three males were capable of natural mating (SB 329 and SB 308, which were wild caught, and SB 394, which is a son of SB 308). Twenty-one of 29 offspring at the Wolong facility (72.4%) – shown in Table 10.4 – are offspring of SB 308 or SB 394. The genetic consequences of this type of over-representation are discussed in more detail by Ballou *et al.* (in Chapter 21).

There is a broader representation of founder individuals in the Chengdu pedigree (see Fig. 10.5) where AI in the absence of natural mating has been more common. However, the dilemma here is that paternity remains unsolved for 17 living offspring and grand-offspring of key dams SB 278, SB 297 and SB 314 (see Fig. 10.5). Breeding records indicate SB 174, SB 201 and SB 202 as potential sires, but SB 202 can be excluded as a potential sire in all but three cases. DNA samples were not available for SB 174 and SB 201 for our study so these key paternities remain unresolved.

An important presumption in designing a maximally outbred *ex situ* breeding programme is that the founder individuals are unrelated or at least not first-order relatives. However, for rare species where wild populations often are under 30 or 40 individuals, the likelihood of recent inbreeding and consanguinity among captive founders may be appreciable. It is possible, however, to assess roughly the degree of relatedness among members of a captive population by comparing their composite microsatellite genotypes and examining the overall distribution (Queller & Goodnight, 1989; Vigilant *et al.*, 2001; Uphyrkina *et al.*, 2002).

We computed the genetic 'relatedness' of all pairs of individual genotypes at the Wolong and Chengdu facilities and plotted their

Table 10.5. Paternity information for 15 pairs of giant panda twins

Twins' studbook numbers	Names	Number of genotypes	Sexes	Current locations	Sire 1/Sire 2	Dam	Date of birth	Potential sires in order of mating/insemination (sire determined in bold)
362/363	Ya Ya/Jun Jun	2	F/M	Chengdu/deceased	202/202	278	1990	**NM 202**[a] AI 201[b] AI 343
400/401	Jian Jian/Eryatou	2	M/F	Wuhan/Chengdu	342/342	278	1993	AI **342** AI 298
407/408	Zhi Qiang/Mei Mei	2	F/F	Chengdu/Wakayama	298/298	312	1994	AI**298** AI342
453/454	Da Shuang/Xiao Shuang	2	F/M	Chengdu/Chengdu	287/287	278	1997	**NM 287** AI 298 AI 231
474/475	You You/none	2[c]	F/M	Wolong/deceased	298/298[d]	382	1998	AI **298**
478/479	Xiu Xiu/Qing Qing	2	F/M	deceased/Wolong	394/394	418	1999	**NM 394** AI 298
484/485	Yuan Yuan/Guo Qing	2[c]	F/M	deceased/Wolong	ND[e]/308	397	1999	**NM 308** NM 394 AI 329 AI 298
490/491	Qi Zhen/SB491	2	F/M	Chengdu/Chengdu	287/287	408	1999	AI **287** AI 306 NM 287
493/494	Yalaoda/Yalaoer	2	F/F	Chongqing/Chengdu	287/287	362	1999	AI 386 **NM 287**
495/496	Ye Ye/Gu Gu	2	F/M	Wolong/Wolong	**329/308**[f]	446	1999	**NM 329** NM 308 AI 298
509/510	Zhu Zhu/Chuang Chuang	2	F/M	Wolong/Wolong	329/329	418	2000	**NM 329** NM 308 AI 357
513/514	Liang Liang/Yang Yang	2	M/F	Wolong/Wolong	394/394	382	2000	**NM 394** AI 305
519/520	Bingxin/Bingdian	2	M/M	Chengdu/Chengdu	287/287	314	2000	**NM 287** AI 287 AI 386
522/523	Chenggong/Chengji	2	F/F	Chengdu/Chengdu	386/386	297	2000	AI **386**
524/525	Long Ten/Long Fei	2	M/M	Wolong/Wolong	394/394	397	2000	**NM 394** NM 308 NM 329 AI 305

[a] NM, natural mating; [b] AI, artificial insemination; [c] Inferred, twins are different sex; [d] Inferred, only one possible parent; [e] ND, not determined; [f] A single case of mixed paternity was observed for twins SB 495 and SB 496 who were sired by SB 329 and SB 308, respectively.

distribution (Fig. 10.6). For each centre, we subdivided the pair-wise comparisons to groups that were presumed 'unrelated' (presuming the founders were unrelated) and to groups with kinship estimates of 0.5 (first-order relatives), 0.25 (grandchildren and grandparents) and 0.125 (first cousins). In both centres, there were two peaks of pair-wise comparisons for each group. In comparing varying kinship (based on pedigree structure determined from Figs. 10.4 and 10.5), mean related-ness values were bimodal with a 'left' major peak with a normal distri-bution of low relatedness, and a second 'more related' peak, likely a reflection of closer kin relationships among some smaller group of individuals not known to be close relatives (see Fig. 10.6). The second, 'right-most' related peak reflects historic kinship not apparent in the pedigree structure. This empirical measure allows zoo managers to identify heretofore unidentified close relatives that occur in pedigrees derived from tiny free-ranging populations.

PRIORITIES FOR THE FUTURE

As discussed by Ballou and colleagues (in Chapter 21), our data indicate that a high priority for the *ex situ* population is to maximise genetic representation across the captive population by planned natural ma-tings and by AI. In the short term this will be an important step, especially when few males are naturally mating. However, if a long-term goal of captive breeding is eventually to contribute to reintroduc-tion, then it will also be important to focus on finding ways to improve the proportion of animals that will naturally mate as well as maternally rear cubs. Furthermore, as it is likely that natural mating *combined* with AI (with multiple males) will continue to occur, paternity testing will be essential for effective programme management. Also, this technology certainly has the potential to improve the effectiveness of AI as a breeding tool. For example, using different males as semen donors on different days of oestrus followed by paternity testing would provide new knowledge on the optimal timing for mating or assisted breeding.

Giant panda DNA has been isolated from blood, faeces, hair and urine (Ding *et al.*, 1997; Fang *et al.*, 1997a); the last three can potentially be obtained from free-living individuals. In turn, the hearty nature of the microsatellites described here will hopefully be incorporated into future studies to evaluate the genetic robustness of the *in situ* popula-tion (e.g. Lu *et al.* 2001). Microsatellite markers are powerful genetic

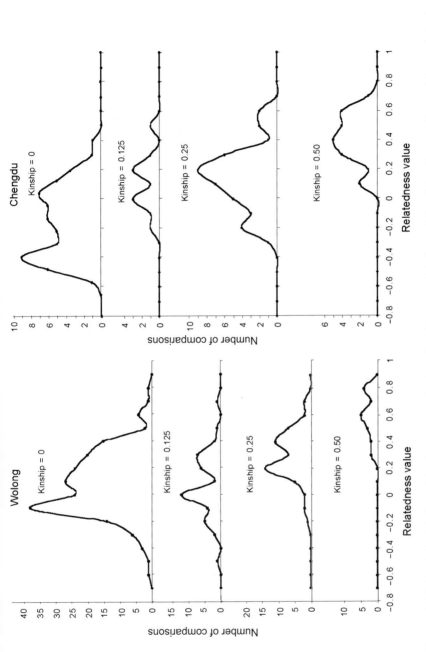

Figure 10.6. Estimated relatedness (R) values among giant pandas of known relatedness at the China Conservation and Research Centre for the Giant Panda and the Chengdu Research Base of Giant Panda Breeding. All wild-caught animals were assumed to be unrelated for this calculation but the complex nature of the curves indicated that this assumption was probably not true. Some animals appeared to be related to the degree expected, whereas others appeared more related than expected. Relatedness among animals can be estimated based on the microsatellite data, and this information could be useful in planning pairings for future matings.

resources that can be used to identify individuals in field studies, to assess parentage and to monitor genetic diversity. Even when cubs are born as a result of AI with semen from a single sire, it would be prudent to confirm the identity of the sire by microsatellite analysis to rule out AI errors (e.g. the mislabelling of vials of frozen semen). Finally, the establishment of genome banks to preserve samples from all captive-born and wild-caught animals could prove to be an invaluable resource for future genetic studies (Wildt, 2000; Ryder *et al.*, 2001), including perhaps assisting with the more accurate censusing of the wild population.

Faecal and hair samples from giant pandas have proven to be good sources of DNA for microsatellite analysis (authors' personal observations). While sightings in the wild are rare, field scientists often encounter giant panda faeces and hair. It has been proposed that microsatellite analysis on DNA isolated from such sources could be used to estimate population size more accurately and determine home ranges as well as the ranges of overlap among animals (M. Durnin, pers. comm.). Based on findings presented in this chapter, the authors support this concept and also suggest that results could be used to estimate genetic relatedness within populations. Faecal samples have the added benefit of (so far) being exempt from CITES regulations and, thus, easier to share across national boundaries.

In summary, today more than 160 giant pandas reside in Chinese and western zoological institutions. Natural and assisted mating of giant pandas has led to the birth of more than 200 cubs in China in the past three decades. However, many of these offspring were born after both natural mating and AI, making the paternity of cubs uncertain. A modern 'state-of-the-art' molecular genotyping centre was established in the Laboratory of Genetics and Reproductive Studies at the Chengdu Research Base of Giant Panda Breeding. A parentage assessment involved determining the composite microsatellite molecular genotype (n = 17 loci) of 71 giant panda DNA samples. Unequivocal paternity and maternity were established for 48 of 50 cubs. In cases where both natural mating and AI were performed, all offspring were derived from the natural mating sire. This largely reflects the delay in AI procedures until after the oestrual peak. Fifteen twin births were determined to be exclusively dizygotic and, in only one case, from more than a single sire. An analysis of microsatellite genotype relatedness suggests that some founders from both centres may share previously undisclosed consanguinity. The full genotype data set is publicly available on http://home.ncifcrf.gov/ccr/lgd and should be useful in managing mating to preserve genomic diversity among captive giant pandas.

ACKNOWLEDGEMENTS

We thank Janet Ziegle and Larry Joe of Celera AgGen for the generous donation of the automated DNA sequencer and to Jianbing Cai of PE China for instrument installation. We are grateful to British Airways for complementary airline tickets for travel to China as a part of this company's continuing support of wildlife conservation research. This publication has been funded with federal funds from the National Cancer Institute, National Institutes of Health, under Contract No. NO1-CO-12400.

REFERENCES

Brownstein, M. J., Carpten, J. D. and Smith, J. R. (1996). Modulation of non-templated nucleotide addition by *Taq* DNA polymerase: primer modifications that facilitate genotyping. *BioTechniques*, **20**, 1004–10.

Bruford, M. W and Wayne, R. K. (1993). Microsatellites and their application to population genetic studies. *Current Opinions in Genetics and Development*, **3**, 939–43.

Callen, D. F., Thompson, A. D., Shen, Y. *et al.* (1993). Incidence and origin of 'null' alleles in the (AC)n microsatellite markers. *American Journal of Human Genetics*, **52**, 922–7.

Ding, B., Zhang, Y. and Ryder, O. A. (1997). DNA preparation, PCR amplification and sequence analysis from scent marks and feces in giant panda. In *Proceedings of the International Symposium on the Protection of the Giant Panda*, ed. A. Zhang and G. He. Chengdu: Sichuan Publishing House of Science and Technology, pp. 157–62.

Ding, B., Shi, P., Xiangyu, J. G. *et al.* (2000). Microsatellite DNA analysis proves nucleus of interspecies reconstructed blastocyst coming from that of donor giant panda. *Chinese Science Bulletin*, **45**, 1883–5.

Driscoll, C. A., Menotti-Raymond, M. and O'Brien, S. J. (2002). Genomic microsatellites as evolutionary chronometers: a test in wild cats. *Genome Research*, **12**, 414–23.

Fang, S., Ding, Z., Feng, W. *et al.* (1997a). A preliminary study of the material resource of DNA in the DNA fingerprinting analysis of giant pandas. *Journal of Sichuan University*, **34**, 98–102.

Fang, S., Feng, W. and Zhang, A. (1997b). The development of giant panda DNA fingerprinting probe F2ZGP96060801 and the comparative test analysis of 5 probes. *Acta Theriologica Sinica*, **17**, 165–71.

Fang, S., Feng, W., Zhang, A. *et al.* (1997c). DNA fingerprinting analysis on the paternity determination of giant pandas. *Acta Theriologica Sinica*, **17**, 92–9.

Frankham, R. (1995). Conservation genetics. *Annual Review of Genetics* **29**, 305–27.

Frankham, R., Ballou, J. D. and Briscoe, D. A. (2002). *Introduction to Conservation Genetics*. Cambridge: Cambridge University Press.

Goldstein, D. B. and Schlotterer, C. (1999). *Microsatellites: Evolution and Applications*. Oxford: Oxford University Press.

Hedrick, P. W. and Kalinowski, S. (2001). Inbreeding depression in conservation biology. *Annual Review of Ecology and Systematics*, **31**, 139–62.

Huang, Y., Wang, P. Y., Zhang, H. M. *et al.* (2001). Efficiency of artificial insemination in giant pandas at the Wolong Breeding Center. *Proceedings: American Society of Andrology, Journal of Andrology*, Suppl, **118** (abstract 079), 146.

Lu, Z. and Kemf, E. (2001). *Giant Pandas in the Wild*. Gland, Switzerland: WWF-World Wide Fund for Nature.

Lu, Z., Wenshi, P. and Harkness, J. (1994). Mother–cub relationships in giant pandas in the Qinling Mountains, China, with comments on rescuing abandoned cubs. *Zoo Biology*, **13**, 567–8.

Lu, Z., Johnson, W. E., Menotti-Raymond, M. *et al.* (2001). Patterns of genetic diversity in remaining giant panda populations. *Conservation Biology*, **15**, 1596–607.

Marshall, T. C., Slate, J., Kruuk, L. E. B. and Pemberton, J. M. (1998). Statistical confidence for likelihood-based paternity inference in natural populations. *Molecular Ecology*, **7**, 639–55.

Menotti-Raymond, M., David, V. A., Lyons, L. A. *et al.* (1999). A genetic linkage map of microsatellites in the domestic cat (*Felis catus*). *Genomics*, **57**, 9–23.

O'Brien, S. J. (1994a). Genetic and phylogenetic analysis of endangered species. *Annual Review of Genetics*, **28**, 467–89.

(1994b). A role for molecular genetics in biological conservation. *Proceedings of the National Academy of Sciences*, **91**, 5748–55.

O'Brien, S. J. and Knight, J. A. (1987). The future of the giant panda. *Nature*, **325**, 758–9.

O'Brien, S. J., Goldman, D., Knight, J. *et al.* (1984). Giant panda paternity. *Science*, **223**, 1127–8.

O'Brien, S. J., Pan, W. and Lu, Z. (1994). Pandas, people and policy. *Nature*, **369**, 179–80.

Peng, J., Jiang, Z. J. and Jinchu, H. (2001a). Status and conservation of giant panda (*Ailuropoda melanoleuca*): a review. *Folia Zoologica*, **50**, 81–8.

Peng, J., Jiang, Z., Liu, W. *et al.* (2001b). Growth and development of giant panda (*Ailuropoda melanoleuca*) cubs at Beijing Zoo. *Journal of Zoology (London)*, **254**, 261–6.

Queller, D. C. and Goodnight, K. F. (1989). Estimating relatedness using genetic markers. *Evolution*, **43**, 258–75.

Roelke, M. E., Martenson, J. S. and O'Brien, S. J. (1993). The consequences of demographic reduction and genetic depletion in the endangered Florida panther. *Current Biology*, **3**, 340–50.

Ryder, O. A., McLaren, A., Brenner, S., Zhang, Y.-P. and Benirschke, K. (2001). DNA banks for endangered animal species. *Science*, **288**, 275.

Uphyrkina, O., Miquelle, D., Quigley, H., Driscoll, C. A. and O'Brien, S. J. (2002). Conservation genetics of the Far Eastern leopard (*Panthera pardus orientalis*). *Journal of Heredity*, **93**, 303–11.

Vigilant, L., Hofreiter, M., Seidel, H. and Boesch, C. (2001). Paternity and relatedness in wild chimpanzee communities. *Proceedings of the National Academy of Sciences*, **98**, 12890–5.

Wildt, D. E. (2000). Genome resource banking for wildlife research, management and conservation. *Institute for Laboratory Animal Research Journal*, **41**, 228–34.

Xie, Z. and Gipps, J. (2001). *Giant Panda International Studbook*. Beijing: Chinese Association of Zoological Gardens.

Zhang, Y., Ryder, O. A., Zhao, Q. *et al.* (1994). Non-invasive giant panda paternity exclusion. *Zoo Biology*, **13**, 569–73.

II

The science of behavioural management: creating biologically relevant living environments in captivity

RONALD R. SWAISGOOD, GUIQUAN ZHANG, XIAOPING ZHOU, HEMIN ZHANG

INTRODUCTION

As for many highly specialised carnivores, breeding giant pandas in captivity has had sporadic gains and setbacks over its 40-year history (see Chapters 1 and 19). Although many husbandry issues have been addressed successfully, we are still learning about behaviour and its relevance to *ex situ* management. This chapter updates the state of captive breeding at the China Conservation and Research Centre for the Giant Panda in the Wolong Nature Reserve (hereafter referred to as the Wolong Centre). We also provide details of our scientifically guided husbandry and management strategies that are contributing to a rapidly growing database of scholarly knowledge as well as to recent improvements in reproductive success.

Even with our limited knowledge about giant pandas in nature, it appears that, in the presence of plentiful natural resources and the absence of human disturbance, giant pandas mate, become pregnant and rear offspring without problem. Thus, reproduction is not a limiting factor to wild population viability (Lu *et al.*, 2000). Because this is not the case for the *ex situ* population, we can surmise that reproductive problems are rooted in the captive environment – a place that fails

Giant Pandas: Biology, Veterinary Medicine and Management, ed. David E. Wildt, Anju Zhang, Hemin Zhang, Donald L. Janssen and Susie Ellis. Published by Cambridge University Press. © Cambridge University Press 2006.

to fully meet the needs of at least some individuals. In principle, and with a proper understanding of species-salient factors, it should be possible to create captive environments that result in, or even surpass, reproductive rates occurring in the wild. Targets for improvement include health, nutrition, husbandry and behavioural management, this chapter concentrating on the latter two factors.

Although human intervention to promote reproduction (including through artificial insemination, or AI) are important back-up tools at the Wolong Centre, the emphasis has been on promoting natural mating and mother-rearing of cubs. This is due to a strong belief that optimal psychological well-being and species-specific behavioural management are prerequisites to a consistently effective propagation programme. It is critical to look to the natural world of giant pandas for guidance, using what is known or inferred from studies of wild individuals (Schaller et al., 1985; Lu et al., 2000). This has been the heart of our research activities (Swaisgood et al., 2000, 2001, 2003a). Although it is impossible to recreate all aspects of nature in captivity, it makes sense to find ways to reproduce functional aspects of natural animal–environmental interactions.

Improved behavioural management, coupled with advancements in health, nutrition and AI, have led to a population explosion at the Wolong Centre, where giant panda numbers increased from 25 to more than 70 from 1996 to 2003 (some of these are now on loan elsewhere). Captive breeding accounts for almost all of this growth, although a few ailing animals are occasionally rescued from the wild. In a previous report (Zhang et al., 2004), we described how behavioural management added new breeders to the Wolong population from 1996 to 2000. This trend continues today with 10 of 12 females aged 6 to 20 mating naturally from 2000 to 2004. Most of these animals mated in multiple seasons, with five of seven to eight of nine oestrual females mating in a given season. Similarly, from 2000 to 2004, six of seven males residing at the centre mated naturally, and another male mated for the first time at the San Diego Zoo just 3 months after shipment from Wolong. Clearly, much progress has been made since Lu and colleagues (2000) reported that, as of 1997, 74% of adult pandas in captivity had failed to breed. Still, reproduction is problematic for certain individuals and is better for the entire population in some years than others. And, as pointed out by Ballou et al. (in Chapter 21), reproduction and management remain inadequate currently to sustain a genetically viable population. Therefore, it is crucial to learn more to reach genetic goals and to increase overall animal numbers, especially if reintroduction of captive-born individuals is on the horizon.

AREAS OF INVESTIGATION

Captive environments and abnormal behaviours

Suboptimal captive environments for wildlife are associated with abnormal behaviours, stress and poor reproduction (Mason, 1991; Carlstead & Shepherdson, 1994; Shepherdson *et al.*, 1998). Besides often being too small, more importantly zoo enclosures lack complexity and opportunities to perform natural behaviours. Unlike wild counterparts, the captive animal has little control over its environment. Attaining 'natural' goals of food and shelter are not required in zoos (e.g. the animal has no mandate to work for food), so there is no need to perform normal behaviours. Additionally, while avoidance behaviours are critical to surviving in nature, a captive animal may have few options to avoid unpleasant situations, such as crowds or noise. As a result, well-being may be compromised.

A prevailing theory, which helps to explain the motivational basis of suboptimal animal well-being in captivity, is the *ethological needs model of motivation* (Hughes & Duncan, 1988). This model proposes that animals are motivated not only to obtain important resources such as food and shelter, but also to *perform* the behaviours that have evolved within the species. Thus, an animal given a bowl of processed food fails to use its evolved behavioural strategies to search for, extract and handle food for ingestion. The lack of opportunity to perform these behaviours may lead to poor psychological well-being and overt abnormal behaviours. Feeding-related behaviours appear to be among the most strongly motivated, and a disproportionate amount of behavioural anomalies are observed while an animal spends time waiting for food delivery (Falk, 1977). A major class of abnormal behaviour is called stereotypies – highly repetitive behaviours that do not vary in form and have no apparent function or goal (Mason, 1991).

According to our empirical evidence (Swaisgood *et al.*, 2001), the ethological needs model of motivation fits the giant panda. It is widely acknowledged that this species has been kept in suboptimal enclosures at many breeding facilities. Interestingly, the giant panda's rather simplistic ecology and behaviours in nature (e.g. solitary, few predators, dietary specialist) should make abnormal behaviours unlikely (Swaisgood et al., 2001). On the contrary, pandas in captivity tend to display a high incidence and impressive array of stereotypies (Swaisgood et al., 2001; Lindburg et al., in press). Of 47 individuals observed at six facilities (in China, Japan, Mexico and the USA), 62% displayed stereotypic behaviours, especially pacing, head tossing, pirouetting and cage climbing. Figure 11.1 demonstrates examples of pacing routines. In extreme cases, these activities persist for hours (pacing) or occur at rates of nearly 80 bouts per hour (head tossing). More males than females are afflicted, and the difference between sexes is statistically significant in the adult, but not the subadult population. Well over half of the wild-caught individuals in the Lindburg et al. (in press) sample (10 of 17) displayed stereotypies (although several of these individuals had been captured from nature at an early age). As might be expected, the incidence was higher among captive-born adults (78% of males and 29% of females). These findings reflect the pervasiveness of these anomalous behaviours and the need for remedial measures.

Environmental enrichment

Environmental enrichment is the behaviourally relevant modification of a captive milieu to optimise physical and psychological well-being. The guiding step in developing an effective enrichment programme is to look at nature (Swaisgood et al., 2003a). How do individuals of a species 'make a living' in nature? In the case of the giant panda, what are the critical resources and how are they distributed, located and obtained? Thus, what is the size of the home range and how far does the animal travel daily? What is the form and frequency of interactions with conspecifics, and how do individuals locate and choose mates? Much of this basic natural history is known, but many gaps in knowledge continue to leave us guessing. We do know that the giant panda spends 98% of its life eating and sleeping, leaving little time for locomotion, social interaction or other behaviours (Schaller et al., 1985). Diet consists almost exclusively of bamboo, which generally is prevalent throughout the home range. Therefore, there is little incentive to search for food proactively, with only occasional movement to an alternative bamboo patch.

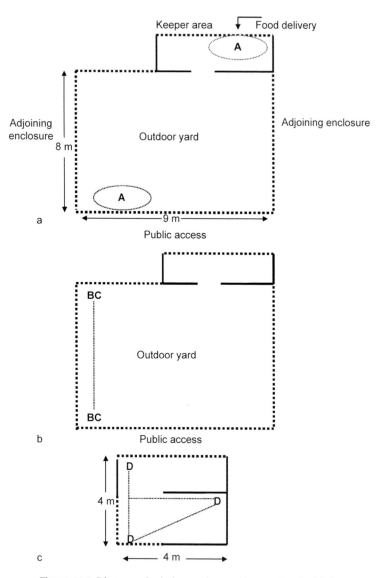

Figure 11.1. Diagrams depicting pacing routines and embedded stereotypies. (a) The routine for an adult female that paced in a circle, using only a small fraction of existing space. With each lap she paused, sat, rolled backward to the ground and then rocked into the seated posture in a movement resembling a 'sit-up' (A). Sit-ups invariably occurred at the cage bars, in one case at the location where food was delivered. The pattern of pacing and sit-ups was often repeated more than 30 times consecutively and was especially prevalent during peri-oestrus.

Our recent studies have involved evaluating and modifying enclosure design to affect the *ex situ* management of giant pandas. Indeed, design of the captive habitat is an important factor, which is not surprising as this issue is important for many wildlife species held in zoos (Forthman *et al.*, 1995). More than 20 enclosures at the Wolong Centre are relatively small, approximately 9 × 11 m (indoor plus outdoor portions). Size is important, but clearly it is not practical or financially feasible to build enclosures nearing home range size in nature (i.e. >400 ha; Schaller *et al.*, 1985). Nonetheless, it is possible to test the value of larger captive pens and, through enrichment, to introduce some of the complexity and variability normally encountered in the wild. Toward this end, the Wolong Centre has constructed seven large (>0.25 ha each), semi-natural (containing native vegetation) enclosures on a slope along the Pitiao River. More such 'enclosed habitats' are planned at this location, and 16 similar enclosures are under construction in another mountain range south of the reserve. Each has the capacity of maintaining up to four or five subadults or two adults. Individual giant pandas from the centre are rotated through these enclosures, but the emphasis is on providing access to young animals, hoping that such special exposure will prevent the behavioural problems seen in pandas housed in smaller pens. An additional development is construction of multiple, even larger (1–30 ha) 'reintroduction pens' on the mountain above the centre. The plan is to release pregnant

(b) The routine of an adult male that paced back and forth by the cage bars adjoining a neighbouring female. At the turnaround point in the corner, he often stood bipedally and rotated the body about 180° (B: a 'pirouette') before pacing back to the opposite corner. The male often head tossed (C), an abrupt exaggerated upward swing of the head, before or instead of the pirouette. This behaviour increased, as did the speed of locomotion, when a neighbouring female was in oestrus. (c) The stereotypical routine of an adult male housed in a small enclosure. This male spent nearly one-third of his time following a specific travel path. He invariably stopped at three distinctive points along the path, sat down, scratched himself with his hind paws and looked outside the enclosure (D) for 30 to 60 seconds. One habitual stopping point was at the site of food delivery, and another overlooked the preparation area where he could observe keepers arriving with food. These behaviours accelerated in frequency as the normal feeding time approached.——, solid wall; - - - -, cage bars, ..., animal's path of travel; A, sit-up; B, pirouette; C, head toss; D, sit and scratch.

females at these sites to allow resulting cubs to be reared in the absence of most human contact. Such offspring are likely to be better candidates in the event of future reintroduction studies.

Since 1996, we have collected quantitative behavioural data on giant pandas living in small versus large enclosures. Preliminary evidence has revealed a marked reduction in stereotypies among individuals maintained in the larger pens. There have been other behavioural changes indicative of enhanced psychological well-being too, including reduced pacing and increased diversity of natural behaviours (Swaisgood, unpublished data). We are learning that even very subtle differences in enclosure design can influence behavioural measures of well-being. For example, the female giant panda maintained at the San Diego Zoo (SB 371) displays more signs of behavioural distress in one pen than another ($p < 0.01$). Both are open to the public, but the larger of the two has less vegetation, which may mean that its design contributes to aversive behaviours. Other giant pandas at this zoo (SB 415 and 487) have shown no behavioural preferences between enclosures.

Presently, less than one-third of the giant pandas at the Wolong Centre are still housed in small enclosures. Initially, these pens were rather barren and contained no structural complexity, manipulable objects or visual barriers. Biologically meaningful aspects of complexity include vertical dimension, variety of textures, mobile and non-mobile furnishings and temperature and light gradients. These, in turn, provide behavioural opportunities for climbing, resting at elevated sites, locomotor play, exploration, resting and digging on different substrates, hiding from conspecifics, humans or disturbances and behavioural thermoregulation. Such enrichments give the animal behavioural options, allowing it some control over its interaction with the environment. That is, an animal's behaviour influences whether or not it gains access to important resources, even if it is as simple as moving from a sunny to a shaded area. To increase behavioural control options for giant pandas at the Wolong Centre, we have begun to equip enclosures with permanent furnishings. Table 11.1 summarises changes for some or all of the pens. These alterations have been successful in that pandas spend a great deal of time interacting with or using the furnishings while displaying more diverse and naturalistic behaviours. Importantly, stereotypic behaviour has been reduced significantly for animals living in these revitalised enclosures, but these measures appear insufficient to eradicate stereotypies altogether. These altered

Table 11.1 *Use and function of structural alterations to small enclosures*

Pen enrichment	Predominant behaviours observed and/or purpose
Trees, logs, stumps, wooden climbing structures (~3 × 3 × 2 m)	Climb; locomotor play; rest; body rub; scent mark; tactile and oral explore; sniff; hide
Stones of various sizes	Object play and manipulation; body rub; scent mark; tactile and oral explore; sniff; hide
Digging pits with sand	Dirt bath; locomotor play; rest; dig; scent mark; tactile explore; sniff
Woven metal baskets in bedroom	Rest/play area secluded from public; locomotor play; rub; scent mark; sniff
Railing outside enclosures	Minimise public interaction with pandas
Bushes, tall bamboo outside enclosures	Minimise public interaction with pandas; temperature and light gradient (shade); rest; hide

environments have also offered a more complex topography for presentation of supplemental enrichment (e.g. hidden or scattered food and the placement of novel objects; see below).

The essence of the giant panda's existence is eating bamboo. This fundamental fact of panda life pervades and consumes a remarkable 14 hours of each individual's day (Schaller *et al.*, 1985). But this is also a valuable piece of information for enrichment, and we have exploited this inherent need to spend a lot of time feeding into the design of our studies. For example, the recent panda population boom at the centre has mandated the need to ensure adequate bamboo provisioning for all animals. This has been no trivial matter as more bamboo harvesters had to be hired and local farmers paid to grow bamboo as a crop. Winter shortages, which plagued the centre's early years, have been reduced and now, regardless of season, each giant panda receives *multiple* feedings of good-quality bamboo daily. This altered husbandry practice has increased opportunities for pandas to engage in behaviours related to bamboo consumption (removing the leaves to make a wad for consumption, stripping the outer inedible shell of the culm and, of course, lots and lots of chewing). Still, because one cannot be certain that all nutrients are available in the provisioned bamboo, the diet must be supplemented. The problem is that dietary supplements are quick, easy-to-process energy sources (see below) resulting in overall less time spent feeding than in wild counterparts. Thus, a captive giant panda under such conditions can have 'empty time' that needs to be filled with other activities, preferably related to feeding and foraging. If this does not occur, the development of stereotypies is likely.

In our studies, this available time has been used to test various enrichment strategies, usually settling on functional analogues. Giant panda foraging strategy entails little searching, but requires a great deal of handling to strip leaves and peel culms before consumption (Schaller *et al.*, 1985) and processing to chew this fibrous food source. We have argued that the ethological needs model does not imply that animals are solely motivated to perform specific behaviours (Swaisgood *et al.*, 2001). Rather, functional analogues may suffice – opportunities that allow animals to engage in goal-directed behaviours to secure resources. In this respect, we have developed multiple feeding-enrichment strategies for adult giant pandas. One of the most important has been replacing milk and low-fibre bread with a high-fibre biscuit (see Chapter 6). This has nutritional benefits, but the hard biscuits also require protracted time to chew and ingest. Mastication also can be promoted by freezing food in a large block of ice. Scatter feeding (hiding food around the enclosure) encourages searching behaviour. Likewise, several feeding devices have been used, all of which involved highly prized food items (apples, carrots and biscuits) and encourage handling and processing before consumption (Swaisgood *et al.*, 2001; Tepper *et al.*, 2001). The strategy can be made even more complex by hiding the food in a puzzle feeder. In one of these, a small hole is cut in a large section of bamboo (about 10 cm in diameter; Fig. 11.2) and apples inserted. The panda must roll the feeder with its paws or head, or pick it up and manipulate to extract the food. Similar devices have been created using a PVC pipe (Fig. 11.3) and hard plastic balls, also designed to increase handling time. Food contained in a cardboard box forces the panda to tear with teeth or claws, whereas food wedged into holes (drilled into logs or hidden in a pile of rocks) requires clawing. Enrichment devices hung with chain or rope from climbing structures will promote a different class of handling behaviours. Even nonfeeding enrichment can promote feeding-like behaviour. For example, fresh evergreen branches are handled, stripped and partially chewed by most giant pandas in a fashion remarkably similar to bamboo consumption (Swaisgood *et al.*, 2001).

Figure 11.2. A panda at the China Conservation and Research Centre for the Giant Panda (Wolong) uses its muzzle to manipulate a bamboo puzzle feeder, rolling it back and forth until apple pieces fall out. Made from a section of large bamboo, the feeder contains a single, small hole from which apple pieces must be extracted. Encouraging pandas to work for food in such ways can lead to improved signs of well-being.

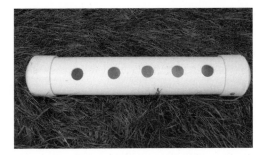

Figure 11.3. A PVC puzzle feeder with several large holes to make access to food treats easier. This design works well with a 'lazy' individual who refuses to work for food using the more challenging bamboo feeder.

Because animals often develop stereotypies due to predictable food delivery schedules, we also have altered the temporal feeding pattern and frequency at the centre. Before 1997, individuals were fed twice per day at 1000 and 1600 hours. At these times, a common sight at the feeding locations was pacing, weaving back and forth and head tossing (Swaisgood, unpublished data). Now, there are as many as seven feedings per day with food given on a more varied schedule. The result has been an apparent reduction in food anticipation and abnormal behaviours.

Our supplemental enrichment programme for giant pandas involves the occasional providing of novel and manipulable objects. Items are chosen on the basis of varying physical properties that facilitate different behavioural opportunities (e.g. sniff, pick up or swat with paws, push with head, bite, shake, throw, roll with or on, jump on, carry, chase) or needs that may well vary with sex, age or individual temperament. In general, the enrichment item is left in the panda's enclosure for 24 hours and then removed. The list of items utilised is continually growing and changing and has included: burlap sacks stuffed with straw, sand or mulch; plastic manipulable objects; piles of evergreen branches, bark, rocks and mulch; boxes; tyres; straw mats; and logs. Such an array of items results in fascinating results. For example, we know that style of presentation can be important. The male (SB 381) at the San Diego Zoo preferred substrate presented in a tall, narrow pile over the same amount of substrate presented in a flat, wide pile (Tepper *et al.*, 2001). We also have discovered that giant pandas in general have a significant preference for a plastic bobbin over a 26-cm-diameter ball (Swaisgood *et al.*, 2000). The bobbin, which is shaped like a thread spool, stands 35 cm high and is 30 cm in diameter at the two flat ends, tapering to 18 cm in the middle. We have consistently observed that a panda can readily pick up the bobbin, which is easy to hold, manipulate and explore it with all four paws while lying on its back. A ball, apparently more conducive to swat and chase, is of less interest. This should not be surprising in that the giant panda is a consummate 'manipulator', using its famous pseudothumb to assist in grasping and handling bamboo.

Enrichment is only as valuable as the ability to measure an animal's responsiveness quantitatively, thereby permitting efficacy to be accurately evaluated. A variety of methods has been used to measure enrichment in the giant panda, some relying on labour-intensive, formal behavioural observations. Others have capitalised on casual observations by keepers going about their daily activities. Regardless of approach, it is evident that enrichment has a major impact on several indices of psychological well-being. Most importantly, these giant pandas are more active and engaged in diverse play and non-play behaviours than when not enriched. They also have an overall reduced incidence of stereotypy while spending less time expressing abnormal behaviours (especially associated with predicted feeding schedules)

(Swaisgood *et al.*, 2001). Particularly exciting is that enriched giant pandas express signs of improved well-being, even in the absence of direct interaction with the enrichment item. This suggests that the effect is not just the result of less time available to engage in stereo-typies. Rather, the animal's *motivation* to perform abnormal behaviour is diminished. This discovery is also consistent with the ethological needs model of motivation. That is, the enrichment objects only provide an opportunity to perform a behaviour without offering the panda an important biological resource. Another important implication is that the animal's time spent interacting with enrichment may not reflect how enriching the experience truly is. Studies at the San Diego Zoo demonstrate that enrichment occupying as little as 2 to 3% of a panda's time significantly improves well-being later in the day (Tepper *et al.*, 2001). Thus, although the enrichment opportunity may be brief, it may be of long-term importance.

Biologically relevant behavioural management for reproduction

In developing breeding management protocols at the Wolong Centre, we capitalised on the scientific studies of wild giant pandas emphasis-ing the solitary nature of the species punctuated by occasional contacts outside the mating season (Schaller *et al.*, 1985; Schaller, 1993; Lu *et al.*, 2000). Although more intensive social housing arrangements may some-times work (Kleiman, 1984; Hoyo Bastien *et al.*, 1985), we ascribe to the philosophy that the best approach is to mimic nature as much as possible and then modify if methods fail. For example, field researchers have noted that males have large home ranges that encompass several females' home ranges. Thus the male in nature has access to these females' scents but is unlikely to spend much time in any given female's range. Direct encounters between male and female giant pandas are relatively rare (Schaller *et al.*, 1985). Lu and colleagues (2000) suspected that these social traits of wild pandas and the lack of similar opportun-ities in captive counterparts likely contribute to reproductive failure in the latter group. Thus, management activities at the Wolong Centre have been altered to ensure that adult pandas are housed individually, as fights can occur outside the mating season even between males and females. These pandas have access to neighbours through cage bars but can avoid proximity and generally most visual, olfactory and acoustic contact if desired. Animals also are moved to a different enclosure several times each year, where they encounter new conspecific odours and neighbours. These conditions probably afford opportunities for social exchange not too disparate from patterns in the wild.

Olfactory management studies

One of the hallmarks of the centre's programme is the use of species-appropriate olfactory management to create a more natural context for intersex communication. We designed a series of studies to elucidate the functions and practical application of the chemical signalling system in the giant panda (Swaisgood *et al.*, 2004). Because these animals live a solitary existence, they must rely heavily on chemical signals to communicate without requiring face-to-face encounters. Chemical signals, including scent marks from the anogenital gland and urine, are unique and persist in the environment after the signaller has left. Anogenital marks last about three months while urine marks are sustained for two weeks (Swaisgood *et al.*, 2004). The species appears to have developed a system of traditional communal scent-mark stations, or 'community bulletin boards', which provide individuals with reliable locations to visit, and where they can deposit signals and investigate signs left by other giant pandas.

By presenting conspecific odours, we have begun experimentally to tease apart the meaning of these chemical signals. First, it is clear that pandas can discriminate between the odours of individuals: each panda's scent mark leaves a 'chemical signature' (Swaisgood *et al.*, 1999). Secondly, pandas 'date' these chemical messages, that is, they are able to distinguish between scents deposited at different times (Swaisgood *et al.*, 2004). Thirdly, recipients extract a great deal of information from these chemical messages, including the sex (Swaisgood *et al.*, 2000; White, 2001), age (White *et al.*, 2003), reproductive status (Swaisgood *et al.*, 2000, 2002) and competitive status of the signaller (White *et al.*, 2002). This information is clearly important. For example, we now know that opposite-sex odours appear instrumental for priming sexual motivation before face-to-face encounters (Swaisgood *et al.*, 2000, 2002). Giant pandas in captivity are notorious for aggressiveness and/or indifference when placed in mating situations (Zhang *et al.*, 2004). We have found that individuals show increased sexual arousal solely in response to odours of the opposite sex. At the Wolong Centre, pandas are exposed to odours of potential mating partners for days to weeks before the mating introduction. The result is less aggression and more sexual interest.

Reproductive behaviour studies

Reproductive behaviour studies have been instrumental not only at the Wolong Centre (Swaisgood et al., 2003b; Zhang et al., 2004), but elsewhere (Kleiman, 1983; Snyder et al., 2004) to improve management. A common approach is to merge behavioural and morphological signs of oestrus through parallel endocrine monitoring to identify the onset of the fertile period (see Chapter 14). In studies at the San Diego Zoo, we have found that behaviour, hormonal profiles and vaginal cytology patterns are all similarly predictive of the timing of impending ovulation (Lindburg et al., 2001; see also Chapter 9).

At the Wolong Centre, two complimentary methods are used to record various behaviours indicative of oestrus beginning in February at the onset of the breeding season. One is a systematic, quantitative protocol for keepers to record casual observations, whereas the other involves formal observations made by trained researchers during five specific 45-minute intervals each day during morning and afternoon periods of high activity. Table 11.2 lists some of the most important behavioural and morphological signs used to track oestrus (adapted from Kleiman, 1983). By monitoring behaviours closely, we can recognise early signs of impending oestrus, and move the female to a pen adjacent to an appropriate male so that communication and familiarisation occur well in advance of the mating introduction. These early signs can often be identified 15 to 20 days before actual introduction for copulation. Visual, acoustic, olfactory and limited tactile contact through cage bars is promoted in this 'protected contact' arrangement. This method is supplemented with frequent 'pen swaps' to expose the pair more directly to each other's odours (Swaisgood et al., 2000). This approach of promoting an optimal level of familiarity between the potential pair before introduction is also advocated at the Chengdu Research Base of Giant Panda Breeding (see Chapter 14). We are now also using supplemental recording protocols to collect more intensive data for the female as well as the neighbouring male (Swaisgood et al., 2003b), eventually allowing enough information to seriously consider making the mating introduction.

Introduction of animals for mating

A useful tool for deciding to attempt or terminate a mating introduction is a 'decision tree' that covers various contingencies. Figure 11.4

Table 11.2 *Behavioural and physical signs used to monitor oestrus*

Behavioural and physical signs	Description
Active	Percentage time spent in non-rest behaviour
Scent mark	Rubbing anogenital gland on a surface
Urinate	Elimination of urine; often accompanies scent marking
Bleat	High-pitched twittering vocalisation of one to three seconds duration, sounding much like a goat's call; 'friendly' and promotes contact with male
Chirp	Short (<1 second) high-pitched, harmonic vocalisation descending in pitch at the end of each call; 'friendly', promotes contact with male
Roll	Lying on back and rolling from side to side
Water play	Splashing and rolling in water
Backwards walk	Walking backwards for more than four steps, sometimes for >20 minutes, often backing into a male or inanimate object with tail up
Tail up	Tail is raised uncovering vaginal area
Lordosis	Sexual posture with tail up, hind quarters raised, back arched and kneeling on forelegs
Proximity to male	Female approaches male and/or spends time located adjacent to male enclosure
Appetite	Consumption of bamboo and dietary supplements declines
Vulvar colour	Coloration of external genitalia changes from grey to red
Vulvar swelling and opening	Enlargement of external genitalia and size of the vaginal opening

Adapted with permission from Kleiman, 1983.

offers a simplified version of such a decision tree, which can be expanded or tailored to the dispositions of individual pairs. The following rules of thumb are for giant pandas housed separately in relatively small enclosures. The ultimate decision to place two individuals together must be based on observations of behaviour indicating

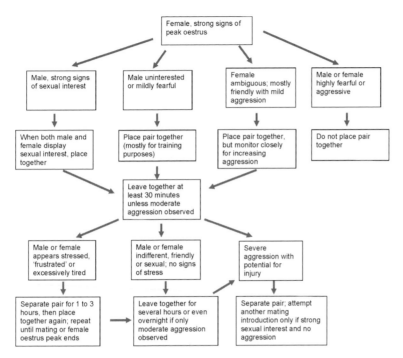

Figure 11.4. Mating introduction decision tree. This simplified version shows how behavioural signs are used by management staff to make decisions to maximise mating opportunities while taking care to avoid injurious aggression. Accompanying such a tree are detailed definitions based, in part, on behaviours included in Table 11.2.

pair compatibility. Other signs (i.e. vulvar colour, vaginal cytology and urinary oestrogen patterns) are important, but more so for identifying the physiological period of highest fertility. Behavioural traits, however, determine the level of sexual interest and the potential capacity for aggressiveness, which can be injurious to one or both individuals. At the Wolong Centre, the staff waits for a strong bout of sexual interaction in protected contact before introducing the animals together. Ideally, this is expressed as a male being excited and bleating very frequently, whereas the female is bleating, chirping and demonstrating a lordotic stance. It has been useful to gently touch the female's genitalia with a piece of bamboo, stimulating lordosis as the male enters the female's pen. If the gate separating the pair is opened at this time (while the female is lordotic), it is more likely that the male will mount and copulate quickly. This seems to minimise fighting.

Figure 11.5. Aggression is not uncommon in mating introductions. Here the male (above) was rough with the female, and she turned in defence. Approximately two hours later, after a brief separation, the pair successfully mated.

Even when not apparent during protected contact, aggression can occur during the mating introduction (Fig. 11.5). Mild hostility is common, is no cause for alarm and should be tolerated for a few minutes before attempting to separate the pair. It is not unusual for the male to bite the female's back and perhaps paws when trying to mount, resulting in some fur removal. A portion of this activity is related to positioning the female for copulation (Fig. 11.6) and is not motivated by aggression. If the biting becomes dangerous and/or the female responds with excessive fear or belligerence, the pair should be separated immediately. Some aggressive vocalisations (bark, moan and squeal) are common for the female during courtship and mating, and are not signals for pair separation. However, excessive and hostile vocalisation from the male is unusual and reflects a potential attack on the female. Under normal conditions, a male should continue bleating frequently while with the female. An absence of this activity may indicate insufficient sexual interest. Similarly, the female should bleat and/or chirp frequently during pairing, although barking, squealing and moaning are also normal. If the pair is separated due to aggression, a re-attempt at introduction should be made within a few

Figure 11.6. Mating in species-typical positions.

hours if adequate sexual interest is apparent and the earlier bout did not result in injuries. Alternatively, a mating introduction with a different male may be attempted.

Aggression is most likely to occur in one of three circumstances and can be dangerous, especially in small enclosures where escape options are limited:

1. Despite signs of sexual arousal before pairing, upon introduction the male may attack the female immediately, especially if she is overly defensive or uncooperative. This situation requires immediate pair separation.
2. After making a series of unsuccessful mounts over more than 15 minutes, the male becomes 'frustrated' and attacks the female, often escalating to a dangerous encounter. In general, both the male and female giant panda appear to tire after 20 or 30 minutes of attempted mounts and failed copulations, a situation that can evolve into aggression. Prior to significant hostility, the pair should be separated. (This stamina issue underscores the need to keep animals in good physical condition, including by providing enrichment opportunities as discussed above.)
3. Immediately after copulation, it is common for the female to be aggressive, which is expressed as breaking away from the male, barking, growling and lunging. The post-copulatory male generally shows little interest in the female, but some individuals may retaliate; serious attacks are unusual. In any case, a pair that has just mated should be separated unless there are indications of additional sexual arousal. For example, very short copulations

(<1 minute) may be followed by another successful copulation within 10 to 30 minutes.

Separating giant pandas that are together during the nonbreeding season is generally easy through conditioned training and the use of favoured treats (e.g. sugar cane or apple). This is not the case during the breeding season when individuals are often uncooperative. Usually it is easier to provoke the male to move from an enclosure because a female in peak oestrus is often unresponsive, assumes the lordotic stance and will not budge even in response to a favourite treat. Alternative techniques for separating or moving individuals include spraying water with a hose or swatting the ground in front of or behind the panda with a large stick of bamboo.

Neonatal management

Our behavioural research has also been targeted to birth and maternal caregiving, which are both critical to successful captive management. It is at this time that many critical decisions are required, often on short notice and with little time. For example, if the mother is deemed inadequate, it may be necessary to remove the infant for hand-rearing (see Chapter 13). Also, weak or sick neonates must be identified as early as possible for human intervention and intensive care. Judicious decision-making at this critical juncture necessitates a clear understanding of the normal range of successful maternal as well as neonatal behaviours.

Although hand-rearing has increased the total number of offspring surviving at the Wolong Centre (and other facilities), it may well be that mother-reared individuals could be more valuable. Although more studies are needed, we can assume that dam-reared infants may have fewer behavioural abnormalities in adulthood and be better candidates for natural mating or reintroduction. Therefore, new methods are being used to encourage reluctant dams to rear their own offspring. In one case, a mother initially abandoned her neonate, which was taken to the nursery for hand-rearing. During the following weeks the mother was exposed to the odours and vocalisations of her cub and was given a surrogate toy panda. These actions evoked normal maternal care behaviours, reduced her fear and gradually allowed the cub to be reintroduced to the dam, who eventually assumed all caregiving activities (Zhang et al., 2000). The importance of investigating the mother–infant relationship does not end with the initial maternal response. In studies similar to those described in Chapter 14 by Snyder and colleagues, we have followed the lives of 39 cubs from birth to weaning, generating a large database on successful and less successful maternal caregiving styles.

Young at the Wolong Centre are routinely weaned from the dam at five to six months of age, which is more than a year earlier than that observed in the wild (Zhu et al., 2001). It has been speculated that precocious weaning may have serious consequences for adult reproductive performance (Lu et al., 2000; Snyder et al., 2003; see also Chapter 14), although our experiences have not revealed such an interaction. Sixty-four cubs have been produced at the centre since 1991. Of these, seven were adults of more than six years of age in 2003 (thus having at least one to two reproductive seasons of experience). An additional female rescued from the wild at less than six months of age and reared in Wolong was also included in our study group. Of these eight adults, six have copulated (more than 70 times). Of the four females that have mated, all have been successful mothers and reared their own young. One explanation is that a mother in nature interacts with her cubs relatively infrequently after the first few months of life, often leaving them behind in a tree while the dam forages at a distance for hours or even days (Lu et al., 1994; Zhu et al., 2001). Captive-born cubs probably grow up to be successful breeders because they actually have more interaction with their dam and also, generally, are reared with same-age peers that provide ample social stimulation (see Chapter 14). Therefore, in this case, the need for imitating nature is probably not essential. Nonetheless, it remains worthwhile to investigate further even potentially subtle impacts of peer exposure on behavioural development and subsequent reproductive success (see Chapter 14). Moreover, mother-rearing may be essential for producing a giant panda capable of surviving reintroduction.

Another challenge to the management of newborns is the production of twins and the rejection of one neonate. Following the practice of the Chengdu Research Base of Giant Panda Breeding, the Wolong Centre has also found that a dam can be encouraged to raise both cubs through the technique of twin swapping (see Chapter 13). In brief, one cub is removed at a time and maintained in the nursery for two to four days, then traded for its twin, which is being mother-reared (a rotation system that results in a combination of dam- and hand-rearing with both offspring receiving maternal milk and colostrum as well as formula). Using this technique at the Wolong Centre, survival to six months has been 100% for 24 offspring (one triplet was lost). Prior to initiating the twin-swapping strategy, only one nursery-reared cub had ever survived for more than a few days (in this case, to five months). We are also exploring the value of using the dam's scent to reduce behavioural stress in the early days post weaning (Swaisgood *et al.*, 2004). We suspect that intense stress at this time could provoke development of stereotypies. More recently, we have experimented with an artificial mother for freshly weaned cubs, using a soft pillow covered with artificial fur and equipped with a nipple. The idea is that the latter will satisfy the cub's need to suckle (much like a human pacifier) and stave off development of oral stereotypies (e.g. paw sucking).

PRIORITIES FOR THE FUTURE

This overview of our experiences with behavioural management and enrichment at a major *ex situ* giant panda centre in China has emphasised the value of science and biological relevance as the most useful tools for developing effective species management. Mating behaviour was once severely impaired by poorly understood behavioural deficits. However, today all but the most reluctant pandas can be encouraged to mate as a result of combined sound scientific data and good judgement. Although there is no universal recipe for success, a decision tree (as discussed above) can guide managers through the contingencies likely to be encountered when attempting to pair and mate two sometimes recalcitrant giant pandas. Regardless, if there is one consistency within this species, it is that no two individuals are alike in behaviour and temperament (Zhang *et al.*, 2004). Therefore it will continue to be necessary to tailor methods to unique individuals.

The contributions of enrichment to reproduction have been more difficult to quantify, but in fact may be most important. Clearly, enrichment leads to short-term reductions in behaviours indicative of poor psychological health. Therefore, continuing to enhance living environments should impact positively on overall well-being on a protracted and perhaps permanent basis. We need not wait to know the precise long-term impact of enrichment – there are enough lessons already learned from other species to satisfy us that a better captive environment is good for the giant panda. Studies of laboratory species have demonstrated a variety of positive *developmental* effects on brain function, with enriched animals having a higher brain weight, more glial cell and synaptic connections between neurones, increased neurotransmitters and more RNA (Renner & Rosenzweig, 1987). These findings may reflect memory consolidation and, therefore, improved learning ability, less emotional reactivity, more exploratory capacity and less hesitancy with novel objects and places. Animals living in enriched environments may also exhibit lower levels of pituitary–adrenal activation and other indices of chronic stress (Shepherdson *et al.*, 1998) which, in turn, can lead to suppressed immune function, reproductive failure and higher disease rates (Hofer & East, 1998). We can infer from these findings that pandas raised in enriched environments will be smarter, healthier, more adaptable and better candidates for future projects, such as reintroduction. Ironically, the scientific determination of the true value of enrichment may never be documented in this species. Due to the preciousness of every individual, we plan to 'leave no giant panda behind', thereby eliminating comparative controls without enrichment.

The future mandates that we scientifically determine the best ways to identify *which* enrichment strategies are ideal. Among the priority challenges is the absolute need to ensure that all giant pandas outside the natural world are living inside high-quality, naturalistic enclosures. Second, all pandas should receive an ample and consistent supply of bamboo and other fibre-rich food. Third, researchers should continue to collaborate with animal caretakers to devise and fine-tune supplemental enrichment programmes, tailoring to different age–sex groups or even to individuals. If conducted carefully, such studies will allow hypothesis testing and data quantification/analysis to identify ideal protocols.

Suppose for the moment that these rather grand goals are indeed realised. That, combined with the need to manage the population genetically to maintain all existing genetic diversity (see Chapter 21) would, we hope, result in a healthy, well-adjusted and viable *ex situ* population. The question then becomes: toward what end is this work with captive animals aimed? Part of the answer has already been addressed in Chapters 1 and 2. Such populations are valuable as insurance for wild counterparts and to inspire and educate the public. But what is most exciting is the opportunity to learn from captive animals – to generate new scholarly knowledge that can increase the population even further and eventually support giant pandas in nature. We believe that the scientific community is well on its way from data gathering with the *ex situ* population to advancing species conservation *in situ*. In the context of the discipline of behaviour, we are advocates for the value of giant panda scents to further field conservation (Swaisgood *et al.*, 2004). By learning how individuals respond to each other's odours in captivity, it is possible to infer how odours could be placed in field situations to encourage pandas to use unoccupied areas in new reserves or habitat corridors. Additionally, conservation initiatives will eventually begin in Wolong and elsewhere which will target animal reintroduction, with the first efforts being both experimental and instructive. Certainly, the success of such a programme will lie not in the sheer quantity of pandas released but in the *quality* of the introduced individuals. Natural behaviours, adaptability and cognitive abilities will be essential, and it has already been demonstrated in other successfully reintroduced species that these traits can be instilled through excellent management and enrichment programmes (Shepherdson, 1994).

Much remains to be learned about the giant panda, but some of the mystery enshrouding this iconic species is being resolved through a host of systematic studies conducted across disciplines, through partnerships and being summarised in texts such as this one and in another recent contribution (Lindburg & Baragona, 2004). This approach is crucial if this beloved species is to be conserved while proving to sceptics that all the effort is indeed worth it. Who would have guessed that the many problems facing this species in captivity and surmised by some to be virtually insurmountable (Lu *et al.*, 2000) would in fact be overcome in a relatively short time? Yet scientifically guided management is making impressive progress, and the near future must be directed at more studies applying all existing new data to more focused conservation efforts in the field. In particular, our new knowledge should be merged with the approaches of panda research pioneers, including George Schaller, Jinchu Hu, Zhi Lu and Wenshi Pan, and joining field biologists in a full, participatory exchange of skills, information and ideals.

ACKNOWLEDGEMENTS

We thank the Zoological Society of San Diego and the State Forestry Administration of China for continued support of our programme. Many individuals have contributed to this work, the most prominent including Suzanne Hall, Hongyin Han, Valerie Hare, Yan Huang, Damin Hu, Desheng Li, Donald Lindburg, Laura McGeehan, Megan Owen, Pengyan Wang, Staci Wang, Rongping Wei, Angela White and Dafu Wu.

REFERENCES

Carlstead, K. and Shepherdson, D. J. (1994). Effects of environmental enrichment on reproduction. *Zoo Biology*, **13**, 447–58.

Falk, J. L. (1977). The origin and functions of adjunctive behaviour. *Animal Learning and Behaviour*, **4**, 325–35.

Forthman, D. L., McManamon, R., Levi, U. A. and Bruner, G. Y. (1995). Interdisciplinary issues in the design of mammal exhibits (excluding marine mammals and primates). In *Conservation of Endangered Species in Captivity: An Interdisciplinary Approach*, ed. E. F. Gibbons, B. S. Durrant and J. Demarest. Albany, NY: State University of New York Press, pp. 377–99.

Hofer, H. and East, M. L. (1998). Biological conservation and stress. *Advances in the Study of Behaviour*, **27**, 405–525.

Hoyo Bastien, C. M., Schoch, J. F. and Tellez Giron, J. A. (1985). Management and breeding of the giant panda (*Ailuropoda melanoleuca*) at the Chapultepec Zoo, Mexico City. *Proceedings of the International Symposium on the Giant Panda. Bongo*, pp. 83–92. Bongo, Berlin.

Hughes, B. O. and Duncan, I. J. H. (1988). The notion of ethological 'need', models of motivation and animal welfare. *Animal Behaviour*, **36**, 1696–707.

Kleiman, D. G. (1983). Ethology and reproduction of captive giant pandas (*Ailuropoda melanoleuca*). *Zeitschrift für die Tierpsychologie*, **62**, 1–46.

(1984). Panda breeding. *International Zoo News*, **31**, 28–30.

Lindburg, D. G. and Baragona, K. (eds.) (2004). *Giant Pandas: Biology and Conservation*. Berkeley, CA: University of California Press.

Lindburg, D. G., Czekala, N. M. and Swaisgood, R. R. (2001). Hormonal and behavioral relationships during estrus in the giant panda. *Zoo Biology*, **20**, 537–43.

Lindburg, D. G., Swaisgood, R. R., Zhang, J., Narushima, E. and Zhou, X. Stereotyped behaviour in the giant panda. In *Making Enrichment a 21st Century Priority: Proceedings, 5th International Conference on Environmental Enrichment*. Sydney: Australia (in press).

Lu, Z., Pan, W. and Harkness, J. (1994). Mother–cub relationships in giant pandas in the Qinling Mountains, China, with comment on rescuing abandoned cubs. *Zoo Biology*, **13**, 567–8.

Lu, Z., Pan, W., Zhu, X., Wang, D. and Wang, H. (2000). What has the panda taught us? In *Priorities for the Conservation of Mammalian Diversity: Has the Panda had its Day?*, ed. A. Entwistle and N. Dunstone. Cambridge: Cambridge University Press, pp. 325–34.

Mason, G. J. (1991). Stereotypies: a critical review. *Animal Behaviour*, **41**, 1015–37.

Renner, M. J. and Rosenzweig, M. R. (1987). *Enriched and Impoverished Environments: Effects on Brain and Behaviour*. New York, NY: Springer-Verlag.

Schaller, G. B. (1993). *The Last Panda*. Chicago, IL: University of Chicago Press.

Schaller, G. B., Hu, J., Pan, W. and Zhu, J. (1985). *The Giant Pandas of Wolong*. Chicago, IL: University of Chicago Press.

Shepherdson, D. J. (1994). The role of environmental enrichment in the captive breeding and reintroduction of endangered species. In *Creative Conservation: Interactive Management of Wild and Captive Animals*, ed. G. Mace, P. J. S. Olney and A. Feistner. London: Chapman & Hall, pp. 167–77.

Shepherdson, D., Mellen, J. and Hutchins, M. (eds.) (1998). *Second Nature: Environmental Enrichment for Captive Animals*. Washington, DC: Smithsonian Institution Press.

Snyder, R. J., Zhang, A. J., Zhang, Z. H. *et al.* (2003). Behavioural and developmental consequences of early rearing experience for captive giant pandas. *Journal of Comparative Psychology*, **117**, 235–45.

Snyder, R. J., Lawson, D. P., Zhang, A. *et al.* (2004). Reproduction in giant pandas: hormones and behaviour. In *Giant Pandas: Biology and Conservation*, ed. D. Lindburg and K. Baragona. Berkeley, CA: University of California Press, pp. 125–32.

Swaisgood, R. R., Lindburg, D. G. and Zhou, X. (1999). Giant pandas discriminate individual differences in conspecific scent. *Animal Behaviour*, **57**, 1045–53.

Swaisgood, R. R., Lindburg, D. G., Zhou, X. and Owen, M. A. (2000). The effects of sex, reproductive condition and context on discrimination of conspecific odours by giant pandas. *Animal Behaviour*, **60**, 227–37.

Swaisgood, R. R., White, A. M., Zhou, X. *et al.* (2001). A quantitative assessment of the efficacy of an environmental enrichment programme for giant pandas. *Animal Behaviour*, **61**, 447–57.

Swaisgood, R. R., Lindburg, D. G. and Zhang, H. (2002). Discrimination of oestrous status in giant pandas via chemical cues in urine. *Journal of Zoology* (London), **257**, 381–6.

Swaisgood, R. R., Ellis, S., Forthman, D. L. and Shepherdson, D. J. (2003a). Commentary: improving well-being for captive giant pandas: theoretical and practical issues. *Zoo Biology*, **22**, 347–54.

Swaisgood, R. R., Zhou, X., Zhang, G., Lindburg, D. G. and Zhang, H. (2003b). Application of behavioural knowledge to giant panda conservation. *International Journal of Comparative Psychology*, **16**, 65–84.

Swaisgood, R. R., Lindburg, D. G., White, A. M., Zhou, X. and Zhang, H. (2004). Chemical communication in giant pandas: experimentation and application. In *Giant Pandas: Biology and Conservation* ed. D. G. Lindburg and K. Baragona. Berkeley, CA: University of California Press, pp. 106–120.

Tepper, E. M., Hare, V. J., Swaisgood, R. R. *et al.* (2001). Evaluating enrichment strategies with giant pandas at the San Diego Zoo. In *Proceedings, Fourth International Conference on Environmental Enrichment*, ed. V. J. Hare, K. E. Worley and K. Myers. San Diego, CA: Shape of Enrichment, pp. 226–39.

White, A. M. (2001). *Chemical Communication in Giant Pandas: The Role of Marking Posture and Age of Signaler*. San Diego, CA: San Diego State University, Masters Thesis.

White, A. M., Swaisgood, R. R. and Zhang, H. (2002). The highs and lows of chemical communication in giant pandas (*Ailuropoda melanoleuca*): effect of scent deposition height on signal discrimination. *Behavioural Ecology and Sociobiology*, **51**, 519–29.

(2003). Chemical communication in giant pandas: the role of age in the signaller and assessor. *Journal of Zoology* (London), **259**, 171–8.

Zhang, G. Q., Swaisgood, R. R., Wei, R. P. *et al.* (2000). A method for encouraging maternal care in the giant panda. *Zoo Biology*, **19**, 53–63.

Zhang, G. Q., Swaisgood, R. R. and Zhang, H. (2004). An evaluation of the behavioral factors influencing reproductive success and failure in captive giant pandas. *Zoo Biology*, **23**, 15–31.

Zhu, X., Lindburg, D. G., Pan, W., Forney, K. A. and Wang, D. (2001). The reproductive strategy of giant pandas: infant growth and development and mother–infant relationships. *Journal of Zoology* (London), **253**, 141–55.

12

Evaluating stress and well-being in the giant panda: a system for monitoring

RONALD R. SWAISGOOD, MEGAN A. OWEN, NANCY M. CZEKALA,
NATHALIE MAUROO, KATHY HAWK, JASON C. L. TANG

INTRODUCTION

Giant pandas are being maintained in captivity largely for the purpose of creating a reproductively viable population that will support conservation of the species in nature. Toward this end, researchers and managers have targeted many aspects of husbandry for improvement through scientific investigations. Among the many priorities is the ability to measure 'well-being' and possibly alleviate 'stress' imposed by a captive environment. Stress research has been increasingly incorporated into captive wildlife breeding programmes, in part because it is widely believed that small enclosures may not allow animals to execute normal escape and avoidance responses to aversive stimuli. Coping mechanisms may be constrained, thus resulting in stress that can compromise psychological and physiological health, including reproduction (Carlstead & Shepherdson, 2000). Among the many deleterious consequences, stress compromises immune function, reproduction, pregnancy sustainability and maternal care (Munck *et al.*, 1984; Baker *et al.*, 1996; Carlstead, 1996; Moberg & Mench, 2000).

How susceptible is the giant panda to stress imposed by *ex situ* environments? The charisma of this species causes it to attract large

Giant Pandas: Biology, Veterinary Medicine and Management, ed. David E. Wildt, Anju Zhang, Hemin Zhang, Donald L. Janssen and Susie Ellis. Published by Cambridge University Press. © Cambridge University Press 2006.

and noisy crowds. Also, giant pandas are commonly held at major institutions that often undertake large construction projects. This chapter deals with the sensitivity of the giant panda to its captive environment. Stress, more than other biological concepts, has limited utility at the population level. In a single species, however, individual animals seem to vary remarkably in response to environmental change. Thus, we have become advocates for assessing stress on the basis of individuals, and not only for minimising it but to identify those frequently occurring or chronic factors to which an animal fails to habituate. Whether due to suboptimal enclosure design, husbandry practices or disturbances, these variables need to be thoroughly understood and then modified so that an optimal environment can be created for that individual. Our 'individual assessment approach' across many giant pandas living in diverse institutions has been useful for developing a list of idiosyncratic stress responses that potentially may well impact on the species or the individual. We suggest that searching out common threads among these factors *post hoc* will provide a more comprehensive understanding of stress responsiveness in this species while helping to manage this phenomenon to promote well-being, health and reproduction.

WHAT IS STRESS AND HOW IS IT MEASURED?

'Stress' is not a well-defined concept (Hofer & East, 1998; Moberg & Mench, 2000; Sapolsky *et al.*, 2000). Most contemporary definitions involve the animal's perception of a threat that challenges internal homeostasis (both motivational and physiological 'set points'), and behavioural and physiological adjustments that the animal undergoes to avoid or adapt to a 'stressor' and return to homeostasis.

Most stress is acute, short-lived and harmless. However, if a perceived threat persists and the animal fails to adapt, chronic stress may develop, homeostasis is not achieved, and hyperadrenal activity results in sustained and problematically elevated glucocorticoid concentrations (Mendoza *et al.*, 2000). Hereafter, we use the term 'stress' loosely as a descriptive concept involving the behavioural and physiological reactions to external threats to homeostasis (stressors) (Selye, 1956). It is generally inferred that an animal's subjective experience is negative during protracted stress. In this light, avoidance or minimisation of *excessive* stress may have implications for psychological and physiological well-being, including health and reproduction.

To understand stress and its implications for individuals, both the stressor and the type of reactive response must be studied. Environmental challenges can be recorded systematically or manipulated experimentally. If their occurrence is associated with a measurable stress response, then one can conclude that they are biologically salient stressors. However, responses can be remarkably variable. As Hofer & East (1998) point out:

> if individuals vary in the extent to which hormonal, resource allocation, immunological and behavioural systems are triggered. . ., and if these systems follow different time courses. . ., then measuring only part of the response may provide incomplete information about the magnitude and consequences of the full response.

Thus, stress studies must examine multiple indices (Cook *et al.*, 2000).

Hypothalamic–pituitary–adrenal (HPA) activity is one of the most useful measures of stress, in part because it can be monitored non-invasively via metabolites excreted in urine or faeces (Wasser *et al.*, 2000). When an animal perceives a threatening stimulus, a relatively predictable series of physiological adjustments, known as the General Adaptation Syndrome, are activated (Selye, 1946). The sympathetic–adrenal response involves epinephrine and norepinephrine secretion, which increases heart rate and mobilises energy reserves for use by the muscles. This fast-acting response is followed by more long-term HPA activation, resulting in glucocorticoid release by the adrenal cortex, which ultimately can be quantified by increased amounts of metabolites in urine or faeces (Sapolsky *et al.*, 2000; see also Chapter 8). A primary function of glucocorticoids is to shut down components of the stress response in a negative feedback loop to keep it from escalating out of control. Short-term HPA activation is adaptive, but prolonged glucocorticoid production can have pathological effects, including immunological and reproductive suppression, disease and even death (Carlstead & Shepherdson, 1994; 2000).

Stressors do not always activate the HPA system, and changes in glucocorticoids can occur for non-stress-related reasons, for example, due to seasonality, in support of reproductive activity or to mobilise energy reserves (Walker *et al.*, 1992; Altemus *et al.*, 1995; Hofer & East 1998; Romero & Wikelski, 2002). Indeed, our data show that glucocorticoid concentration fluctuate seasonally in the great panda, potentially obfuscating or confounding HPA activity related to stress (Owen *et al.*, in press). Thus, there is not always a clean cause and effect. For example, one recent study of cheetahs demonstrated that paired females experienced evidence of behavioural stress and suppressed ovarian activity, but no elevated corticoid excretion (Wielebnowski *et al.*, 2002). In cattle, individuals that appeared to habituate physiologically (in terms of HPA activity) still showed behavioural distress (Cook *et al.*, 2000). And, in nonhuman primates, chronic stress – or failure to restore homeostasis – has been associated with *reduced* corticoids (Mendoza *et al.*, 2000). Therefore the accurate measurement of a response to stress must go beyond simply tracking fluctuations in adrenal hormone activity (Cook *et al.*, 2000).

One other commonly used and viable index is behaviour, an easily measured, sensitive indicator of an animal's perception of environmental change (Weary & Fraser, 1995; Baker & Aureli, 1997; Boinski *et al.*, 1999; Swaisgood *et al.*, 2001; Wielebnowski *et al.*, 2002). Altered behaviours to a stressor may include stimulus avoidance, formation of stereotypies and changed activity levels associated with feeding, exploratory and sexual behaviours (Mason, 1991; Carlstead & Shepherdson, 2000). However, behavioural signs of stress may also be unreliable because inappropriate behaviours may be selected for measurement, they may be subtle or correlate poorly with physiological indicators, or they may reflect a permanent adaptive response to an environmental change (Hofer & East, 1998; Cook *et al.*, 2000). One frequently cited argument for the unreliability of behaviour is the dichotomous nature of the behavioural component to the stress response – the animal may show heightened activity suggestive of agitation and escape motivation (fight–flight response) or, alternatively, withdraw, hide and become sluggish (conserve–withdraw response) (Moberg, 1985). These differing response modes do not invalidate each other. On the contrary, they highlight the importance of examining *species-specific* or even *individual* responses to stressors rather than expecting universal monotypy.

We have found that behaviour is a valuable window into the perceptual processes of the animal, useful for inferring if its state has changed to address a specific environmental challenge. This deviation from 'behavioural homeostasis' can offer insight into the animal's wellbeing, especially given that a suite of behavioural variables is evaluated in concert. Cook *et al.* (2000) have suggested that behavioural indices illustrative of a stress response include measures of the startle or defence response, the time required to resume normal activity after a stress, aggression, stereotypy, lack of responsiveness or apathy and decreased complexity of behaviour. Assessments can also be carried out more experimentally, for example, by giving an animal a choice between two environments, one presumably less evocative of stress than the other. Better yet, one could determine how much effort an animal is prepared to expend to gain access to a less stressful environment.

POTENTIAL IMPACT OF ANTHROPOGENIC NOISE ON THE GIANT PANDA

In recent years and through partnerships (amongst the San Diego Zoo, the China Conservation and Research Centre for the Giant Panda – Wolong Nature Reserve – and Ocean Park-Hong Kong), we have documented the beneficial effects of enrichment on giant panda well-being (Swaisgood, 1996; Swaisgood et al., 2001; Tepper et al., 2001; see also Chapter 11). Despite our best efforts, sometimes individuals held in enriched environments are still exposed to stressors. Indeed, we have tentatively identified stress as a contributing factor in reproductive failure of at least three female giant pandas at the Wolong facility (Zhang et al., 2004). One significant factor in most zoo environments is noise, such as loud talking by the public, vehicular traffic and construction activities. In 1997, we began to document the impact of noise on the two resident pandas at the San Diego Zoo, an adult male (Studbook, SB, number 381) and an adult female (SB 371) (Owen et al., 2004). Using a sound meter (Model 573; Casella/CEL, Inc., Amherst, NH), we recorded noise levels throughout the day. Behavioural observations were made for two hours (1300 to 1500 hours each day) at a time when noise levels peaked. We recorded behaviours potentially indicative of stress, including locomotion, honking vocalisation, excessive scratching and waiting restlessly at the outside enclosure door, apparently anxious to leave the exhibit. Overnight urine samples were collected daily and analysed for corticoid concentrations, a reflection of the previous day's adrenal activity (see Chapter 8 for methods and explanation).

In an early, crude analysis, we compared the 20 'loudest' and 20 'quietest' days for differences in behavioural and hormonal indices. These preliminary data suggested no major impact of noise on the giant panda (Swaisgood & Borst, 1998). A more detailed analysis of a larger four-year data set did document several significant noise impacts (Owen et al., 2004). First, we defined loud days as the loudest 25% of days (as determined by recorded amplitude levels) and quiet days as the quietest 25%. In one analysis, we examined the average amplitudes for the entire day (loud days = 72 decibels or dB; quiet days = 65 dB). In another, we examined the very loudest sounds heard during the loudest 1% of one-minute intervals (loud days = 81 dB; quiet days = 70 dB). Therefore, this latter analysis examined the impact of the loudest 15 minutes of noise exposure on a given day. During loud days, this maximum noise level was approximately 15 times higher than maximum noise on quiet days. Because decibels are measured on a logarithmic scale, a 3-dB change reflects a doubling of noise energy, i.e. it sounds twice as loud. Although duration of individual noise events was not examined, it is helpful to use the average decibel level for the entire day as a measure of 'chronic' noise exposure and the 15 minutes of the loudest 1% of noise as an index of 'acute' noise.

In general, loud days were associated with high rates of locomotion and honking, restless manipulation of the enclosure door, increased scratching and/or increased concentrations of urinary glucocorticoids. Data also suggested that acute exposure to very loud noise reliably induced behavioural distress, whereas chronically moderate noise was required to activate the HPA and to increase urinary corticoid excretion. Indices of stress were also stronger for low-pitched sound (16–64 hertz or Hz) compared to medium (125–500 Hz)- or high-pitched (1000–16000 Hz) noise. The female that we studied showed greater behavioural sensitivity to noise during oestrus and lactation than during nonreproductive periods and pregnancy/pseudopregnancy. In contrast, urinary glucocorticoid excretion profiles were more dynamic during nonreproductive periods, suggesting that adrenal responsiveness to noise was suppressed during intervals of heightened ovarian activity (as has been found for other species; Cook, 1997). If this observation holds true, using urinary corticoids to monitor stress during crucial reproductive periods may be misleading. Finally, it was clear from our analyses that the two giant pandas differed markedly in noise responsiveness, especially hormonally. For example, only the female produced significantly different glucocorticoid patterns in the urine in response to this particular disturbance.

Although there was a significant noise effect on behaviour and adrenal hormone response, the overall impact on what might be considered well-being appeared admittedly minor. We found that only 15% of a giant panda's total time was spent expressing these mild behavioural indicators of stress. Additionally, when excreted corticoids were elevated, increases were only about 10 to 40% above baseline. Nonetheless, we were able, through an assessment of multiple variables, to document some change in behavioural and physiological function to an environmental factor. Thus, it makes sense that more detailed studies be explored to determine if different (and perhaps more severe) perturbations can influence health and well-being, especially at sensitive intervals, such as at implantation and parturition. Even commonly encountered noise in a zoo atmosphere clearly can activate the stress response system in the giant panda and, therefore, should be monitored and mitigated; and, if noise can be aversive to pandas that have had years to habituate in captivity, then their wild counterparts may be even more likely to respond negatively to such a disturbance. Potentially, noise (caused by encroaching people via mining, farming or ecotourism) could cause pandas to abandon good habitat, dens and cubs.

LESSONS LEARNED

A new stress-monitoring system for the giant panda

Our noise effect studies motivated our interest in examining other potentially stress-inducing environmental challenges. We also realised that our initial methods had several limitations. For example, although the duration, pitch and amplitude of noise were known, we did not always have an idea of the source. Our data collection also was not ideal. Our behavioural observations were based on only a few hours of observation on each panda each week. Although keeper records provided useful information about what occurred outside our observation period, it was not recorded in a manner that allowed statistical analysis. Therefore, we decided to explore a better monitoring system which would capitalise on the observations of animal care staff that work with and observe the pandas intensively each day. Our goal was to provide a way for the keepers to record observations quantitatively as they went about their daily routines which, in turn, would allow statistical evaluations of environmental influences on animal well-being.

We also concluded that this new approach needed to be useful across institutions while addressing a much larger set of factors that may cause stress and aversive reactions. We first made a list of independent variables with the potential of causing stress, including:

- number of visitors;
- visitor behaviour (e.g. arm waving);

- noise;
- novel or strong odours (e.g. vehicular exhaust);
- exposure to conspecifics;
- aggressive interactions with conspecifics;
- new enrichment items;
- interactions with keepers;
- conditioning/training;
- illness;
- mucous stool;
- veterinary procedures;
- medications.

We also identified a host of husbandry variables, including enclosure modifications, confinement in different or smaller portions of customary enclosures, diet modifications and schedule changes (e.g. time of feeding). We recognised that the list of behavioural responses to these potential stressors was similarly diverse. For example, a panda may consume less bamboo, or it may vocalise, pace, increase general movement or engage in more abnormal, stereotypical or door-directed behaviour. Finally, the panda may attend to, retreat or hide from the disturbance or adopt some other idiosyncratic response. As an example of the latter, a 12-year-old female at San Diego Zoo (SB 371), when troubled, would typically climb a tree, 'huff' and bounce up and down while looking at the source of the disturbance.

The new monitoring system has been designed to reflect two discrete modes of the stress response, both of which require the use of a detailed ethogram. The first deals with behaviours that may change immediately upon stressor exposure. In this fight–flight response (Selye, 1946), the HPA may not be activated and, thus, adrenal corticoid production may not change substantially. Similarly, behavioural activity may quickly return to baseline if the disturbance is not protracted or if accommodation occurs rapidly. The second assessed mode involves situations when a disturbance continues unabated, with the animal's activities affected for hours or days, accompanied by a concomitant rise in corticoids. An ability to monitor both types of response simultaneously is critical to understanding what environmental dimensions cause aversion. While a record-keeping system has been created for caretakers to track environmental changes and the behavioural responses that could be provoked, urine samples continue to be collected twice daily to be analysed for corticoids as a reflection of physiological status and adaptive ability.

Experiences from daily monitoring of stress and well-being

Keepers record all changes in the panda's environment and its behaviour and health on a day-to-day basis. It is important to record this information even when no environmental changes are evident and when pandas show no signs of behavioural agitation. For example, a keeper may note that the animal appeared 'anxious' today, and that 'there were many visitors'. But by making this same anxiety observation on another day when there are few visitors, then it will eventually become possible to test statistically the impact of visitor numbers on giant panda 'anxiety'. The success of this approach, however, requires adherence to specific definitions of observable behaviours. The giant panda ethogram used by the authors and with operational definitions and guidelines for recording frequency or severity is depicted in Table 12.1.

Throughout the day, one designated keeper observes the panda's behaviour and makes notes on different behaviours from the ethogram. At the day's end, the keeper estimates the amount of time spent monitoring each animal's activity, a reflection of actual awareness of that person for what that particular animal was doing on a given day. Once this time-frame has been established, the keeper records the approximate percentage of this time the panda spent engaged in behaviours defined in the ethogram. Directions are provided to guide the keeper's rough estimates (see Table 12.1). In conjunction with the keeper's detailed records, these data are later analysed to determine if changes in the environment have an impact on stress and well-being. For example, data *may* reveal that substantial changes to the enclosure increase pacing, that changes to the feeding schedule promote restless door-directed behaviour or that mucous stools are associated with elevated honking vocalisations. By contrast, it may be determined that novel odours and the number of visitors have no effect.

Monitoring acute responses to disturbances

Table 12.2 illustrates some of the variables being tested at the San Diego Zoo and Ocean Park-Hong Kong to assess a giant panda's acute response to disturbance – a sudden change in environment. An animal may

Table 12.1. *Examples of behaviours recorded in the daily monitoring ethogram for the giant panda*

Behaviour	Definition
Appetite-bamboo	Consumption of provisioned bamboo. If food is weighed, record weight of consumed food in kg. If not, estimate percentage of normal diet consumed. For example, if animal eats half of its normal amount, record '50%'
Pace/quasi-pace	Back and forth, or perimeter, travel in a repetitive, sustained, predictable pattern. Animal follows the same or a similar path repeatedly, using only a small portion of the space available for locomotion
Anxious walk	Locomotion in an anxious manner which is reminiscent of pacing but does not meet the repetitive, predictable pattern required for pace or quasi-pace
Door-directed behaviour	Panda at a door or gate or any area where food or keeper interaction may occur. Panda's behaviour is oriented toward food, keeper or gate; waiting with mild to extreme restlessness. Behaviours may include sniffing, pawing or scratching, pushing, head butting or other manipulations
Honk	A short, tonal, nasal call, falling in pitch. Honking almost always occurs in a series lasting at least several minutes
Uncooperative	Participation when shifting from one enclosure to another or during training sessions, or less formal keeper interactions. Unlike other behavioural measures in this section, rate degree of uncooperativeness on a scale from 0 to 3; 3 is most cooperative

become startled and run in response to a loud noise or appear agitated and move away from a noisy, large group of people near the exhibit. Again, the keeper (or other observer) can record the exact nature of the disturbance according to defined categories, as well as the severity of response on a scale of 0 to 3 (Table 12.3).

Such an approach adds a more detailed and temporal dimension to data collection. Furthermore, it has the potential of identifying the precise source of the disturbance while pairing it with the immediate behavioural response, thus providing better information on cause and effect. It also allows us to 'catch' ephemeral responses. For example, a brief honking bout after a visitor waves his arms may not influence honking rate for that day, yet the fact that it occurred suggests that it

Table 12.2. *Examples of environmental disturbances or husbandry changes recorded for monitoring responses to acute disturbances*

Variable	Definition/comments
Strong novel odours	Describe odour or known source. Sources may include fresh paint or cleaning products
Enclosure restrictions	Panda restricted to a smaller portion of the enclosure than normal
Activity of visitors	Visitor(s) wave arms, jump up and down, lunge forward, move rapidly or throw objects
Monitoring noise with sound meter	Methods vary according to equipment used
Subjective noise monitoring	Evaluate noise level based on his/her experience during course of the day. Describe noise events and note approximate duration
• Source(s) of noise	If noise source is outside the panda facility (e.g. construction), the proximity of the disturbance is noted
• Approxi- mation of pitch	*Low* Examples: large truck or deep human voice *Medium* Example: group of adults talking in normal speaking voices *High* Examples: shrill, piercing, metal grinding against metal, a small dog barking or a small child's scream
• Approxi- mation of amplitude	*Quiet* Not louder than normal speaking voice *Moderate*: Not louder than voices in a heated argument (not quite yelling)
	Loud Ambient noise such that it would be difficult to hear someone talking at a normal speaking voice
Mucous stool	Panda passes faeces containing large quantities of mucus, often accompanied by signs of physical discomfort
Exposure to conspecific	Any substantial *change* in sensory access to other panda individuals. Examples include exposure to the odours of another panda, visual access through cage bars, visual access with limited physical access through cage bars or direct physical access in the same enclosure. Indicate frequency with which panda is subjected to this exposure

may have been an important event. Similarly, an animal retreating from a noise disturbance would not be recorded under the daily monitoring system. The acute response system does not, however, allow us to determine whether a panda displays a chronic stress response, has a

Table 12.3. *Examples of a behavioural ethogram for the response to acute disturbance*

Behaviour	Definition/scoring
Attends	Panda orients towards the disturbance
	0 = Not observed
	1 = Animal calmly focuses attention in the direction of the disturbance (e.g. sniffs air or looks toward disturbance). May be brief or last several seconds
	2 = Animal intensely focuses attention in the direction of the disturbance (e.g. eyes riveted). May stand or otherwise reorient body position. Posture may appear rigid, and response will be held for at least several seconds
	3 = Animal is focused on, and moves towards, the disturbance in a deliberate manner (at least two body lengths or approaching the limit of the enclosure)
Hides	Behaviour that places visual barrier between panda and source of disturbance
	0 = Not observed
	1 = Animal covers eyes/head with paws
	2 = Animal withdraws to a corner or crouches behind available cover, such as a log pile that offers a partial visual barrier
	3 = Animal withdraws to another space, including behind a full visual barrier (e.g. wall or enters another room)
Retreats	Behaviour that increases distance between panda and source of disturbance
	0 = Not observed
	1 = Animal moves less than three body lengths away in a calm manner
	2 = Animal moves greater than three body lengths away in a calm manner *or* less than three body lengths in rapid fashion
	3 = Animal moves greater than three body lengths away from disturbance rapidly or climbs a tree or other vertical structure

delayed response to a stressor or if it responds to less abrupt environmental changes. The two methods together offer complimentary information and a more complete picture of the potential effects of stress on well-being in captivity as well as simply more biological information about this fascinating species.

Our data collection for the 'acute response system' is not yet complete, but we foresee its potential to provide concrete answers to questions about which aspects of the typical zoo environment pandas may find aversive. Examples of potential findings that these data collection protocols could produce include:

1.   A loud noise (especially at a high pitch) commonly startles a giant panda causing it to retreat and hide, often followed by pacing and/or honking.
2.   A giant panda often honks after an aggressive interaction with a conspecific.
3.   A giant panda often attends to arm-waving visitors but shows no other overt response.
4.   Most episodes of stereotypic behaviour are in response to groups of large visitors (i.e. more than 100 people).

Preliminary findings from Ocean Park-Hong Kong are already providing useful feedback for making management changes. The giant panda enclosures at this facility are entirely indoors, primarily for temperature regulation. While this allows the animals to enjoy the benefits of appropriate temperature and decreases external noise, indoor crowd noise may be amplified. Although we have found no dramatic physiological evidence indicating severe distress, factors such as husbandry changes, noise, crowd size and illness appear to be associated with slightly elevated levels of stress-related behaviours throughout the day. More intensive monitoring has helped identify the source of noise disturbances. For example, on four occasions, this particular panda pair responded to high-pitched, moderately loud drilling from inside the facility with increased pacing, restless behaviour at the door leading to the night quarters and, in one case, immediate rapid retreat from the disturbance. Similar responses were also observed following low-pitched, medium-amplitude noises from outside the facility.

PRIORITIES FOR THE FUTURE

We believe that the value of our observations to date are to re-emphasise the importance of intensive and interdisciplinary assessment of giant pandas – behaviourally and physiologically, as well as among individuals. We have only begun to 'mine' the amount of information that is needed to optimise captive habitats which, in turn, will maximise health and reproduction. This chapter has provided some tantalising evidence that the giant panda is indeed sensitive to its environment. It definitely reacts to acute disruptions, but as yet we have only a limited understanding of the impact on psychological and physiological well-being, acutely or long term. Yet it is possible using a combination of tools and approaches described in this chapter to address this topic experimentally. At the same time, it is obvious that highly detailed monitoring will be necessary to identify explicit causes and effects. Resulting data will be highly valued by captive managers, allowing them to design or recreate (e.g. through enrichment) *ex situ* environments that promote naturalistic behaviours and reproductive functions. Additionally, close tracking through the use of both keepers and researchers, armed with systematic data collection protocols, will allow the earliest possible detection of illness for treatment.

Finally, we predict that stress-related studies of giant pandas living in captivity will have application to studying and conserving wild counterparts. A final example is recent work in our laboratory that has addressed the impact of a radiotelemetry collar placed on a giant panda. For years, controversy has swirled in China about radiocollars being problematic for free-living pandas. Without substantive evidence to the contrary, the central government now bans radiocollars on wild giant pandas, which greatly compromises ecological studies of, for example, movement and foraging activities. To begin to generate real, quantifiable data, we immobilised and fitted four giant pandas at the Wolong *ex situ* facility with standard radiocollars (Durnin *et al.*, 2004). Behaviours and daily urine samples (for glucocorticoid assessments) were measured for two weeks before and two weeks after collar placement. Despite intensive monitoring, we observed no differences between the control (pre-collar) and treatment (post-collar) interval in any of the observed behaviours or in hormonal profiles or concentrations. The 'absence of stress' after anaesthesia and collar placement offers potentially important new information to decision-making authorities in China. These specific results could be critical in helping to consider lifting the current moratorium on radiocollaring wild giant pandas, a policy that hinders obtaining crucial knowledge about wild populations and individuals. Lastly, it may be that these types of study could also eventually play a role in identifying and remediating a growing list of potential stressors that may well be influencing giant pandas in nature. Many anthropogenic activities, ranging from pollutants, traffic noise and ecotourism to research and conservation activities themselves, may activate stress response systems in wild animals, negatively affecting population fitness (Hofer & East, 1998). It is our hope that these research methods can eventually be adapted for evaluating and managing stress in wild living giant pandas.

ACKNOWLEDGEMENTS

We are grateful to the Zoological Society of San Diego for financial support. A portion of this research was funded by a grant from the Hong Kong Society for Giant Panda Conservation. We also thank observers who contributed to these projects, including Kim Bacon, Susan Ewing, Suzanne Hall, Valerie Hare, Reimi Kinoshita, Nathalie Kwok, Barry Kwok and Dallas Ripsky.

REFERENCES

Altemus, M., Deuster, P. A., Galliven, E., Carter, C. S. and Gold, P. W. (1995). Suppression of hypothalamic–pituitary–adrenal axis responses to stress in lactating women. *Endocrinology*, **80**, 2954–9.

Baker, A. J., Baker, A. M. and Thompson, K. V. (1996). Parental care in captive mammals. In *Wild Mammals in Captivity*, ed. D. G. Kleiman, M. E. Allen, K. V. Thompson and S. Lumpkin. Chicago, IL: University of Chicago Press, pp. 497–512.

Baker, K. C. and Aureli, F. (1997). Behavioural indicators of anxiety: an empirical test in chimpanzees. *Behaviour*, **134**, 1031–50.

Boinski, S., Swing, S. P., Gross, T. S. and Davis, J. K. (1999). Environmental enrichment of brown capuchins (*Cebus apella*): behavioral and plasma and fecal cortisol measures of effectiveness. *American Journal of Primatology*, **48**, 49–68.

Carlstead, K. (1996). Effects of captivity on the behavior of wild mammals. In *Wild Mammals in Captivity*, ed. D. G. Kleiman, M. E. Allen, K. V. Thompson and S. Lumpkin. Chicago, IL: University of Chicago Press, pp. 317–33.

Carlstead, K. and Shepherdson, D. J. (1994). Effects of environmental enrichment on reproduction. *Zoo Biology*, **13**, 447–58.

(2000). Alleviating stress in zoo animals with environmental enrichment. In *The Biology of Animal Stress: Basic Principles and Implications for Animal Welfare*, ed. G. P. Moberg and J. A. Mench. Cambridge, MA: CABI, pp. 337–54.

Cook, C. J. (1997). Oxytocin and prolactin suppress cortisol response to acute stress in both lactating and non-lactating sheep. *Journal of Dairy Research*, **64**, 327–9.

Cook, C. J., Mellor, D. J., Harris, P. J., Ingram, J. R. and Mathews, L. R. (2000). Hands-on and hands-off measurement of stress. In *The Biology of Animal Stress: Basic Principles and Implications for Animal Welfare*, ed. G. P. Moberg and J. A. Mench. Cambridge, MA: CABI, pp. 123–46.

Durnin, M. E., Swaisgood, R. R., Czekala, N. and Zhang, H. (2004). Effects of radiocollars on giant panda stress-related behavior and hormones. *Journal of Wildlife Management*, **68**, 987–92.

Hofer, H. and East, M. L. (1998). Biological conservation and stress. *Advances in the Study of Behavior*, **27**, 405–525.

Mason, G. J. (1991). Stereotypies: a critical review. *Animal Behaviour*, **41**, 1015–37.

Mendoza, S. P., Capitanio, J. P. and Mason, W. A. (2000). Chronic social stress: studies in nonhuman primates. In *The Biology of Animal Stress: Basic Principles and Implications for Animal Welfare*, ed. G. P. Moberg and J. A. Mench. Cambridge, MA: CABI, pp. 227–47.

Moberg, G. P. (1985). Biological response to stress: key to assessment of animal well-being? In *Animal Stress*, ed. G. P. Moberg. Bethesda, MD: American Physiological Society, pp. 27–50.

Moberg, G. P. and Mench, J. A. (2000). *The Biology of Animal Stress: Basic Principles and Implications for Animal Welfare*. Cambridge, MA: CABI.

Munck, A., Guyre, P. M. and Holbrook, N. J. (1984). Physiological functions of glucocorticoids in stress and their relation to pharmacological actions. *Endocrine Reviews*, **5**, 25–44.

Owen, M. A., Czekala, N. M., Swaisgood, R. R., Steinman, K. and Lindburg, D. G. Seasonal and diurnal dynamics of glucocorticoids and behavior in giant pandas: implications for monitoring well-being. *Ursus* (in press).

Owen, M. A., Swaisgood, R. R., Czekala, N. M., Steinman, K. and Lindburg, D. G. Monitoring stress in captive giant pandas (*Ailuropoda melanoleuca*): behavioral and hormonal responses to ambient noise. *Zoo Biology*, **23**, 147–64.

Romero, L. M. and Wikelski, M. (2002). Exposure to tourism reduces stress-induced corticosterone levels in Galapagos marine iguanas. *Biological Conservation*, **108**, 371–4.

Sapolsky, R. M., Romero, L. M. and Munck, A. U. (2000). How do glucocorticoids influence stress responses? Integrating permissive, suppressive, stimulatory and preparative actions. *Endocrine Reviews*, **21**, 55–89.

Selye, H. (1946). The general adaptation syndrome and the diseases of adaptation. *Journal of Clinical Endocrinology*, **6**, 117–230.

(1956). *The Stress of Life*. New York, NY: McGraw-Hill.

Swaisgood, R. R. (1996). Behavioral monitoring at the San Diego Zoo: implications for captive enrichment. In *Proceedings, Giant Panda Technical Conference*. Chengdu.

Swaisgood, R. R. and Borst, J. (1998). Sound advice for giant pandas. *Sound and Vibration*, **32**, 6–12.

Swaisgood, R. R., White, A. M., Zhou, X. *et al.* (2001). A quantitative assessment of the efficacy of an environmental enrichment programme for giant pandas. *Animal Behaviour*, **61**, 447–57.

Tepper, E. M., Hare, V. J., Swaisgood, R. R. *et al.* (2001). Evaluating enrichment strategies with giant pandas at the San Diego Zoo. In *Proceedings, Fourth International Conference on Environmental Enrichment*, ed. V. J. Hare, K. E. Worley and K. Myers. San Diego, CA: Shape of Enrichment, pp. 226–39.

Walker, C. D., Lightman, S. L., Steele, M. K. and Dallman, M. F. (1992). Suckling is a persistent stimulus to the adrenocortical system of the rat. *Endocrinology*, **130**, 115–25.

Wasser, S. K., Hunt, K. E., Brown, J. L. *et al.* (2000). A generalized fecal glucocorticoid assay for use in a diverse array of non-domestic mammalian and avian species. *General and Comparative Endocrinology*, **120**, 260–75.

Weary, D. and Fraser, D. (1995). Calling by domestic piglets: reliable signs of need? *Animal Behaviour*, **50**, 1047–55.

Wielebnowski, N., Ziegler, K., Wildt, D. E., Lukas, J. and Brown, J. L. (2002). Impact of social management on reproductive, adrenal and behavioural activity in the cheetah (*Acinonyx jubatus*). *Animal Conservation*, **5**, 291–301.

Zhang, G., Swaisgood, R. R. and Zhang, H. (2004). An evaluation of behavioral factors influencing reproductive success and failure in captive giant pandas. *Zoo Biology*, **23**, 15–31.

13

The neonatal giant panda: hand-rearing and medical management

MARK S. EDWARDS, RONGPING WEI, JANET HAWES, MEG SUTHERLAND-SMITH,
CHUNXIANG TANG, DESHENG LI, DAMING HU, GUIQUAN ZHANG

INTRODUCTION

Among eutherians, ursids have a significant disparity between maternal weight and neonatal weight (Leitch *et al.*, 1959). The giant panda also produces a smaller litter mass relative to maternal body mass than, for example, the American black bear (Oftedal & Gittleman, 1989; Ramsay & Dunbrack, 1996; Zhu *et al.*, 2001). The giant panda neonate is particularly altricial (i.e. highly dependent on parental care), requiring 24-hour care during the first weeks of life. This chapter deals with the issues and intricacies associated with the newborn giant panda cub, including hand-rearing and medical management.

NEONATAL CARE AND HAND-REARING: METHODS, RESULTS AND RECOMMENDATIONS

Indications for hand-rearing

Although maternal care is always preferred for the giant panda cub, there are situations when human care-giving is mandatory. The most obvious is maternal abandonment, which usually becomes apparent

Giant Pandas: Biology, Veterinary Medicine and Management, ed. David E. Wildt, Anju Zhang, Hemin Zhang, Donald L. Janssen and Susie Ellis. Published by Cambridge University Press. © Cambridge University Press 2006.

315

within the first five to ten minutes of birth. A female that abandons her cub will typically leave it on the ground and move away, showing little or no interest. Intervention is also required when the dam holds the cub improperly (malpositioning). Such a cub can neither nurse nor rest, often moves about excessively (in an attempt to achieve proper positioning on its own) and then can fall to the ground. A third complication is the common production of two or more cubs (mean litter size is 1.7; range 1–3) (Schaller *et al.*, 1985). Despite the frequency of multiple births, a female giant panda rarely cares for more than one offspring, probably because of the intensity of maternal attention required. One mitigating strategy has been hand-rearing one of the twins or, more recently, implementing 'twin swapping' (see below). Some females also produce inadequate amounts of milk, thereby requiring human assistance. In this case, the neonate often shows excessive activity searching for the female's nipple. Finally, dam or cub illness or injury may require separating the female from neonate to provide treatment.

Criteria indicating the need for hand-rearing are most obvious during the first three days postpartum and include neonate vocalisation, activity levels and skin colouration. The significance of these criteria declines with the cub's increasing age. Of the three, vocalisation is perhaps most important. Loud calls, emitted a few times each hour, indicate a healthy cub. However, increased frequency of loud vocalisations suggests that the neonate is uncomfortable, e.g. being held in an awkward position or consuming too little milk. A decline in call intensity and frequency can mean reduced vitality. If vocalisations are not heard for more than one hour, the cub should be examined. It may be useful to awaken the female, prompting her to reposition the cub so that caregivers can more easily monitor its movements and calls. It is also important to see skin with a healthy pink colour. The skin of a sickly cub will increase in pallor, suggesting the need for hands-on evaluation.

Admittance to the nursery

A cub transferred from the dam's enclosure to a nursery should be examined and treated as needed (see 'Medical management of neonates', p. 330). Once completed, the neonate should be placed in a prewarmed incubator (Fig. 13.1; Plate XI) or a similar controlled environment. Most healthy cubs will be moving actively within two to three hours, seeking a nipple. Animals showing this behaviour should be fed

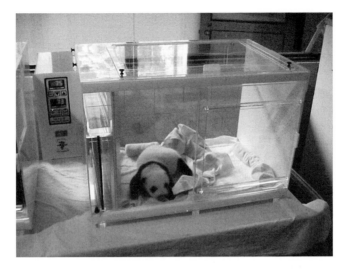

Figure 13.1. A controlled microenvironment (Animal Intensive Care Unit) used for housing giant panda cubs from Day 0 to Day 50 (photograph by Mark S. Edwards). (See also Plate XI.)

(see 'Feeding and nutrition', p. 320 below) and stimulated to urinate and defecate (see 'Feeding regimes and concerns', p. 325). Each of these activities should be conducted within the confines of the incubator to avoid sudden shifts in environmental temperature. If 'nipple seeking' is not observed within five hours, the neonate should be encouraged to take its first bottle. Those that are weak or unable to suck may require enteral feeding via a stomach tube (see 'Feeding and nutrition').

Personnel, housing and environmental conditions

The number of personnel working with a hand-reared cub should be limited during the first month – fewer caregivers increase the chance to detect early problems in appetite, behaviour and overall vitality. A schedule of 12-hour shifts around the clock may be necessary during this critical time.

Due to its altricial nature, the supporting microclimate for a giant panda cub is critical, especially for the first 50 days of life. A summary of required housing and environmental conditions on the basis of early age is provided in Table 13.1.

Table 13.1 *Housing equipment, environmental temperature and humidity suggested for hand-rearing the giant panda cub from day of birth (Day 0) to Day 50*

Day	Housing	Temperature	Humidity (%)
0	Incubator	35.0°C (95.0°F)	80
2	Incubator	34.0°C (93.2°F)	80
5	Incubator	32.5°C (90.5°F)	80
10	Incubator	30.0°C (86.0°F)	70
15	Incubator	29.0°C (84.2°F)	70
20	Incubator	26.5°C (79.7°F)	67
25	Incubator	25.5°C (77.9°F)	68
30	Incubator	24.5°C (76.1°F)	69
35	Incubator	23.5°C (74.3°F)	73
40	Begin weaning from incubator	22.0°C (71.6°F)	70
45		21.5°C (70.7°F)	73
50	Incubator discontinued	19–20°C (66.2–68.0°F)	77

The incubator environment (either human infant incubator or animal intensive care unit) should be kept sanitary with any soiled bedding changed immediately. The incubator should provide warmth, softness and security. The neonate needs supplemental heat to maintain its body temperature in the early weeks of life. Since heat lowers humidity, both should be monitored inside the incubator using a thermometer and hygrometer, respectively. Low humidity causes dry skin (most notable on the feet and at the tail base) and may lead to dehydration. The incubator should be adjusted to achieve a desired humidity (see Table 13.1) and, if necessary, a small sponge or cloth soaked with clean water may be placed in the rear of the incubator. A warm or cool mist humidifier should also be added to the nursery room to increase overall humidity. Humidifiers and incubators should be cleaned routinely to prevent mould and mildew growth.

While in the incubator, the giant panda cub should be provided with a 'surrogate' for the maternal body. Unlike other bears, the mother panda does not place the infant on the ground, floor or substrate unless she leaves the den. Instead, she continually holds the cub ventrally (similar to many primates) while using a paw to steady and support the infant. Surrogates developed at the San Diego Zoo include a human infant car seat lined with fur as well as a 30-cm round bean-bag covered with fur, and long flat pillows to simulate the mother's paws (Fig. 13.2;

Figure 13.2. Bean-bag support used to feed giant panda cubs in an appropriate position (photograph by Mark S. Edwards). (See also Plate XII.)

Plate XII). Simulated fur hairs should be about 2.5 cm long (similar to that found on an adult female's chest) and should be high quality to avoid shedding and accidental ingestion. Because rough edges can catch on bedding (especially in the case of towels with loops), the animal's claws should be clipped and filed. Ripped or frayed bedding should be discarded to avoid entanglement. Cage furnishings should not block the air circulation openings in the incubator unit.

At about six weeks of age, the cub can begin a transition over about a seven-day interval from the incubator to a draft-free room environment (at room temperature). Based on observations of mother-reared giant pandas, dams may leave cubs on the ground for long periods by Day 30; thus, some surrogate weaning should be attempted at this age. Durability and safety are important in designing transition housing for young pandas. The interim enclosure should be safe while providing adequate space for stimulation, investigation, climbing and enrichment. Favourite comfort items from the previous incubator environment should be incorporated into the new enclosure to provide continuity and security. Any toys should be durable and disinfected, as a young giant panda can easily destroy and ingest fragile materials.

Visits to adult enclosures or larger exercise areas are important to develop locomotor skills and desensitise the hand-reared animal to noise, commotion and the larger world. Visits should begin as soon as the panda is able to self-ambulate. However, these exercise areas require frequent inspection for hazards and for the animal's time to be supervised by a familiar care-giver. A young panda will gradually become comfortable in such surroundings, which will allow increased exercise time.

Feeding and nutrition

When preparing formula, all equipment must be kept sanitary, and personnel must wash their hands properly and frequently. Measuring cups, mixing utensils and containers used for the storage or preparation of milk formula should be washed in hot water with mild detergent and a 2% bleach solution. All equipment should be soaked, scrubbed and then thoroughly rinsed and air-dried. Bottles, nipple rings and formula containers should be steam-sterilised over boiling water for three minutes before refilling. Latex nipples should never be heat-sterilised as this can cause premature hardening or cracking. Formula ingredients must be accurately weighed and kept only for the length of time indicated by the manufacturer. Containers of prepared formula should be dated, and no formula fed after the recommended expiration date.

Colostrum

Concentrations and relative transfer of maternal immunoglobulins (Ig) and active phagocytic cells via first milk produced during lactation (colostrum) have not been quantified in the giant panda. However, practical experience and information from other carnivores (i.e. cat, dog and mink) suggest that this secretion is important for establishing the neonate's passive immunity. How long the gastrointestinal tract remains permeable to IgG is unknown, but this interval ranges from eight days in mink to 170 to 200 days in altricial marsupials (Hanson & Johansson, 1970; Yadav, 1971). The potential importance of other Ig, such as IgA, in protecting against infection at the level of the gastro-intestinal tract, even after the period of intact protein absorption, should not be overlooked.

Four strategies have been considered for providing colostrum, or its physiological equivalent, to a giant panda cub needing supportive care. The first – colostrum collection from a donor – requires extensive pre-partum conditioning and a strong relationship between the panda and its human care-giver. In the absence of quantified concentrations of Ig in early stage milk, samples collected two to three days postpartum may be used in feeding the neonate. The mammary gland and nipple are massaged four to five times per day for the first three days postpartum, with each collection typically providing 1 to 3 ml of colostrum-rich milk. Colostrum has a light green colour for six or seven days, sometimes longer (G. Zhang, pers. comm.); however, it is important to recognise that the association of this colour with secretions of higher Ig content has not been quantified. Harvested colostrum is poured through a coarse filter to remove particulate (hair, dirt) contamination, then placed in a hot water bath at no less than 63°C for 30 minutes to kill bacteria (low temperature, long-time treatment pasteurisation). It can be used immediately for feeding, be refrigerated until needed or frozen for future use.

The remaining strategies are all theoretical. One could rely on using Ig from serum that has been aseptically collected from a healthy donor. Handling, processing and administration of such serum has been outlined for other mammals (Levy et al., 2001). Harvested serum should be administered on the day of birth (Day 0) and the day after (Day 1) based on 1 ml per 20 g of body weight or 3650 mg IgG per kg body weight. Another strategy could include using a commercial colostrum substitute, such as those available for the dog and cat. These products may provide Ig for passive transfer, as well some protection at the gut mucosa. Finally, there is the potential of using concentrated giant panda Ig, a procedure described by Loeffler et al. (see Chapter 16).

Maternal milk

Samples of opportunistically collected maternal milk may be available in quantities sufficient for feeding a neonate. If used for feeding, these samples would be handled in a sanitary fashion similar to colostrum (as described above).

Formula

In the absence of maternal milk, formula may be offered as the primary nutrient and energy source. A formula that has supported giant panda cubs as part or all of the diet from birth to weaning is provided in Table 13.2. The principal components of this formula are a human milk

Table 13.2 *Ingredients and selected nutrient composition of a formula for feeding giant panda cubs*

Ingredient	Amount (g)	
Esbilac® powder	12.5	
Enfamil® (low iron[a]/with iron[b]) powder	12.5	
Water	75	
Composition	As-fed basis	Dry-matter basis
Total solids (%)	24.6	100.0
Energy (kcal metabolisable energy g^{-1})	1.39	5.64
Crude protein (%)	5.7	23.2
Whey (%)	0.84	6.90
Casein (%)	2.19	17.90
Whey:casein	28:72	
Crude fat (%)	8.9	36.1
Total carbohydrates (%)	8.9	36.2
Ash (%)	1.1	4.5
Calcium (%)	0.18	0.73
Phosphorus (%)	0.13	0.52

[a] Day 0–30; [b] Day 31: Weaning.

replacer (Enfamil®, Mead Johnson Nutritionals, Evansville, IN; www.meadjohnson.com) and a domestic dog milk replacer (Esbilac®, PetA9®, Hampshire, IL; www.petag.com). The blending of these two commercially available products delivers micronutrients consistent with estimated requirements of growing carnivores from the Esbilac® component, along with a lower casein, higher whey fraction from the Enfamil® component. The latter characteristic appears to be important in regulating curd formation in the stomach of the giant panda. High-casein milks tend to form a strong clot, or curd, in the neonate's stomach. This is an adaptive strategy in some species (e.g. bovids) which regulates the digestion and absorption of energy and nutrients between feedings. In other species (e.g. humans) that consume high-casein milk, these curds can become indigestible, forming obstructions (lactobezoars) in the gastrointestinal tract. Morbidity associated with the formation of lactobezoars has been reported in other hand-reared species, including tropical bears. This formula was developed to minimise the potential of indigestible curd forming.

These two products are available in many different forms, including powders, concentrated liquids and ready-to-use liquids. The use of powdered ingredients is preferred as they provide a higher total whey content, which is less likely to form lactobezoars. However, using powders complicates mixing. Blenders should never be used to mix powders with water as the force of the process can separate the lipids from the remainder of the formula. After water is added, it is important to strain the mixture once by stirring or pressing the formula through a kitchen strainer, incorporating all powder lumps. This should be repeated, as necessary, to eliminate larger particles and thus avoid clogging the nipple of the feeding bottle. Additionally, it is important to include Enfamil® low iron powder in the formula from Days 0 to 30, due to the tendency of the higher iron product to cause constipation and dyspepsia. On Day 31, Enfamil® low iron should be replaced with Enfamil® with iron. Finally, in the case of a weak or dehydrated infant, an oral electrolyte solution, such as Pedialyte® (Abbott Laboratories/Ross Products, Columbus, OH; www.similac.com) may also be offered as the sole food item for the first few feedings. The quantities offered are dependent on the extent that the infant is dehydrated; however, fluid maintenance requirements for other carnivores (e.g. the domestic dog and cat) range from 50 to 90 ml kg^{-1} body weight per day (Haskins, 1988).

There are alternative ingredients for some of the products shown in Table 13.2. One is Enfamil® Concentrated Liquid, which has 50% less water than the ready-to-use liquid. Using Enfamil® Concentrated Liquid requires adding an equal volume of water to make a formula equivalent to Enfamil® ready-to-use liquid. Enfamil® Powder is also available and requires mixing with two volumes of water to generate a product equivalent to ready-to-use Enfamil®. Esbilac® is also commercially available as a liquid or dry powder. For the latter, powder is mixed with two volumes of water, which gives the equivalent of the liquid counterpart.

Although it is recommended that all components, including water, of any hand-rearing formula be measured and allocated by weight, some facilities may be lacking appropriate weighing equipment. As an interim solution, the weight of various formula components per 100 ml volume is summarised in Table 13.3.

Feeding apparatus

The type of nipple used is crucial, especially during the early days of life because the neonate's swallowing reflex is incomplete. Heat sterilising nipples causes premature degradation, and worn, hardened or cracked

Table 13.3 *Weights per 100 ml of products used in formula for feeding giant panda cubs*

Product	Volume (ml)	Weight (g)
Enfamil®, ready-to-use liquid	100	–
Enfamil®, concentrated liquid	100	105
Enfamil®, powder	100	63
Esbilac®, liquid	100	105
Esbilac®, powder	100	43
Water	100	100

nipples increase the risk of aspiration and must be discarded. Soft nipples with large holes require less vigour for successful nursing, but can also allow excessively rapid formula flow. The preferred apparatus for the newborn giant panda for the first two weeks is the Ross/Abbott Infant Special Care Nipple (Abbott Laboratories/Ross Products), which is short, soft and originally developed for the premature human infant. The end of each of these nipple units has a pre-stamped, appropriately sized hole. Even so, a dilute formula can flow too rapidly from the nipple. Securing the nipple ring tightly to the bottle and holding the bottle horizontally during feeding will slow formula flow to avoid aspiration. From Day 14 until four months of age, an Evenflo® Standard Single Hole Nipple (Evenflo® Products, Piqua, OH; www.evenflo.com) is recommended. Either of the above nipples can be affixed to a Gradu-feed® Nurser (Evenflo® Products), a 60-ml graduated bottle which allows accurate monitoring of the amount of formula consumed. From four until eight months of age, a standard lambing nipple can be used in combination with a Nasco® Feeding Bottle (8 or 16 ounce; Nasco®, Modesto, CA; www.nascofa.com).

Giant pandas may feed from a bottle up to 18 months of age, but should be encouraged to consume formula from a shallow bowl by eight months.

Feeding a 'compromised' neonate

In the case of a cub that is injured, weak or extremely small (<80 g body weight), the suckling response may be insufficient to ingest adequate formula from the Ross/Abbott nipple. Tube feeding should be considered. Then, once the swallowing/suckling reflex is developed, the infant should be graduated to a soft marsupial nipple (WXICOF,

Wentzville, MO; www.wxicof.com) before advancing to the Ross/Abbott nipple. However, extreme care should be used when first using this marsupial nipple, ensuring that the nipple hole is as small as possible and that the nipple be placed in the panda's mouth at a horizontal angle to avoid too-rapid fluid flow and aspiration of formula into the lungs. At the breeding centre in the Wolong Nature Reserve in 1999, a 53-g cub accepted a short, rounded teat (Pet Nurser®, Pet A9®) after the initial first days of tube feeding. Within ten days, he was nursing from the Ross/Abbott nipple. By Day 22, his muzzle had grown sufficiently to accommodate the standard Evenflo® nipple.

Feeding regimen and concerns

The semi-hairless, newborn panda can easily become chilled when removed from its incubator for feeding. Therefore, the feeding area should be both warm and humid, using a portable space heater and humidifier, if necessary. Personnel should begin by washing hands thoroughly or wearing clean gloves. The formula should be warmed to body temperature (38 to 39°C) in a cup of warm clean water. A microwave should not be used for warming because uneven heating can cause serious burns.

The amount of formula per bottle should not exceed gastric capacity, which is projected as 50 ml kg^{-1} body weight. For the first week of life, total daily intake should not exceed 20% body weight. The total volume offered is equally divided among the number of feedings offered per day. If the formula is tolerated well by the end of the first week, the total daily volume can be increased to 25% of body weight. Suggested feeding frequencies and volumes for giant panda cubs of various ages are provided in Table 13.4.

The neonatal giant panda will not urinate or defecate without stimulation. To mimic stimulation by licking from the dam, warm water should be gently applied to the urogenital region with a cotton ball or damp cloth. This is typically done before each feeding; stimulation does not always produce urine and faeces. However, frequent opportunities to eliminate should be given because mother pandas frequently lick the genital region of the cub during the first few weeks postpartum.

The cub should be fed in a sternal position with the head slightly elevated, although some individuals appear more comfortable with their entire body (head up) at 45 degrees. Rolled-up artificial fur or a surrogate add comfort. Begin by feeding only small amounts, gradually

Table 13.4 *Suggested feeding frequency, number of feedings and total formula volume (% body weight) offered during a 24-hour period for giant panda cubs of various ages (Day 0 = day of birth)*

Age	Feeding frequency	Per 24 hours	Total % body weight
Days 0 to 7	Every 2 hours	12	20
Days 8 to 28	Every 2 hours	10 to 12	25
Days 29 to 56	Every 3 hours	8	25
Days 57 to 98	Every 3 to 4 hours	6	25
Days 99 to 140	Every 4 hours	5	24
Days 141 to 182	Every 5 hours	4	23
Months 6 to 12	Every 7 hours	3	23
Months 12 to 13	Every 12 hours	2	20
Months 13 to 14	Every 12 hours	2	18.5
Months 14 to 15	Every 12 hours	2	14
Months 15 to 16	Every 12 hours	2	9
Months 16 to 17	Every 12 hours	2	4.5
Months 17 to 18	Every 12 hours	1	2.5

increasing volume over time. Refusing a feeding is normal, and a cub should never be forced to finish all the formula being offered. Neonates ingest air during nursing bouts. Therefore, post-feeding pats and massage should be provided until air is expelled. This may be necessary repeatedly in some individuals during and after the same feeding.

Feeding transitions (e.g. changing the dilution of formula) can cause gastric upset and diarrhoea. Formula changes should be gradual to ensure that the neonate is not suddenly overwhelmed by a fluctuating diet. Signs of underfeeding include restlessness, searching, frequent crying and poor growth. Indications of overfeeding include fussiness at feeding time, regurgitation, bloating, excessive weight gain and diarrhoea. The latter, even if dietary in origin, can lead to serious dehydration or hypoglycaemia. In this case, parenteral fluid support should be provided by diluting formula with water to one-third to one-half concentrations or by offering electrolyte solutions orally until the stool returns to normal.

Introduction of solid foods

The weaning process for all herbivores is difficult, from a fibre-free, easily digestible diet (milk) to a highly fibrous, poorly digestible diet (bamboo). Removal of milk and introduction of new food items should be slow and gradual. At six months of age, some bamboo shoots and bamboo may be offered. At this age, most individuals play with these foods, but occasionally undigested bamboo fragments can be seen in faeces. Additional solids may be introduced by ten months, including the concentrate portion of the adult diet. Weaning from formula should be complete by 24 months of age.

Monitoring

Body weight, growth rate and development

Body weight is a critical metric for monitoring successful growth in relation to same-age cohorts or earlier-produced individuals. Absolute body mass is also used to determine stomach capacity that, in turn, is used to calculate amount of formula to provide (see above). For the hand-reared panda, body weight should be collected once daily, preferably in the morning before first feeding. Newborns may lose weight in the first few days after birth. Thereafter, the neonate should gain weight regularly; a contrary observation indicates the need for immediate medical attention. Typical mean body weights for newborn giant pandas are provided in Table 13.5. Although there was a tendency from this small sampling for newborn females to be slightly heavier than males, the difference was not significant ($p > 0.05$).

At birth, giant pandas are pink with a light white coat of lanugo. Various other development characteristics of young giant pandas (mother- or hand-reared) are provided in Table 13.6. Body measurements should be taken regularly (e.g. every five days) as an indicator of normal growth. However, more so than absolute weight at a given age, postnatal growth (i.e. average daily gain) is perhaps most indicative of development. Although information is available on both mother- and hand-reared neonates, care should be taken in evaluating mother-reared data

Table 13.5 *Birth weights of giant pandas born in captivity*

	Number of cubs	Mean ± SD (g)	Range (g)
Combined gender	12	152.0 ± 47.4	54 to 216
Male	6	125.3 ± 53.1	157 to 216
Female	6	178.7 ± 20.6	54 to 207

Table 13.6 *Developmental events documented in mother-reared and hand-reared giant pandas (Day 0 = day of birth) (Zhang et al., 1996)*

Developmental event	Age (days)
Black skin pigmentation (initial)	7 to 10
Hair coat (initial)	7 to 10
Eyes open	35 to 48
Ears open	31 to 50
Deciduous dentition (eruption)	82 to 121
Permanent dentition (eruption)	350

Table 13.7 *Mean (± SEM) growth rate of mother-reared giant panda cubs of various ages from day of birth (Day 0)*

Days	Mother-reared, captive[a]	Mother-reared, free-ranging[b]	Mother-reared, captive[c]
0 to 30	47.1 ± 18.6 (n = 4)	–	42.0 ± 18.8 (n = 2)
31 to 60	62.0 ± 13.9 (n = 4)	60 (n = 2)[d]	69.0 ± 22.2 (n = 2)
61 to 90	62.2 ± 13.2 (n = 5)	–	53.4 ± 10.5 (n = 2)

[a] Zhang *et al.*, 1996; [b] Zhu *et al.*, 2001; [c] Edwards, unpublished data; [d] Days 28 to 52.

as these cubs are often supplementally fed by human care-givers. Rates of growth published in the literature (Zhang *et al.*, 1996), as well as preliminary data on two giant panda cubs that were entirely mother-reared (i.e. no supplementation), (Edwards unpublished data) are provided in Table 13.7.

Temperature

The rectal temperature range for the giant panda cub regardless of age is 38 to 39°C (100.5 to 102.5°F), with an average external temperature of 37°C (range 36.0 to 37.5°C). The ability of the giant panda neonate to thermoregulate is minimal and variable at birth, but increases with age. Taking at least one daily body temperature is an effective means of monitoring a cub's status. Although a rectal measure is the most reliable index of core body temperature, this method poses some risk to the neonate. A specialised thermometer with a thin, flexible probe is required (e.g. Mon-a-Therm Temperature System Model 4070 with a size 8 or 9 flexible French probe; Mallinckrodt, Inc., St Louis, MO). The probe tip should be lubricated with medical-grade, sterile lubricant and inserted carefully to reduce irritation and avoid mechanical injury. An alternative method of monitoring body temperature is taking a reading from the external surface from the animal using a standard or laser thermometer. Regardless of the type of method, the location of taking the measure and the duration should be recorded to allow relative comparisons over time. Daily temperature monitoring should continue beyond the time that the cub is removed from the incubator (at 50 days) and up to about 100 days when it leaves the nursery. Illness, of course, may suggest more frequent temperature assessments.

Excreta

Urine and faecal characteristics are important indicators of an individual's health or response to a particular supportive care approach. In particular, the appearance of these 'end-products' is essential for identifying an animal that may be entering, or already in, a marginal status.

Table 13.8 *Progressive changes in faecal consistency typical of giant panda cubs from day of birth (Day 0)*

Day	Description
0 to 1	Meconium, appears brown/black and dry; in small bead-like shapes; green/yellow stools produced from colostrum, also with dry, bead-like shapes
2	Blue/yellow with mucus (colostrum origin)
3	Blue/yellow (colostrum) or brown/yellow with mucus, bead-like or tubular
4 to 7	Light yellow, tubular soft with mucus
8 to 20	Light yellow, bead-like, sometimes in a string or series, with mucus; floats on water; the middle of the bead is empty; sometimes a fermented odour
21 to 45	Yellow, brown/yellow or deep yellow with mucus; shaped like a bead in either a separate bolus or together; not like a necklace; sometimes a fermented odour
55	Brown/yellow in a tubular shape; dry with mucus
100 to 120	Some cubs can pass faeces on their own; tubular shape, yellow and soft with mucus

High-moisture faeces are not typical for a panda cub nursing from the dam or a bottle. Faecal moisture content also appears to decline with age. A progressive change in faecal consistency typical of giant panda cubs up to 120 days of age is provided in Table 13.8.

Twin swapping

Facilities in China with extensive giant panda experience have developed a protocol that increases the survivability of litters with multiple cubs. Although a dam producing multiple offspring will normally care for only a single cub, it is possible with some females to rotate the cubs from the mother to the nursery (see Chapter 14). While one neonate is nursing, the other is maintained in an incubator and attended and supplemented by humans. This tactic obviously requires a serene dam that tolerates the switching of her cubs. The frequency of cub exchange is highly variable, and a function of the status of each individual, including the dam. This approach offers an enormous advantage to the cub in providing both maternal milk and social access, but requires a high level of monitoring and scrutiny of caretakers familiar with the species.

MEDICAL MANAGEMENT OF NEONATES: METHODS, RESULTS AND RECOMMENDATIONS

Most neonatal mammals born with inadequate Ig concentrations depend upon transfer of maternal Ig to protect against infectious diseases. This transfer occurs through transplacental movement, colostrum ingestion or a combination of both, and is determined by placentation type. For several bear species, placentation has been described as 'endotheliochorial' (Rau, 1925; Wimsatt, 1974; Michel, 1984), and the giant panda is presumed to follow this model. Because this placentation permits only very limited transfer of maternal Ig (Tizard, 1982), colostrum ingestion by the giant panda neonate is critical. In a review of eight giant panda neonatal deaths (1980 to 1987), bacterial infections caused mortality in five (Montali *et al.*, 1990). Of the latter, three cubs were being hand-reared. Another (stillborn) was believed to have acquired an infection *in utero*, and histological examination revealed bacterial pneumonia, enteritis, meningitis and sepsis. *Escherichia coli* and *Pseudomonas* sp. were isolated in two cases, with *Staphylococcus intermedius* septicaemia in another. Inadequate colostrum ingestion was believed to have contributed to all of these cub mortalities.

More contemporary advancements in cub rearing, especially supplementing colostrum via hand feeding and/or the practice of twin swapping, have markedly decreased newborn mortalities; all indirect evidence for the panda neonate being particularly sensitive to an immunodeficiency. Therefore, an essential component of managing a potentially pregnant female is being adequately prepared to care for a rejected or sick neonate. The clinician needs to be concerned not only with the need to boost immunity but also common hypothermia, hypoglycaemia and/or dehydration.

To develop an appropriate database, veterinary staff at the San Diego Zoo have conducted regular physical examinations of giant panda cubs (mother-reared) at about two-week intervals. This usually begins at about 14 days postpartum, when the female reliably begins to leave the den, allowing the cub to be removed for a physical assessment. The first examinations are performed quickly to minimise separation time from the dam (no more than five minutes is recommended). In addition to a routine physical examination and body measures, faecal examination (cytology, Gram stain and culture) and urinalyses can easily be carried out to generate valuable data.

The authors do not have experience with intravascular catheterisation of the giant panda neonate but suspect it would be difficult due to the size limitations of the infant. However, intraosseous and subcutaneous or intraperitoneal routes of fluid administration would be feasible. Treatment with antibiotics and other medications should be based on guidelines for other neonatal carnivore species (Macintire, 1999; Boothe, 2001).

PRIORITIES FOR THE FUTURE

In the last decade, there have been remarkable advances in reducing newborn giant panda mortality, largely through improvements in husbandry, especially in feeding the neonate and salvaging both offspring (in the case of twins). In the past, the inevitable loss in one twin was a colossal waste of a valuable resource. The technique of twin swapping has been one of the greatest contributors to the recent substantial growth in the *ex situ* population.

We predict that progress could be further accelerated if giant panda milk composition and lactation characteristics could be determined throughout the natural lactation curve. There is also a need to establish the significance of colostrum for this species as well as to develop colostrum alternatives for establishing or boosting passive immunity. Finally, to better understand requirements associated with developmental nutrition, there is a need to understand more about young giant pandas in nature. For example, a particularly high priority should be elucidating the age at which solid foods become the primary nutritional source as well as foods that are selected during that transitional period.

REFERENCES

Boothe, D. M. (2001). Drug therapy. In *Veterinary Pediatrics: Dogs and Cats from Birth to Six Months,* 3rd edn, ed. J. D. Hoskins. New York, NY: W. B. Saunders Co., pp. 35–45.

Hanson, L. A. and Johansson, B. G. (1970). Immunological studies of milk. In *Milk Proteins, Chemistry and Molecular Biology. Volume I,* ed. H. A. McKenzie. New York, NY: Academic Press, pp. 45–125.

Haskins, S. C. (1988). A simple fluid therapy planning guide. *Seminars in Veterinary Medicine and Surgery,* 3, 227–36.

Leitch, I., Hytten, F. E. and Billewicz, W. Z. (1959). The maternal and neonatal weights of some Mammalia. *Proceedings of the Zoological Society of London,* 133, 11–28.

Levy, J. K., Crawford, P. C., Collante, W. R. and Papich, M. G. (2001). Use of adult cat serum to correct failure of passive transfer in kittens. *Journal of the American Veterinary Medical Association,* 219, 1401–5.

Macintire, D. K. (1999). Pediatric intensive care. In *Veterinary Clinics of North America, Pediatrics: Puppies and Kittens,* ed. J. D. Hoskins. 4, New York, NY: W. B. Saunders Co., pp. 971–88.

Michel, V. G. (1984). On the structure of the placenta of the bear. *Anatomischer Anzeiger,* 155, 209–15.

Montali, R. J., Bush, M., Phillips, L. G. *et al.* (1990). Neonatal mortality in the giant panda (*Ailuropoda melanoleuca*). *Proceedings of the Second International Symposium on Giant Pandas,* Tokyo, pp. 83–94.

Oftedal, O. T. and Gittleman, J. G. (1989). Patterns of energy output during reproduction in carnivores. In *Carnivore Behavior, Ecology and Evolution,* ed. J. G. Gittleman. Ithaca, NY: Cornell University Press, pp. 355–78.

Ramsay, M. A. and Dunbrack, R. L. (1996). *Nutritional Constraints On Life History Phenomena: Bears Revisited.* Leesburg, VA: Comparative Nutrition Society, pp. 121–5.

Rau, A. S. (1925). On the structure of the placenta of Mustelidae, Ursidae and Sciuridae. *Proceedings of the Zoological Society of London B,* 1027–69.

Schaller, G. B., Jinchu, H., Wenshi, P. and Jing, Z. (1985). *The Giant Pandas of Wolong.* Chicago, IL: University of Chicago Press.

Tizard, I. R. (1982). Immunity in the fetus and newborn animal. In *An Introduction to Veterinary Immunology,* 2nd edn, ed. I. R. Tizard. New York, NY: W. B. Saunders Co., pp. 165–77.

Wimsatt, W. A. (1974). Morphogenesis of the fetal membranes and placenta of the black bear, *Ursus americanus* (Pallas). *American Journal of Anatomy,* 140, 471–96.

Yadav, M. (1971). The transmissions of antibodies across the gut of pouch-young marsupials. *Immunology,* 21, 839–51.

Zhang, G., Zhang, H., Chen, M. *et al.* (1996). Growth and development of infant giant pandas (*Ailuropoda melanoleuca*) at the Wolong Reserve, China. *Zoo Biology*, **15**, 13–19.

Zhu, X., Lindburg, D. G., Pan, W., Forney, K. A. and Wang, D. (2001). The reproductive strategy of giant pandas (*Ailuropoda melanoleuca*): infant growth and development and mother–infant relationships. *Journal of Zoology (London)*, **253**, 141–91.

14

Consequences of early rearing on socialization and social competence of the giant panda

REBECCA J. SNYDER, MOLLIE A. BLOOMSMITH, ANJU ZHANG, ZHIHE ZHANG,

TERRY L. MAPLE

INTRODUCTION

For more than five decades, various nonhuman primate species have been studied to determine how early rearing experiences influence behaviour in later life. Because of this wealth of information, the non-human primate literature is extremely useful for application to the giant panda in developing appropriate methodologies, testing hypotheses and understanding the breadth of behavioural outcomes that might result from different types of early socialisation. Although we recognise the limitations of comparing these distantly related taxa, we believe that the depth of controlled nonhuman primate studies makes comparisons worthwhile and of scholarly interest. Given the close phylogenetic relationship between the giant panda and other carnivores within the superfamily Canoidea (Ewer, 1973; O'Brien et al., 1985), other species within this group may also be useful comparative models, and these are also briefly reviewed.

Giant pandas in captivity can experience inadequate sexual behaviour, maternal behavioural deficits and severe aggression, which is also common to bears, other carnivores and nonhuman primates. It is our general hypothesis that socialisation (particularly the early relationship between mother and cub) is important in the ontogeny of

Giant Pandas: Biology, Veterinary Medicine and Management, ed. David E. Wildt, Anju Zhang, Hemin Zhang, Donald L. Janssen and Susie Ellis. Published by Cambridge University Press. © Cambridge University Press 2006.

normal social behaviour. Our long-term goal is to develop and evaluate management interventions that will overcome behavioural inadequacies and contribute to creating a naturally reproducing, self-sustaining and genetically viable population (Lindburg *et al.*, 1997; Zheng *et al.*, 1997; Zhang *et al.*, 2000; see also Chapter 21).

Most captive giant pandas are housed, bred and raised in breeding centres and zoos in China. Standard management practices include permanently separating cubs from their mothers before six months of age, which enables the adult female to cycle every year, reducing the interbirth interval and, presumably, increasing the number of offspring produced over an individual's life-time. However, in the wild, giant panda cubs remain with their mothers for 1.5 to 2.5 years (Schaller *et al.*, 1985). Given the long-term, detrimental effects that other species experience with early social life disruptions (Carlstead, 1996), it is essential to examine the influence of contemporary rearing practices on the giant panda, especially on subsequent sexual, maternal and agonistic behaviours.

METHODS

Our research on giant panda social behaviour over the past seven years has compared cubs that remained with their mothers for 12 to 13 months with cubs that have been permanently removed from their mothers before six months of age. Studies include five mothers and 18 cubs (nine males, nine females). Ten cubs (six males, four females) remained with their mothers for 12 months, whereas the remaining eight youngsters were removed from their dams at four to five months of age and reared with peers. Nine subjects in this long-term study are currently 4.5 to 6.5 years of age and, thus, at or near reproductive maturity. Most study animals have been housed at the Chengdu Research Base of Giant Panda Breeding and the Chengdu Zoo (see Snyder *et al.*, 2003, for details on housing and data collection).

RESULTS AND DISCUSSION

Early social experience and subsequent sexual behaviour (in other species)

In general, early social deprivation negatively influences nonhuman primate sexual behaviour, but more strongly in males than females

(Mason *et al.*, 1968; Reisen, 1971; Sackett, 1974). Male and female macaques (*Macaca* spp.) raised without mothers experience a variety of deficiencies in sexual behaviour, both in form (e.g. improperly oriented mounting) and sequence (e.g. male genital examination is not followed by male positioning or female presentation) (Mason 1960; Harlow, 1969; Testa & Mack, 1977). The mechanism by which this occurs seems to be through insufficient coordination and communication rather than an inability to perform any individual behavioural component (Capitanio, 1986). In the chimpanzee (*Pan* spp.), both male and female offspring that are removed from the mother and then reared by humans are less proficient at copulating (Davenport & Rogers, 1970; Reisen, 1971). However, it appears that the level of nursery management can influence this outcome. With more intensive nursery practices involving the presence of peer youngsters and extensive human care, these chimpanzees can normally copulate and reproduce, although they still express less frequent sexual behaviours than mother-reared counterparts (Bloomsmith & Baker, 2001).

Rearing domestic carnivores in socially deprived conditions also results in sexual behaviour deficiencies. Female domestic cats hand-reared and isolated from conspecifics show fewer courtship, amicable and copulatory behaviours than mother-nurtured females or those reared with siblings (Mellen, 1989, 1992). Additionally, female kittens that grow up with siblings are more sexually competent than females reared in isolation but less competent than mother-reared counterparts. Male dogs developed in isolation have sexual behaviour deficiencies unlike males raised in groups or with limited peer contact (Beach, 1968).

There are well-recognised breeding challenges in a host of wild carnivores maintained in captivity, including the cheetah (Caro, 1993; Wielebnowski, 1999), clouded leopard (Yamada & Durrant, 1989), small-sized felids in the genus *Felis* (Mellen, 1991) and the maned wolf (Rodden *et al.*, 1996). Although there has been little experimental research in these or other wild carnivores, preliminary findings to date have been unable to link early social deprivation to current reproductive problems (Mellen, 1991; Wielebnowski, 1999).

Early social experience and subsequent sexual behaviour (in the giant panda)

For the giant panda, a primary behavioural challenge is that many females do not display behaviours normally associated with oestrus

(Zheng *et al.*, 1997). This is especially important given that a female has only one period of receptivity per year (one three-day oestrus) to copulate for producing offspring. In one *ex situ* population survey, 21 of the 43 breeding-age females (6 to 20 years of age) had not produced any surviving offspring (J. Ballou, pers. comm.). Nine of these were wild-born and, thus, critical recruits to the breeding programme as founders (Zheng *et al.*, 1997; see also Chapter 21). Nonbreeding males either lack sexual interest or show aggression toward oestrual females (Zheng *et al.*, 1997; see also Chapter 3).

As of 1997, fewer captive males (29.3%; n = 12) than females (45.7%; n = 21) had produced offspring (Lindburg *et al.*, 1997; Zheng *et al.*, 1997). As of 2003, the situation had improved with 50% (17 of 34) of males and 51.2% (22 of 43) of females producing surviving young. However, a more recent examination revealed that, of 56 adults (19 males, 37 females), 36.8% of males had copulated versus 48.6% of females (S. Sun, pers. comm.). Therefore, a lower proportion of males are reproductively successful, largely due to an inability to exhibit successful sexual behaviours. Lindburg *et al.* (1997) examined a variety of factors potentially influencing reproductive success of male pandas housed in Chinese breeding centres and zoos. Although the importance of early rearing history per se was not examined, these investigators stated that "early rearing deficiencies could be one of several factors contributing to mating dysfunction in male adults".

The lower reproductive success of captive males compared to females indicates that early removal from the mother may impact young male pandas more than young females, as has been found for primates. In nature, male and female giant pandas play very different social roles. Males have larger home ranges, spend more time travelling, compete and fight for access to oestrual females and do not care for young (Schaller *et al.*, 1985). By contrast, females are solely responsible for offspring rearing, thereby concentrating their activities over a more restricted habitat range that has the required resources to contribute to successful cub rearing (Schaller *et al.*, 1985).

If there are fundamental differences in the social requirements for young males to develop the necessary skills to reproduce, then one would expect to find within species sex differences in early behavioural development. In fact, young male pandas spend significantly more time play-fighting with their mothers than do young females (Snyder *et al.*, 2003). We have not found sex differences for other general categories of behaviour, including locomotion or self-motion play (i.e. a variety of

Figure 14.1. A mother play-fighting with a seven month old cub
(photographs reproduced with permission from Sarah Bexell).

nondirected body movements, including rolling, headstand and somer-
saulting that do not result in travel from one point to another). Sexual
differentiation of play is most common in juveniles of species in which
males and females have different adult roles. Certainly, this is the
case for giant pandas. Thus, the play-fighting of young male pandas
may prepare them for aggressive encounters with rivals in adulthood
(Fig. 14.1).

Although giant panda males are reproductively mature at five to
six years of age (Schaller *et al.*, 1985), some captive-held males do not
begin breeding until much later. For example, one male (SB number
342) at the Chengdu Research Base did not copulate until 14 years of age
(Snyder, unpublished data) despite being given access to at least one
oestrous female each year from the time he was four to seven years of
age. He was aggressive during some of these early introductions, and,
because he mildly injured two females, was not given the opportunity
to breed again until age 14, when he was suddenly successful. Another
male (SB 287) at Chengdu Zoo was placed with prime age females at
seven and eight years of age but no breeding occurred. When given
the opportunity again at age 13, he mated successfully. Both SB 342 and
287 were captured from the wild at less than one year old and, thus,
may have had a few more months of mother-rearing than typical captive-born counterparts. In retrospect, both may have been capable of

breeding at earlier ages than when mating actually occurred (at 13 and 14 years, respectively). However, because these males were labelled as hyper-aggressive or nonbreeders, there were limited (or no) further opportunities for mating until later years. Therefore, we believe that caution should be used in qualifying animals with labels such as 'non-breeder'. Males (even if not successful at the first attempt) need to learn breeding behaviours, and this is possible only through regular exposure to oestrual females. We recommend allowing a male to interact with females though a barrier, which provides olfactory, visual and limited tactile contact. The male's and female's behaviour can be safely assessed and used to decide if free access should be given. If possible, a male should be given this type of exposure to several different females in each breeding season, because some pairs are more compatible than others (Snyder *et al.*, 2004). This strategy has been used successfully at the Chengdu Research Base since 2001 and has been facilitated by the construction of a new facility with enclosures specifically designed for these types of breeding introduction.

Some captive females also do not mate until well into their adult years. For example, two females (SB 297 and 314) from the Chengdu institutions copulated for the first time at 12 years of age. These females were introduced to an experienced male every year during their oestrous cycles for six to seven years before finally mating. Although both displayed obvious behavioural changes associated with oestrus and were characterised by the Chengdu staff as having strong oestrus, each failed to hold a lordotic posture when the male mounted. Rather, each female would sit, walk away or put her tail down during the copulatory attempt. Interestingly, these observations were similar to Mellen's (1992) findings for the isolate-reared female domestic cat which can behave in a similar manner. Also, SB 297 and 314, which were slow to acquire copulatory behavioural skills, were captive born and reared similarly to other Chengdu-born females that copulated at four to six years old. Therefore, we conclude that early rearing history is not the only variable predictive of later sexual behaviour.

Applying knowledge about breeding interactions in wild giant pandas may offer insight into the problem of delayed breeding success in captive individuals. Wild male giant pandas congregate around a given oestrous female and fight for breeding access, and a female sometimes mates with more than one male from such assemblages (Schaller *et al.*, 1985). It has been hypothesised that the largest, most experienced male dominates these interactions, breeds with a female early in oestrus (ensuring that his sperm are in the oviduct at the time of ovulation) and then moves on in search of other females (Schaller *et al.*, 1985). The female continues to be receptive for a short time thereafter, mating with other, possibly younger males who were a part of the original congregation. This system is probably one way young males acquire the motor skills necessary for successful copulation. This process also has the advantage of allowing a young male the opportunity to mate with an experienced partner, which might assist with learning correct positioning and sequencing of behaviours. Young males in a particular area might encounter several receptive females each year offering multiple opportunities to practice copulation. The female also benefits from the multiple matings by ensuring the presence of plenty of viable sperm at the time of her ovulation.

Applying this multi-male mating strategy in captivity is not as simple as it might seem, largely because of the risk of injury. Unlike in nature, where there is plenty of room to escape in the presence of behavioural incompatibility, captive situations (by definition) do not provide the same opportunities to flee. Thus, injuries can and do occur during breeding encounters. Nonetheless, it is possible to give a young, inexperienced male access to a receptive female, especially after she has mated with a more experienced male. This opportunity has not been a routine part of captive giant panda management in China but should be tested and evaluated. If successful in enabling more males to copulate at a younger age, then it may be necessary to rethink the current model, especially for smaller institutions and those outside China that usually hold only a single pair. Perhaps it makes more sense to hold a minimum of three or four individuals, where there is the opportunity for a young male to have frequent encounters with an older, experienced female. Therefore, a female is maintained not only as a breeder (with a compatible, older adult), but also as a mentor for an inexperienced, young male.

Early social experience and subsequent maternal behaviour (in other species)

The early social environment can influence adult maternal behaviour. Experimental and anecdotal reports provide a consistent picture of infant neglect, abuse and infanticide by nonhuman primates that lacked normal social experiences (Ruppenthal *et al.*, 1976; Capitanio,

1986; Beck & Power, 1988) with rearing in social isolation causing the most severe later maternal deficits. Chimpanzee mothers raised by humans in a nursery are less likely to care for their own infants than mother-raised counterparts (Brent et al., 1996; Bloomsmith et al., 2003). For example, 65% of nursery/peer-reared chimpanzee females were incompetent primiparous mothers (Bloomsmith et al., 2003). The result is a cycle of parental incompetency that is passed from one generation of nursery-reared primates to another (Maple, 1980). The relatively severe consequences for apes may be the result of motor immaturity at birth, a prolonged period of infancy and the long duration of dependency on the dam.

Depending upon litter size, solitary carnivores spend most of their early development associated with the mother alone with or without siblings (Ewer, 1973). Carnivore offspring are invariably altricial, that is highly dependent on parental care (although no other carnivore is so much so as the giant panda). The period of dependence of many large carnivores is long (Ewer, 1973), and the age of independence is greatest in the ursids (Ewer, 1973; Gittleman, 1986).

Although zoo professionals widely believe that mother-rearing increases the likelihood of producing individuals with good parenting skills, very little experimental or archival data are available for species other than primates and rodents (Baker et al., 1996). There are many accounts of captive, wild carnivore mothers abandoning or cannibalising their young. Most of these reports cite disturbance by humans and noisy environments as possible proximate causes (Flint, 1975; Vogt et al., 1980; Aquilina, 1981). Experimental evidence indicates that carnivores reared in socially deprived settings also exhibit later deficiencies in maternal behaviour. Two female domestic kittens removed from their mothers at seven weeks of age and subsequently reared in isolation cannibalised their first litters (Baerends-van Roon & Baerends, 1979). The maternal behaviour of one of these females improved with experience (she successfully raised later litters), but the other continued to cannibalise litters. Mellen (1989) attempted to examine the effects of early rearing on maternal behaviour by raising domestic kittens in three conditions: mother-reared with sibling; hand-reared with sibling; and hand-reared in isolation. Little could be deduced about maternal behaviour because sexual behaviour was so detrimentally affected by hand-rearing – only one female in each of the hand-reared groups ever conceived. Both of these females were adequate mothers and successfully raised their litters to independence.

Early social experience and subsequent maternal behaviour (in the giant panda)

Giant pandas give birth to the most altricial offspring of all placental mammals (Zhu *et al.*, 2001). According to Grand (1992), more than 99% of giant panda brain and body growth occurs postnatally. Thus, giant pandas have greater motor immaturity and prolonged infancy and dependency on the mother than any other carnivore. Most solitary species give birth to multiple young, and littermates play an important role in sociosexual development. Such development in the giant panda is unusual because the female normally rears only one offspring. Thus, in the wild, a cub's first several months of development are spent entirely in the company of the mother, whose role in offspring socialisation may be especially critical. Giant panda cubs have a relatively long period of maternal dependency similar to that of other bear species whose cubs are not fully weaned until two or more years of age (Gittleman, 1986). Therefore, we might expect giant panda maternal behaviour to be strongly impacted by early rearing deficits.

In the past at the Chengdu institutions, females were raised by their dams until four to six months of age, after which they were separated and placed with at least one peer until the age of three to four years. After this time, each female was typically housed with one to three other adult females. Six of the nine females examined were competent first time mothers, naturally breeding at 4 to 12 years. This high proportion of success probably indicates that mother-rearing for the first few months followed by peer-rearing may be sufficient to develop adequate mothering skills in this species (X. M. Huang & S. Zhong, pers. comm.). There also is evidence that females with less than four months of mother-rearing may not become competent mothers. One female (SB 425) was hand-raised alone for her first year of life, after which she was placed with peers. Although she has never mated, she has given birth three times via AI and rejected all three of her cubs. Another female (SB 314), with only one month of mother-rearing, displayed inappropriate maternal care with her first three births (all twins). She held her cubs too tightly (two cubs died from injuries that may have been the result of squeezing by the mother), did not hold them near her nipples and licked her own rather than the cubs' fur (X. M. Huang & S. Zhong, pers. comm.). However, it is worth noting that another female (SB 401), raised similarly to the maternally competent females, also rejected her first cub. Therefore, more data are needed to determine whether females hand-reared from a very early age have a higher incidence of cub rejection than counterparts that were removed from their mothers at a few months of age and subsequently lived with peers. A high priority is determining the time-frame or critical period after which removal from the mother does not impair subsequent maternal behaviour.

It is important to note that female giant pandas that initially reject cubs or exhibit other inadequate mothering behaviours can later acquire these skills. This was demonstrated by Zhang and colleagues (2000) at the China Conservation and Research Centre for the Giant Panda (Wolong Nature Reserve). A primiparous female (SB 446) was provided with controlled periodic access to a cub that she had rejected at birth. These investigators documented the dam mastering specific care-giving behaviours until the cub could be fully returned to its mother at 73 days of age. Corroborating anecdotal evidence is also available from the Chengdu institutions where females are normally categorised on a 1 to 10 scale for mothering skills (10 being the best rating). Female (SB 314) was rated as a 4 for mothering skills for her first three births, and her cubs died or were removed by six days of age for hand-rearing. However, the staff continued to give her the opportunity to raise subsequent cubs, and by her fourth birth she rated a 7 on the 'mothering meter' and cared for her cub for five months according to standard management in Chengdu.

Cross-fostering

Female giant pandas exhibit tremendous flexibility in their maternal behaviour and associated tolerances, including extending their mothering behaviour to multiple young not their own (see below). The ability to cross-foster increases infant survivorship and the quality of life for cubs under an intensive management system. It also probably provides a unique socialisation opportunity for cubs while allowing some females to return more quickly to the breeding population. For example, it is well known that some dams will tolerate offspring swapping – the periodic switching of infants from the dam to the nursery thereby allowing a given female to nurse multiple offspring.

In theory, genetically valuable females could be reserved for offspring production but not cub rearing – with these females then allowed to recycle and rebreed on an annual basis. Schaller and colleagues (1985) speculated about the phenomenon of twinning in the giant panda. They suggested that wild female pandas abandon one twin cub because of energy constraints, as well as the high level of caregiving required by two extremely altricial neonates (Zhu *et al.*, 2001). A captive female's willingness to care for multiple cubs simultaneously may indicate that she retains the behavioural flexibility to raise a litter and/or the realisation that energy requirements for raising more than one offspring are being readily met in the captive environment.

The first successful cross-fostering of a giant panda cub from one mother to another at the Chengdu institutions occurred in 1990. This strategy is now practised routinely here and elsewhere, most often for a rejected cub or often when twins are produced. Mothers accepting cross-fostered cubs do not hesitate to accept another female's cub, and there is no indication that they distinguished the unrelated cubs from their own. Cross-fostering has been successful with cubs varying in age from a few hours to several months. In Chinese institutions, giant panda females typically give birth within a few weeks of each other. Thus, all the cubs born at the Chengdu institutions in a particular year are similar in age and size, which may contribute to the likelihood that a cub from one mother will be accepted by another. Four dams (SB 278, 297, 362 and 314) at the Chengdu Research Base have been successful foster mothers, including nursing and playing with all cubs (Fig. 14.2). When cubs are less than four months old, a mother is usually given only one cub at a time. However, when cubs are more than four months of age (when they are walking, urinating and defecating on their own), a mother may care for two or more cubs concurrently. Amazingly, some dams have cared for as many as four cubs simultaneously. No attempt has yet been made in the Chengdu institutions to introduce a cub to a nonlactating female without a cub of her own.

Early social experience and subsequent aggressive behaviour (in other species)

Many nonhuman primates raised in isolation from conspecifics (especially from their mothers) are hyperaggressive when placed in social situations (Mason, 1960; Baker & Aureli, 2000), with males being more aggressive than females (Meder, 1989). It may be that the nursery/peer-rearing experience (which offers little or no understanding about the consequences of inappropriate hostility) produces animals that display aggression at abnormally high rates. Nonhuman primates with histories

Figure 14.2. Female caring for her own twins and unrelated twins simultaneously (photograph reproduced with permission from Megan Wilson).

of social isolation also can be hyperfearful, sometimes engaging in almost no positive social behaviour (Mitchell *et al.*, 1966). While there is some variation in onset and degree of hyperaggression and fearfulness (Capitanio, 1986), it is clear that early social deprivation disrupts the incidence of agonistic interactions.

Similar findings occur in the domestic cat. Kuo (1960) found that kittens raised with rats and in isolation from conspecifics were hyperaggressive toward conspecifics later in life. Age of separation from the mother affects the development of antagonistic behaviour. Seitz (1959) reared kittens in isolation after removing them from their mothers and littermates at 2, 6 or 12 months of age. Individuals isolated from 2 weeks of life were most aggressive and fearful toward conspecifics and humans. Mellen (1989, 1992) found that female kittens reared with their mother and one sibling exhibited the least aggression toward males during their eventual oestrus. She also reported that cats reared alone by humans were excessively hostile toward humans compared to mother-reared counterparts exposed to a sibling. The latter (unlike the former) were calm and friendly. Kittens raised in isolation from one week (Weiss, 1952) or seven weeks (Baerends-van Roon & Baerends, 1979) of age also try to escape when later placed with conspecifics.

For captive wildlife species, hyperaggression has been documented in certain carnivores. The clouded leopard is notoriously difficult to breed in captivity because of intersexual aggression by males, which has resulted in serious injury or death of females at 22 of 44 institutions (Yamada & Durrant, 1989).

Managing aggressive behaviour in the giant panda

Managing aggression in captive giant pandas can be challenging, particularly during breeding introductions involving unfamiliar individuals. Male hyperaggression is a main reason cited for breeding failure in captivity (Zheng *et al.*, 1997; see also Chapter 3). In the wild, aggression may be valuable for competing with other males. Captive males, however, live in isolation because of the potential for injury if contact was allowed. Hyperaggression in captivity may be an expression of a species-typical behaviour with no outlet and, thus, is misdirected toward females. Furthermore, the intensity of the interaction, or perhaps inappropriate responses to interactions that escalate rather than tempering aggression, may be a result of early rearing practices common in managing giant pandas.

Variations in male assertiveness may be driven by seasonal patterns in testosterone (a recently discovered rhythm in this species; see Chapter 8). Giant pandas living in nature are probably familiar with other nearby residents (via chemical communication or even occasional chance meetings or possibly involvement in previous breeding congregations). Captive individuals are often not given much opportunity to become familiar with each other prior to breeding introductions. Thus, it is possible that some males do not recognise a female that is in oestrus when introduced for breeding, and first reactions are aggressive rather than sexual.

Because the less aggressive males are easier to manage in breeding introductions, these giant pandas are used most often for mating to the exclusion of more assertive individuals. The former's genes then become common in the population, whereas the latter males become more genetically valuable (see Chapter 21). Therefore, the current management regimen selects against aggression, a trait that is probably essential to survival and the spreading of one's genes in nature. This may not be an immediate problem for the *ex situ* population, given continued improvements in AI. However, if the reintroduction of captive individuals to the wild is ever to be successful, behaviourally competent individuals with the social skills necessary to interact appropriately with wild conspecifics will be essential.

Approaches have been developed for other species (Burks *et al.*, 2001) which allow aggression to be managed or circumvented to permit breeding and to minimise injury. This usually involves a step-wise process that controls exposure between adults to allow developing familiarity before the breeding introduction. Individuals first have visual and olfactory contact, then limited tactile contact and finally full access in a controlled setting. This strategy for resocialising individuals with hyperaggressive histories warrants attention for the giant panda, especially as a means of recruiting genetically, underrepresented males into the population.

PRIORITIES FOR THE FUTURE

Because our studies are longitudinal and long-term, so far we have incomplete data for making definitive conclusions about the impact of early weaning and hand-rearing on long-lasting social and behavioural competency in the giant panda. It has been standard practice in Chinese breeding institutions to wean a giant panda cub permanently from its mother at least one year earlier than weaning occurs in the wild. Certainly, removing a cub at a younger age reduces the interbirth interval by enabling females to re-cycle (and potentially conceive every year), thus increasing total offspring number in the *ex situ* population. However, we would argue that depriving a young panda of a full and protracted social experience with its mother may have long-term detrimental consequences on its social competence and its eventual value and contribution to the population. To date, we know that allowing at least four months of mother-rearing does not appear to significantly compromise a female's later reproductive ability. However, weaning earlier in life appears to produce females that lack critical mothering skills. Nonetheless, if these females do reproduce, their offspring can survive by the remarkable ability of unrelated dams to accept cubs – often multiple offspring – through cross-fostering. Additionally, some behaviourally deficient females appear to be able to learn to become good mothers over time. The impact of early separation from the dam with the onset of the young's eventual reproductive ability is less clear and confounded by the usual practice of labelling an individual as a nonbreeder. These individuals have historically been ignored, with little chance to contribute to the gene pool. But when given a later opportunity to breed, many of these pandas reproduce. Finally, as yet, we have been unable to relate the severe problem of male hyperaggression to any early social deficiency. Rather, this behaviour may be a natural trait that is being misdirected to females because of inadequate captive management techniques. Regardless, we will continue to monitor the male cohort in our study (a few of which have now entered adulthood) for their reproductive proficiency in the context of their earlier life experiences.

Meanwhile, we continue to see value in exploring the vast amount of data available from other species on the influence of early rearing on later reproductive competence. This approach will be useful for identifying future research priorities for the giant panda. Our highest overall priority will continue to focus on determining the impact of longer periods of mother-rearing on an offspring's subsequent reproductive viability and parental behaviour. The latter appears especially important in an altricial species such as the giant panda.

More specifically and based on experiences to date, we would suggest that high priorities in basic research include:

1. additional studies on the significance of sex differences in play-fighting of cubs and its role in subsequent reproductive and social success;
2. identifying the earliest time (or window) of weaning that will allow offspring to become fully socially competent as an adult (in parallel, such efforts should closely monitor the potentially adverse health effects of premature separation from the dam on traits such as growth rate and disease susceptibility);
3. exploring the significance of size and sex of a cub's peer group post-weaning on subsequent development;
4. detailed investigations of the relationship between testosterone concentrations, behaviour and possible influences of social histories on male behaviours, especially aggression.

From an applied management perspective, our findings would strongly suggest that more attention be focused on providing 'learning opportunities' for breeding. This is relevant to both sexes, but especially for placing young males with older, experienced females. This indeed may require re-analysing the existing paradigm for most small holding institutions where two individuals is the usual number. Rather the management norm may need to be shifted to maintaining a small population, perhaps at least two experienced adults and one to two younger adults to be mentored by their older counterparts. In a related matter, it would be interesting to study further the ways to train behaviourally deficient dams to be better mothers. It is exciting to know that dysfunctional females can acquire these skills, but how can the process be made more efficient? Certainly another management-related research target should be investigating the value of allowing both males and females more opportunities to explore each other's scents, largely to improve familiarity, reduce aggression and promote safe breeding. Similarly, any technique that has proven useful for reducing antagonistic behaviours in other species, especially carnivores, should be tested in the giant panda.

Finally, there is value in being more diligent in recording individual animal behaviours and predilections, including adding this information to the studbook. Indeed, the studbook for this species should contain relevant information on rearing methods for every animal, data that will allow tracking the importance of mother-rearing on overall species' well-being in captivity. Eventually it will be necessary to assess the overall benefits (or detriments) of premature mother–cub separation. Although it makes sense that a practice that allows a cub to be produced annually (rather than every two to three years) is valuable, ultimately the questions need to be asked:

1.   Are giant pandas biologically built to produce healthy cubs annually?
2.   Even if so, are our management practices sufficient to ensure that prematurely weaned offspring can be raised into competent adults that will contribute to a successful, self-sustaining *ex situ* population?

The experiments to address these queries are ongoing, with answers expected in the near future.

ACKNOWLEDGEMENTS

We thank China's Ministry of Construction for its support and cooperation. We greatly appreciate assistance provided by staff from the

Chengdu Research Base of Giant Panda Breeding and the Chengdu Zoo. Financial support for the described studies was provided by Zoo Atlanta, the Morris Animal Foundation (grant D00ZO-42) and the Georgia Tech Center for Conservation and Behaviour.

REFERENCES

Aquilina, G. D. (1981). Stimulation of maternal behaviour in the spectacled bear at Buffalo Zoo. *International Zoo Yearbook*, **21**, 143–5.

Baerends-van Roon, J. and Baerends, G. (1979). *The Morphogenesis of the Behaviour of the Domestic Cat.* Amsterdam: North-Holland.

Baker, A. J., Baker, A. M. and Thompson, K. V. (1996). Parental care in captive mammals. In *Wild Mammals in Captivity*, ed. D. G. Kleiman, M. E. Allen, K. V. Thompson and S. Lumpkin. Chicago, IL: University of Chicago Press, pp. 497–512.

Baker, K. C. and Aureli, F. (2000). Coping with conflict during initial encounters in chimpanzees. *Ethology*, **106**, 527–41.

Beach, F. A. (1968). Coital behaviour in dogs. III. Effects of early isolation on mating in males. *Behaviour*, **30**, 218–38.

Beck, B. and Power, M. (1988). Correlates of sexual competence in captive gorillas. *Zoo Biology* **7**, 339–50.

Bloomsmith, M. A. and Baker, K. C. (2001). Social management of captive chimpanzees. In *The Care and Management of Captive Chimpanzees*, ed. L. Brent. New York, NY: Wiley-Liss, pp. 204–41.

Bloomsmith, M. A., Baker, K. C., Ross, S. R. *et al.* (2003). Social rearing conditions and later maternal performance of primiparous chimpanzees. *American Journal of Primatology*, **60**, 38–9.

Brent, L., Williams-Blangero, S. and Stone, A. M. (1996). Evaluation of the chimpanzee breeding program at the Southwest Foundation for Biomedical Research. *Laboratory Animal Science*, **46**, 405–9.

Burks, K. D., Bloomsmith, M. A., Forthman, D. L. and Maple, T. L. (2001). Managing the socialization of an adult male gorilla (*Gorilla gorilla gorilla*) with a history of social deprivation. *Zoo Biology*, **20**, 347–58.

Capitanio, J. P. (1986). Behavioural pathology. *Comparative Primate Biology*, 2(A), 411–54.

Carlstead, K. (1996). Effects of captivity on the behavior of wild mammals. In *Wild Mammals in Captivity*, ed. D. G. Kleiman, M. E. Allen, K. V. Thompson, and S. Lumpkin. Chicago, IL: University of Chicago Press, pp. 317–33.

Caro, T. M. (1993). Behavioral solutions to breeding cheetahs in captivity: insights from the wild. *Zoo Biology*, **12**, 19–30.

Davenport, R. K. and Rogers, C. M. (1970). Differential rearing of the chimpanzee: a project survey. In *The Chimpanzee: A Series of Volumes on the Chimpanzee: Immunology, Infections, Hormones, Anatomy and Behavior of Chimpanzees*, ed. G. H. Bourne. Baltimore, MD: University Park Press, pp. 337–60.

Ewer, R. F. (1973). *The Carnivores.* New York, NY: Cornell University Press.

Flint, M. (1975). Hand-rearing the small-spotted genet at Randolph Park Zoo, Tucson. *International Zoo Yearbook*, **15**, 244–5.

Gittleman, J. L. (1986). Carnivore life history patterns: allometric, phylogenetic and ecological associations. *American Naturalist*, **127**, 744–71.

Grand, T. I. (1992). Altricial and precocial mammals: a model of neural and muscular development. *Zoo Biology*, **11**, 3–15.

Harlow, H. F. (1969). Age-mate or peer affectional system. In *Advances in the Study of Behaviour*, ed. D. S. Lehrman, R. A. Hinde and E. Shaw. New York, NY: Academic Press, pp. 333–83.

Kuo, Z. Y. (1960). Studies on the basic factors in animal fighting: interspecies co-existence in mammals. *Journal of Genetic Psychology*, **75**, 363–77.

Lindburg, D. G., Huang, X. M. and Huang, S. Q. (1997). Reproductive performance of male giant pandas in Chinese zoos. In *Proceedings of the International Symposium on the Protection of the Giant Panda*, ed. A. Zhang and G. He. Chengdu: Sichuan Publishing House of Science and Technology, pp. 67–71.

Maple, T. L. (1980). *Orang-utan Behavior*. New York, NY: Van Nostrand Reinhold.

Mason, W. A. (1960). The effects of social restriction on the behaviour of rhesus monkeys. I. Free social behavior. *Journal of Comparative Physiological Psychology*, **53**, 582–9.

Mason, W. A., Davenport, R. K. and Menzel, E. W. (1968). Early experience and the social development of rhesus monkeys and chimpanzees. In *Early Experience and Behavior: the Psychobiology of Development*, ed. G. Newton and S. Levine. Springfield, IL: Charles Thomas, pp. 440–80.

Meder, A. (1989). Effects of hand-rearing on the behavioral development of infant and juvenile gorillas (*Gorilla g. gorilla*). *Developmental Psychology*, **22**, 357–76.

Mellen, J. D. (1989). *Reproductive Behavior of Small Captive Exotic Cats (Felis spp.)*. Davis, CA: University of California, doctoral dissertation.

(1991). Factors influencing reproductive success in small captive exotic felids (*Felis* spp.): a multiple regression analysis. *Zoo Biology*, **10**, 95–110.

(1992). Effects of early rearing experience on subsequent adult sexual behavior using domestic cats (*Felis catus*) as a model for exotic small felids. *Zoo Biology*, **11**, 17–32.

Mitchell, G. D., Raymond, E. J., Ruppenthal, G. C. and Harlow, H. F. (1966). Long-term effect of total social isolation upon behavior of rhesus monkeys. *Psychological Reports*, **18**, 567–80.

O'Brien, S. G., Nash, W. G., Wildt, D. E., Bush, M. and Benveniste, R. E. (1985). Riddle of the giant panda's phylogeny: a molecular solution. *Nature*, **317**, 140–4.

Reisen, A. H. (1971). Nissen's observations on the development of sexual behaviour in captive-born, nursery-reared chimpanzees. In *The Chimpanzee*, ed. G. H. Bourne. Basel: S. Karger, pp. 1–18.

Rodden, M. D., Sorenson, L. G., Sherr, A. and Kleiman, D. G. (1996). Use of behavioral measures to assess reproductive status in maned wolves (*Chrysocyon brachyurus*). *Zoo Biology*, **15**, 565–85.

Ruppenthal, G. C., Arling, G. L., Harlow, H. F., Sackett, G. P. and Suomi, S. J. (1976). A 10-year perspective of motherless monkey behaviour. *Journal of Abnormal Psychology*, **85**, 341–49.

Sackett, G. P. (1974). Sex differences in rhesus monkeys following varied rearing experiences. In *Sex Differences in Behavior*, ed. R. C. Friedman, R. M. Richart and R. L. van de Wiele. New York, NY: Wiley, pp. 99–112.

Schaller, G. B., Hu, J., Pan, W. and Zhu, J. (1985). *The Giant Pandas of Wolong*. Chicago, IL: University of Chicago Press.

Seitz, P. (1959). Infantile experience and adult behaviour in animal subjects: age of separation from the mother and adult behavior in the cat. *Psychosomatic Medicine*, **21**, 353–78.

Snyder, R. J., Zhang, A. J. and Zhang, Z. H. (2003). Behavioral and developmental consequences of early rearing experience for captive giant pandas. *Journal of Comparative Psychology*, **117**, 235–45.

Snyder, R. J., Lawson, D. P., Zhang, A. J. *et al.* (2004). Reproduction in giant pandas: hormones and behaviour. In *Giant Pandas, Biology and Conservation*, ed. D. Lindburg and K. Baragona. Berkeley, CA: University of California Press, pp. 125–32.

Testa, T. J. and Mack, D. (1977). The effects of social isolation on sexual behavior in *Macaca fascicularis*. In *Primate Bio-Social Development: Biological, Social and Ecological Determinants*, ed. S. Chevalier-Skolnikoff and F. E. Poivier. New York, NY: Garland, pp. 407–38.

Vogt, P., Schneidermann, C. and Schneidermann, B. (1980). Hand-rearing a red panda at Krefeld Zoo. *International Zoo Yearbook*, **20**, 280–81.

Weiss, G. (1952). Beobachtungen an zwei isoiert aufgezogenen Hauskatzen. *Zeitschrift für Tierpsychologie*, **9**, 451–62.

Wielebnowski, N. C. (1999). Behavioural differences as predictors of breeding status in captive cheetahs. *Zoo Biology*, **18**, 335–49.

Yamada, J. K. and Durrant, B. S. (1989). Reproductive parameters of clouded leopards (*Neofelis nebulosa*). *Zoo Biology*, **8**, 223–31.

Zhang, G. Q., Swaisgood, R. R., Wei, R. P. *et al.* (2000). A method for encouraging maternal care in the giant panda. *Zoo Biology*, **19**, 53–63.

Zheng, S., Zhao, Q., Xie, Z., Wildt, D. E. and Seal, U. S. (1997). *Report of the Giant Panda Captive Management Planning Workshop*. Apple Valley, MN: IUCN–World Conservation Union/SSC Conservation Breeding Specialist Group,

Zhu, X., Lindburg, D. G., Pan, W., Forney, K. A. and Wang, D. (2001). The reproductive strategy of giant pandas: infant growth and development and mother–infant relationships. *Journal of Zoology (London)*, **253**, 141–55.

15

Medical management of captive adult and geriatric giant pandas

DONALD L. JANSSEN, PATRICK MORRIS, MEG SUTHERLAND-SMITH, MARK
GREENBERG, DESHENG LI, NATHALIE MAUROO, LUCY SPELMAN

INTRODUCTION

The medical management of giant pandas has advanced significantly in recent years due to cooperative programmes between Chinese and western institutions, specifically zoos and breeding centres. Key to these partnerships have been veterinarians who have become committed to understanding the diseases affecting this species. Progress has emanated from efforts such as the Biomedical Survey (see Chapter 4) and international personnel exchanges related to giant panda loans to western zoos (see Chapter 22). The result has been many opportunities for veterinarians working with giant pandas to share philosophies, tools, expertise and knowledge which, in turn, have vastly improved medical care of this species in captivity.

There are unique as well as overlapping medical issues impacting the giant panda according to age. For example, Chapter 13 has already addressed health-related topics facing neonates and juveniles. After four years of age, however, the giant panda has matured physically and sexually, leaving behind many of the diseases associated with its youth. Then, after the age of 20 years and during the period of reproductive senescence, another set of potential problems face managers

Giant Pandas: Biology, Veterinary Medicine and Management, ed. David E. Wildt, Anju Zhang, Hemin Zhang, Donald L. Janssen and Susie Ellis. Published by Cambridge University Press. © Cambridge University Press 2006.

and veterinarians – degenerative changes related to the geriatric condition. Because health and reproduction are improving so rapidly in the *ex situ* panda population, it is a given that more animals will live longer, requiring more sophisticated veterinary management to ensure well-being for up to 25 years of age or beyond.

This chapter describes the authors' medical experiences with adult and aged giant pandas living in zoos, especially in the USA. We believe that our observations and findings will be a useful guide for the medical management of this species throughout this portion of its life cycle. Because restraint and anaesthesia are prerequisites for many preventive medical activities as well as performing diagnostic and therapeutic procedures, one section is devoted to this topic. This is followed by a discussion of preventive medicine, which then turns to our experiences with commonly encountered clinical challenges in adults and geriatric individuals.

RESTRAINT AND ANAESTHESIA

Restraint with no sedation

Operant conditioning can be used to train giant pandas to enter and remain quiet in a restraint device. The steel-constructed restraint cage used at the San Diego Zoo is rectangular (2 m long by 93 cm wide by 118 cm high) (Fig. 15.1) and on wheels for mobility. It has a solid bottom with bars on the sides and top. Along one side, some of the bars are removable to allow better access to the animal. The opposite wall is moveable allowing the available internal cage space to be decreased (i.e. the restraint function). Similar cages are being used successfully at other institutions in the USA as well as China.

This type of restraint device can facilitate various medical procedures. For example, at both the San Diego Zoo and the Smithsonian's National Zoological Park, females have been trained to enter the device and lay down in dorsal recumbency or sit upright with the front paws gripping the top bars. Usually, the cage is 'squeezed' only modestly to limit, but not prohibit, animal movement. This has allowed ultrasound examinations (see Fig. 15.1), recovery of vaginal swabs (see Chapter 9) and milk collection (when in the upright position). Some individuals will allow blood pressure measurements to be taken from the upper front foreleg (see below). Others, including those at the Smithsonian's National Zoo and Ocean Park-Hong Kong have been trained to place a

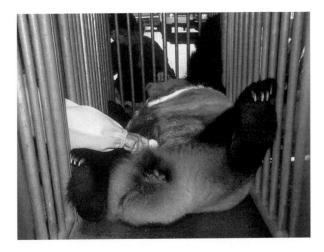

Figure 15.1. A female giant panda, trained for resting in dorsal recumbency in a restraint device, undergoing transabdominal ultrasound (photograph reproduced with permission by Ken Bohn).

foreleg into a sleeve for blood sampling. Still others have remained so still as to permit radiography. Finally, certain individuals have been trained to place the lower jaw on a metallic bar to allow ocular examinations. In short, this approach allows multiple, important procedures to be conducted in nonsedated giant pandas that otherwise would require repeated anaesthetic events.

Anaesthesia

The following description is designed to aid in planning and implementing anaesthesia when either multiple procedures have to be accomplished quickly and efficiently, or a few complex procedures are required. There are four steps: planning; anaesthesia induction; maintenance; and recovery.

Pre-procedural planning

Preparation for anaesthesia must include thinking ahead about every impending event and required resource from how to transport a large and potentially dangerous animal to making sure that all necessary drugs and tools are at hand. Because of the high profile of giant pandas, it is common that everyone wants to observe the anaesthetic event, from the animal's keepers to the zoo's director. This can result in a room that is crowded (Fig. 15.2) and noisy which, in turn, can be distracting for the veterinarians. It is especially important to have a quiet room during the intubation period, when the veterinarians must be able to hear the monitors and pay close attention to maintaining a patent airway. Thus the issue of who must be present needs to be addressed pre-emptively to limit people in the examination room to those playing an essential role. Once the animal is stable and the procedure long underway, it may be possible for non-essential personnel to get a closer look.

A pre-induction planning meeting that includes all relevant personnel is an excellent time to review everyone's roles and responsibilities. Besides ensuring that all necessary tasks are organised, such preparation allows the various components of the overall procedure to be adjusted to reduce the overall anaesthesia interval. With appropriate planning, proper techniques and rigorous monitoring, the anaesthesia interval can be minimised and, if necessary, extended safely to several hours for necessary procedures.

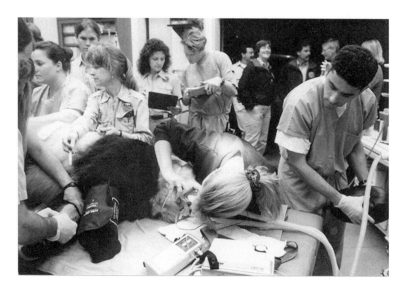

Figure 15.2. Anaesthetic scene at the San Diego Zoo.

Induction and intubation

Anaesthesia induction in the giant panda is predominately accomplished by intramuscular (i.m.), remote delivery dart or by hand injection in a restraint cage. Food and water are withheld for at least 12 hours before induction to reduce the risk of regurgitation and pulmonary aspiration of gastric contents. At induction onset, the panda is separated from its mate or offspring (even if nursing). In addition to the primary anaesthetic drug, other essential items to have available include reversal and resuscitation drugs, endotracheal tubes and stylets, a self- or flow-inflating bag with mask (to deliver oxygen), a pulse oximeter (to measure blood oxygen saturation), a stethoscope, intravenous (i.v.) catheters and fluids and a tarpaulin (to carry the animal to the examination room).

Ketamine hydrochloride (Ketaset®; Fort Dodge, Fort Dodge, IA) is the primary induction agent used for the giant panda. Other drugs, such as tiletamine/zolazepam (Telazol®; Fort Dodge, IA) and medetomidine hydrochloride (Wildlife Pharmaceuticals, Fort Collins, CO) are also options but are no more effective than ketamine. A 6–9 mg kg^{-1} dose of ketamine given in a single dart or injection is usually effective although supplemental ketamine doses (1–1.5 mg kg^{-1}) provided i.m. or i.v. may be necessary to induce an anaesthetic plane that allows the animal to be moved safely. Initial post-ketamine effects (e.g. sitting, head drooping and uncoordination) generally occur within about five minutes, with the animal being tractable and accessible within 15 minutes. Once it has been determined that the panda can be handled, vital signs are obtained quickly and the eyes covered with a blindfold to minimise external stimuli. During transport, oxygen is provided by face-mask and an attempt made to monitor oxygen saturation by pulse oximetry, although the latter is challenging. Readings are best secured by affixing the pulse clip to the tongue, cheek or genitals (other extremities, including the ear, give poor readings).

Once the panda has been relocated to the main examination area, full physiological monitoring can begin, including using electrocardiography, pulse oximetry, measuring end-tidal carbon dioxide and tracking blood pressure noninvasively (using a cuff around a fore-limb). Systemic hypertension often occurs initially, with blood pressure usually returning to normative values within 10 to 15 minutes. Noninvasive blood pressure readings are helpful to determine when the animal is sufficiently relaxed for intubation. Giant pandas anaesthetised with ketamine are usually physiologically stable with excellent respiratory function as determined by arterial blood-gas sampling.

An isoflurane delivery system is one method that can be used to deepen anaesthesia and optimise conditions for intubation. This is usually required if the jaw muscles remain tense, respiration is rapid and/or the animal responds to stimulation (e.g. repositioning or a skin pinch). For supplemental anaesthesia before intubation, we have occasionally used propofol (Diprivan; AstraZeneca Pharmaceuticals, Wilmington, DE) in 0.5–1 mg kg^{-1} boluses i.v. or ketamine in 50–100 mg boluses i.v. Care is exercised using propofol because, even at therapeutic doses, the animal may become apnoeic. During this time, the panda is supported using supplemental oxygen mixed with the volatile anaesthetic given by mask. The most useful mask is a plastic jug with the bottom cut away to fit the panda's muzzle and face while avoiding the eye region and ocular injury. The mask is attached using adhesive tape, which provides an adequate seal, although a towel can be placed around the mask's edge to further reduce air leakage.

Endotracheal intubation is desirable to assure control of the animal's airway. This is particularly important for performing procedures in the oral cavity or for any prolonged procedure. It also decreases the risk of pulmonary aspiration from oral secretions or regurgitated stomach contents. For the giant panda, direct laryngoscopy and intubation are most easily accomplished with the animal in lateral recumbency. To aid in opening the jaws and for obtaining the best possible view of the upper airway, two assistants apply ropes anchored around the upper and lower incisors behind the canine teeth (Fig. 15.3). Topical anaesthesia is applied to the larynx to decrease the incidence of laryngospasm during intubation. The preferred drug and dose are lidocaine at about a 2–5 ml volume of a 2% solution sprayed on the vocal cords and periglottic structures. Intubation is usually accomplished with a long, straight blade (Miller 4 or 5; Welch Allyn Medical Products, Skaneateles Falls, NY) and a well-lighted laryngoscope (Welch Allyn Medical Products) (see Fig 15.3). Most adult giant pandas can be intubated with a size 12, 13 or 14 mm endotracheal tube with placement often facilitated by a stylette within the tube. The full length of the tube is generally used for proper positioning within the trachea. However, it should be noted that intubation into the right primary bronchus has occurred with a 63-cm long tube in an adult male. Therefore, it is

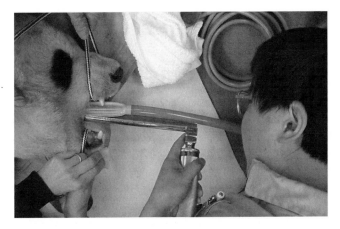

Figure 15.3. Proper positioning of the head and jaws during laryngotracheal intubation. The larynx and vocal folds are visualised using a laryngoscope and long blade.

important to estimate the position of the tracheal bifurcation. On most occasions, an ideal view of the vocal cords is obtained, the tube placed and the cuff inflated to prevent air leakage. Placement and appropriate ventilations are further confirmed by listening to bilateral breath sounds over the anterior lung fields and by placing the end-tidal carbon dioxide monitor in the circuit. Simply seeing the chest rise and fall is not an accurate indicator of proper tube placement. This, in fact, can result from an upper airway occlusion or laryngospasm.

Once the tube is appropriately placed, it is secured with a tie around the animal's neck. An alternative method is holding the tube with a plastic prop placed between the teeth, a technique that also serves as a block to prevent accidental bite obstruction of the endotracheal tube. Throughout all of these procedures, care is taken to avoid hyperextending the jaw.

While the endotracheal procedure is underway, an i.v. catheter is simultaneously placed. The medial cephalic vein is most accessible and can be catheterised using an 18-gauge, 2-cm long catheter. The external jugular vein can also be used for multipurpose access relying on a 7 French, double-lumen catheter. Once vital signs are satisfactory via the various monitoring approaches, the airway is secured and there is i.v. access, then the planned medical procedures may commence.

Anaesthetic maintenance techniques to ensure effective procedures

Volatile agents, including isoflurane and sevoflurane, provide a stable base for anaesthetic maintenance delivered by the endotracheal tube. Propofol and ketamine supplementation have also been useful for sustaining effective anaesthesia. However, the medical procedure itself can influence animal stimulation which, in turn, requires modifying the anaesthesia protocol. For example, electroejaculation involves periods of no or low activation followed by more intensive stimulation, which can evoke acute tachycardia and hypertension. Therefore, when semen collection is scheduled for sperm banking, a background volatile agent (isoflurane or sevoflurane) combined with bolus doses of propofol or ketamine can be used. In contrast, for simple examinations without electroejaculation, volatile anaesthesia can be used alone.

Throughout the interval of anaesthesia maintenance, vital signs are continuously and carefully monitored. This involves recording blood pressure, pulse, oxygen saturation, end-tidal carbon dioxide and respiratory rate every five minutes. Normative values are found elsewhere in this book (see Chapter 4). We have found excellent correlation between noninvasive blood pressure monitoring via a cuff on a limb extremity and values measured directly and invasively via intra-arterial means. End-tidal carbon dioxide values likewise correlate well with direct arterial blood-gas analysis.

During this period, it is also essential to consider fluid and temperature management (normal temperature values for this species are presented in Chapter 4). The target is usually to provide a fluid bolus to replace the deficit resulting from the 12-hour fast. Approximately 5–10 ml kg^{-1} of a balanced electrolyte solution (e.g. lactated Ringer's) are given i.v. over the first 30 minutes after the animal has been stabilised under anaesthesia. Blood-gas analysis of fasted giant pandas has revealed a metabolic acidosis of unclear aetiology, which is usually corrected by this initial fluid bolus. Maintenance fluids are then given to sustain adequate hydration, generally lactated Ringer's at a rate of about 200 ml per hour, which prevents electrolyte abnormalities.

Adequate personnel during the anaesthetic episode are critical, especially given individual animal variations and the complexity of these anaesthetic drugs. Ideally, there should be a team of three qualified individuals focusing only on anaesthesia and with the following assignments.

One person is responsible for airway management, including mask oxygenation (immediately after the animal becomes tractable), aiding or performing the intubation, securing the endotracheal tube, confirming (and reconfirming) tube placement and 'guarding' the airway. This team member may want to keep a hand physically on the tube, whenever possible, and at all times during any animal movements. The latter includes (but is not limited to) animal repositioning or activities associated near the head or neck (e.g. ophthalmic or dental procedures).

The second anaesthesia team member assumes the role of monitoring and adjusting the anaesthetic drug, as necessary, to maintain an adequate plane of anaesthesia. This includes closely observing the animal and its reaction to various stimuli, with the ability then to anticipate and react by changing anaesthetic delivery. This usually means simply adjusting the percentage of volatile agent provided. However, this person is also prepared to give required i.v. drugs, including vasopressors and beta-blockers perhaps needed to control blood pressure or heart rate.

The third person keeps detailed records, documenting not only anaesthetic events, but also procedural activities to allow future correlations. A standardised record form is used and formatted to allow recording anaesthetic drugs, doses and times of delivery as well as anaesthesia and procedural start and end times. The forms are designed so that the times the agents are given can be correlated with changes in vital signs, which are also recorded. Additional useful information includes that associated with monitoring data, i.v. site information, amounts of fluids provided, airway status and general comments.

Emergence and recovery

Emergence from general anaesthesia is fraught with potential complications such as laryngospasm, vomiting and ingesta aspiration. This is yet another time to be highly vigilant and to take precautions to ensure smooth recovery with minimal physiological stress. One potential prophylaxis to reduce the incidence of vomiting is the use of metoclopramide (0.1 mg kg^{-1}, i.v.; Faulding Pharmaceutical Co., Parmus, NJ) given as an anti-emetic about 30 minutes before emergence. Lidocaine (1 mg kg^{-1}) administered i.v. also helps diminish laryngospasm and gagging on the endotracheal tube or during its withdrawal.

When all procedures that required anaesthesia are complete, the panda is moved to a transport or recovery cage. At this time, the endotracheal tube is left in place with the animal spontaneously breathing 100% oxygen. Oxygen saturation is monitored continuously while the panda is observed transitioning from unconsciousness to full awareness. Once the swallowing and protective airway reflexes return (just prior to arousal), the cuff on the endotracheal tube is deflated and the tube carefully withdrawn. An alternative approach is to transport the panda back to the home enclosure with the tube in place after which, upon arrival, it is extubated. The former option is preferable, that is, it is safer to remove the endotracheal tube with the animal in the transport cage held at or near the examination room where all medical resources are readily available.

Even after returning to consciousness, the giant panda is monitored continuously until ambulatory in order to identify complications or signs of discomfort. If the animal still seems to be under the influence of the anaesthetic, pulse oximetry continues to be used while oxygen is provided via the face-mask. Detailed record-keeping also continues, marking the times from the end of the procedure to extubation, head lift, sternal recumbency, attempting to stand and full standing. Throughout, attention is paid to limiting stimulation, especially noise. When judged to be in a satisfactory condition, the giant panda is returned to its enclosure.

PREVENTIVE MEDICINE

Routine examination

Routine examinations are key to a comprehensive and proactive preventive medicine programme for adult and geriatric giant pandas. Such evaluations are made prior to animal transport, during quarantine and on a periodic basis. Each assessment is carried out under anaesthesia (using the principles and methods described above); the benefits gained far outweigh the minor risk of anaesthesia. Each holding institution should strive to develop a thorough, consistent database of information. The types of desirable data and a systematic assessment protocol have been presented in Chapter 4. In addition to standard health factors, the routine examination always considers opportunities to collect data on morphometry, ultrasound observations and reproduction (e.g. semen collection or vaginal smears). Each animal is also checked for its permanent identifier (tattoo or transponder chip code). Other pieces of information may also be important depending on the particular medical needs of a given giant panda. Collectively, securing the same type data consistently over time is invaluable in tracking the health of an individual as it ages or in comparing it to conspecifics in the same institution or across facilities. Important medical and disease trends of individuals or groups also are identified in this way.

The ideal time of year to perform a routine examination in the giant panda is late winter, immediately preceding the breeding season (i.e. January or February in the northern hemisphere). At this time, adult females with young are near weaning or can easily be separated from a cub for evaluation. Assessment of both males and females allows checking overall health, as well as breeding soundness (see Chapters 4 and 7) in anticipation of impending reproductive activity. This timing also works for juveniles who have often had no previous comprehensive examination but are at a convenient age and stage of development.

Quarantine procedures

Giant pandas that arrive from another facility should be isolated physically from other animals (especially conspecifics) for at least 30 days. This minimises the inadvertent transmission of infectious disease(s) from the new individual to other specimens in the collection. The quarantine interval also provides a 'controlled' time for the panda to adjust to the new environment and variations in diet, all under close observation. Ideally, the newly arrived individual should undergo a thorough quarantine examination following the systematic approach (described above for routine examinations). The scope of this evaluation can be reduced if a pre-transfer examination was performed by the previous holding facility.

Vaccinations

The relative risks of acquiring a given disease need to be considered when formulating a vaccination programme for giant pandas in captivity. A serological survey of eight wild giant pandas from a reserve in China revealed antibody titres to several canine viral diseases, but clinical disease information is limited (Mainka *et al.*, 1994). Antibody titres were found in two of eight pandas against canine distemper virus, three of eight against canine coronavirus, six of eight against canine parvovirus and four of eight against canine adenovirus. Local domestic dogs were suspected as one source of viral exposure.

Morbidity and mortality due to canine distemper has been documented in captive giant pandas, so vaccination against this disease is important (Qiu & Mainka, 1993), although challenging. Many wild carnivore species have developed vaccine-induced canine distemper from using modified live vaccines (Deem *et al.*, 2000). Killed distemper vaccines, although safe, provide only short-lived humoral immunity. The development of a monovalent, canary pox-vectored, canine distemper vaccine (PUREVAX™ Ferret Distemper Vaccine; Merial Limited, Inc., Athens, GA) has been shown to be safe while stimulating a significant humoral immune response (R. Montali, pers. comm.). This vaccine is currently used in giant pandas at 6- to 12-month intervals at several USA and Chinese holding institutions.

Parvovirus has been suspected, but not proven, as a cause of gastrointestinal disease in the giant panda (Qiu & Mainka, 1993). A coronavirus has been isolated from a giant panda with diarrhoea in China (Chen & Pan, 1991). The giant pandas at Ocean Park-Hong Kong are vaccinated yearly against parvovirus using a killed vaccine.

Rabies is known to occur in bears and, although the risk of rabies exposure may be low in some facilities, giant pandas have been vaccinated every one to three years against this disease using a killed vaccine.

In all cases, it is advisable to monitor antibody titres to ensure adequate immune responses to vaccinations, as well as to track exposure to those diseases for which vaccines are unavailable or not used. Such titres can be analysed from blood samples taken during routine annual examinations or while in quarantine.

Parasites

Intestinal parasitism is not uncommon in giant pandas. In one study, which reviewed mortality in this species, parasites were found in 39% of the cases, with the majority of parasitic infections being in wild individuals (Qiu & Mainka, 1993). *Balisascaris (Ascaridia) schroederii* is the main intestinal parasite reported (Mainka, 1999). Because ascarid eggs are highly resistant to environmental degradation, it is difficult for an animal to avoid exposure once a physical area is contaminated. Due to the large size and fibrous nature of the giant panda faecal bolus, a microscopic assessment can easily miss parasitic ova, especially if only a small portion of a sample is examined. Whole, adult ascarids can be found in giant panda faeces and vomitus, especially in juveniles.

The authors recommend routine faecal screening for parasites and anti-helmintic treatment in the giant panda. Upon arrival at the San Diego Zoo and the Smithsonian's National Zoological Park, giant pandas from China had ascarid ova observed in faeces. Monthly faecal ova parasite examinations and regular anti-helmintic treatment were initiated. Beginning four months after the first treatment and through to the present time, no further parasites have been detected. Anti-helmintics used include:

- fenbendazole (Panacur®; Hoechst-Roussel Agri-Vet Co., Somerville, NJ) at 50 mg kg^{-1} orally daily for five days;
- ivermectin (Ivomec®; Merial) at 0.2–0.5 mg kg^{-1} orally for 1 day;
- milbemycin (Interceptor®; Ciba-Geigy Animal Health, Greensboro, NC) at 1 mg kg^{-1} orally once;
- pyrantel pamoate (Strongid®; Pfizer, Inc., New York, NY) at a standard carnivore dose.

Due to the presence of *Dirofilaria immitis* in Hong Kong, the giant pandas at Ocean Park-Hong Kong are treated with ivermectin monthly (0.01 mg kg^{-1}).

Demodectic mange has been detected in some captive pandas in China (see Chapters 4 and 16). Other mites of an unknown species have also been seen in skin scrapings from captive giant pandas in China with dermatological manifestations.

COMMON MEDICAL PROBLEMS OF THE ADULT GIANT PANDA

There is a paucity of well-documented information about the common medical problems of giant pandas; often what is known is presented in Chinese journals. From 1998 to 2000, the CBSG Biomedical Survey helped to improve substantially the body of reference knowledge on this species, as well as to survey for medical problems (see Chapter 4). The only other comprehensive reference on medical problems in the giant panda was produced by Qiu & Mainka (1993). Regardless, it is important to emphasise that, if a natural diet is available and adequate veterinary care is given, the giant panda appears to be relatively free of major medical problems in captivity.

Gastrointestinal disorders

Gastrointestinal disease is the most common cause of mortality in both captive and free-living giant pandas (Qiu & Mainka, 1993). In the latter population, pancreatitis secondary to ascaridiasis was the most frequently described gastrointestinal lesion found at necropsy. For giant pandas in captivity, gastric, intestinal and colonic disorders predominate as determined at necropsy (see Chapter 16). Endoscopy has been used recently to identify a gastric ulcer in a living individual (see Chapter 18). Alimentary lymphoma has been observed in some captive giant pandas in China (Qiu & Mainka, 1993) as well as animals held at Zoológico de Chapultepec (Mexico City, Mexico; F. Gual-Sil, pers. comm.). Although hepatic neoplasia has been reported (Qiu & Mainka, 1993), the prevalence is not as high as in other species of Ursidae.

Colic

Colic is commonly observed in adult pandas held at various facilities. These episodes usually are spaced at approximately one week to several week intervals. Affected animals characteristically lie on the abdomen with the legs splayed laterally and are anorexic and lethargic. Vocalisations (especially 'honking') are typically associated with colic. Bouts usually last one to two days (but sometimes up to five days) and are followed by the passage of large amounts of mucous stool. Mucoid stools, evaluated cytologically using Wright–Giemsa and Gram stains, generally contain inflammatory cells and (less commonly) red blood cells and sloughed enterocytes. Large, spore-forming Gram-positive rods are seen, but inconsistently.

Extensive bacteriological evaluations through the aerobic and anaerobic culture of normal and mucoid panda stools have failed to reveal a clear association between bacteria and colic. Bacterial pathogens associated with this condition include *Campylobacter lari*, *Plesiomonas shigelloides*, *Clostridium perfringens* enterotoxigenic strains and *Aeromonas hydrophila*. *Aeromonas* probably results from storing bamboo browse in water-filled containers. Hanging and misting bamboo browse before feeding (rather than storing it submersed in water troughs) greatly reduces *Aeromonas hydrophila* prevalence in the gastrointestinal tract of these animals. Episodes of colic can also be managed conservatively by withholding food for 24 hours and providing drinking water *ad libitum*. Relief and recovery to normal activity, appetite and stool production tend to be rapid. In general, we believe that the incidence of colic and mucoid stools can be prevented (or at the least greatly reduced) through proper dietary management (see Chapter 6).

Ascites

Ascites is an incidental finding in many normally healthy giant pandas. The exact cause of ascitic fluid accumulation in this species is unknown. Aspirates of this fluid reveal a clear to slightly yellowish, non-turgid fluid characterised as a modified transudate. Ascitic fluid, as monitored by ultrasound imaging, is intermittent and changing in volume. The presence of ascites can complicate diagnosis of other, more serious, conditions such as Stunted Development Syndrome (see Chapter 4) or renal and liver disease.

Breeding trauma

Bites and scratches around the head, neck, trunk and extremities occur in both the male and female giant panda following a breeding inter-action. Most wounds are superficial (punctures or minor lacerations) that do not require primary closure. However, there are exceptions, including serious, life-threatening wounds. In most cases, documenta-tion of the presence and location of these lesions, conservative treat-ment with topical wound care (in tractable individuals through protected contact) and oral antibiotics are sufficient management options.

MEDICAL MANAGEMENT OF THE GERIATRIC GIANT PANDA

The giant panda has the capacity to have a relatively long lifespan in captivity, up to almost 30 years. Since reproduction generally ends after 20 years, we consider animals beyond this age to be 'geriatric'. Condi-tions that have been recognised in the aged panda include osteoarth-ritis, renal insufficiency, epistaxis, hypertension and dental disease. However, information available for each of these problems is often limited to studying a single individual. This is because few geriatric giant pandas have undergone routine examinations that have included clinical pathology, radiology and ultrasonography. Until recently, few giant pandas have lived beyond 20 years in captivity but with rapidly improving husbandry and medical management, numbers of geriatric pandas requiring care will skyrocket, so this age group needs more attention now.

Most health problems can be managed in the aged giant panda but never completely resolved during the animal's last years. Thus, it is important to balance concerns for 'quality of life' with the disturbances that normally result from diagnostics and treatments. Here again, advances in operant conditioning (see 'Restraint with no sedation', p. 354) offer many diagnostic and treatment opportunities as part of the animal's daily routine while avoiding chemical restraint.

Osteoarthritis

Osteoarthritis is a frequent problem in old animals, regardless of species. Diagnosis is often based initially on clinical lameness specific to the affected limb or joint. However, the condition can also be manifested by non-specific lethargy and reduced appetite, especially since considerable strength and mobility is required to eat bamboo. Giant pandas with vertebral spondylosis spend less time in a sitting position (without back support) to eat bamboo and rather often use one limb (rather than all four) to manipulate bamboo.

Radiographic indicators of osteoarthritis in aged giant pandas have been documented and include bridging spondylosis in the thoracolumbar vertebrae (Fig. 15.4). One geriatric giant panda (SB 121) developed a lameness shifting from leg to leg and stiffness associated with spondylosis and severe humeroulnar osteoarthritis (Fig. 15.5). Another geriatric female (SB 230) had obvious radiographic lesions of osteoarthritis in the carpal joints, although clinical signs were mild.

One challenge in the giant panda is that many medical conditions are associated with nonspecific signs of variable activity and partial anorexia. As a result, it may be impossible to distinguish suspected arthritis-related pain and stiffness from other clinical conditions without a thorough examination, including radiography. One alternative is to evaluate a trial course of therapy with anti-inflammatory agents when signs of stiffness or lameness are apparent. The goal here is to relieve pain and improve overall quality of life.

The decision to treat osteoarthritis chronically or intermittently (i.e. only when pain and stiffness worsen) must be based upon the

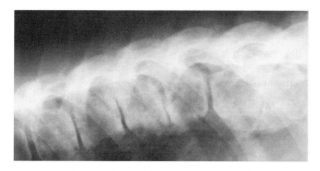

Figure 15.4. Lateral radiograph of the thoracolumbar spine in a geriatric giant panda (SB 12; at 28 years old) showing bridging spondylosis and disc space.

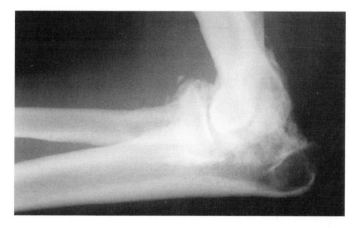

Figure 15.5. Lateral radiograph of the left elbow joint in a giant panda (SB 12; at 26 years old) showing degenerative osteoarthritis.

animal's other known health conditions or its responsiveness to therapeutic agents such as nonsteroidal anti-inflammatory drugs (NSAIDs). In one short-term case, vedpaprofen (Quadrisol®; Intervet International, Boxmeer, The Netherlands; 150 mg given orally every 48 hours for two weeks and repeated as needed) or carprofen (Rimadyl; Pfizer Animal Health, New York, NY; 500 mg orally once daily for ten days) was effective in relieving pain and stiffness in a 27-year-old male giant panda (SB 140) (F. Ollivet, pers. comm.). In terms of long-term NSAID therapy in this species, such work has been restricted to only a single individual. In this case, carprofen was used successfully for 2.5 years in a 26-year-old male giant panda (SB 121) at an initial dosage of 2.2 mg kg^{-1} orally once per day or 250 total mg per day mixed in sweet potato. After one month, the dosage was lowered to 1.5 mg kg^{-1} given once per day for the next two years. Injectable carprofen (1 mg kg^{-1} subcutaneously or i.m.) was also used intermittently in this individual. A dosage of less than 1.5 mg kg^{-1} appeared to be ineffective. Mobility and appetite increased in this giant panda during therapy, with even resumption of play behaviour. During the second year of NSAID treatment, SB 121 developed chronic renal insufficiency, but there was evidence that this animal had pre-existing renal disease. When carprofen was discontinued there was rapid deterioration of the panda's overall condition. Upon carprofen treatment resumption, the animal's quality of life was restored, and parallel efforts were initiated and conducted for the next six months to address the renal insufficiency problem. During the last six months of this panda's life, the carprofen dosage was increased gradually to 3 mg kg^{-1} per day, divided into three oral doses to control arthritic pain (as measured by appetite maintenance). During the 2.5 years of NSAID treatment, the animal's vertebral spondylosis worsened steadily, and yet carprofen effectively maintained an adequate quality of life.

Monitoring weight and activity levels is also important in the geriatric giant panda because excessive body mass, sedentary behaviour and overexertion will contribute to degenerative joint diseases. Thus, it is essential to maintain proper body condition by ensuring adequate diet, sufficient exercise space and daily enrichment activities.

Renal insufficiency

Clinical signs of renal insufficiency in the giant panda include lethargy, inappetance, polyuria and polydypsia. Clinical pathology findings generally are associated with azotaemia, anaemia and hyperphosphataemia. Ultrasonography is useful, although obtaining images of the kidney can be difficult, especially if anaesthesia time is limited. Renal insufficiency has been definitively documented in three geriatric giant pandas (cases summarised as follows).

A 20-year-old female (SB 112) developed acute renal failure associated with pyelonephritis. A striking feature of this individual's illness was severe anaemia (haematocrit, 10%) and azotaemia (blood urea nitrogen, BUN, 145 mg dl^{-1}; creatinine 95 mg dl^{-1}). This animal responded to aggressive therapy, including a blood transfusion and antibiotic therapy (Bush *et al.*, 1984).

A 24-year-old female (SB 230) developed mild renal insufficiency associated with one episode of epistaxis and intermittent anorexia. Initially, she was found to be hypertensive, but clinical pathology results were within normal limits. Subsequently, polydypsia, polyuria, mild nonregenerative anaemia (haematocrit, 31%) and persistent hyposthenuria (specific gravity, SG, 1.002–1.008) developed (D. Spielman, pers. comm.).

A 29-year-old male (SB 121) developed moderate, compensated renal insufficiency that progressed over six months to renal failure associated with degenerative cystic renal disease. This giant panda had a long history of infrequent epistaxis (nosebleed) episodes that became more severe as hyposthenuria (SG 1.005–1.010), moderate azotaemia (BUN 42 mg dl^{-1}; creatinine 6.8 mg dl^{-1}) and moderate nonregenerative anaemia (haematocrit, 25%) developed. This individual was managed successfully with intermittent fluid therapy and human recombinant erythropoietin despite worsening azotaemia (BUN, 34–101 mg dl^{-1}; creatinine, 6.3-22.8 mg dl^{-1}). After six months of therapy, the animal was euthanised due to worsening epistaxis (nosebleed) and retinal haemorrhages associated with secondary renal hypertension. In retrospect (and based upon the clinical course of the second case described above), the mild anaemia (haematocrit, 37%), intermittent mild nosebleeds and occasional anorexia noted in this male two years earlier may have been early indicators of renal insufficiency.

Treatment of renal failure in the giant panda using fluid therapy is limited by the requirement for anaesthesia. Nonetheless, i.v. and subcutaneous fluid therapy has been effective in stabilising ill giant pandas that were episodically anaesthetised. However, it is possible to use behavioural training to condition the giant panda to accept fluid therapy while fully awake.

Blood transfusions or treatment with human recombinant erythropoietin have been effective for treating anaemia in this species. In the case of the geriatric arthritic giant panda on carprofen (SB 121; see 'Osteoarthritis', p. 368), erythropoietin (Epogen®, Amgen Pharmaceuticals, Thousand Oaks, CA) was effective in addressing chronic renal insufficiency and was associated with improved haematocrit, appetite and attitude. In that case, the risk of the giant panda developing anti–erythropoietin antibodies and, therefore, paradoxically accelerating the anaemia was outweighed by anticipated potential benefits. The efficacy of anaemia treatments is also enhanced in this species by the naturally high water and relatively low protein content of the bamboo diet. Offering high-water-content bamboo shoots or young bamboo could probably further facilitate treatment.

Epistaxis and hypertension

Epistaxis, characterised by intermittent and sometimes severe nosebleeds, is the most dramatic clinical sign associated with hypertension in the giant panda. Spattered blood is often found in the animal's den overnight, or dried blood may be present on the nose and fur. Unless the nosebleed is severe, most pandas eat and appear otherwise normal.

Clinical signs of hypertension, including epistaxis, lethargy and inappetance, have been noted in several older giant pandas. In one geriatric male (SB 381), epistaxis occurred on multiple occasions. Hypertension was noted using direct and indirect methods in this animal during anaesthesia as well as when it was fully conscious. Treatment in this case was apparently successful with propranolol starting at 20 mg given orally twice daily and then increasing the dose slowly over six weeks to 60 mg orally twice daily. Treatment continued for seven weeks. No further episodes were observed after treatment ended. Hypertension was also noted in a 24-year-old female (SB 230) using daily, indirect blood pressure measurements via operant conditioning (Mauroo *et al.*, 2003). Average values for this female were 241/35 mmHg with a mean arterial pressure of 176. The only clinical sign reported perhaps related to systemic hypertension was an episode of epistaxis. A daily dose of 7.5 mg of amlodipine (Norvasc[®]; Pfizer, West Ryde, NSW, Australia; equivalent to 0.093 mg kg^{-1} BW) was administered orally and was found to reduce systolic blood pressure, diastolic blood pressure and mean arterial pressure, with no adverse effects. In one male (SB 121), episodes of epistaxis occurred throughout his life and increased in severity at the onset of severe and chronic renal failure at 26 years of age. In one 24-year-old female (SB 112), one occasion of epistaxis was noted prior to a diagnosis of chronic renal failure.

To rule out the potential hypertensive impacts of the anaesthetic ketamine hydrochloride, some giant pandas have been trained to accept indirect blood pressure cuffs, allowing readings to be taken while fully awake (Fig. 15.6). Observations have revealed similar

Figure 15.6. A giant panda trained to accept an automatic indirect blood pressure measurement.

episodes of hypertension in the relaxed, fully conscious condition as well as under ketamine anaesthesia.

Dental disease

Giant pandas, like other browsing herbivores, require healthy dentition to masticate a diet of coarse, fibrous plant material. Excessive wear of the premolar and molar grinding surfaces can result in under-nutrition. The most significant dental problems observed in one geriatric panda (SB 381 at about 26 years of age) was extreme dental wear (Fig. 15.7; Plate XIII) and associated traumatic gingivitis secondary to the animal's insistence on eating bamboo. Interdental erosive gingivitis was observed in the premolars at about the level where the animal appeared to chew bamboo stalks. This individual was coincidentally receiving long-term antibiotic treatment for cystitis. The antibiotic treatment improved the bacterial gingivitis that had resulted from repetitive trauma of ineffective bamboo consumption. Because of the chronic nature of his urinary tract infection, and the apparent benefit of

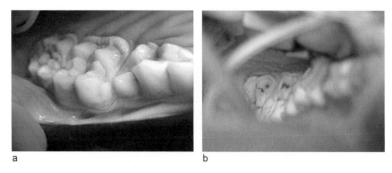

a   b

Figure 15.7. View of the mandibular premolars and molars of (a) a
2.5-year-old giant panda with minimal wear of the grinding surfaces and
(b) a 26-year-old giant panda (SB 381) with extreme wear of the grinding
surfaces. (See also Plate XIII.)

long-term antibacterial therapy in minimising secondary bacterial gingivitis, the animal was placed on long-term antibacterial therapy with
the result being generally improved quality of life. Routine examinations revealed that the focal gingivitis was adequately managed with
a combination of a dietary 'bread' made up from dry, ground bamboo
leaves, dry biscuits and long-term antibacterial chemotherapy.

PRIORITIES FOR THE FUTURE

In this chapter, we have described aspects of managing anaesthesia as a
critical first step for ensuring effective health care for the giant panda.
Preventive medicine practices are also essential for deterring disease,
detecting early indicators of a problem and revealing how a given
animal fares compared to others in the same facility or even in the
genetically valuable, worldwide *ex situ* population. Effective prevention
means that high, *routine* priorities include regular health examinations,
quarantine, vaccinations and parasite control. It does appear that a
giant panda maintained under good husbandry and veterinary care
generally remains healthy, although occasional problems remain with
gastrointestinal disorders, colic, ascites, dermatophytosis and breeding
traumas. More and more giant pandas will live to 'old age', and this
process needs to be managed humanely. There is early evidence that the
common medical problems of the geriatric giant panda include predominately osteoarthritis, renal insufficiency, epistaxis, hypertension
and dental disease.

In terms of priorities, we first recommend the need for more efforts devoted to operant conditioning. Behavioural training can allow veterinarians to provide noninvasive diagnostics and treatments to giant pandas throughout their long life-span. This will permit routine ultrasonography, blood sampling and blood pressure monitoring (among other techniques) which, in turn, will allow rapid expansion of medical databases, earlier diagnosis of potential problems and more effective treatments.

Second, it must be realised that there will always be the occasional need to use anaesthesia in this species. Although many giant panda managers are comfortable with using injectable ketamine as a short-term anaesthetic option, this is not an adequate drug for performing many necessary long-term veterinary procedures, such as surgery or endoscopy. There is a need for more capacity building in China in the arena of alternative and advanced anaesthetic approaches. This particularly includes training in the use of intubation and gaseous anaesthesia. The transfer of such technology and basic knowledge will permit addressing serious medical issues that require prolonged, hands-on access to a safely anaesthetised giant panda.

Finally, geriatric medicine is booming in zoos that are now managing animal populations that are able to live longer due to modern advances in husbandry and veterinary care. To date, dealing with geriatric giant pandas has been largely occasional, due to the few individuals reaching advanced age. This is changing dramatically. The good news is that the few older individuals that have been studied have been so quite thoroughly. Not surprisingly, giant pandas, like all living creatures, eventually suffer from age-related organ system failures. However, it also is evident based on our limited information, that this species can respond nicely to therapies that have been developed for other aged animals. Now there is a need to focus attention specifically on the more systematic collection of hard data, the problems to be expected and especially the solutions to ensure that the growing giant panda population ages gracefully with a high quality of life.

ACKNOWLEDGEMENTS

The authors thank David Fagan DDS for his professional expertise in helping to understand and treat dental problems in the geriatric giant panda.

REFERENCES

Bush, M., Montali, R. J., Phillips, L. G. *et al.* (1984). Anemia and renal failure in a giant panda. *Journal of the American Veterinary Medical Association*, **185**, 1435–7.

Chen, Y. and Pan, X. (1991). Etiological studies on acute enteritis in giant pandas. *Proceedings of the Third Asian Bear Conference*, Harbin, China, p. 193 (abstr.).

Deem, S. L., Spelman, L. H., Yates, R. A., and Montali, R. J. (2000). Canine distemper in terrestrial carnivores: a review. *Journal of Zoo and Wildlife Medicine*, **31**, 441–51.

Mainka, S. A. (1999). Giant panda management and medicine in China. In *Zoo and Wild Animal Medicine V,* ed. M. E. Fowler and E. Miller. Philadelphia, PA: W. B. Saunders Co., pp. 410–14.

Mainka, S. A., Qiu, X., He, T. and Appel, M. J. (1994). Serologic survey of giant pandas (*Ailuropoda melanoleuca*) and domestic dogs and cats in the Wolong Reserve, China. *Journal of Wildlife Diseases*, **30**, 86–9.

Mauroo, N. F., Routh, A. and Hu, W. (2003). Diagnosis and management of systemic hypertension in a giant panda (*Ailuropoda melanoleuca*). *Proceedings of the American Association of Zoo Veterinarians*, ed. C. K. Baer. Minneapolis, MN pp. 289–90.

Qiu, X. and Mainka, S. A. (1993). Review of mortality of the giant panda (*Ailuropoda melanoleuca*). *Journal of Zoo and Wildlife Medicine*, **24**, 425–9.

16

Diseases and pathology of giant pandas

I. KATI LOEFFLER, RICHARD J. MONTALI, BRUCE A. RIDEOUT

INTRODUCTION

The study and control of diseases have not been traditional priorities in giant panda management, even though neonatal mortality, chronic and debilitating disease, compromised reproduction and premature death have been problems. Recent years have seen an increased awareness of the role of diseases in captive and free-living wildlife populations, with pathology integral to both diagnosis and creating new scholarly knowledge.

Growing concerns in the zoo community about the stress of captivity, pathogen transmission and the emergence of novel infectious agents are driving a rising interest in wildlife disease. It is also critical to understand diseases in *ex situ* populations of animals that may be released into the wild. The reintroduction of giant pandas into native habitats has been a focus of several conservation proposals, including the National Conservation Management Plan for China (MacKinnon *et al.*, 1989). The recommended course of action in this plan failed to emphasise the importance of veterinary care and pathological investigations of illness and mortality in the captive population. Ten years later, the CBSG Giant Panda Biomedical Survey (1998 to 2000; Zhang *et al.*, 2000; see Chapters 4 and 15) recognised that a clear understanding of health and disease must be a priority in the plan to

Giant Pandas: Biology, Veterinary Medicine and Management, ed. David E. Wildt, Anju Zhang, Hemin Zhang, Donald L. Janssen and Susie Ellis. Published by Cambridge University Press. © Cambridge University Press 2006.

secure a viable *ex situ* giant panda population. The next step then would be to integrate new information with mitigating approaches to optimise health, which, in turn, would promote reproduction.

A workshop on veterinary medicine and nutrition (again associated with CBSG) was held in Chengdu in 1999 to provide baseline training in these areas. Later in that year, another CBSG workshop was held in the Wolong Nature Reserve that focused on identifying priorities for conserving giant pandas *in situ* (Yan *et al.*, 2000). A working group on captive management at this meeting placed substantial emphasis on health and pathology. It pointed out that information about diseases in captive giant pandas is not only necessary for the controlled reintroduction of individuals into the wild, but also may influence disease management and prevention in giant pandas *in situ*. The working group also recommended the need for training zoo veterinarians, advice that resulted in the first Workshop on Diagnostic and Clinical Pathology in Zoo and Wildlife Species (June 2002, Beijing). Participants at this workshop recognised the importance of pathology in the veterinary management of zoo collections, and expressed a personal need for more training. The international giant panda conservation community has also realised the need for training in veterinary medicine, diagnostics and pathology in China, in part to reach the overall goal of a self-sustaining, *ex situ* panda population. Although much remains to be learned about this species, a remarkable amount of rudimentary information is available. This chapter summarises what is known about diseases in this species, with a special emphasis on pathology.

SPECIAL CONSIDERATIONS OF GIANT PANDA ANATOMY AND PHYSIOLOGY

The giant panda is an enigmatic carnivore, adapted to a highly specialised ecological niche (see Chapter 1). Although the species looks, moves and is genetically similar to bears (see Chapter 10), many of its phenotypic traits are unlike those of other ursids. For a thorough understanding of the anatomy of giant pandas, the reader is referred to the classic publication of Davis (1964). Many of the species' distinctive features are related to its highly specialised reliance on bamboo. For example, the bone structure and musculature of the head differ markedly from that of other bears. The cheek teeth are enlarged (as in herbivores) to provide large grinding surfaces (see Chapter 6), and an

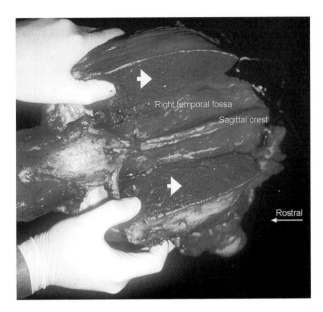

Figure 16.1. Dorsal aspect of the adult giant panda skull displaying the large temporal muscles (arrows). The masticatory apparatus is massive relative to other carnivore species. The volume of the temporal fossa and the associated temporalis muscle are particularly large. Similar adaptations are seen in the red panda (a procyonid with a similar diet) and the hyaena (Davis, 1964). The relative mass of bone in the giant panda cranium and mandibles is also much greater than in other ursids. (See also Plate XIV.)

articulated jaw with massive masticatory muscles and skull bones are adapted to the panda's herbivorous diet (Davis, 1964; Fig. 16.1; Plate XIV).

Given its bamboo diet, the giant panda's short and seemingly unspecialised intestinal tract is a physiological enigma (Fig. 16.2; Plate XV). Most mammals maintain a correlation between gut length and body length such that the greater the herbivorous proportion of the diet, the longer the gut length. Intestinal length of four to five times body length is typical of a true carnivore (e.g. cat), whereas six to eight times is representative of an omnivore (e.g. raccoon), and the ratio in a fully herbivorous ungulate (e.g. deer) is about 10 to 22 (Davis, 1964). The giant panda diverges from this pattern entirely in that, despite its wholly herbivorous diet, its gut is only four times as long as its body

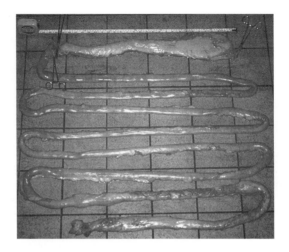

Figure 16.2. Intestinal tract of the adult giant panda illustrating
uniformity in diameter along its entire length. There is no discernible
demarcation between the small and large intestine and no caecum. The
extent of the tract is only four to five times the body length, which is
more characteristic of a true carnivore than a species that survives on a
fibrous bamboo diet. (See also Plate XV.)

length, such as in true carnivores. The panda has a simple stomach
(precluding the possibility of foregut fermentation) and no caecum (precluding the possibility of hindgut fementation). There is no recognisable
demarcation between the small and large intestine and no anatomical or
histological clue for its ability to utilise bamboo as a primary energy
source. The high transit rate of food and an intestinal flora comprised
largely of facultative anaerobes (Hirayama *et al.*, 1989) preclude the
microbial-enhanced utilisation of fibrous plants (as occurs in true herbivores). The giant panda presumably secures its major nutrient requirements by selecting leaves and shoots rather than stems, ingesting large
quantities of food, digesting plant cell contents (rather than cell walls)
and rapidly excreting undigested residues.

The unique diet of the giant panda and its lack of natural predators may have influenced skeletal structure evolution. For example,
proficiency in locomotion is reduced as much as that in handling
bamboo is increased, and the animal appears well adapted to sitting
on its hindquarters for 10 to 12 hours per day, stripping and chewing
bamboo. Modification of the radial sesamoid bone of the wrist to

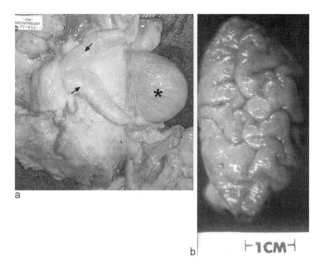

Figure 16.3. (a) Bladder and ductus deferens of an adult giant panda. The glandular ductus deferens (arrows) opens into the neck of the bladder (*). (b) Ovary of an adult giant panda with its characteristic convoluted surface. (See also Plate XVI.)

function as a sixth prehensile 'digit' (see Chapter 6 for depiction) allows the panda to strip leaves from bamboo stalks with remarkable dexterity (Davis, 1964). Other modifications include the concentration of the panda's mass toward the front of the body (with a concomitant gradient in bone density), relatively broad and heavy vertebrae, fewer thoraco-lumbar vertebrae (four compared to five in other carnivores) and unique pelvic size and shape (Davis, 1964).

The male giant panda lacks seminal vesicles and bulbourethral glands but has well-developed glands at the terminus of the ductus deferens which supply seminal fluid (Davis, 1964; Fig. 16.3a; Plate XVI). Davis (1964) indicated that the species lacks a prostate gland, an asser-tion disputed by Hildebrandt et al. (see Chapter 17) who have observed a bilobal prostate by ultrasound. The penis is relatively small, located ventral to the anus, is directed posteriorly (like in the cat) and is normally withdrawn into the prepuce. The baculum is small and differs from the rod shape in other carnivores in that it projects ventrolateral wings (Davis, 1964). The female reproductive tract is bicornuate and unusual only in the highly convoluted surface structure of the ovary (see Fig. 16.3b), an appearance that seems retained into the breeding season.

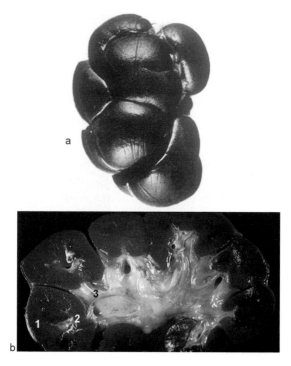

Figure 16.4. The renculate kidney of the giant panda. (a) Whole kidney and (b) sagittal section. Each renculus has a thick cortex (1) and small medulla (2) with the papillae of each joining into a common minor calyx (3). In the renal fossa, the minor calyces unite into two major ones which, in turn, join to form the proximal end of the ureter. There is no renal pelvis in this species. (See also Plate XVII.)

Giant panda kidneys share the lobular or renculate structure of other bears, although each kidney is comprised of fewer lobes and contains two or three papillae rather than one (Davis, 1964; Fig. 16.4; Plate XVII). In place of a renal pelvis, two major calyces drain the minor calyces of each renculus and join at the proximal end of the ureter.

Giant panda vocalisations more closely resemble bleats, barks and honks than the roars or growls of typical bears, which may be related to differences in the anatomy of the epipharyngeal pouches. These highly elastic, tubular diverticulations of the caudodorsal pharyngeal wall are lined with respiratory epithelium and are unique to Ursidae. Most bear species have two juxtaposed pouches, but the relative sizes and positions

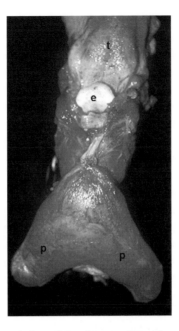

Figure 16.5. Dorsal view of the pharyngeal pouches of the adult giant panda. These paired sacs (p, shown here filled with water to facilitate viewing) open into the dorsal nasopharynx and may play a role in vocalisation. Some authors (Davis, 1964) describe the left sac as being much smaller than the right (15 mm vs. 130 mm in length), while others (Weissengruber *et al.*, 2001 and references therein) describe just a single medially located pouch. e, epiglottis; t, tongue. (See also Plate XVIII.)

of the pouches vary greatly (Weissengruber *et al.*, 2001 and references therein). The anatomy of the panda's epipharyngeal pouches is somewhat controversial. Some authors describe two pouches of which one is much smaller than the other, while others found only one medially located pouch (Davis, 1964; Weissengruber *et al.*, 2001 and references therein; Fig. 16.5; Plate XVIII).

NEONATAL PATHOLOGY

Twinning is common in the giant panda, with one neonate usually rejected by the dam at birth. Hand-rearing the rejected cubs is challenging but has become more successful in recent years (see Chapter 13).

The species is highly altricial. A newborn cub is generally 10 to 15 cm long and weighs 60 to 200 g. The per cent body weight ratio of cub to mother is only about one-tenth of one per cent, which is more typical of a marsupial than eutherians. Necropsy and histological assessments of giant panda neonates have revealed an underdeveloped immune system (compared with domestic carnivores), including a haemolymphatic organisation that contains few lymphocytic cells. Any interference with colostrum intake within the first few hours of life renders the infant extremely vulnerable to sepsis and death (Knight *et al.*, 1985; Montali *et al.*, 1990).

Research into the neonatal immunity of giant pandas and immunoglobulin (Ig) supplementation is just beginning, but its importance is well recognised. One impediment to progress in this area has been the unavailability of stored colostrum or sufficient volumes of panda serum for supplementation. In the USA, concentrated Ig of the horse, cow and cat are commercially available for oral or parenteral use in hypoimmune neonates (generally <400 mg dl^{-1} of Ig) of those species. The Smithsonian's National Zoological Park has begun to bank serum from the resident female giant panda (studbook, SB, 473) and is collaborating with a commercial producer of purified Ig to generate a lyophilised product in preparation for future giant panda offspring. Methods of evaluating the immune status of neonates will continue to be challenging due to the small size of cubs and the hesitation to perform invasive procedures such as venipuncture.

The factors responsible for the deaths of eight giant panda neonates born at the zoos in Washington, DC, Mexico City and Madrid were evaluated in a study by Montali *et al.* (1990). Of the eight deaths, six were associated with infections involving frequently identified, ubiquitous organisms such as *Escherichia coli*, *Staphylococcus* spp. and *Pseudomonas* spp. (Fig. 16.6; Plate XIX). The frequency of fatal opportunistic infections in neonatal giant pandas suggests that failure of passive transfer may be a significant problem in this species (Montali *et al.*, 1990).

Neonatal bacterial infections were also the leading cause of death in a study of 17 cubs less than 30 days of age in Chinese giant panda breeding facilities (B. Rideout, unpublished data). The umbilicus appeared to be the primary portal of bacterial entry regardless of whether the cub was mother- or hand-reared. Accidental maternal trauma was the second most common cause of death in this survey, but some traumatised neonates also had evidence of pre-existing bacterial infections.

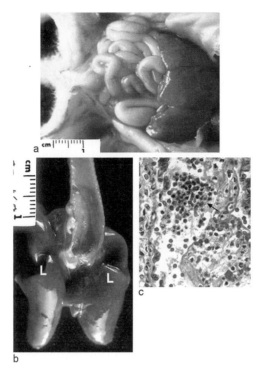

Figure 16.6. (a) Gross necropsy (liver) of neonate with septicaemia and
hepatitis caused by *Staphylococcus intermedius*. (b) Lungs (L) of a neonate
that acquired pneumonia prenatally from a dam with an ascending
urogenital tract infection that caused chorioamnionitis. (c)
Microscopically the pulmonary alveolar spaces contained inflammatory
cells (arrow). *Pseudomonas aeruginosa* was cultured from the lungs and
P. fluorescens from heart blood. (See also Plate XIX.)

    Chronic, subclinical infections in the dam may also compromise
the newborn giant panda. Four of the five cubs produced by SB 112 (Ling
Ling) at the Smithsonian's National Zoological Park over a six-year
interval died of bacterial infections within the first three days after
birth (one cub was a stillborn, a twin). Her second cub acquired an
infection *in utero* due to bacterial chorioamnionitis (see Fig. 16.6b). It
died three hours postpartum (Montali *et al.*, 1990), presumably due to an
ascending infection caused by the dam masturbating during late pregnancy (Bush *et al.*, 1984). SB 112 developed an acute pyelonephritis four

months after parturition (Bush *et al.*, 1984), but responded well to treatment and became pregnant again the next breeding season. As all of SB 112's live-born cubs died shortly after birth of sepsis, it is possible that she maintained a chronic, low-grade urogenital tract infection that carried over to all offspring at parturition.

THE DISEASES BY ORGAN SYSTEM

The digestive system

Digestive disorders represent by far the most prevalent and chronic form of disease in captive giant pandas (Bush *et al.*, 1985; Göltenboth, 1985a,b; Villares *et al.*, 1985; Pan *et al.*, 1991; Gual-Sil *et al.*, 2000; F. Ollivet, unpublished data, pathology report, Paris Zoo). Gastrointestinal disease is the leading reported cause of death in captive pandas and also may be so in free-ranging counterparts (Qiu & Mainka, 1993). Cubs often develop bouts of diarrhoea within the first month of life, particularly if hand reared on artificial milk replacer (I. K. Loeffler, unpublished observations). Cubs raised on milk replacer and recent weanlings often develop diarrhoea that can become a chronic condition. Episodes of abnormal stool, abdominal pain, passage of mucous faeces, vomiting and inappetance occur frequently, if not chronically, throughout the lives of many giant pandas *ex situ*.

Pathology of the oral cavity

A condition identified during the CBSG Biomedical Survey described as Stunted Development Syndrome (Zhang *et al.*, 2000; see Chapters 4 and 15) is characterised by slow growth, small body size, ascites, infertility, general unthriftiness and chronic digestive problems. The individuals affected with this syndrome typically also have dental abnormalities characterised by severely worn, discoloured teeth, dentine exposure and enamel hypo- or dysplasia (Fig. 16.7; Plate XX). The cause of the dental abnormalities in these pandas is unknown, although possible aetiologies include pre- or neonatal exposure to canine distemper virus, malnutrition, fluorosis or early exposure to other infectious agents or toxins. Worn and broken dentition and open root canals are also a fairly common finding in giant pandas ten years old or older (Zhang *et al.*, 2000; see Chapters 4 and 15).

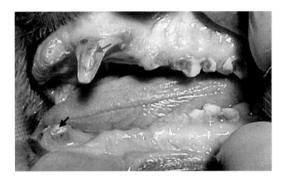

Figure 16.7. Giant pandas with Stunted Development Syndrome have dental abnormalities characterised by severely worn, stained teeth with dentine exposure (black arrow) and enamel hypo- or dysplasia (blue arrow). (See also Plate XX.)

An oropharyngeal fistula and sublingual abscess were diagnosed in a 28-year-old male giant panda (SB 121) at the Smithsonian's National Zoological Park (R. Montali, unpublished data, pathology report). The lesion appeared to be associated with a bamboo fragment which had migrated under the tongue. This animal was severely ill with chronic renal failure at the time and, despite two surgical treatments, the lesion failed to heal. Post-mortem examination revealed a fistulous pocket of abscessed material found caudolateral to the base of the tongue which was lined with fibrous connective tissue that extended into the oropharynx.

Gastrointestinal distress and faecal mucus excretion

Giant pandas are often observed to experience episodes of gastrointestinal discomfort, apathy and inappetance (particularly for bamboo) which are followed by faecal mucus excretion and then immediate relief. The frequency of these bouts of catarrhal colitis varies among institutions and individual animals. The male giant panda (SB 140) at Paris Zoo had a history of chronic abdominal pain with anorexia, groaning, faecal mucus and occasional vomiting from the age of 11 years. The condition was managed symptomatically with spasmolytic and demulcent therapy. This individual died at 27 years of age, and necropsy findings included chronic ulcerative and necrotising,

suppurative colitis with local colonic perforations and chronic peritonitis (F. Ollivet, unpublished data, pathology report, Paris Zoo). To what
extent the 16 years of intermittent colitis or the nonsteroidal anti-
inflammatory drug (NSAID) treatment the panda received for musculo-
skeletal pain toward life's end contributed to the colonic ulcers and
perforations is not known.

Cytological examination of excreted mucus often reveals mild
granulocytic inflammation, but does not point to obvious aetiologies
(B. Rideout, unpublished observations). Mucus excretion and gastrointestinal pain may be associated with insufficient dietary bamboo.
However, the association between mucus excretion and bamboo may
be entirely coincidental, and other aetiological possibilities have not
been fully explored. For example, pathogenic organisms have occasionally been cultured from mucous faeces (Göltenboth, 1985a; Villares et al.,
1985; Gual-Sil et al., 2000), but not to an extent that could strengthen a
causative association. Spore-forming, Gram-positive rods are also occasionally seen in smears of excreted mucus but likewise have not been
correlated with clinical disease or mucoid stools (see Chapter 15).

Parasites

Frequent, unexplained episodes of gastrointestinal pain in the giant
panda may be associated with parasites or (sometimes consequent)
intestinal obstruction. Mortalities due to pancreatitis in the free-
ranging giant panda have been associated with parasite burdens and
blockage of pancreatic ducts by ascarids (Qiu & Mainka, 1993). Although
there was no reference to bile-duct obstruction in the latter citation,
this could be an associated pathological finding. *Baylisascaris schroederi* is
the only endoparasite that appears to be consistently recognised in the
giant panda (Fig. 16.8). Larvae of this ascarid hatch in the intestine and
penetrate the intestinal wall to enter the portal circulation to the liver
and lungs where their migration may cause extensive inflammation
and scarring. Eggs of this roundworm are notoriously difficult to observe upon faecal examination, probably because they are shed intermittently and because the high roughage content of the faeces interferes
with routine laboratory detection. Pandas whose faecal examinations
have produced repeated negative egg counts can suddenly vomit
bundles of ascarids (D. Janssen, pers. comm.). Treatment of individuals
(including those with possible false-negative results for *B. schroederi* eggs)

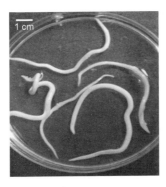

Figure 16.8. *Baylisascaris schroederi* is a common intestinal nematode of the giant panda. The parasite has a life-cycle similar to other roundworms in which infective eggs are ingested by the host, the larvae hatch and penetrate the intestinal mucosa where they mature and return to the small intestine as adults 2.5 to 3 months later. Evidence of visceral larval migrans is often apparent upon necropsy.

with antiparasitic drugs may require multiple doses until the animal ceases to expel worms or shed eggs in faeces (Göltenboth, 1985a; Leclerc-Cassan, 1985; L. Spelman, unpublished observations, clinical records, Smithsonian's National Zoological Park). Yang (1998) provides one of the few descriptions of other intestinal parasites in the giant panda, including a hookworm (*Ancylostoma caninum*), another roundworm species (*Toxocara selenactis*) and a protozoan (*Sarcocystis* sp.).

The incidence of parasitism and its clinical significance to captive giant pandas is unknown. Analysis of haemograms during the CBSG Giant Panda Biomedical Survey (Zhang *et al.*, 2000; see Chapters 4 and 15) frequently demonstrated an eosinophilia, which suggests parasitism or an allergic response. Protozoa such as *Giardia*, coccidia and *Cryptosporidia* may cause subclinical disease that could become important in the context of other stressors. Ascarids may compromise nutritional status by directly interfering with nutrient absorption or by causing pathological changes to intestinal tissue. Hookworms cause mucosal haemorrhage or blood extraction that, in turn, may lead to anaemia and critical illness in young animals. Migrating larvae may cause extensive damage to hepatic or pulmonary parenchyma, which may be especially harmful to young, developing individuals. High parasite loads in females may be detrimental to a nursing mother (due to the high energy demand of lactation) as well as to her offspring because of the presumed lactational transmission of recrudescent larvae (as occurs with roundworms in the dog and cat). The impact of an endoparasitic infection on cub development is unknown but is an important topic for further investigation.

Enteritis and diarrhoea

Haemorrhagic enteritis and bacterial diarrhoeas occur relatively frequently in giant pandas and appear to be associated with a variety of potential aetiologies (Bush et al., 1985; Villares et al., 1985; del Campo et al., 1990a; Qiu & Mainka, 1993; Gual-Sil et al., 2000). Bacterial pathogens such as E. coli (del Campo et al., 1990a), haemolytic and nonhaemolytic Streptococcus, Salmonella (Villares et al., 1985; Gual-Sil et al., 2000) and Clostridium (Dämmrich, 1985) have been implicated in various reports, although definitive diagnoses of the primary or underlying problems have not been made.

Faecal culture of an adult male giant panda (SB 187) at the Madrid Zoo who suffered frequent gastrointestinal disturbances demonstrated E. coli proliferation nearly every time he was ill (del Campo et al., 1990a). The predisposing factors of the recurrent infection remain unknown. Three endoscopies over four years revealed (at various times) pseudomembranous colitis, haemorrhagic gastritis and oesophagitis (del Campo et al., 1990a). Madrid Zoo veterinarians performed an experimental oral inoculation on this panda with flora from normal faeces of another giant panda (SB 249) who also had suffered symptoms of gastrointestinal distress, but less frequently. The oral inoculations were combined with large quantities of lyophilised Lactobacillus, mucosal protectants, activated charcoal, dimethylpolisiloxane (another carminative) and spasmolytics, a combination that appeared somewhat effective. Dietary modifications included the removal of milk products and the reduction of roughage by feeding only stripped bamboo leaves (del Campo et al., 1990a).

It is worth noting that haemorrhagic enteritis in giant pandas in China has sometimes been diagnosed solely on the basis of a gross reddening of the intestine (B. Rideout, pers. obs.). Some of these cases could be misclassified and may in fact be due to shock-related congestion or intra-intestinal haemorrhage. The chronic enteritis experienced by many giant pandas poses the added risk of intestinal ulceration and megacolon (Fig. 16.9; Plate XXI).

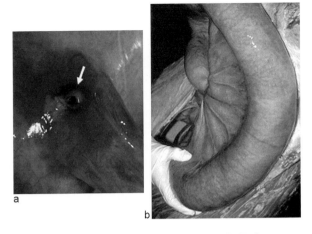

Figure 16.9. (a) Chronic enteritis in a giant panda. Endoscopy demonstrating ulceration (arrow) of the intestine. (b) Megacolon found at necropsy. (See also Plate XXI.)

Some cases of acute hemorrhagic diarrhoea, although recurrent, resolve within a few days with no medical intervention (Bush *et al.*, 1985; Göltenboth, 1985a). Clinical pathology in these cases indicates localisation of disease to the intestinal tract, and histopathological changes are limited to inflammatory cell infiltration in areas of sub-mucosal haemorrhage without evidence of infectious organisms (Bush *et al.*, 1985). Other cases of haemorrhagic intestinal disease in this species have been severe and occasionally fatal. An adult female (SB 169) at Madrid Zoo died after an acute episode of haemorrhagic enteritis that lasted only 24 hours (Villares *et al.*, 1985). Clinical signs included bloody faeces, abdominal pain, respiratory distress and progressive weakness. Necropsy revealed that the lesions were restricted to the small and large intestines. Mucosal surfaces were congested, and the wall of the large intestine was intermittently oedematous and friable. Histopathology indicated varying degrees of serosal and mucosal congestion, haemorrhage, necrosis, oedema and inflammatory cellular infiltration. Regional lymph nodes were similarly affected, but there was no evidence of mesenteric vascular occlusion. Villares *et al.* (1985) considered the lesions to be consistent with an acute toxicosis, allergic episode or thrombosis of a major mesenteric vessel, but not with a primary bacterial enteritis. The authors suggested that the giant panda died from toxic shock associated with an acute ischae-mic or anaphylactic intestinal necrosis.

Some reports describe enteritis as part of a multisystemic illness. A case from the Berlin Zoo involved a six-year-old female panda (SB 210) who died after an illness characterised by intestinal haemorrhage and severe central nervous system disturbance (Göltenboth, 1985b). The clinical course of illness progressed from what was believed to be enterotoxaemia with neurological signs to intestinal haemorrhage, sepsis, coagulopathy and hypovolaemic shock. Post-mortem examin-ation revealed extensive oedema throughout the body (including the head) and a haemorrhagic diathesis along the full length of the intestinal tract (Dämmrich, 1985). Histologically, the primary lesions affected the intestinal mucosa and lymphoreticular tissue and resembled those associated with enteropathogenic viral disease in the domestic cat and dog (Dämmrich, 1985). This giant panda had been vaccinated against parvovirus annually for the preceding three years. Antibody titres were 1:256 and 1:480, the latter of which was higher than a typical vaccination titre in the cat or dog (Dämmrich, 1985). Parvovirus, coronavirus and rotavirus could not be isolated from this individual's tissues, and efforts to detect other common canine and feline viruses by immunofluorescence also failed. An un-identified *Clostridium* sp. was isolated from the stomach and small intestine, which may or may not have been responsible for the signs of enterotoxaemia.

Viral diseases

Disease associated with parvovirus or other common canine and feline viruses in giant pandas is a largely unexplored issue in China. One report describes a survey for canine parvo-, corona-, adeno- and distemper virus in five captive and three recently rescued giant pandas at a Chinese breeding centre (Mainka *et al.*, 1994). None of these animals had been vaccinated, and all had been in captivity either for some time or had been recently captured. The titres to coronavirus (in two captive individuals 1:160), distemper (one captive 1:50; one recently wild caught 1:30) and parvovirus (three captive 1:20, 1:320, 1:1280; two recently captured-individuals 1:160, 1:50) indicated prior exposure to these viruses. Seven (of seven) dogs and two (of three) cats surveyed in the study area had high parvovirus titres (1:250 to >1:10240). Positive titres to all viruses in all sampled dogs suggested that these pathogens were endemic. An outbreak of distemper that affected giant pandas and red pandas at one Chinese zoo has been described in the literature (Qiu & Mainka, 1993).

Vaccines approved for the giant panda are not currently available in China. Some zoos vaccinate with attenuated multivalent canine vaccines produced in China, but antibody titres have not been evaluated and vaccine efficacy is unknown. Some carnivores (ferret, red panda, maned wolf, African wild dog, kinkajou, grey and fennec fox and European mink) are highly sensitive to distemper virus, and vaccination with modified live (canine) product has resulted in clinical distemper and death (Montali et al., 1983). Anecdotal accounts of clinical cases in China suggest that giant pandas are susceptible to canine distemper virus, but the degree of vulnerability and epidemiology of the disease are unknown. Coronavirus particles have been observed in giant panda faeces, but a causative relationship between the virus and clinical signs has not been established (Pan et al., 1991; X. Xia, pers. comm.).

Vomiting

Frequent vomiting or regurgitation has been observed in some giant pandas who appear otherwise clinically normal (Zhang et al., 2000). One female (SB 297, born in captivity and classified as a Prime Breeder during the CBSG Biomedical Survey) regularly regurgitates and re-ingests her food while otherwise being in good health. This may be some form of stereotypical behaviour. Another young male (SB 394) described during the survey as robust and a proven breeder also had a history of postprandial vomiting.

Allergies

Food allergy was implicated in a case of generalised gastrointestinal illness in an adult female giant panda (SB 127) at the London Zoo (Knight et al., 1982). This individual presented with anorexia, polydipsia, lethargy, ascites and progressive, severe weight loss. Following weeks of supportive therapy, the panda made a clinical recovery while maintaining a fluctuating eosinophilia that periodically rose to 40% of the total white cell count. She also had a peritoneal exudate with a specific gravity up to 2.4 g dl^{-1}. The possibility of food allergens was considered, and, on the basis of a rising antibody titre to ovalbumin, chicken eggs were eliminated from the diet. The antibody titre to ovalbumin decreased over the next 14 months and the panda's clinical condition improved, although the undulating eosinophilia and low-grade peritonitis persisted. This individual also demonstrated antibody responses to monkey chow and wheat, but the titres fell after the acute illness episode without removal of these foodstuffs. Investigations into a separate cause for the persistent peritonitis and recurrent eosinophilia were initiated at the time the article was written, and preliminary findings localised a mass in the upper epigastric region.

A possible allergic reaction to soya-bean meal may have been responsible for acute joint pain in two giant pandas (SB 208 and 210) at the Berlin Zoo (Göltenboth, 1985a). The soya-bean meal ratio in the pandas' gruel had been doubled to encourage weight gain. Two days later, both individuals moved as though in great pain. The animals were treated with NSAIDs, and the soya-bean meal was omitted from the gruel. Both animals recovered fully within three days. Diagnostic evaluations were not reported.

Ascites

Abdominal fluid is a frequent, incidental observation in physical and post-mortem examinations of the giant panda (Zhang et al., 2000; see Chapters 4 and 15; R. Montali & B. Rideout, pers. obs.). In some individuals, ascites has been associated with a disease process, such as heart failure (R. Montali, unpublished data, pathology report, Smithsonian's National Zoological Park), chronic ulcerative colitis or peritonitis (F. Ollivet, unpublished data, pathology report, Paris Zoo) or as part of Stunted Development Syndrome (see Chapter 4). Giant pandas with chronic gastrointestinal disease often have ascites (Zhang et al., 2000; I. K. Loeffler, unpublished observations). In most cases, the aetiology is unclear. The condition does not appear to be associated consistently with clinical illness (e.g. cardiac or hepatic failure) or abnormal serum chemistry values (e.g. hypoproteinaemia) (Zhang et al., 2000; B. Rideout, pers. obs.).

The reproductive system

Most diseases of the reproductive tract of giant pandas are poorly described, and their impact on reproduction in males or females is unknown.

Embryonic mortality and abortion

Because it is currently impossible to distinguish pregnancy from pseudopregnancy in giant pandas (see Chapter 8), the incidence of embryonic death is difficult to assess. Anecdotal accounts of late-term abortion have been noted at two Chinese giant panda facilities but have not been investigated (I. K. Loeffler, unpublished observations).

Infectious diseases of the reproductive tract

A case of metritis and cervicitis was diagnosed in a 6.5-year-old female (SB 404) during the CBSG Biomedical Survey (Zhang et al., 2000). In the previous year, this individual had a sporadic urinary oestrogen profile, had been ill most of the year and was described during the survey as 'unhealthy'. Laparoscopy revealed ascites, and an ultrasound examination showed fluid in the uterus. A purulent vaginal discharge and suppurative cervicitis (diagnosed on cytology and biopsy) confirmed a diagnosis of endometrial infection. This panda was treated with antibiotics and conceived in the same season to produce a healthy cub that year (see Chapter 4).

Knight et al. (1985) reported a presumed Trichomonas infection of the vaginal vestibule (in an unspecified individual) that was treated with metronidazole (20 mg kg^{-1} orally; duration and outcome not reported). The same report indicated a laparoscopic finding of chronic endometrial hyperplasia, bilateral oviductal occlusion and diffuse peritonitis in another giant panda (SB 127) at the London Zoo. The authors described clearing the obstructions by catheterisation and oviductal lavage in two successive years, but did not report the panda's subsequent reproductive success.

Uterine lesions

Incidental observations of uterine lesions may indicate a potential cause of infertility in some giant pandas. A 22-year-old female (SB 112) who died of heart failure at the Smithsonian's National Zoological Park was found at necropsy to have three leiomyomas distributed throughout the uterus (Fig. 16.10a; Plate XXII). Leiomyomas are benign tumours of smooth muscle cells that may (in high numbers) adversely affect fertility, particularly in older animals.

Testicular anomalies

A giant panda at the Smithsonian's National Zoological Park (SB 121) developed an enlarged testis at 25 years of age (see Fig. 16.10b). Testicular biopsy revealed a neoplasm that was confirmed postcastration to be a seminoma. Although vascular involvement was noted microscopically, there was no evidence of metastasis upon pathological examination at this male's death three years later. The testicular tumour in this giant panda resembled seminomas commonly observed in the domestic dog (R. Montali, unpublished data, pathology report, Smithsonian's National Zoological Park). Undescended testes or testicular hypoplasia (see Fig. 16.10c–f) were detected in three male giant pandas during the CBSG Biomedical Survey (see Chapter 7). One testis in a 14-year-old, wild-born male (SB 298) had hyperechoic foci at ultrasound that were believed to represent fibrosis or calcification (Zhang et al., 2000; see Chapter 4). He had successfully sired seven litters (11 total cubs) over the five years prior to the examination, and the lesions did not appear to interfere with his fertility.

The renal system

Clinical signs of renal dysfunction have been documented in young, adult and geriatric giant pandas. Aetiology and pathophysiology of uraemia in this species are usually undetermined (Knight et al., 1982; Nakazato et al., 1985; Li et al., 2001).

Renal failure in a young individual

Li et al. (2001) described a case of renal failure in an eight-month-old female (SB 484). She had been weaned a month prior to the onset of illness and was reported to have been less thrifty than her peers. The course of her six-month illness was marked with frequent vomiting,

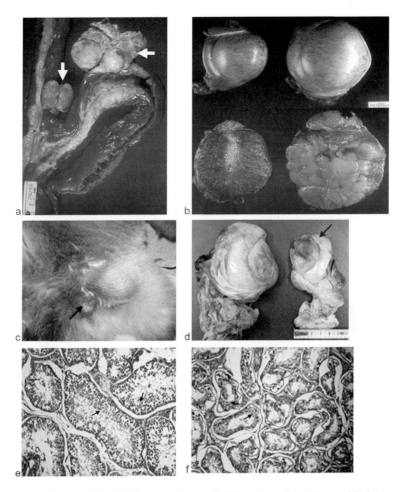

Figure 16.10. (a) Leiomyoma (arrows) of the uterus in a 22-year-old giant panda. Leiomyomas are benign tumours of smooth muscle tissue and are common in Carnivora. (b) Seminoma (right) found incidentally in a 25-year-old individual. Both testes were surgically excised, and the neoplasm did not recur or metastasise. The top two items in the image depict the gross specimens, and the bottom two their cut surfaces. (c) Gross and (d) excised testes of an adult giant panda (SB 323) with unilateral testicular hypoplasia (arrows). The photograph in (c) was taken during the CBSG Biomedical Survey when the panda was 13 years old; he died one year later. Histopathology of a normal (e) versus hypoplastic (f) testis demonstrating the absence of germ cells in the seminiferous tubules of the latter. Arrows indicate the germ cell layer. (See also Plate XXII.)

diarrhoea, anorexia and abdominal discomfort. She was also pruritic and developed alopecia on the head and abdomen. Physical examination and blood analysis performed a month after illness onset revealed uraemia (blood urea nitrogen, BUN, >140 mg dl^{-1}; normal is <15 mg dl^{-1}), acidosis, anaemia and enlarged kidneys. Severe renal disease was confirmed by biopsy, ultrasonography and scintigraphy. Each kidney was twice normal size. Scintigraphy and urinalysis revealed reduced renal perfusion, hyposthenuria, glucosuria and proteinuria. A biopsy revealed renal tubular necrosis and mineralisation and renal tubule dilation. Although clinical signs improved gradually in response to allopathic and Chinese medicines, BUN and creatinine concentrations remained elevated, anaemia persisted, and she never recovered normal activity levels. SB 484 died six months after the initial uraemia diagnosis following an acute episode of haemorrhagic enteritis and anorexia. Necropsy revealed extensive kidney calcification, hepatomegaly and mucosal haemorrhage of the stomach and intestines. Further diagnostic tests were unavailable, and the aetiology of this young animal's renal failure remains undetermined. One possible consideration is the treatment of the initial diarrhoea with gentamicin, although the author reported that this drug was used frequently in giant pandas at this institution without complication.

Pyelonephritis and anaemia in an adult

Acute pyelonephritis occurred in a 14-year-old female giant panda (SB 112) at the Smithsonian's National Zoological Park four months after parturition (Bush *et al.*, 1984). Her cub had died within a few hours of birth due to aspiration of infected amnionic fluid *in utero* (see 'Neonatal pathology',). SB 112 developed signs of acute renal failure with azotaemia, haematuria and hyperphosphataemia. She also had a profound macrocytic, hyperchromic anaemia and hyperbilirubinaemia of unexplained origin, although the possibility of a haemolytic crisis was considered. *Enterococcus* sp., *Klebsiella pneumoniae* and *E. coli* were cultured from the urine and biopsied renal tissue, and *Enterococcus* was cultured from blood. The renal biopsy revealed neutrophils and proteinaceous casts within dilated proximal tubules (Fig. 16.11a; Plate XXIII). The medullary architecture was abnormal, with oedema and interstitial mononuclear inflammatory cells.

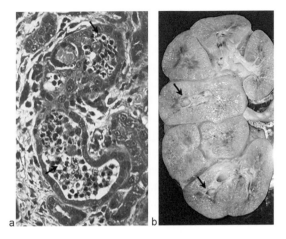

Figure 16.11. (a) Renal biopsy of a 14-year-old female giant panda with acute renal failure who was diagnosed with ascending coliform pyelonephritis. (Arrows designate neutrophils in the renal tubules.) (b) Renal cystic glomerular disease (arrows) in a 28-year-old male giant panda that died of chronic renal failure. (See also Plate XXIII.)

The severity of this female's anaemia warranted aggressive emergency therapy, which included a blood transfusion from her mate (SB 121). SB 112 responded to therapy with a return of clinical chemistry values to normal levels, and she became pregnant at her next oestrus the following spring. However, she may have retained a persistent, subclinical urogenital tract infection, which was offered as a possible explanation for the failure of all the live-born cubs produced in succeeding years to survive beyond a few days (see 'Neonatal pathology'.

Renal failure in a geriatric individual

The need to manage chronic renal failure is not uncommon in captive collections as good husbandry and veterinary care allow animals to live to ages that are unusual in the wild. The male giant panda SB 121 at the Smithsonian's National Zoological Park developed signs of renal failure at the age of 28 years. The clinical course of his six-month illness was characterised by azotaemia, anaemia and episodic epistaxis that was considered to be associated with renal hypertension. He was treated with fluids, erythropoietin (Epogen; Amgen, Thousand Oaks, CA; 10 IU kg^{-1} given subcutaneously at a tapering frequency), injectable iron supplementation (2 mg kg^{-1}) and later amilodipine (a human anti-hypertensive drug; Pfizer, La Jolla, CA; 0.025–0.05 mg kg^{-1} orally once daily) (L. Spelman, unpublished observations, clinical records, Smithsonian's National Zoological Park). SB 121 had also been effectively treated for degenerative osteoarthritis for several years, but with the deterioration of his renal condition the arthritis became refractory to treatment. He was euthanised when advancing uraemia also became resistant to treatment (see Chapter 15; L. Spelman, unpublished observations, clinical records, Smithsonian's National Zoological Park).

At necropsy, both kidneys were enlarged, pale, scarred, mineralised and contained numerous cysts throughout the cortices and medullae (R. Montali, unpublished data, pathology report, Smithsonian's National Zoological Park; see Fig. 16.11b). Histologically, the kidneys showed diffuse cystic changes in the glomeruli, with tubular atrophy and interstitial fibrosis. Additionally, there was evidence of cardiomegaly and congestive heart failure (possibly hypertension induced) and uraemic metastatic calcifications of blood vessels and soft tissue. A macroaneurysm and thrombosis in the left retina were probably also associated with hypertension. Haemosiderosis in the lymph nodes, liver, spleen, adrenals and pancreas was consistent with anaemia secondary to end-stage renal failure. There was no evidence of the involvement of infectious agents. Despite SB 121's long-term treatment with NSAIDs to manage osteoarthritis, there was no evidence of renal papillary necrosis or other toxic changes. Cortical cysts are commonly seen in animals with end-stage kidneys.

Diabetes insipidus

An unusual case resembling diabetes insipidus in a subadult giant panda (SB 249) is reported in the literature (del Campo et al., 1990b). The three-year-old male was described as apathetic, inappetant and polydipsic (drinking up to 221 of water per day; normal is around 2.51 per day). Based on clinical signs and haematological analysis, diagnostic tests were undertaken for psychogenic polydipsia. A water deprivation test was discontinued after 12 hours due to a 4% loss in body weight. Urine osmolality revealed no significant variation in the 24 hours following initiation of water deprivation. Two injections of porcine vasopressin tannate (5 and 10 IU on consecutive days; source unspecified) and a few days of treatment with synthetic antidiuretic hormone (0.05 and then 0.09 $\mu g\ kg^{-1}$ daily; source unspecified) resulted in no change in urine osmolality or water intake. Plasma concentrations of antidiuretic hormone were not detectable using standard laboratory assays. Treatment with chlorothiazide (0.2 mg kg^{-1}; source unspecified) was initiated as a diagnostic indicator and then continued at 0.4 mg kg^{-1} every three days as a therapeutic measure. Like other saluretic drugs, chlorothiazide has paradoxical effects in patients with diabetes insipidus in that it actually reduces diuresis and polydipsia, and SB 249 responded accordingly.

However, this individual remained severely depressed and anorexic, particularly after the chlorothiazine treatments. Staff observed that the panda appeared nearly 'euphoric' after recovering from ketamine hydrochloride anaesthesia (this male was tranquilised frequently for blood sampling and supportive therapy) and that he responded most positively to one particular keeper. Efforts then concentrated on the panda's psychological well-being. The keepers adopted a more positive attitude and were trained to brush the animal as a form of massage. Although water intake remained relatively high (about 6l daily), SB 249's weight and behaviour returned to normal over the course of a year. A behavioural and psychological evaluation would have been interesting, especially in light of this male's concomitant chronic gastrointestinal illness (see above). Regardless, this case illustrates the potentially profound influence of a giant panda's social environment on its health and well-being.

The neurological system

Seizure disorders in the giant panda have been reported occasionally (Hime, 1976; Keymer, 1976; Qiu & Mainka, 1993; I. K. Loeffler, collected anecdotal accounts), but aetiological or pathological origins have not been studied. Qiu & Mainka (1993) reported that five of 18 known neurological cases in the Chinese literature have been associated with: 1) a nasal sinus abscess with extension to the brain; 2) low blood levels of potassium and calcium; 3) adverse reaction to medication for tuberculosis; 4) a hormonal problem; or 5) heavy metal toxicosis (arsenic and mercury).

The details of the clinical course and diagnosis of these cases are unavailable. There has been no investigation into the incidence and pathology of microorganisms or other parasites that may affect the giant panda's central nervous system.

It is worth noting that five of six giant pandas maintained at the London Zoo over a 30-year span were reported to have had 'fits' (Keymer, 1976). In most cases, post-mortem examinations were not sufficiently detailed to establish a diagnosis or aetiology. The case of a 15-year-old female (SB 18) who died at the London Zoo following a five-month period of increasingly frequent seizures was carefully examined (Keymer, 1976). This individual had no histological lesions of the central nervous system or biochemical evidence for heavy metal toxicity or vitamin deficiency (Hime, 1976; Keymer, 1976). Histological changes in the brain were consistent with prolonged, repeated seizures, and no specific cause was identified.

The cardiac and respiratory systems

Cardiomyopathy (including dilated cardiomyopathy and endomyocardial fibrosis) has been found in some older giant pandas post-mortem (Keymer, 1976; Nakazato, 1985; R. Montali, unpublished data, pathology report, Smithsonian's National Zoological Park). The male at the Paris Zoo (SB 140), who died at 27 years of noncardiac causes (see 'The digestive system'), was found to have moderate, chronic endocarditis with dystrophic or senile valvular sclerosis, but no myocarditis (F. Ollivet, unpublished data, pathology report, Paris Zoo). SB 121, the 28-year-old giant panda who died at the Smithsonian's National Zoological Park of chronic renal failure (see 'The renal system'), had enlarged left heart chambers that were attributed to chronic hypertension and anaemia (R. Montali, unpublished data, pathology report).

Other cases are reported as having had no signs of cardiac insufficiency prior to sudden collapse. For example, the 15-year-old female (SB 18), who died with convulsions at the London Zoo (see 'The neurological system'), had a dilated right ventricle and evidence of aortic and heart valve fibrosis (Keymer, 1976). SB 112, the 22-year-old female at the Smithsonian's National Zoological Park, who died of sudden heart failure, had extensive cardiac fibrosis with an endomyocardial distribution observed in both ventricles (R. Montali, unpublished data)

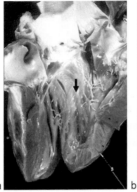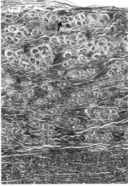

Figure 16.12. Endocardial fibrosis in a 22-year-old female giant panda that died suddenly of cardiac arrhythmia. (a) Gross necropsy demonstrating fibrosis of endocardial tissue (arrow). (b) Histopathology of endocardial tissue. Fibrosis disrupts the course of electrical signalling through conducting fibres (arrow). (See also Plate XXIV.)

(Fig. 16.12; Plate XXIV). The fibrosis appeared to entrap Purkinje fibres, particularly near the bundle of His and may have led to dysrhythmias and a heart block. Some pericardial and peritoneal effusions were present, but the histological appearance of the liver and lungs was inconsistent with long-standing congestive heart failure.

There is limited information on respiratory disease in giant pandas. One report describes pulmonary hyaline membrane disease in an 18-year-old female, SB 341, who apparently died of chronic pancreatic disease and respiratory failure (Chen & Pan, 1991). As already discussed, a leading cause of cub mortality is pneumonia, generally due to sepsis, failure of passive transfer and an immature or weak immune system. Pulmonary injury caused by migrating *Baylisascaris* larvae may also predispose young pandas to secondary bacterial pneumonia, although this has not yet been reported. There is one case (in the Chinese literature) of tuberculosis in the giant panda (see Qiu & Mainka, 1993 for a list of citations).

Skin

Alopecia and pruritus have been observed in giant pandas in association with renal failure (Li *et al.*, 2001), stereotypical behaviours (Zhang *et al.*, 2000) and ectoparasitism (Leclerc-Cassan, 1985). At the workshop on

Diagnostic and Clinical Pathology in Zoo and Wildlife Species (Beijing, June 2002), demodecosis was recognised as a significant skin disease in the giant panda. Periocular and generalised demodecosis were also observed during the CBSG Giant Panda Biomedical Survey, with one animal being severely affected with alopecia, erythema, pyoderma and skin lichenification (Zhang *et al.*, 2000; see Chapters 4 and 15). References to *Demodex* in the giant panda also appear in the literature (Zhu, 1991). Although no data are available, we suspect that demodecosis in the giant panda is associated with a compromised immune status, as it is in the domestic dog.

A squamous cell carcinoma was observed in a 16-year-old male (SB 305) during the CBSG Biomedical Survey (Zhang *et al.*, 2000; see also Chapters 4 and 15). The lesion was 50 cm in diameter and its pathology well advanced. It had been treated topically as a superficial skin infection for a chronic but unspecified period of time. The panda died four months after the survey examination.

The musculoskeletal system

Like most animals, the giant panda develops osteoarthritis with age. SB 121 of the Smithsonian's National Zoological Park began to show episodes of lameness by 11 years of age, particularly during the winter and breeding seasons. Progressive stiffness and lameness in the forelimbs and then in the rear limbs and back prompted the initiation of long-term NSAID therapy (see Chapter 15; L. Spelman, unpublished observations, clinical records, Smithsonian's National Zoological Park). SB 121's osteoarthritic pain was managed medically for 2.5 years until euthanasia due to renal failure. At necropsy, the articular surfaces of the left elbow were severely eroded, and the epicondyles of the ulna were marked by multiple exostoses; both carpal joints were similarly affected. Progressive osteoarthritis affected the thoracolumbar vertebrae.

Ocular system

Ocular changes described for the giant panda include conjunctivitis, corneal ulcers, cysts, retinal degeneration, neoplasms and lesions involving the whole eye that may have resulted from disease or trauma (Ashton, 1976; Hime, 1976; Lopez *et al.*, 1996; Zhang *et al.*, 2000; McLean *et al.*, 2003; R. Montali, unpublished data, pathology report, Smithsonian's National Zoological Park). A recurrent case of keratoconjunctivitis in a young female (SB 18) at the London Zoo finally resolved with topical and injectable penicillin treatment (Hime, 1976). Upon this animal's death eight years later after a period of increasingly frequent seizures (see 'The neurological system'), cysts were identified histologically in the iris, retina and pars plana. This panda was also found to have bilateral retinal degeneration and lenticular sclerosis attributed to old age, but there was no evidence of cataract formation (Ashton, 1976).

A mass determined histopathologically to be a low-grade haemangiosarcoma was removed from the outer ocular limbus of a 13-year-old male giant panda (SB 249) at Madrid Zoo (Lopez et al., 1996). The tumour appeared to be painless and was confined to the conjunctiva with no invasion of the underlying tissues. Haemangiosarcomas are not uncommon in a variety of species, and their development is potentially stimulated by sunlight (see citations in Lopez et al., 1996). It is not unusual for captive pandas to be exposed to ultraviolet radiation far in excess of levels in their native habitat of foggy, wet mountain forests. The panda in Madrid was moved to a more shaded enclosure, which is probably a good suggestion for all giant pandas housed in sunny areas.

A 14-year-old male (SB 298) examined during the CBSG Biomedical Survey in China had corneal and lenticular opacities in the left eye but without evidence of a cataract. The anterior chamber of the affected eye was distended, and the iris and pupil appeared abnormal. The lesions were suggestive of trauma, glaucoma or retrobulbar disease (Zhang et al., 2000). This male was noted for his aggressive behaviour, which may have been precipitated, at least in part, by the irritating ocular lesions or, alternatively, the trauma could have resulted from fighting with other pandas.

SB 121 at the Smithsonian's National Zoological Park had periodic episodes of bilateral ulcerative keratitis that began at 25 years of age and progressed markedly in the last two weeks of his life (R. Montali, unpublished data, pathology report). During the original bouts of keratitis, this male had positive titres against Herpesvirus canis (or another herpesvirus species that cross-reacted with H. canis; R. Montali, unpublished data). However, a direct relationship between herpesvirus and the ocular lesions could not be determined. Histologically, the ulcerated lesion showed superficial corneal inflammatory calcification associated with Gram-negative bacilli (McLean et al., 2003). Some of these changes were attributed to the panda's chronic renal failure and debilitated condition. Both retinas had small nodules of proliferated astrocytes, indicative of hamartomas and similar to the congenital lesions found in humans which arise from abnormal tissue development and maturation (McLean *et al.*, 2003). Human hamartomas are usually associated with a multisystemic disease complex known as tuberous sclerosis (characterised by seizures, mental retardation, skin and eye lesions and neurobehavioural problems). SB 121 had none of the hallmarks of the human disease complex, and the hamartomas were believed to have arisen spontaneously, as they do occasionally in humans (McLean *et al.*, 2003).

PRIORITIES FOR THE FUTURE

Although this chapter has reviewed diseases and pathology in the giant panda, it is clear that we have only scratched the surface of what needs to be learned to maintain a healthy, viable captive population effectively. The foremost priority is the institution of systematic, pathological evaluations of diseased giant pandas and certainly of each mortality. This includes the documentation of gross and microscopic pathological examinations, the results of which need to be entered into a computerised, internationally accessible database. As discussed throughout this book, one of the primary reasons for such rapid success in giant panda *ex situ* management is past and current cross-institutional and multidisciplinary partnerships. In the spirit of this philosophy, discussions are ongoing about an international partnership that would address the most important aspects of building the knowledge and technical resources necessary for improved health and veterinary management of captive giant pandas in China.

There is clear evidence for such a need, and there is willingness among the appropriate parties to collaborate. As acknowledged in this chapter's introduction, two training workshops associated with veterinary medicine and clinical pathology have already been conducted in China, the first in Chengdu (1999) and the second in Beijing (2002). The need for annual workshops dealing with the diagnosis and understanding of giant panda diseases has been expressed by the Chinese themselves. Such opportunities would allow Chinese and international colleagues to share observations and expertise. After all, none of the problems identified and discussed in this chapter can be resolved by any one person or institution. For this reason, and to begin to understand the pathogenesis and epidemiology of diseases, there is an urgent need to develop a comprehensive database for this species that consolidates clinical information, laboratory data and pathology findings. As most of the information will be generated in China, it is imperative that this resource be developed and maintained within that country.

The concerted effort required to build and maintain a giant panda medical and pathology database would improve communication among veterinary staff at different giant panda institutions. The shared information would greatly improve the application of available information for treating and preventing diseases in giant pandas and would enhance research opportunities and activities. Issues of particularly high priority include neonatal immunity, digestive disorders, reproductive failure and developmental abnormalities of cubs associated with nutrition, parasitology, infectious disease and toxicology. These are complex topics, and advances in the veterinary care and disease prevention of giant pandas – as in all wildlife conservation – will depend on functional partnerships within the international community.

REFERENCES

Ashton, N. H. (1976). The eye. Chi-Chi, the giant panda (*Ailuropoda melanoleuca*) at the London Zoo 1958–1972: a scientific study. *Transactions of the Zoological Society of London*, **33**, 127–31.

Bush, M., Montali, R. J., Phillips, L. G. *et al.* (1984). Anemia and renal failure in a giant panda. *Journal of the American Veterinary Medical Association*, **185**, 1435–7.

Bush, M., Montali, R. J., Phillips, L. G., Alvarado, T. P. and Ravich, W. J. (1985). Acute hemorrhagic enteritis in a male giant panda. *Proceedings of the International Symposium on the Giant Panda. Bongo*, **10**, 153–8.

Chen, Y. and Pan, X. (1991). Pulmonary hyaline membrane disease of the giant panda. In *The Pandas: A Conservation Initiative*. Washington, DC: Smithsonian's National Zoological Park, p. 4.

Dämmrich, K. (1985). Post-mortem report on the female giant panda of the Berlin Zoo. *Proceedings of the International Symposium on the Giant Panda. Bongo*, **10**, 33–138.

Davis, D. D. (1964). *The Giant Panda. A Morphological Study of Evolutionary Mechanisms. Fieldiana: Zoological Memoirs. Volume 3*. Chicago, IL: Chicago Natural History Museum Press.

Del Campo, A. L. G., Monsalve, L. S., Villares, M. C. and Taylor, D. C. (1990a). Gastrointestinal disorders in two giant pandas at Madrid Zoo. In *Giant Panda: Proceedings of the Second International Symposium on Giant Panda*, ed. S. Asakura and S. Nakagawa. Tokyo: Tokyo Zoological Park Society, pp. 77–8.

Del Campo, A. L. G., Monsalve, L. S., Villares, M. C. and Taylor, D. C. (1990b). Diabetes insipidus-like syndrome in a subadult male giant panda. In *Giant Panda: Proceedings of the Second International Symposium on Giant Panda*, ed. S. Asakura and S. Nakagawa. Tokyo: Tokyo Zoological Park Society, pp. 79–82.

Göltenboth, R. (1985a). Some notes on the veterinary care of giant pandas (*Ailuropoda melanoleuca*) at the Berlin Zoo. Proceedings of the International Symposium on the Giant Panda. *Bongo*, **10**, 127–8.

(1985b). Clinical progress report on the fatal illness of the female giant panda (*Ailuropoda melanoleuca*) Tian Tian. Proceedings of the International Symposium on the Giant Panda. *Bongo*, **10**, 129–32.

Gual-Sil, F., Muñoz, I. Y., Morales, R. C., Romahan, C. L. and Gomez-Llata, P. R. (2000). Disease and cause of death in giant pandas at Chapultepec Zoo in Mexico City. *Proceedings of Panda 2000: Conservation Priorities for the New Millennium* (Abstract 22). San Diego, CA: Zoological Society of San Diego.

Hime, J. M. (1976). Clinical history, serology and chemistry: Chi-Chi, the giant panda (*Ailuropoda melanoleuca*) at the London Zoo 1958–1972: a scientific study. *Transactions of the Zoological Society of London*, **33**, 95–8.

Hirayama, K., Kawamura, S., Mitsuoka, T. and Tashiro, K. (1989). The faecal flora of the giant panda (*Ailuropoda melanoleuca*). *Journal of Applied Bacteriology*, **67**, 411–15.

Keymer, I. F. (1976). Pathology: Chi-Chi, the giant panda (*Ailuropoda melanoleuca*) at the London Zoo 1958–1972: a scientific study. *Transactions of the Zoological Society of London*, **33**, 103–18.

Knight, J. A., Brostoff, J., Pack, S. and Sarner, M. (1982). A possible allergic illness in a giant panda. *The Lancet*, **2**, 1450–1.

Knight, J. A., Bush, M., Celma, M. *et al.* (1985). Veterinary aspects of reproduction in the giant panda (*Ailuropoda melanoleuca*). Proceedings of the International Symposium on the Giant Panda. *Bongo*, **10**, 93–126.

Leclerc-Cassan, M. (1985). The giant panda Li-Li: history and pathological findings. Proceedings of the International Symposium on the Giant Panda. *Bongo*, **10**, 169–74.

Li, D., Hu, D., Tang, C., Sutherland-Smith, M. and Rideout, B. (2001). Diagnosis and treatment of acute renal failure in an 8 month old giant panda (*Ailuropoda melanoleuca*). In *Proceedings of the American Association of Zoo Veterinarians*, pp. 239–41.

Lopez, M., Talavera, C., Rest, J. R. and Taylor, D. (1996). Haemangiosarcoma of the conjunctiva of a giant panda. *The Veterinary Record*, **138**, 24.

MacKinnon, J., Bi, F., Qiu, M. *et al.* (1989). *National Conservation Management Plan for the Giant Panda and its Habitat. Joint Report of the Ministry of Forestry and World Wide Fund for Nature*. Hong Kong: China Alliance Press.

Mainka, S. A., Qiu, X., He, T. and Appel, M. J. (1994). Serologic survey of giant pandas (*Ailuropoda melanoleuca*) and domestic dogs and cats in the Wolong Reserve, China. *Journal of Wildlife Diseases*, **30**, 86–9.

McLean, I. W., Bodman, M. G. and Montali, R. J. (2003). Retinal astrocytic hamartomas: an unexpected finding in a giant panda. *Archives of Ophthalmology*, **121**, 1786–90.

Montali, R. J., Bartz, C. R., Teare, J. A. *et al.* (1983). Clinical trials with canine distemper vaccines in exotic carnivores. *Journal of the American Veterinary Medical Association*, **183**, 1163–7.

Montali, R. J., Bush, M., Phillips, L. G. *et al.* (1990). Neonatal mortality in the giant panda (*Ailuropoda melanoleuca*). In *Giant Panda: Proceedings of the Second International Symposium on the Giant Panda*, ed. S. Asakura and S. Nakagawa. Tokyo: Tokyo Zoological Park Society, pp. 83–94.

Nakazato, R., Sagawa, Y., Tajima, H. *et al.* (1985). Giant pandas at Ueno Zoo. Proceedings of the International Symposium on the Giant Panda. *Bongo*, **10**, 33–42.

Pan, X., Chen, Y. and Chen, Y. (1991). Etiological studies on acute enteritis of the giant panda. In *The Pandas: A Conservation Initiative*. Washington, DC: Smithsonian's National Zoological Park, p. 5.

Qiu, X. and Mainka, S. A. (1993). A review of mortality in the giant panda (*Ailuropoda melanoleuca*). *Journal of Zoo and Wildlife Medicine*, **24**, 425–9.

Villares, M. C., del Campo, A. L. G., Greenwood, A., Torraca, L. S. M. and Taylor, D. (1985). Health problems and clinical aspects of the giant panda at Madrid Zoo. In Proceedings of the International Symposium on the Giant Panda. *Bongo*, **10**, 159–68.

Weissengruber, G. E., Forstenpointner, G., Dubber-Heiss, A. *et al.* (2001). Occurrence and structure of epipharyngeal pouches in bears (Ursidae). *Journal of Anatomy*, **193**, 309–14.

Yan, X., Deng, X., Zhang, H. *et al.* (2000). *Giant Panda Conservation Assessment and Research Techniques Workshop, Final Report*. Apple Valley, MN: IUCN/SSC Conservation Breeding Specialist Group.

Yang, G. (1998). Advances on parasites and parasitology of *Ailuropoda melanoleuca*. *Chinese Journal of Veterinary Science*, **18**, 206–8.

Zhang, A., Zhang, H., Zhang, J. *et al.* (2000). *1998–2000 CBSG Giant Panda Biomedical Survey Summary*. Apple Valley, MN: IUCN/SSC Conservation Breeding Specialist Group.

Zhu, C. (1991). Pathological changes of demodecosis of giant panda. *Sichuan Journal of Zoology*, **10**, 39.

17

Ultrasonography to assess and enhance health and reproduction in the giant panda

THOMAS B. HILDEBRANDT, JANINE L. BROWN, FRANK GÖRITZ, ANDREAS OCHS, PATRICK MORRIS, MEG SUTHERLAND-SMITH

INTRODUCTION

Ultrasonography is a routine diagnostic procedure used for assessing soft tissue characteristics in the human and veterinary medical fields of ophthalmology, cardiology, neurology, nephrology, obstetrics, oncology and orthopaedics. Because various forms of ultrasonography have existed for more than 50 years, it is surprising that this technology has only recently been applied to the study and management of wildlife species (Hildebrandt & Göritz, 1998; Hildebrandt et al., 2003). Nonetheless, there already is enough evidence making it clear that ultrasonography, combined with other technologies, can address issues that directly impact the health and reproductive welfare of wildlife species. This chapter focuses on the relevance of this technique for assisting in the assessment of medical and reproductive health in the giant panda.

The struggle to propagate and maintain viable wild animal populations in captivity is often related to information gaps that limit our ability to develop breeding and health strategies that are species appropriate. As demonstrated throughout this book, the giant panda presents some significant challenges to *ex situ* managers, which are

Giant Pandas: Biology, Veterinary Medicine and Management, ed. David E. Wildt, Anju Zhang, Hemin Zhang, Donald L. Janssen and Susie Ellis. Published by Cambridge University Press. © Cambridge University Press 2006.

exacerbated by a lack of basic biological knowledge about the species. Ultrasonographical studies are helping fill these physiological and anatomical voids by allowing the:

1.   characterisation of reproductive tract morphology;
2.   description of reproductive events;
3.   monitoring of foetal development;
4.   documentation of progression and treatment of pathologies.

Ultrasonography is also playing a significant role in developing and using artificial insemination (AI), which plays a critical role in the genetic management of this species (see Chapters 20 and 21).

This chapter is a compendium of information on experiences using ultrasound at two institutions, the Berlin Zoo and the San Diego Zoo. In 1981, the Berlin Zoo acquired a male giant panda (studbook, SB, 208) at the approximate age of three years. He was without a mate for 13 years before the arrival of a ten-year-old female (SB 378) in 1995. In multiple attempts to breed these individuals, ultrasonography was used extensively in reproductive and health monitoring. Ultrasonography has also been used similarly in four giant pandas at the San Diego Zoo (male SB 381 and female SB 371 who arrived from China in 1996; female SB 487 who was born at the zoo in 1999; and male SB 415 who arrived from China in 2003).

OVERVIEW OF ULTRASOUND HEALTH ASSESSMENTS OF MALE AND FEMALE GIANT PANDAS

The Berlin Zoo

Ultrasound examinations have been conducted since 1997 in the male and female giant panda at the Berlin Zoo, with and without anaesthesia. Anaesthesia induction was achieved with ketamine hydrochloride (10 mg kg^{-1} body weight, BW) administered via blowpipe dart injection. Anaesthesia was then maintained using isoflurane gas delivered via a face-mask (2.5–3.0 volume percentage with oxygen flow of 2–4 l per minute). Transabdominal assessments were carried out using a portable ultrasound unit (Sonosite, Inc., Bothell, WA) equipped with a 4-2 MHz convex transducer (Fig. 17.1a). When the animal was assessed fully conscious, the scan head was mounted on an extension rod for safety. The female was behaviourally conditioned to fixate on a handheld

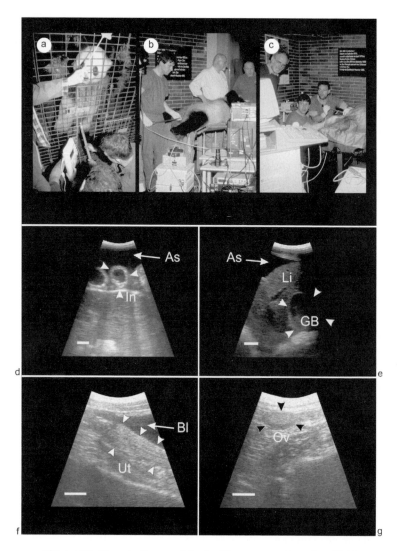

Figure 17.1. Use of ultrasound for health assessments and artificial insemination (AI) of giant pandas at the Berlin Zoo. (a) Transabdominal ultrasound examination for pregnancy diagnosis; (b) transcutaneous and transrectal ultrasound of the female; (c) endoscopic and ultrasound-guided AI; (d) transcutaneous sonogram (3.5 MHz) of the lower abdominal region of the male showing accumulated ascites (As) and free-floating intestinal loops (In, arrow heads); (e) transcutaneous sonogram (3.5 MHz) of the upper abdominal region of the male showing ascites (As) surrounding a liver lobe (Li) containing an anechoic-appearing blood

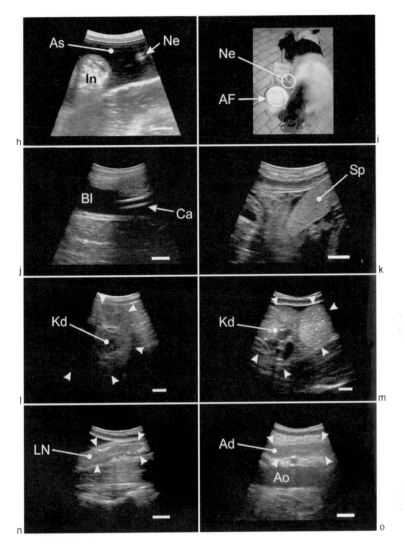

Figure 17.1. (Cont.)

vessel and part of the gall bladder (GB, arrow heads); (f) transcutaneous sonogram (3.5 MHz) of the female pelvic region showing part of the urinary bladder (Bl) and the uterine body (Ut, arrow heads);

(g) transcutaneous sonogram (3.5 MHz) of the lower abdominal region of the female showing an inactive ovary (Ov, arrow heads);

(h) transcutaneous (3.5 MHz) ultrasound-guided abdominal puncture with needle (Ne) to drain the ascites (As) in the male panda (In, intestine);

(i) drainage of abdominal fluid (AF) from the male panda; (j) transrectal

target in the standing position, thus allowing the veterinarian to conduct several consecutive, two-minute ultrasound scanning sessions.
This approach was particularly useful for assessing the status of the
uterus and confirming no pregnancy after AI.

Transcutaneous and transrectal ultrasound examinations were
performed on both sexes during annual physical examinations while
each animal was anaesthetised (see Fig. 17.1b). These procedures relied
on a portable ultrasound system (Hitachi EUB 405; Physia GmbH, Nei-
Isenburg, Germany) equipped with a 3.5-MHz convex transducer (EUP-C
318T) and a miniaturised intraoperative 7.5-MHz ultrasound probe
(EUP-F 334). The specific curved linear array transducer was 35 mm long
and 12.5 mm wide and had a minimum head of 9 mm at the top and 14
mm at the end. Additionally, a stationary colour-flow Doppler ultrasound unit (HDI 1000; ATL Ultrasound, Inc., Bothell, WA) equipped with
a miniaturised linear transducer (LI9-5, 5.0–9.0 MHz) was used for imaging the ovarian blood supply. The dimension of this transducer was
slightly larger (48 × 18 × 25 mm) but could be introduced easily into
the rectum. Both scan heads were inserted using three probe extensions
(Arno Schnorrenberg Chirurgiemechanik, Inc., Schönewalde, Germany)
as shown in Figure 17.2. The TR-500 (500 mm in length by 12 mm
diameter) was developed for use with large carnivores to visualise the
uterine horn, ovary, kidney and adrenal gland. The TR-250 (250 mm in
length by 12 mm diameter) was used to image the caudal urogenital
tract. The TR-150 (150 mm in length by 12 mm diameter) was designed
to view scent glands in large felid and ursid species.

Successful AI of giant pandas in China typically relies on a transcervical approach under anaesthesia, whereby catheter passage is aided

Figure 17.1. (Cont.)
(7.5 MHz) ultrasound-guided catheterisation (Ca) of the urinary bladder
(Bl) of the male; (k) transrectal sonogram (7.5 MHz) of the caudal part of
the elongated spleen (Sp, dotted line) surrounded by omentum;
(l) transrectal sonogram (7.5 MHz) of part of the lobulated kidney (Kd);
(m) transrectal sonogram (7.5 MHz) of part of a kidney (Kd) showing
generalised focal parenchyma degeneration. Arrow heads mark the outer
border of three renculi and the central renal blood vessels; (n) transrectal
sonogram (7.5 MHz) of an altered iliac lymph node (LN); (o) sonogram (7.5
MHz) of a lymph node near a healthy left adrenal gland (Ad) and the aorta
(Ao). The white bar represents 10 mm.

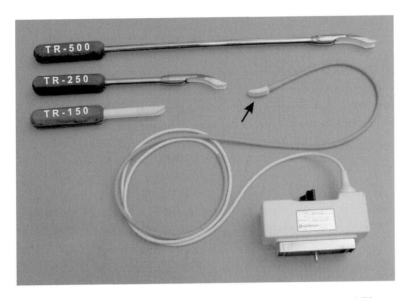

Figure 17.2. Probe extensions (TR-150, TR-250, TR-500) and the 7.5-MHz convex transducer (arrow) used for the transrectal ultrasound examinations.

by using a speculum and insemination tube. Ideally, sperm are deposited into the uterus, but in many cases it is possible to pass the catheter only into the cervix or at the external os. For AI of SB 378 at the Berlin Zoo, a technique using both endoscopy and ultrasound permitted semen deposition directly into the uterus (see Fig. 17.1c). Endoscopic imaging of the cervical portio was used to insert a catheter (5.5 French; Jansen-Anderson Intratubal Transfer Set; Cook Deutschland GmbH, Weyarn-Holzolling, Germany) into the cervix, while transrectal ultrasound facilitated catheter passage through the cervical canal and sometimes into the uterus. As described in more detail below, this approach provided a safe and easy way to place the semen deep within the giant panda's reproductive tract.

Ultrasound also aided reproductive procedures in the male. Transrectal ultrasound was used to monitor catheterisation of the urinary bladder (see Fig. 17.1j) to prepare for semen collection by electroejaculation. Urine was removed using a flexible 10-French embryo-flushing catheter (Cook Veterinary Products Australia, Inc.,

Banyo, QLD, Australia) and the bladder refilled with 50 ml of culture medium (Medium 199; Sigma Chemical Co., St Louis, MO). This was done to prevent sperm death (resulting from urine contamination) and to allow collecting any retrograde ejaculated sperm from the bladder. In 1991 (when male SB 208 was about 16 years old), a good quality electroejaculate (with 85% sperm motility) was cryopreserved and used to artificially inseminate SB 214 at the London Zoo in 1992 and again in 1993; both inseminations were unsuccessful. When SB 208 was electro-ejaculated again in January and March 1996, no sperm were obtained. The following year, just before the breeding season (February 1997), transcutaneous ultrasound revealed a pathological accumulation of ascites in the lower abdominal region which resulted in free-floating intestines (see Fig. 17.1d). As shown in Figure 17.1e, the ascites was observed near a liver lobe containing an anechoic-appearing blood vessel and part of the gall bladder. Transcutaneous ultrasound was used to monitor an ultrasound-guided abdominal puncture with a 100 mm long, 16-gauge needle to drain the accumulated fluid (see Fig. 17.1h); 8 l of ascites fluid were removed (see Fig.17.1i). The echogenic needle tip was easily visualised, optimising needle positioning and preventing the accidental puncture of other abdominal organs, such as the free-floating intestinal loops (see Fig. 17.1d). The cause of the progressive ascites build-up appeared to be acute pancreatitis. Administering diuretics and broad-spectrum antibiotics daily for one week brought about a rapid recovery that was confirmed by a follow-up ultrasound examination five weeks later. Interestingly, a normal spermic ejaculate was obtained at that time, and this male then continued to exhibit normal seminal traits.

Transcutaneous ultrasound examinations of the female pelvic region were conducted yearly, usually outside the February breeding season, allowing multiple views of the urinary bladder and uterine body (see Fig. 17.1f). However, image quality and resolution were insufficient to characterise endometrial activity, especially compared to sonograms generated by transrectal ultrasound. Transcutaneous sonography was effective for determining that the ovaries were inactive during the nonbreeding season (see Fig. 17.1g). This finding was supported by noninvasive urinary steroid monitoring, which also indicated a lack of ovarian cyclicity (Meyer *et al.*, 1997).

As part of the yearly health assessments, other internal organs, including the spleen, kidney and lymph nodes, were evaluated using transrectal ultrasound. In Figure 17.1k, the border of the caudal part of the elongated spleen was clearly distinguishable from the omen-

tum and was surrounded by anechoic abdominal fluid indicating ascites. The splenic parenchyma was characterised by a homogeneous,
moderate echogenicity. The absence of major blood vessels, except in
the hilus region, easily distinguished it from the liver, which was more
vascularised. The giant panda kidney is lobulated with 15 to 18 lobes
(renculi) (Davis, 1964; see also Chapter 16). A sonogram of part of the
lobulated kidney, with the outer border marked by white arrowheads, is
shown in Figure 17.1l. Each single lobe measured 20 to 30 mm in
diameter and had its own capsule with a common minor calyx. The
calyx centre was fluid filled and anechoic, whereas the periphery was
more echoic. The echogenicity of the cortex and medulla did not differ
sonographically in a healthy individual, and overall was less echoic than
the spleen. In Figure 17.1m, the kidney of male SB 208 showed clear
signs of generalised focal parenchyma degeneration. The white arrows
mark the outer border of three renculi and central renal blood vessels.
White spots noted in the sonogram were interpreted as scar tissue
replacing necrotic parenchyma. This was the first sonographic evidence
of chronic kidney degeneration found in connection with a severe case
of pancreatitis in 1997 (see above). Although the latter condition was
successfully resolved with antibiotics, the kidney problem was not.
It is believed that an attack of acute nephritis together with the pancreatitis (and possibly a bacterial infection as indicated by an altered iliac
lymph node noted during the same period) probably resulted in the
chronic kidney degeneration. As shown in Figure 17.1n, there were
white echogenic spots in the cortical region of the lymph nodes that
indicated replacement of parenchyma by scar tissue due to extensive
cell death. Several altered abdominal lymph nodes were observed
during the sonographic examination in 1997, but none since. Lastly,
Figure 17.1o shows a healthy, normal-sized left adrenal gland located
near the aorta. Characteristic of the adrenal gland are the relatively
large nutritive blood vessels shown between the adrenal gland (top)
and the aorta (bottom) in cross-section. However, it was difficult to
differentiate sonographically between the adrenal medulla and cortex
in a normal-sized gland.

The San Diego Zoo

Transabdominal ultrasound has been used to evaluate abdominal
organs in the giant pandas at the San Diego Zoo. Anaesthesia was also

induced with ketamine hydrochloride (10 mg kg^{-1}) administered via remote dart (Telinject USA, Inc., Saugus, CA) and then maintained using isoflurane (IsoSol; Vedco, Inc., St Joseph, MO) or sevoflurane (SevoFlo™; Abbott Laboratories, North Chicago, IL) administered via a face-mask. Examinations were conducted using an Aloka ultrasound unit (Aloka Co. Ltd., Wallingford, CT) and a 3.5- or 7.5-MHz transducer. Varying amounts of free abdominal fluid (ascites) were observed in all of the zoo's adult animals (Fig. 17.3). Fluid samples were collected via ultrasound-guided abdominocentesis using a 16- or 18- gauge, 7-cm long spinal needle. Cytological examination of the fluid was consistent with a transudate to modified transudate with eosinophils as the primary cell type. No organisms were isolated after bacteriological culture of this fluid. However, the presence of eosinophils suggested a possible link to gastrointestinal parasitism, although this was not proven. The significance of ascites needs to be considered in more detail. However, based on our observations and those from the CBSG Biomedical Survey (see Chapter 4), it is apparent that small to moderate amounts of ascites

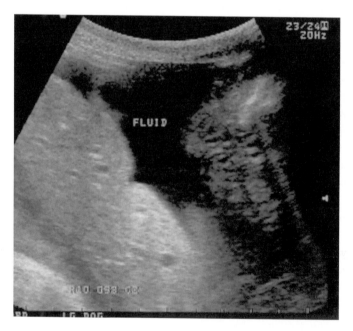

Figure 17.3. Transabdominal ultrasound image illustrating abdominal fluid around the liver of an adult female giant panda at the San Diego Zoo.

occur, even in apparently healthy animals. Thus, at this time, this phenomenon must be considered incidental.

Investigations at the San Diego Zoo also revealed that renal architecture was easier to evaluate in younger than older giant pandas. Figure 17.4 depicts the kidneys of SB 487 at 28 months of age.

REPRODUCTIVE ULTRASOUND ASSESSMENTS IN THE FEMALE GIANT PANDA

Berlin Zoo

After arrival of the wild-caught, ten-year-old female (SB 378) at the Berlin Zoo in 1995, noninvasive urinary hormone analysis was initiated to characterise ovarian activity. Unchanging patterns of excreted oestrogen and progesterone metabolites for the next two years, as well as a lack of any sexual behaviour, suggested that she was acyclic. A transrectal ultrasound examination in February 1997 indicated that the genital tract was normal but that the ovaries were inactive. Since then, she has undergone yearly ultrasound examinations for health assessments and to monitor ovarian activity after exogenous gonadotrophin treatment (equine chorionic gonadotrophin, eCG, with or without follicle stimulating hormone, FSH, priming) to induce oestrus and to stimulate ovulation (with human chorionic gonadotrophin, hCG) for AI (Table 17.1). In general, gonadotrophic treatments stimulated follicular development based on ultrasound and urinary oestrogen analyses, although oestrous behaviour was never observed.

The genital tract schematic (Fig. 17.5) indicates regions where ultrasound images were generated. Figure 17.6 (Plate XXV) presents examples of ultrasonographic and endoscopic evaluations of the female reproductive tract. Figure 17.6a is an endoscopic image of the external os or portio cervicalis after oestrus induction with eCG and prior to AI. The cone-shaped portio extends into the vagina and is characterised by a typically pink mucosa around the time of oestrus. A concurrent ultrasonographic image of the portio identified mucus accumulation among the portial folds typically found during oestrus (see Fig. 17.6b). The cervix is slightly convoluted and measures about 4 cm in length (see Fig. 17.6c). A longitudinal view of the uterus generated by transrectal ultrasound using the TR-250 probe extension clearly revealed the two endometrial layers (see Fig. 17.6d). The small size of the uterus and the thin endometrium indicated an inactive status before the onset of this

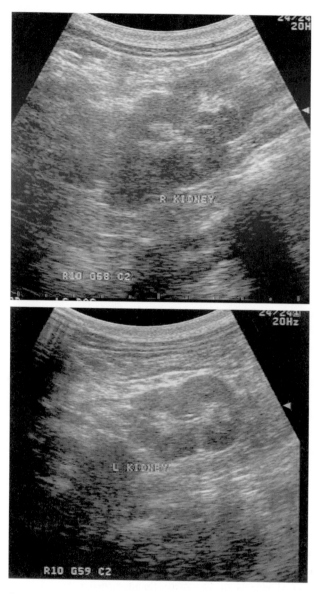

Figure 17.4. Transabdominal ultrasound images of the kidneys from a young (28-month-old) female giant panda at the San Diego Zoo.

Table 17.1 *Summary of artificial insemination (AI) trials conducted on the female giant panda SB 378 at Berlin Zoo*

Year	Oestrous induction/Natural oestrus	Anaesthesia	Outcome
1997	Days 1–5, 100 IU eCG[a] daily; Day 6, 1000 IU hCG[b] followed by AI	Yes plus ultrasound	No pregnancy
1998	Days 1–9, 100 IU FSH[c] daily; Day 10, 150 IU FSH; Days 11 to 13, 200 IU FSH; Day 13, 500 IU eCG; Day 18, 5000 IU hCG plus anti-eCG and AI	Yes plus ultrasound	Possible pregnancy with embryonic absorption
1999	Days 1–11, 200 IU FSH daily; Day 12, 250 IU eCG; Day 17, 2000 IU hCG plus anti-eCG and AI	Yes plus ultrasound	No pregnancy
2001	Natural cycle monitoring via urinary steroids followed by 2000 IU hCG and AI	Yes plus ultrasound	No pregnancy
2002	Natural cycle monitoring via urinary steroids followed by AI after the oestrogen peak	Yes plus ultrasound	Weak oestrus, no pregnancy
2003	Natural cycle monitoring via urinary steroids followed by AI after the oestrogen peak	No	Weak oestrus, no pregnancy

[a] eCG, equine chorionic gonadotrophin;
[b] hCG, human chorionic gonadotrophin;
[c] FSH, follicle-stimulating hormone.

breeding season in February. Each ovary was also quiescent as visualised using the TR-500 probe extension, although there was a clear distinction in echogenicity between the cortical and medullar regions (see Fig. 17.6e). Blood vessels in the medulla were visualised by colour-flow Doppler using an ultrasound system ATL HDI 1000 equipped with an intraoperative microscanner (7.0 to 11.0 MHz), which was inserted into the TR-500 probe extension (see Fig. 17.6f). After ovulation induction and AI, the ovary contained at least three corpora lutea, each with a diameter of 5 to 7 mm (see Fig. 17.6g). This finding corresponded to the high urinary progestin concentrations measured on this same day (data

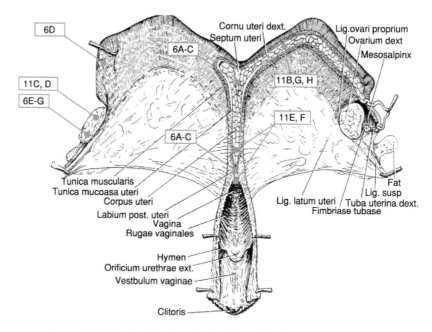

Figure 17.5. Modified drawing (Davis, 1964) of the female urogenital tract showing the regions where the images in Figure 17.6 were generated.

not shown). At this time, a large volume of abdominal fluid (ascites) was found surrounding the ovary (see upper part of Fig. 17.6g). This ascites accumulation was not unlike that observed for the male panda. Although no full-term pregnancies resulted from any of the six AI attempts of the female at the Berlin Zoo, ultrasound imaging suggested a conception followed by embryonic absorption in 1997 (Table 17.1). As seen in Figure 17.6h, cystic degeneration of the endometrium in the cranial part of the uterus resulted from foetal resorption. Cyst size varied from 2 to 8 mm at the time of examination (late September). However, ultrasound conducted during the following breeding season revealed significant healing of the endometrial lesions, with only minor signs of the collapsed endometrial cysts and scar tissue present (see below).

The San Diego Zoo

Both transrectal and transabdominal ultrasound have been used at the San Diego Zoo to assess reproductive integrity and to assist in performing AI (see below). As shown in Figure 17.7, transrectal ultrasound using

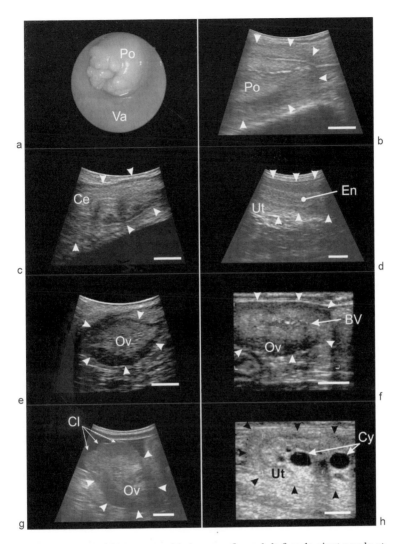

Figure 17.6. Ultrasonographic images of an adult female giant panda at
the Berlin Zoo. (a) Endoscopic image of the portio cervicalis (Po) and
vagina (Va) during AI; (b) ultrasonographic image (TR-150, 7.5 MHz) of the
portio (Po, bordered by arrow heads); (c) transrectal sonogram (7.5 MHz) of
the cervix (Ce, arrow heads); (d) longitudinal aspect of the uterus (Ut,
arrow heads) generated by transrectal ultrasound (TR-250, 7.5 MHz)
showing the two layers of the endometrium (En); (e) sonographic image
(TR-500, 7.5 MHz) of an inactive ovary (Ov, arrow heads) outside the
breeding season; (f) colour-flow Doppler imaging (TR-500, 5.0–9.0 MHz) of
the central blood vessels (BV, including veins and arteries) in the medulla

a 7.5-MHz rectal transducer was effective in visualising the uterus. Transabdominal visualisation of the reproductive tract was more difficult when conducted prior to the breeding season. In contrast, tract integrity was easier to evaluate at the time of AI (conducted in late March/early April) because of an increased uterine wall thickness associated with oestrus.

REPRODUCTIVE ULTRASOUND ASSESSMENTS IN THE MALE GIANT PANDA

The Berlin Zoo

By the time Berlin Zoo acquired female SB 378 in 1995, the male SB 208 was already 16 years old, so it was deemed appropriate to monitor testicular integrity frequently via ultrasound. The genital tract schematic (Fig. 17.8) indicates regions where ultrasound images were generated. The sonogram in Figure 17.9a is a midsegment view of the testis during the breeding season (April). The testes were larger in size, and the parenchyma was less echoic at this time, as compared to other times of the year, presumably because of increased spermatogenic activity. Compared to other bear species (Knauf *et al.*, 2003), the rete testis system in the giant panda appeared more pronounced, especially during the breeding season. Although SB 208 always exhibited seasonal changes in testicular activity, there were indications of age-related parenchymal degeneration. A cystic structure (see Fig. 17.9b) and the presence of two echogenic spots close to the caput epididymis as the result of apoptosis and scar tissue replacement (testicular degeneration) in the parenchyma (see Fig. 17.9c) were observed in the ultrasound examination conducted in 2002. A lesion in the left testis caused by an accidental dart injection while attempting immobilisation can also be seen in Figure 17.9d. The lesion was still detectable near the rete testis one year later, although it was reduced in size by half and presumably had little effect on overall testicular function.

Figure 17.6. (Cont.)

of an inactive ovary (Ov, arrow heads); (g) sonogram of an ovary (Ov, arrow heads) containing at least three active corpora lutea (Cl); (h) degeneration of endometrial cysts (Cy) in the cranial part of the uterus (Ut, arrow heads) as a result of foetal resorption. (See also Plate XXV.)

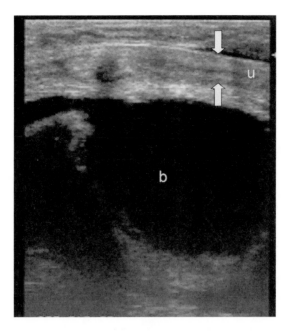

Figure 17.7. Transrectal ultrasound of the uterus (between the arrows) of the adult giant panda SB 371 during a pre-breeding evaluation at the San Diego Zoo. b, bladder; u, uterus.

The sonogram in Figure 17.9e depicts the right scent gland during the breeding season (April). Scent glands, visualised transrectally about 2 cm cranial to the anal sphincter, contain a creamy, yellowish secretion used for scent marking, which can easily be collected by manual palpation. There was a size difference between the left and right gland in this male.

The classical anatomical study of Davis (1964) suggested that the giant panda does not have a prostate gland. However, transrectal ultrasound evaluations clearly revealed that this male had a distinctive prostate comprised of two lobes laterally situated to the urethra with a central dorsal isthmus above the urethra (see Fig. 17.9f). It was located 2 to 4 cm caudal to the neck of the urinary bladder. The prostate parenchyma was less echoic than that of the scent glands and had a homogeneous appearance. Dorsal to the urinary bladder and lying in parallel were the cigar-shaped, ductal glands of the ductus deferens (see Fig. 17.9g; central aspect of the left gland). They were moderately

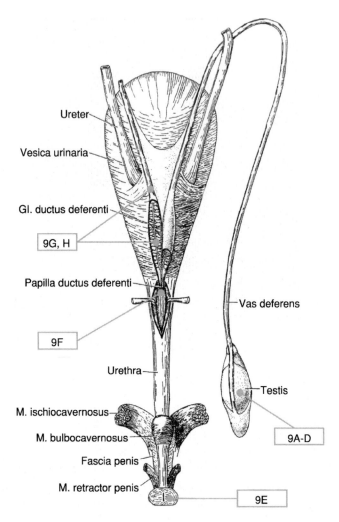

Figure 17.8. Modified drawing (Davis, 1964) of the male urogenital tract showing the regions where the images in Figures 17.9 and 17.10 were generated.

echogenic and had a solid parenchyma and a distinctive echogenic gland capsule approximately 1.5 mm thick. The ductus deferens entered the cranial end of a ductal gland, was slightly coiled and situated partly above the urinary bladder (see Fig. 17.9h). Although easily imaged by transrectal ultrasound (7.5 MHz) during the breeding season (March to May) and presumably because of increased spermatogenic

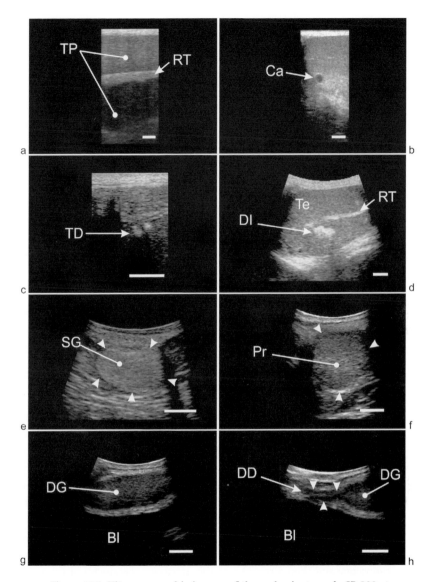

Figure 17.9. Ultrasonographic images of the male giant panda SB 208 at
the Berlin Zoo. (a) Middle segment of the testis during the breeding
season showing a pronounced rete testis (RT) and testicular parenchyma
(TP); (b) cystic alteration (Ca) of the testicular parenchyma; (c) echogenic
spots close to the caput epididymis associated with testicular
degeneration (TD); (d) lesion in the left testis (Te), near the rete testis (RT),
caused by an accidental dart injection (DI); (e) right scent gland (SG) in a
male panda during breeding season; (f) transrectal sonogram (7.5 MHz) of

activity, the ductus deferens were nearly undetectable outside this time of year.

The San Diego Zoo

Transcutaneous ultrasound was used to evaluate the testicles of both males at San Diego Zoo. Several hyperechoic foci were noted consistently in the testes of SB 381 (Fig. 17.10). Because these same foci were observed over a five-year time-span with no change, we presumed that they represented permanent fibrotic areas. However, there was no apparent adverse impact on male testis function as viable sperm were collected via electroejaculation at times when the foci were obvious.

ULTRASOUND-GUIDED AI AND POST-BREEDING MONITORING IN THE FEMALE GIANT PANDA

The Berlin Zoo

At no time during the eight years that the giant pandas were together at the Berlin Zoo did natural breeding occur. Therefore, AI was developed, and from 1997 to 2003, six ultrasound-guided inseminations were conducted (see Table 17.1).

Follicular development and ovulation were induced in the female with exogenous gonadotrophins for four of the six inseminations. Ovarian responses were relatively uniform, with one to three graafian follicles (>6 mm in diameter) and more than ten smaller follicles (1–3 mm) produced after each treatment (eCG with or without FSH). There was also a urinary oestrogen metabolite rise associated with follicular stimulation, but no oestrous behaviour. Ovulation was induced with hCG given on the day of AI and confirmed by ultrasonography and urinary progestin analysis. For each insemination, a disposable catheter (2 mm diameter) was guided into the external cervical os using an endoscope and then directed into the uterus by transrectal ultrasound.

Figure 17.9. (Cont.)

the prostate (Pr, arrow heads); (g) central region of the left ductal gland (DG) of the ductus deferens located dorsal to the urinary bladder (Bl); (h) transrectal sonogram (7.5 MHz) of the ductus deferens (DD, arrow heads) running into the cranial end of a ductal gland (DG) situated partly above the urinary bladder (Bl).

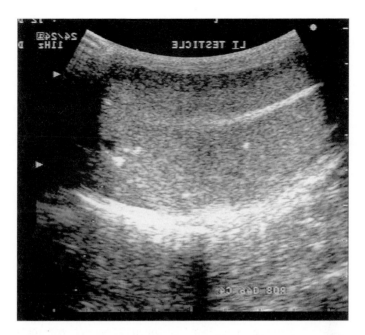

Figure 17.10. Transcutaneous ultrasound image illustrating several hyperechoic foci in the left testis of the adult male giant panda SB 381 at the San Diego Zoo.

Fresh or frozen semen from SB 208 (1–2 ml; 0.35–7.5 × 10^9 spermatozoa ml^{-1}, 85–95% sperm motility) was deposited after the end of eCG treatment. Although no full-term pregnancies resulted, there was ultrasonographic evidence of a pregnancy and embryonic resorption after the 1998 AI.

Images in Figure 17.11 (Plate XXVI) illustrate the concurrent use of endoscopy and transrectal ultrasound to facilitate AI. After gonadotrophin treatment, the vulva enlarged under the influence of oestrogen (see Fig. 17.11a). Near the time of ovulation, ultrasound revealed a thickened uterus, including a proliferated endometrium (see Fig. 17.11b). A Graafian follicle with a diameter of 5 mm on the active ovary was observed shortly before ovulation (see Fig. 17.11c). Although several small follicles were often observed, there were never more than three Graafian follicles present. A colour-flow Doppler image of the same ovary illustrates Graafian follicle characteristics, including the large size (≥5 mm in diameter), location within the ovarian periphery and independent blood supply (see Fig. 17.11d).

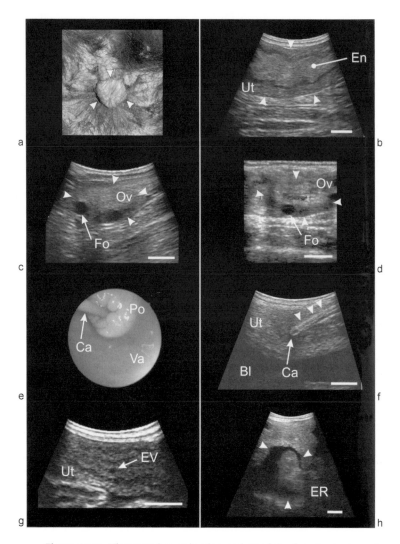

Figure 17.11. Ultrasound examination and AI of the female giant panda
SB 378 at the Berlin Zoo. (a) Mildly swollen vulva (arrow heads) near
oestrus; (b) transrectal sonogram (TR-500, 7.5 MHz) of the uterus
(Ut, arrow heads) and endometrium (En) at the time of ovulation;
(c) transrectal sonogram (TR-500, 7.5 MHz) of a Graafian follicle (Fo) on the
ovary (Ov, arrow heads) shortly before ovulation; (d) colour-flow Doppler
image (TR-500, 5.0–9.0 MHz) of the same ovary (Ov, arrow heads) showing
the independent blood supply of the Graafian follicle (Fo); (e) endoscopic
image of the transcervical passage of the AI catheter (ca, 2 mm in
diameter) showing the swollen portio (Po) containing opaque viscous

The uterus was located just dorsal to the urinary bladder. For AI, the passage of the catheter (2 mm in diameter) through the vagina and into the cervix was visualised by endoscopy (see Fig. 17.11e). The vagina contained mucus of a deep pink colour and was highly oedematised, whereas the cervical portio appeared swollen and contained opaque viscous mucus. Transrectal ultrasound was subsequently used to verify catheter placement for semen deposition within the caudal part of the uterine body (see Fig. 17.11f). After the AI in 1998, the sonogram at one month post insemination revealed evidence of a non implanted embryonic vesicle (see Fig. 17.11g). The hatched blastocyst was less than 1 mm in size and had no visible effect on the uterus at this stage. Six months after AI, an embryonic resorption site was visualised in the sonogram, which caused a temporary cystic degeneration of the endometrium (see Fig. 17.11h).

The San Diego Zoo

Artificial insemination of the female SB 371 was performed under anaesthesia during oestrus in 1998, 1999, 2001 and 2002. Ultrasound was used to confirm the placement of the AI rod within the cervix during each procedure. Transabdominal ultrasound without the use of anaesthesia was used to monitor postbreeding changes in the reproductive tract. The latter examination of the ventral abdomen was facilitated by training this female to lie in dorsal recumbency in a cage. Designated bars were removed from the cage, and a thick fabric sleeve was used to protect the ultrasonographer's arm and ultrasound transducer cord (see Chapter 15 and Fig. 17.1). Since 2001, real-time B-mode ultrasound examinations using the Aloka SSD-900V ultrasound machine with 3.5- and 7.5-MHz sector transducers have been performed. Examinations were performed

Figure 17.11. (Cont.)
mucus, and the mucosa of the vagina (Va); (f) sonographic verification (TR-250, 7.5 MHz) of the intrauterine position of the insemination catheter (Ca, arrow heads). Semen was deposited in the caudal part of the uterus (Ut), which was located dorsal to the urinary bladder (Bl); (g) sonographic evidence (TR-250, 7.5 MHz) of a nonimplanted embryonic vesicle (EV) in the enlarged uterus (Ut) one month after AI; (h) sonogram (TR-250, 7.5 MHz) showing the embryonic resorption site (ER, arrow heads) six months after AI. (See also Plate XXVI.)

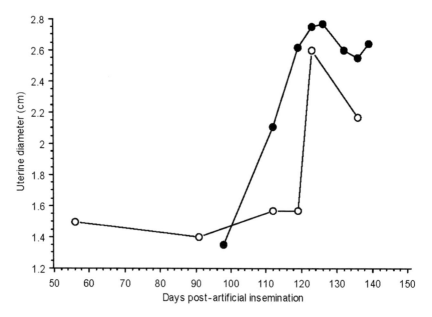

Figure 17.12. Change in uterine diameter over time (in 2002) as assessed by transabdominal ultrasound after AI of the giant panda SB 371 at the San Diego Zoo. •, right uterine horn; ○, left uterine horn.

outdoors in the early morning, which afforded better visualisation of the ultrasound monitor due to low ambient light, and improved viewing of abdominal structures due to less food in the gastrointestinal tract. These examinations, which lasted about 15 minutes each, were generally conducted every one to three weeks beginning in May and continued until the average gestation length had passed, and the female's behaviour and appetite had returned to normal. In 2002, an enlarged uterine diameter was observed (Fig. 17.12); however, foetal structures were never imaged. The increase in uterine diameter combined with the presence of an amorphic structure (30 mm by 15 mm) surrounded by uterine fluid indicated an early embryonic loss and was most likely the resorption site.

In 2003, female SB 371 bred naturally on 22 March (Day 0) with male SB 415. Weekly ultrasound examinations began on Day 32. An increase in the diameter of the right uterine horn was noted by Day 109 (Fig. 17.13), with the left horn not as consistently visualised. At 134 days (16 days before parturition), a twin pregnancy was diagnosed via

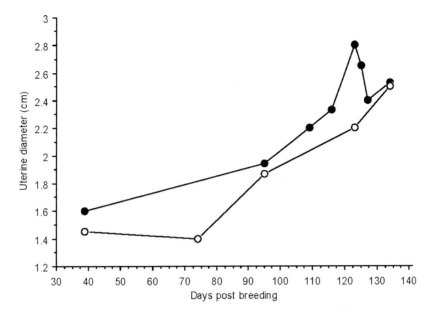

Figure 17.13. Change in uterine diameter over time (in 2003) as assessed by transabdominal ultrasound after natural breeding of the giant panda SB 371 at San Diego Zoo. ●, right uterine horn; ○, left uterine horn.

ultrasound. An anechoic gestational sac containing an elongated echogenic foetus was seen in both the right and left uterine horns, and the placenta was visible. A heartbeat was evident in each foetus; however, other foetal structures were not well developed. The uterine wall, foetus and placenta were isoechoic to slightly hyperechoic relative to the surrounding tissues. The foetuses were located just proximal to the inguinal regions at a depth of 3 to 4 cm. Subsequent ultrasound examinations were performed on Days 137, 138, 140 and 142 (13, 12, 10 and 8 days before parturition).

Metrics associated with the gestational sac, foetus and placenta are provided in Table 17.2. The gestational sac measure was taken from the innermost uterine layers. Foetal length was assessed as the longest length that could be imaged. The appearance of the placenta was discoidal, an elliptical homogeneous mass protruding into the gestational sac. Although it was not possible to image the complete placenta in a single view, it was possible to measure the largest placental dimensions

Table 17.2 *Foetal, gestational sac and placenta measurements in a giant panda at 134, 137, 138, 140 and 142 days post-mating (adapted with permission from Sutherland-Smith et al., 2004)*

Days post-mating	Foetal length (cm)		Gestational sac (cm)		Placenta (cm)	
	Left	Right	Left	Right	Left	Right
134	1.42	0.93	2.58 × 2.14	1.53 × 1.13	[a]	1.28 × 0.93
137	2.32[b]	1.76[b]	3.19 × 3.07	2.99[b]	2.35 × 1.2	1.8 × 1.06[b]
138	[c]	1.79[b]	3.71 × 2.81	2.81 × 2.67	2.1 × 1.19	1.64 × 0.96[b]
140	3.32	2.92	5.33 × 2.99	3.85 × 3.44	3.4 × 1.6	[d]
142	4.8	[e]	4.0 × 6.6	[e]	4.3 × 2.6	[e]

[a] Placenta was visualised, but not measured.

[b] Represents an average measurement.

[c] Measurement was inaccurate so was excluded.

[d] Placenta was difficult to visualise and not measured.

[e] Ultrasound evaluation of the right uterine horn was not conducted.

visualised. The foetuses were located at a depth of 3 to 6 cm within the abdomen.

Figure 17.14 depicts the ultrasound images of the developing foetuses (Sutherland-Smith *et al.*, 2004), which were clearly visible by Day 134 and 137 (see Fig. 17.14a,b). Linear hyperechoic areas along the spine compatible with skeletal ossification were evident at Days 138 and 140 (12 and 10 days before parturition) for the left foetus (see Fig. 17.14c,d). At Day 140, similar changes consistent with ossification of the limb buds and cranium in the left fetus were visible (see Fig. 17.14d). Thin hyperechoic lines consistent with foetal membranes were observed in the left uterine horn at Day 140. Foetal movement within both gestational sacs was noted on Day 140. Heartbeats were evident in both foetuses, but could not be accurately measured because of interference from the dam's respiratory motions. Only the left fetus and placenta were visualised on Day 142 (8 days before parturition), showing both further size development and heartbeats (see Fig. 17.14e, f).

A hypoechoic area was noted adjacent to the right placenta and within the right myometrium on Days 134 and 138 (12 days prior to parturition) (see Fig. 17.14g,h). A hypoechoic area was also evident in

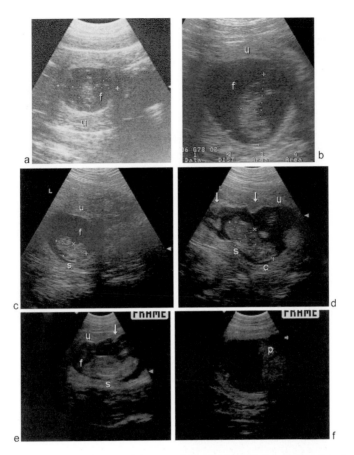

Figure 17.14. Ultrasound images of the right and left gestational sacs, foetus and placenta from the giant panda SB 371 at the San Diego Zoo (adapted with permission from Sutherland-Smith *et al.*, 2004). (a) Day 134, left uterine horn and foetus; (b) Day 137, left uterine horn and foetus; (c) Day 138, left uterine horn and foetus with ossification of the spine evident; length measurement inaccurate due to curled foetal position; (d) Day 140, left uterine horn containing foetus with ossification of the spine and cranium evident; note tent-like projections (arrows); (e) Day 142, left uterine horn and foetus with progressive spinal ossification; tent-like projections still visible (arrow); (f) Day 142, left placenta; (g) Day 134, right uterine horn, foetus and placenta; abdominal fluid is present; (h) Day 138, right uterine horn and foetus; hypoechoic areas adjacent to the placenta and within the uterine wall (arrows); (i) Day 140, right uterine horn and foetus with ossification of the spine and cranium apparent as well as free abdominal fluid; (j) Day 140, right uterine horn and foetus; hypoechoic

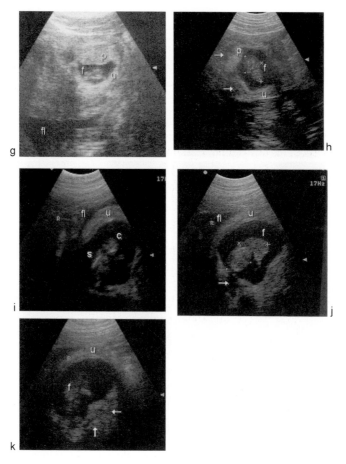

Figure 17.14. (Cont.)

area in uterine wall visible (arrow); free abdominal fluid present; (k) Day 140, right uterine horn and foetus; increased heterogeneity in the placenta and uterine wall present (arrows). c, cranium; f, foetus; fl, abdominal fluid; p, placenta; s, spine; u, uterus.

the right myometrium on Day 140, as well as a foetus with ossification of the spine and cranium (see Fig. 17.14i,j). The placenta was difficult to discern in the right gestational sac at Day 140 and appeared more heterogeneous than on previous evaluations (see Fig. 17.14k). Similar changes were not observed in the left myometrium or placenta. Tent like projections were observed within the left gestational sac protruding into the amniotic fluid on Days 140 and 142 (see Fig. 17.14d,e). These were not apparent within the right gestational sac. The female did not cooperate for additional ultrasound examinations, and on 19 August (8 days after the last ultrasound and 150 days post mating) the female gave birth to a single cub. Neither a second cub nor foetal remnants were observed. The first post-partum ultrasound was conducted opportunistically 45 days after birth but the uterus was not identifiable. Evidence of retained foetal structures was not apparent on subsequent ultrasound examinations, and the uterus remained difficult to visualise.

PRIORITIES FOR THE FUTURE

This chapter has illustrated the value of ultrasonography as a practical tool for studying and managing giant pandas. The technology is still vastly under-utilised, and its many advantages (noninvasiveness, reproducible real-time images, cross-sectional images of tissues/organs and the ability to measure morphometry) certainly argue for more widespread use. Applying ultrasound to the systematic examination of giant panda biology (especially in case-related investigations) will undoubtedly help find new solutions to physiological problems. It will also provoke the design of new and more specialised, species-specific instruments. Continued miniaturisation of transducer technology and the development of microprobes with compact, contoured shapes will improve the scanning of less accessible anatomical regions. Many new ultrasound systems offer colour-flow imaging that allows direct evaluation of blood-vessel architecture in reproductive organs. This is a promising area of research because the blood supply to the genital tract undergoes marked changes related to sexual maturation, gonadal activity and pregnancy. Three-dimensional ultrasonography will also further expand our ability to characterise reproductive soundness while facilitating a better understanding of complex physiological processes and creating near life-like images.

A high priority should be to incorporate this technology into more comprehensive, integrated investigations involving other scientific disciplines such as endocrinology and behaviour. The information in this chapter is based on only a few individuals, so there is a need for a large-scale application to more giant pandas to fully understand the utility of ultrasonography for improving the management of the captive population. Given the importance of artificial insemination for achieving genetic management goals (see Chapters 20 and 21), ultrasonography will be critical for learning more about the time of ovulation with respect to hormonal cues to ensure appropriate timing of AI. It is also clear that more work is needed to identify a consistently reliable ovulation induction regimen in this species. Although noninvasive hormonal monitoring can be valuable in assessing ovarian responses after gonadotrophin therapy (see Chapter 8), ultrasound offers the added advantage of visualising the exact number of ovulation sites. This could be a critical tool for determining the number of ovulations that occur during a natural oestrus as compared to those induced by hormonal therapies. Ultrasound technology can be used to diagnose pregnancy in the giant panda as well as for monitoring foetal development. Finally, it could also be beneficial for addressing other unique health challenges of the species, for example, the unexplained but rather common observation of waves of ascites accumulation, and the aetiology of developmental abnormalities, such as Stunted Development Syndrome (see Chapter 4).

There are undoubtedly more examples of how ultrasound technology could be combined with other reproductive technologies and behavioural assessments to answer questions that directly impact on the health and reproductive welfare of giant pandas. Given sufficient resources and access to these very special animals, there are no limitations to what can be learned. It is our hope that more researchers will accept the challenge of conducting integrated studies that cross disciplines not only to enhance scientific discovery but also to facilitate developing practical solutions for managing this endangered species.

REFERENCES

Davis, D. D. (1964). *The Giant Panda. A Morphological Study of Evolutionary Mechanisms. Fieldiana: Zoology Memoirs. Volume 3*. Chicago, IL: Chicago Natural History Museum.

Hildebrandt, T. B. and Göritz, F. (1998). Use of ultrasonography in zoo animals. In *Zoo and Wild Animal Medicine. Current Therapy 4*, ed. M. E. Fowler and R. E. Miller. Philadelphia, PA: W. B. Saunders Co., pp. 41–54.

Hildebrandt, T. B., Brown, J. L., Hermes, R. and Göritz, F. (2003). Ultrasound for the analysis of reproductive function in wildlife. In *Reproduction and Integrated Conservation Science*, ed. W. V. Holt, A. R. Pickard, J. C. Roger and D. E. Wildt. Cambridge: Cambridge University Press, pp. 166–82.

Knauf, T., Jewgenow, K., Dehnhard, M. *et al.* (2003). Comparative investigations on reproductive biology in different bear species – anatomical, ultrasonographic and endocrine investigations. *Verhandlungsbericht des 41. Internationalen Symposium über Erkrankungen der Zoo- und Wildtiere, Rome*, **41**, 287–96.

Meyer, H. H. D., Rohleder, M., Streich, W. J., Göltenboth, R. and Ochs, A. (1997). Sexual Steroidprofile und Ovaraktivitäten des Pandaweibchen Yan Yan im Berliner Zoo. *Berliner Münchener Tierärztliche Wochenschrift*, **110**, 143–7.

Sutherland-Smith, M., Morris, P. J. and Silverman, S. (2004). Pregnancy detection and fetal monitoring via ultrasound in a giant panda (*Ailuropoda melanoleuca*). *Zoo Biology*, **23**, 449–61.

18

Gastrointestinal endoscopy in the giant panda

AUTUMN P. DAVIDSON, TOMAS W. BAKER, CHENGDONG WANG,
RONG HOU, LI LOU

INTRODUCTION

Endoscopy is a minimally invasive method of evaluating the gross appearance of the mucosal surfaces of the gastrointestinal, respiratory and urogenital tracts (Guilford, 1996). Efficient methods of performing endoscopic evaluation of these systems have been developed in small and large animal medicine (Jones, 1997). Besides providing direct visualisation, endoscopy permits obtaining representative biopsy specimens for subsequent histopathological tissue assessments. Video endoscopy allows recorded observations (for retrospective evaluation) as well as group participation by investigators and students, thereby improving both diagnostics and training opportunities.

The original CBSG Biomedical Survey of the giant panda made minimal use of endoscopy, although laparoscopy was used in a few individuals and found to be effective for evaluating abdominal organs, including the uterine cornuae and all ovarian surfaces (see Chapter 4). Other medical findings from this initial survey suggested the need to test other forms of endoscopy for more advanced diagnostic evaluations. Thus the present study was conducted in 2004 at the invitation of the Chengdu Research Base of Giant Panda Breeding. It was a

Giant Pandas: Biology, Veterinary Medicine and Management, ed. David E. Wildt, Anju Zhang, Hemin Zhang, Donald L. Janssen and Susie Ellis. Published by Cambridge University Press. © Cambridge University Press 2006.

component of a more thorough set of evaluative procedures that included case histories, physical examinations, haematology, blood chemistry, ultrasonography, serology, toxicology, histopathology and faecal analysis of 11 giant pandas. The subject of this chapter exclusively involves the effectiveness of using endoscopy at this same time to examine the gastrointestinal tract of this species.

METHODS FOR GASTROINTESTINAL ENDOSCOPY

Animals

Eleven giant pandas (four males and seven females) were evaluated and ranged in age and weight from nine months to 19 years and 19 to 165 kg, respectively.

Preparation

Effective endoscopy of the gastrointestinal tract requires proper patient preparation. The presence of ingesta and faecal material can inhibit entry into the oesophagus, stomach, duodenum and colon, contribute to damaged tissue or instrumentation and prevent clear visualisation of luminal surfaces. For companion animals (dogs and cats), a minimal 24-hour fast is accompanied by administering an enteric cleansing solution (i.e. Golytely; Braintree Laboratories, Braintree, MA; oral dose, 30 ml kg^{-1} divided over a 24-hour period) (Willard, 2001; Zoran, 2001).

Preparation of the giant panda for upper gastrointestinal endoscopy was limited to fasting due to poor palatability and reluctance to voluntarily accept an oral cleansing solution. Fortunately, the rapid transit time of food through the giant panda gastrointestinal tract (see Chapter 6) enabled adequate preparation with only a 12- to 20-hour fasting interval. Evaluation of the stomach mucosa was improved by preventing access to water during the four to six hours immediately preceding anaesthesia and endoscopy. Without such fasting, partially digested ingesta and watery fluid commonly obstructed visibility of the gastric mucosa and precluded successful entry into the duodenum (Fig. 18.1; Plate XXVII).

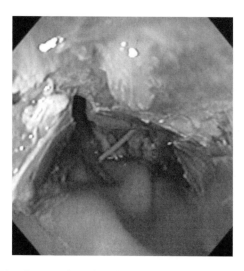

Figure 18.1. Ingesta obscuring visualisation and penetration of the pylorus. (See also Plate XXVII.)

In companion animals, a plain water enema is used to cleanse the distal colon of faecal material before endoscopy. Treatment usually occurs several hours before procedure onset. Again, due to intractability, each giant panda received an enema after general anaesthesia already had occurred. Administration of 1 to 3 litres of plain water into the colon by gravity flow stimulated the elimination of most residual faeces, certainly sufficiently to permit safely introducing an endoscope to permit good visibility of the colonic mucosa. A plain water enema without additives (e.g. soap) is important to minimise iatrogenic inflammation of the mucosa. Enema administration can cause chilling, inhibiting accurate monitoring of rectal temperature, so water temperature should be similar to the panda's normal body temperature (average 37.2°C; see Chapter 4).

Anaesthesia

Typically, gastrointestinal endoscopy requires patient relaxation and analgesia for 30 to 60 minutes. Without adequate anaesthesia, the procedure can be uncomfortable for the animal while putting expensive equipment at risk of damage (e.g. chewing or gagging with an endoscope in the oral cavity or oesophagus). Adequate anaesthesia is also required because of the need to distend the viscera with insufflated room air. Without insufflation, it is impossible to assess organ distensibility accurately (failure of which correlates to pathology; Zoran, 2001) and view the luminal surfaces. For these reasons, inhalant anaesthesia with endotracheal intubation providing airway protection and a surgical plane of anaesthesia are highly recommended. Reduction in saliva production and accumulation also facilitates gastrointestinal endoscopy so using anticholinergics (e.g. atropine or glycopyrrolate) is indicated as part of the pre- or early-anaesthetic regimen.

In this study, giant pandas were sedated with ketamine hydrochloride (Ketaset; Fort Dodge Laboratories, Inc., Fort Dodge, IA; 6–9 mg kg^{-1} i.m.) and then induced by mask inhalation of isoflurane/oxygen until an endotracheal tube could be placed, followed by surgical anaesthesia maintained with isoflurane (see Chapter 15 for more details). Atropine was administered (0.05 mg kg^{-1} i.m.) during anaesthesia induction. Due to its anti-emetic properties, metoclopramide (Faulding Pharmaceutical Co., Parmus, NJ; 0.2 mg kg^{-1} i.v.) was administered just prior to initiation of upper gastrointestinal endoscopy in all animals. Subjectively, the use of this drug appeared to facilitate endoscope passage through the pyloric antrum (in contrast to what has been reported in the veterinary literature for small mammals) (Zoran, 2001). For gastrointestinal endoscopy, each panda was placed in left lateral recumbency, which facilitated traversing the angulus and entering the duodenum by positioning the antrum and pylorus above the gastric contents.

Equipment

Oesophagoscopy, gastroscopy, duodenoscopy and colonoscopy were accomplished using a flexible 13 mm by 168 cm videoscope (Olympus CF-Q140L Video Colonoscope; Endoscopy Support Services, Inc., Brewster, NY) in ten of the 11 individuals. The exception was the smallest panda (studbook, SB, 575; 19 kg BW) where a 9.8 mm by 1 m videoscope (Olympus XQ140 Video Gastroscope; Endoscopy Support Services) was used. All procedures benefited from the use of a video processor (Olympus CV-100; Endoscopy Support Services), a xenon light source (Olympus CLV-U20; Endoscopy Support Services), suction and air insufflation equipment (Insufflator 20; Karl Storz Veterinary Endoscopy, Inc., Goleta,

GA), suction (Gomco 3020; Karl Storz Veterinary Endoscopy) and a 35.6-cm monitor (Sony 14 Inch Monitor; Karl Storz Veterinary Endoscopy). It was determined that the 168-cm endoscope was too short to visualise more than the most proximal duodenum in the largest male giant pandas (those weighing in excess of 100 kg). It was possible to obtain multiple biopsies of the stomach, duodenum and colon using standard, short, oval cup forceps (QFC-1296, 2.4 mm by 225 cm Biopsy Forceps; Endoscopy Support Services). Biopsies were taken of representative areas within these organs, as well as in any area with a grossly abnormal appearance. The oesophagus was not biopsied because of the inherent difficulty in obtaining pinch samples of oesophageal tissues and because no gross pathology was observed. Biopsies were transferred into biopsy tissue cassettes (Endoscopy Support Services) and placed in 10% buffered formalin for preservation.

To avoid pathogen transmission and ensure sample integrity between animal examinations, it was necessary to clean and sterilise the equipment thoroughly as in the manufacturer's recommendations. Immediately after a procedure was completed, a copious amount of water (100–200 ml) interspersed with air was brought through the suction channel to remove gross debris. The water button was activated for 15 seconds to clear debris from the air and water port, and the exterior of the scope was wiped thoroughly with 70% alcohol. The scope was disassembled and cleaned with an enzymatic/mild detergent solution following the owner's manual, then rinsed with clean water and disinfected using a glutaraldehyde-based solution. This was followed by a final rinse and drying according to the manufacturer's recommendations.

RESULTS AND DISCUSSION

Oesophagoscopy

In general, it was readily possible to pass the lubricated (e.g. with K-Y Jelly; Johnson and Johnson, Inc., New Brunswick, NJ) endoscope through the left oropharyngeal region and into the distal oesophagus. Neither ingesta nor watery fluid were found in the oesophageal lumen of any giant panda, and typically there was minimal mucus. The oesophagus of this species is characterised by the presence of numerous strong aboral waves of peristalsis which periodically (every 30 to 120 seconds)

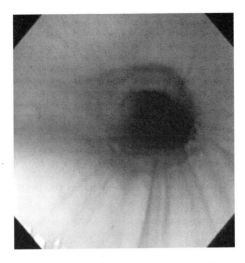

Figure 18.2. Appearance of the oesophageal lumen between peristaltic waves. (See also Plate XXVIII.)

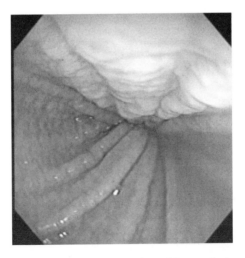

Figure 18.3. Appearance of the oesophageal lumen during a peristaltic wave. (See also Plate XXIX.)

and markedly reduce the size of the oesophageal lumen. Despite the use of general anaesthesia, the oesophagus did not appear dilated in any individual (Figs. 18.2, 18.3; Plate XXVIII and XXIX), as is common in companion animals (Gualtieri, 2001). Air insufflation was mandatory for securing good visibility of the oesophageal luminal structures. The trachea made a curved impression on the ventral oesophageal wall, and pulsation was observed where the base of the heart and the aortic arch were adjacent to the oesophageal wall. Oesophageal mucosal coloration was pale pink/tan with smooth surfaces distally (near the oral cavity), transitioning to a markedly cobblestone or corrugated appearance proximally (adjacent to the gastroesophageal junction) (Fig. 18.4; Plate XXX). Submucosal vessels were not prominent, and the gastroesophageal junction was not typically difficult to negotiate, even when closed.

The oesophageal mucosa was evaluated for erythema, erosions, ulceration, gastric reflux and strictures. Minimal gross abnormalities were observed. In one giant panda (SB 387), the texture of the distal oesophageal mucosa was smooth, but lymphoid aggregates (appearing as small, focal cavitations) were evident adjacent to the gastroesophageal junction (Fig. 18.4; Plate XXXI). Interestingly, and in contrast to all other pandas, the gastroesophageal junction was difficult to negotiate in this individual. One of the study animals (SB 575, a nine-month-old individual) differed from its adult counterparts by having a smooth, 'uncorrugated' proximal oesophageal mucosa (Fig. 18.6; Plate XXXII). Thus the cobblestone appearance (see Fig. 18.5) may be distinctive to adults.

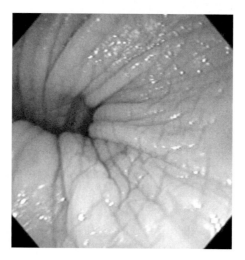

Figure 18.4. Corrugated appearance of the oesophagus adjacent to the cardia. (See also Plate XXX.)

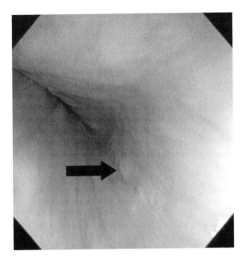

Figure 18.5. Focal lymphoid aggregates (arrow) in the oesophagus near the cardia. (See also Plate XXXI.)

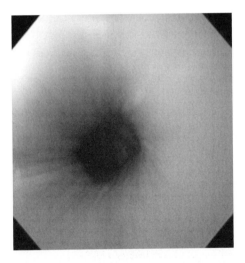

Figure 18.6. Smooth oesophageal mucosa observed throughout the oesophagus in a nine-month-old, juvenile giant panda. (See also Plate XXXII.)

Gastroscopy

Following oesophagoscopy, all giant pandas in this cohort underwent gastroscopy, with it being readily possible to insufflate the stomach with room air. There was a moderate amount of watery fluid in the gastric lumen, with flecks of white mucoid debris throughout, which prevented good viewing of the gastric mucosa in three individuals (SB 387, 522 and 530). In each of these cases, there was also an ingesta mass in the pyloric antrum that prevented observations and penetrating the pyloric with the endoscope. Each of these pandas had undergone minimal fasting and water removal before anaesthesia, which apparently contributed to this difficulty. No technical challenges occurred in the remaining animals where evaluations were completed for the greater and lesser curvatures of the stomach, reflex evaluation of the cardia and assessment of the pyloric antrum. The gastric mucosa appeared moderately pink and had a smooth texture. Large vessels were visible in the submucosa, rugal folds were distinctive and oriented in a direction leading to the pyloric antrum, and moderate to marked peristalsis was evident. Multiple regional (cardiac, body, antral) gastric biopsies were taken (with minimal bleeding) in all pandas, except SB 575 which was not sampled because of young age (nine months old).

The gastric mucosa was evaluated for discoloration, erosions, ulcerations and gastroduodenal reflux. A large oval gastric ulcer, approximately 1.5 cm in diameter, was observed in SB 454 (a seven-year-old male; Fig. 18.7; Plate XXXIII). The ulcer extended into the submucosa, with its edges appearing contracted, but no gross haemorrhage was

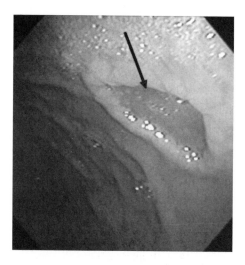

Figure 18.7. Gastric ulceration (arrow) near the cardia in giant panda SB 454. (See also Plate XXXIII.)

evident. Pinch biopsies were obtained adjacent to the ulcer with no increased bleeding as compared to grossly normal gastric tissue. Another giant panda (SB 515) had small fragments of foreign material evident in the gastric lumen, whereas another (SB 387) had a single polypoid mucosal mass in the pyloric antrum, which was also biopsied.

Duodenoscopy

Inspection of, and entry through, the pylorus into the proximal duodenum was facilitated by adequate fasting – there was more technical difficulty in the pandas that were fasted for the shortest intervals. The pylorus of the adult appeared as a puckered rosette of mucosa protruding into the gastric lumen (Fig. 18.8; Plate XXXIV). This structure in the juvenile was similar in appearance but had an increased pink coloration. The pyloric outflow tract was rarely open but was readily penetrated with the endoscope when not obscured by ingesta. The duodenal mucosa was pale pink with a textured, grainy surface that was distinctive from gastric mucosa (Fig. 18.9; Plate XXXV). The duodenum was typically dilated, and minimal insufflation was required for visualisation, which often resulted in observing small amounts of bilious mucoid fluid in the cavernous lumen (Fig. 18.10; Plate XXXVI). Moderate

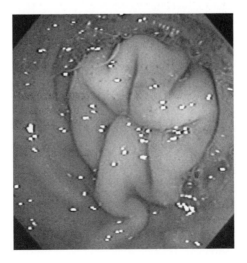

Figure 18.8. Pylorus of an adult panda. (See also Plate XXXIV.)

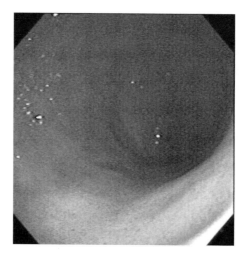

Figure 18.9. Grainy appearance of the duodenal mucosa. (See also Plate XXXV.)

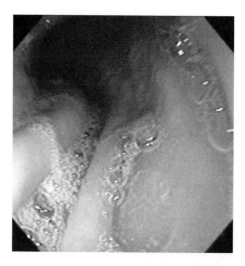

Figure 18.10. Cavernous appearance of the duodenum with bilious mucoid fluid apparent. (See also Plate XXXVI.)

peristalsis was generally evident, which obscured the lumen for short periods (of about three to five seconds). Numerous pinch biopsies were taken at various points within the length of the duodenum, resulting in negligible haemorrhage. Duodenal biopsies were mucoid and friable as compared to similarly sized samples collected from stomach tissue.

When entry into the pylorus was prevented by residual ingesta (SB 387, 522 and 530) or by panda size or age (SB 342 and 454), blind biopsies were made successfully in the duodenum. This was accomplished by passing the pinch biopsy forceps through the pylorus into the duodenum, proceeding until mild resistance was met (indicating mucosal contact) and then taking a biopsy from that location. On two occasions, it was possible to pass the endoscope into the duodenum over the inserted biopsy instrument (using the instrument as a guide). One panda (SB 297) exhibited patchy areas of thick-appearing, pale mucosa in the duodenum, which was highly friable at biopsy as compared to counterpart pandas (Fig. 18.11; Plate XXXVII).

Colonoscopy

Colonoscopy was performed 30 or more minutes after completing the enema procedure. The colonic preparation described on pp. 441–442 was generally adequate for achieving good visibility of mucosa, which was commonly folded into the lumen (reducing its size). The colonic mucosa was dark pink and had a coarse texture (Fig. 18.12; Plate XXXVIII), whereas minimal mucus was present, and blood vessels were not prominent. Numerous pinch biopsies were taken along the length

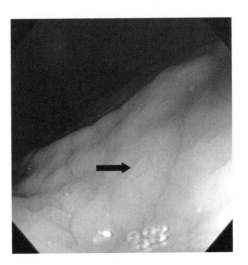

Figure 18.11. Abnormal duodenal mucosa with patches of pale tissue (arrow). (See also Plate XXXVII.)

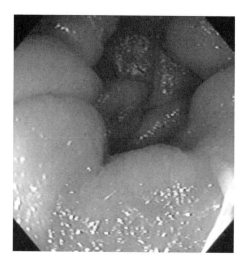

Figure 18.12. Typical colouration of the colonic mucosa. (See also Plate XXXVIII.)

of the descending colon which resulted in minimal bleeding. Inspection for erythema, erosions, ulcerations and strictures revealed only one abnormal incidence. The colonic mucosa in giant panda SB 297 appeared erythemic in coloration and had an even coarser texture than counterpart individuals.

PRIORITIES FOR THE FUTURE

This is the first evidence that endoscopy of the gastrointestinal tract is safe and efficacious in the giant panda. However, it was clear that effectiveness is tightly linked to a sufficient fasting time as well as adequate anaesthesia and availability of suitable equipment. Under appropriate conditions, we were able to examine thoroughly and document the status of the oesophagus, stomach, proximal duodenum and colon. We also suspect that the relatively easy entrance into the small bowel was facilitated by the pharmacological manipulations described. There was a remarkable lack of gastrointestinal problems in this panda cohort, the exception being a stomach ulcer in a seven-year-old male (who interestingly had a history of vomiting). More minor abnormalities were detected in other individuals, and in all cases it was possible to secure biopsied tissue safely. Although the purpose of this chapter was not to correlate biopsy and endoscopic assessments, it is important

to note that all endoscopic observations were supported by histopathological findings. Clearly gastrointestinal endoscopy is another effective
and safe tool available for monitoring the health of the *ex situ* giant
panda population.

There appear to be three high priorities. Although initial observations here seemed to indicate no consistently serious anomalies to the
gastrointestinal system, it is now necessary to apply this technology to
other individuals, including those that display nutritionally related
problems, for example, mucous stools (see Chapter 6). In the case of
the present chapter, there is also a need to further correlate the histories of specific individuals with endoscopic observations. Each of the
adult animals in this cohort had experienced challenges to normal
health and/or reproduction. This is a particularly important concern
for individuals who may be clinically compromised (e.g. SB 454) and
could benefit from potential therapies, with success monitored by additional endoscopy. A final high priority is the training of Chinese veterinarians who have the ultimate responsibility of managing most of the
giant pandas living *ex situ*. Although this technology is becoming quite
common in the west (especially for companion animals), there is little
fundamental expertise for wildlife, including for highly specialised
species such as the giant panda. While its safety and utility have been
demonstrated here, endoscopy requires a batch of combined skills in
general anaesthesia with volatile gas, the availability and effectual use
of sophisticated fibre-optic equipment and especially the ability to
diagnose what is normal versus abnormal. For the giant panda to
benefit fully will require additional partnerships and the sharing of
information and skill between appropriate Chinese and western
institutions.

ACKNOWLEDGEMENTS

We are grateful for the invaluable assistance and participation of the
veterinary and animal care staff of the Chengdu Research Base of Giant
Panda Breeding. The endoscopy equipment was generously loaned by
James F. Burns, Endoscopy Support Services, Inc., and Linda Scott and
Christopher Chamness, Karl Storz Veterinary Endoscopy. The senior
author is indebted to Darlene Riel (RVT), School of Veterinary Medicine,
University of California, Davis, for transcontinental technical assistance
and previous endoscopic training.

REFERENCES

Gualtieri, M. (2001). Esophagoscopy. *Veterinary Clinics of North America Small Animal Practice*, ed. L. Melendez, **31**, 605–30.

Guilford, W. G. (1996). Gastrointestinal endoscopy. In *Strombeck's Small Animal Gastroenterology*, 3rd edn, ed. W. G. Guilford, S. A. Center, D. R. Strombeck, D. A. Williams and D. J. Meyers. Philadelphia, PA: W. B. Saunders, p. 114.

Jones, B. D. (1997). Incorporating endoscopy into veterinary practice. *Compendium on Continuing Education for the Practicing Veterinarian*, **20**, 307–12.

Willard, M. D. (2001). Colonoscopy, proctoscopy and ileoscopy. *Veterinary Clinics of North America Small Animal Practice*, ed. L. Melendez, **31**, 657–69.

Zoran, D. L. (2001). Gastroduodenoscopy in the dog and cat. *Veterinary Clinics of North America Small Animal Practice*, ed. L. Melendez, **31**, 631–56.

19

Historical perspective of breeding giant pandas ex situ in China and high priorities for the future

ZHIHE ZHANG, ANJU ZHANG, RONG HOU, JISHAN WANG, GUANGHAN LI, LISONG FEI, QIANG WANG, I. KATI LOEFFLER, DAVID E. WILDT, TERRY L. MAPLE, RITA MCMANAMON, SUSIE ELLIS

INTRODUCTION

The giant panda is one of the national treasures of China. Many factors, related primarily to increased human activity, have caused a marked decline and geographic fragmentation of the wild population. To preserve this endangered species, the Chinese government, in partnership with many nongovernmental organisations (inside and outside China), has invested significant human and material resources to benefit *in situ* conservation. These collective efforts have resulted in the establishment of more than 40 nature reserves in southwest China in the provinces of Sichuan, Gansu and Shaanxi.

Giant pandas have been sporadically maintained in captivity since the Han Dynasty (206 BC to AD 226) (see Chapter 1). However, it was not until the 1940s that there was serious interest in exhibiting the species in China. It took more than 20 years of giant panda husbandry experience to produce the first cub in captivity, at the Beijing Zoo in 1963. Much progress has been made in the subsequent years in understanding basic giant panda biology and making it possible for the species

Giant Pandas: Biology, Veterinary Medicine and Management, ed. David E. Wildt, Anju Zhang, Hemin Zhang, Donald L. Janssen and Susie Ellis. Published by Cambridge University Press. © Cambridge University Press 2006.

to reproduce consistently in captivity. This chapter reviews the brief history and significance of *ex situ* breeding efforts for the giant panda.

The giant panda is particularly vulnerable to external pressures, in part because of an inherently slow rate of reproduction. In nature, a giant panda female reproduces only every second or third year and, although half the births are twins, only one cub is raised by the dam at the expense of the other. In captivity, the focus has been on finding ways to consistently promote reproduction in every genetically valuable individual, largely through improved husbandry, veterinary health and the use of assisted reproductive technologies, when necessary. Reproductive management is now also beginning to address the need to maintain genetic viability (see Chapter 21). Reduced gene diversity is of concern for the *ex situ* as well as *in situ* populations as this may compromise both health and reproduction through inbreeding depression. It has also become clear among Chinese managers that captive giant pandas are indeed an important research resource and insurance policy against catastrophes that might affect wild counterparts whose population is highly fragmented.

Our goal is simple: to continue to use scientific advances in husbandry, reproductive biology, veterinary medicine, genetics, nutrition and behaviour to build a self-sustaining captive population. We believe that this population could contribute to *in situ* conservation by enhancing biological knowledge about the giant panda, increasing public awareness about species uniqueness and, if necessary, providing healthy, viable individuals for reintroduction. From a Chinese perspective, learning how to manage giant pandas effectively in captivity lays a foundation for the initiation of similar programmes with other rare Chinese species.

A steep learning curve: the first 26 years (1963 to 1989)

Early efforts to breed giant pandas in captivity were frustrated by the animals' strong selectivity in mate preference, the failure of many individuals (particularly males) to breed, low conception rates and high neonatal mortality. The first cub produced in 1963 at the Beijing Zoo [Studbook, SB, 60] resulted from the natural breeding of two wild-born pandas that had been captured in Baoxing four and five years earlier. The cub's sire (SB 31) died within a few months of this milestone birth, and the dam (SB 25) reproduced only two more times before she died 18 years later. During his 26 years of life, this 'inaugural' cub never bred. This, unfortunately, was a preview of things to come. In the first quarter century of *ex situ* giant panda reproduction, only 30% of the animals reproduced, with a neonatal mortality of more than 60% (Fig. 19.1). The number of reproductively mature pandas in the captive population rose in that period from 12 to 88, primarily through the capture of wild-born

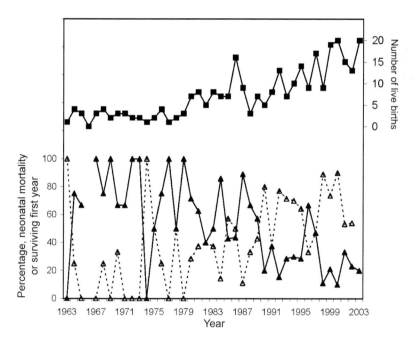

Figure 19.1. Neonatal mortality (<1 month of age) and survival to 1 year of age in captive-born giant panda cubs, 1963 to 2003 (first year survival includes cubs born through 2002). Data are from Xie & Gipps (2003). ■, Number live births in captivity; ▲, neonatal mortality (%); △, surviving first year (%).

animals (Figs. 19.2 and 19.3). Of the 115 cubs born in those 26 years, only 13.9% (i.e. 16 cubs) lived to breeding age (i.e. more than four years of age).

During these early years of attempted breedings, it was apparent that reproductive success in both sexes was compromised. Females failed to enter oestrus, and of those that did, many failed to mate, conceive or sustain a pregnancy. Males often refused to copulate, generally because of serious aggression toward females (see Chapter 14). The latter was the incentive to test the feasibility of assisted breeding, virtually always by artificial insemination (AI). However, in many cases it was difficult to obtain high-quality ejaculates, and the sperm often died soon after collection.

These frustrations motivated managers to develop more systematic efforts to learn about the biology of the species and to take an applied approach. For example, to circumvent behavioural incompatibilities that prevented natural mating, serious experiments with assisted breeding began in the mid 1970s. The first cub produced by AI with fresh sperm was born at Beijing Zoo in 1978. Two years later, the first cub from AI using frozen–thawed sperm was produced at Chengdu Zoo. Thereafter, success with AI was somewhat sporadic but gradually improved with the advent of detailed sperm biology studies, including creating methods to maintain good sperm viability *in vitro* before deposition into the female. Successful AI was an important milestone because it overcame one of the primary challenges to consistent reproduction: behavioural incompatibility between the sexes. This accomplishment was significant because it provided the technology to maintain gene diversity, which is currently one of the highest priorities of the modern breeding programme (see Chapter 21). In this context, it was realised early on that moving genes via sperm could be much more cost-effective and less stressful than moving animals from one breeding facility to another. Furthermore, the ability to cryopreserve viable sperm could maintain valuable panda genes in 'suspended animation' indefinitely, to be used even in future generations. Throughout the 1980s and 1990s, this type of research progressed, as did the routine use of AI at various Chinese breeding centres. Nonetheless, the refusal of individual giant pandas to mate remained an important issue, as it still does today. And no manager has been naïve enough to believe that the captive population should be sustained by assisted breeding alone. On the contrary, a substantial number of studies have been initiated over the years to understand the behaviour of this species, largely for the purposes of promoting natural breeding (Maple *et al.*, 1997; see also Chapters 11 and 14).

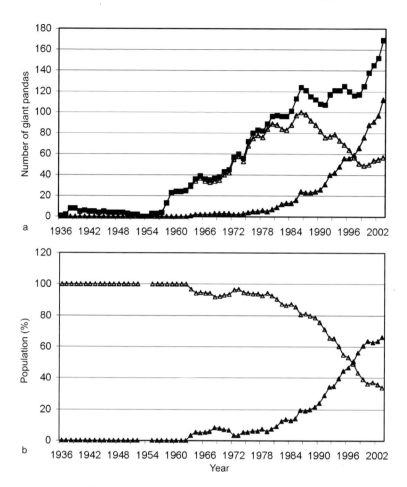

Figure 19.2. *Ex situ* giant panda population from 1936 to 2003, represented by the number of individuals alive on 31 December of each year (including the cubs produced in that year). Data are from Xie and Gipps (2002). (a) The total number of giant pandas and the proportions of the population that are captive born versus wild-caught. (b) The percentage of captive-born and wild-caught individuals in the population from 1936 to 2003. Δ, wild caught; ▲, captive born; ■, total.

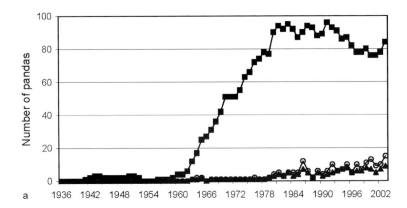

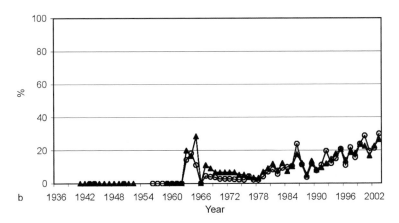

Figure 19.3. (a) The number of reproductive-age giant pandas (five years or older on 1 August of each year) in the *ex situ* population. (b) The percentage of reproductive-age males and females producing live-born cubs (1936 to 2003). Data are from Xie and Gipps (2003). ▲, males; ○, females; ■, total.

The Focus on improving cub survival, assisted breeding and maximising inter-institutional communication (1990 to the present)

During the first 26 years, 115 cubs were born from 77 pregnancies with a neonatal survival rate of 37%. From 1990 to 2002, 179 cubs were born from 126 pregnancies, with 71% of neonates surviving. The captive-born population has grown from one panda born in 1963 to 112 at the end of 2003 (including surviving cubs born in the autumn of 2003; see Fig. 19.2). In 1997, the number of captive-born animals outnumbered wild-born (and captured) animals in the *ex situ* population (see Fig. 19.2). Since then, divergence in the size of these two groups has increased dramatically and reflects the improvements in captive breeding efficiency. This increase in the captive-born population has been attributable to more females having the chance to reproduce (due to the use of AI) and enhanced cub survival (due to improved husbandry).

Improvements in cub survival have been particularly profound. In those early years, much of the early cub loss was due to the almost 50% incidence of twinning and the 'natural' loss of one cub through maternal rejection of both cubs and/or a variety of dam-related incompetencies (see Chapter 13). Efforts to assist dams by supplemental feeding and the partial hand-rearing of cubs first succeeded at Chengdu Zoo in 1990, when a set of twins survived to adulthood. Two years later, the Beijing Zoo hand-reared a cub that had been rejected by its mother and had never nursed from the dam (complete hand-rearing). This was a milestone in that the cub, although presumably never obtaining immunological protection afforded by colostrum, did not succumb to neonatal sepsis which causes most neonatal mortalities in this species (see Chapters 13 and 16). The use of commercial milk formulae, improved sanitary conditions and the novel technique of twin swapping (see Chapter 13) have resulted in a steady rise in neonatal survival over the past decade (see Fig. 19.1). These efforts have also been bolstered by partnerships between panda breeding centres that had little experience with hand-rearing and institutions (often international) whose personnel were skilled in these methods (see Chapter 13). This resulted in a rapid transfer of knowledge and technologies. At the same time, similar cooperation occurred in reproductive biology studies, including those that focused on improving AI using fresh or thawed spermatozoa (see Chapter 20).

Although international partnerships were critical to progress, there was also more communication among Chinese breeding facilities. To oversee and coordinate breeding recommendations and research priorities within China, the Chinese Committee of Breeding Techniques for Giant Pandas was established in 1989. This committee, comprised of scientists and administrators from zoos, nature reserves, universities, research facilities and government agencies, meets annually (usually in November or December), and prior to the impending breeding season.

Information is shared in scientific presentations and through small working groups that allow face-to-face interaction. This annual meeting has recently been opened to international holders of giant pandas. The Committee has also worked closely with the Conservation Breeding Specialist Group (CBSG) of the IUCN–World Conservation Union's Species Survival Commission. In fact, it was through this annual committee meeting that CBSG worked with various stakeholders to initiate the Giant Panda Biomedical Survey (discussed throughout this text). CBSG has also been instrumental in catalysing various contemporary actions involving genetic management, from helping organise paternity testing (see Chapter 10) to annual breeding recommendations based on mean kinship (see Chapter 21).

PRIORITIES FOR THE FUTURE

We now have a wealth of tools available to address the challenges associated with understanding giant panda biology and contributing to conservation. Certainly, the highest and most visible priority is to ensure that viable populations remain in nature. Also of importance is a concerted and coordinated effort involving scientists, managers, veterinarians, husbandry personnel and administrators to build a hedge (insurance) population in captivity that can provide new biological information, inspire and educate the public, and help to generate funds to support giant pandas in nature. Although strides have been made in captive breeding, many of the difficulties first discovered more than 40 years ago are still problematic today. It is the authors' opinion that the following issues warrant attention immediately.

Enhance reproductive representation of the adult population

About 15 years ago, only 7% of breeding age males had copulated (Hu, 1990). Although this has increased significantly, the proportion of adult giant pandas that actually reproduce (especially through natural mating) still remains quite low (see Fig. 19.3). One estimate has only about 58.3% of males (14 of 24) in the contemporary population being able to mate naturally (R. Hou, unpublished data). And, despite advanced technologies that enable sophisticated assisted breeding, genetic analysis, enhanced communication among centres and more feasible animal transport, only about 25% of the breeding population is now reproducing. Such data mean that even more research and cooperation are needed as well as perhaps re-thinking what constitutes a 'breeding population'. For example, Snyder *et al.* (in Chapter 14) suggest that the current strategy of sending pairs of giant pandas to cooperating zoos may work against success. An incompatible pair results in the loss of two valuable individuals indefinitely, especially as zoo staff (having developed affection for the pair) resist an animal exchange and rather decides to wait until next year for a success, which may be unlikely to occur. These authors propose several solutions, which include maintaining an absolute minimum number of animals (more than a pair) in a given institution. Especially important might be maintaining younger adult males in facilities with proven breeder females. The receptive female is then permitted to mate with an older, genetically appropriate male, but then also 'mentors' a younger, inexperienced male, allowing him to develop appropriate breeding behaviours. In this scenario, the concept of maintaining a single pair of giant pandas over many years is rejected in favour of establishing groups on the basis of genetic and behavioural compatibilities.

Most of the behavioural and physiological factors that now limit reproductive success in the captive programme can be resolved scientifically. But the time required to achieve this success will be based largely on our ability to share resources and knowledge. For example, the Biomedical Survey and the many studies that emerged from this collaboration were excellent examples of the cooperative use of intellectual and physical resources. Thus, a high priority is more such collaboration, but not only in research. Particularly important is the sharing of adult offspring (or their gametes) to maintain high levels of genetic diversity. The Chinese giant panda population numbered 161 at the end of 2003 (Xie & Gipps, 2003). Table 19.1 describes the current distribution of giant pandas in the *ex situ* population. Although giant panda institutions are separated by considerable geographic distance, current technology makes it possible to transport (often rapidly) frozen sperm or the animals themselves. Lastly, while significant studies have been made in past decades, there is a need for more cooperation between the Chinese regulatory agencies responsible for the *ex situ* populations (i.e. the Ministry of Construction, MoC, and the State Forestry Administration, SFA). There is already a good relationship among the breeding centres and holding zoos, but this 'positive' is often based on informal arrangements and personal relationships without agreement and action by decision-makers at the highest agency levels.

Table 19.1. *Number and distribution of* ex situ *giant panda population (data current to 31 December 2003)*

Location	Total	Wild caught	Captive born	% of total population
China				
Baoding Zoo	1	0	1	0.6
Baoxing Nature Reserve	0	0	0	0.0
Badaling Wild Animal World (Beijing)	1	1	0	0.6
Baishuijing Reserve	1	1	0	0.6
Beijing Zoo	11	1	10	6.8
Chengdu Panda Base[a] & Zoo	28	3	25	17.4
Chongqing Zoo	5	1	4	3.1
Dalian Zoo	1	0	1	0.6
Fuzhou Giant Panda Research Centre	4	4	0	2.5
Guangzhou Zoo	1	1	0	0.6
Guilin, Guangxi	1	1	0	0.6
Hanzhou Zoo	1	1	0	0.6
Hangzhou Wildlife Park	2	0	2	1.2
Hefei Xialoyaojin Park	1	1	0	0.6
Jinan Zoo	2	2	0	1.2
Lanzhou	1	0	1	0.6
Louguanta	9	6	3	5.6
Kunming Zoo	1	1	0	0.6
Panyu Xiangjiang Park	2	2	0	1.2
Shanghai Safari Park	2	1	1	1.2
Shanghai Zoo	1	1	0	0.6
Shengzhen Safari Park	2	0	2	1.2
Tian Jing Zoo	1	1	0	0.6
Wolong[b]	39	11	28	24.2
Wuhan Zoo	1	0	1	0.6
Xian Zoo	1	0	1	0.6
Ya'an Bifengxia[c]	12	4	8	7.5
International				
Berlin Zoo, Germany [Louguanta][d]	2	2	0	1.2
Chiangmai Zoo, Thailand [Wolong]	2	0	2	1.2
Kobe Park, Japan [Wolong]	2	0	2	1.2
Memphis Zoo, USA [Beijing, Shanghai]	2	0	2	1.2
Chapultepec Zoo, Mexico	3	0	3	1.9
Smithsonian's National Zoo, USA [Wolong]	2	0	2	1.2
Ocean Park, Hong Kong [Wolong]	2	2	0	1.2
Pyongyang Zoo, North Korea [Beijing]	1	1	0	0.6
San Diego Zoo, USA [Wolong]	2	1	1	1.2

Table 19.1. (cont)

Location	Total	Wild caught	Captive born	% of total population
Uneo Zoological Park, Japan	1	0	1	0.6
Adventure World, Japan [Chengdu]	6	0	6	3.7
Vienna Zoo, Austria [Wolong]	2	0	2	1.2
Zoo Atlanta, USA [Chengdu]	2	0	2	1.2
TOTALS (number of institutions: 40)	**161**	**50**	**111**	**100.0**

[a] Chengdu Research Base of Giant Panda Breeding.
[b] China Conservation and Research Centre for the Giant Panda (Wolong Nature Reserve).
[c] Facility associated with the China Conservation and Research Centre for the Giant Panda.
[d] For international facilities, names in square brackets are the source of animals.

Therefore, the MoC and the SFA could advance progress significantly by facilitating compliance with breeding recommendations that are made on the basis of genetic priorities. This mostly involves providing formal governmental approvals to allow the timely shipment of animals (or germ plasm) from one institution to another.

Increase cub survival

Although declining precipitously in the last ten years, mortality rates for neonates remain alarmingly high – at 20 to 40% (see Fig. 19.1). There is a need to emphasise improved care for the dam prenatally and for the cub up to one year of age. This involves a complex array of issues that requires highly skilled personnel in nutrition, reproductive physiology, behaviour, nursery husbandry, neonatal medicine, infectious disease and immunology. The target is to increase offspring survival to one year of age, in part because some cubs survive the neonatal interval only to die before reaching their first birthday. There is a vast area of research into the cause(s) of this phenomenon which remains unexplored.

Improve overall knowledge of animal health

There is a paucity of information in the veterinary sciences for the giant panda, particularly on reproductive dysfunction, infectious disease and digestive disorders. The impact of the latter two on reproductive health is virtually unstudied and is in need of urgent attention. Research on infectious diseases is particularly limited because the extant *ex situ* giant panda population is dangerously limited in geographical scope. Of all the captive animals currently in China, 49% are held in three Sichuan institutions (24% at the centre in the Wolong Nature Reserve, 17% at the two Chengdu institutions and 7% at the new centre in Ya'an (Xie & Gipps, 2003)). A concentration of animals in a few locations places the overall population at extreme risk in the event of an infectious disease outbreak or other disaster. The potential devastation of a disease epidemic has been demonstrated frequently in the history of animal care and human societies, most recently by the outbreaks of severe acute respiratory syndrome (SARS) in Asia, influenza virus in China's tiger farms and avian influenza throughout South-East Asia. The role of subclinical, chronic disease in compromising the reproductive health and survival of young animals is unknown. Loeffler, Zhang and Wildt have launched a serological survey of giant pandas in China to investigate the species' risk of exposure to various infectious diseases. This is an important first step towards filling this knowledge gap. A subsequent action would involve the major giant panda holding facilities in China (and respective government agencies) to support a single infectious disease laboratory. Such a core facility would conduct research and monitor wildlife diseases in China's captive and free-living wildlife. A related need is a national pathology repository that includes the expertise to identify and track diseases as well as maintain information databases that can be readily accessed and shared. This priority requires significant capacity building, including workshops for groups and advanced, highly specialised training for individuals.

Implement enrichment and behavioural management

Zoos throughout the world are recognising the importance of enriched animal environments (Bloomsmith *et al.*, 2003). Enrichment activities not only enhance the quality of life in captivity, but also improve physical and mental health, reproductive capacity and maternal care that, in turn, boost neonatal survival and cub health (see Chapter 11). Operant conditioning techniques are now especially important for facilitating certain medical procedures and open many possibilities for more thorough biomedical investigations and improved preventive medicine and health monitoring.

Therefore, we endorse the need for more behavioural enrichment for giant pandas in captivity. Additionally, a high priority is training more giant pandas via operant conditioning to accept routine physical examinations, blood collection and ultrasonography, thereby avoiding the stress associated with restraint or anaesthesia (see also Chapter 15). We expect that the amount of new information to be gleaned from a closer, noninvasive but 'hands-on' approach could be significant. This will also require more interaction and training between behavioural scientists in China and their Western colleagues. The goal should be to build much more capacity in Chinese zoos in the disciplines of animal behaviour and environmental enrichment.

Step up the pace of activities

Although enormous progress has been made since the first cub was produced in captivity in 1963, we should not become too comfortable with our accomplishments. Given what we do not know about fundamental giant panda biology and accounts about the situation in the wild (especially habitat fragmentation), we cannot be too hasty in our scientific studies.

This priority requires approval by holding facilities and government agencies, commitment by experts and funding. Historically, giant panda breeding centres and zoos have been highly cooperative in conducting conventional studies as well as applying novel concepts and tools to benefit this endangered species. Given a sound, safe and scientific protocol, animals can be made available for study. Towards this point, there have also been adequate numbers of senior scientists wanting to study this species. In fact, much of the recent progress has been due to the dedication of many highly skilled investigators from holding zoos, in China and in the West. But it is critical that scientists from the West do not lose their enthusiasm for working in China, and there is a constant need to train more biologists in China (and in the West) to ensure plenty of investigative expertise.

The last priority – funding – will always be a challenge even for a charismatic species such as the giant panda. Historically, the US Fish and Wildlife Service has endorsed the use of giant panda loan funds (from the USA) only for supporting projects that conserve giant pandas *in situ*. This implies, therefore, that giant pandas in zoos do not contribute to conservation. However, we (as well as others, e.g. Ellis *et al.* in Chapter 1) would argue that they do so at many levels. Much of the critical biological information being generated about the species will never be collected on wild individuals that are rare and elusive. Further, given the shaky and fragmented status of wild habitat, the captive population may well serve as the source for future reintroductions. So, although loan funds should always focus on sustaining giant pandas in nature, it makes sense that support should also be directed for solid scientific studies of giant pandas regardless of their location, *in situ* or *ex situ*. In the end, our goals are the same: sustaining healthy populations and providing the best information and data possible so that decision-makers can ensure the long-term survival of this species. More intensive integration of these three factors – breeding centre cooperation, available scientists and funding – would increase the speed of necessary progress.

REFERENCES

Bloomsmith, M. A., Jones, M. L., Snyder, R. J. *et al.* (2003). Positive reinforcement training to elicit voluntary movement of two giant pandas throughout their enclosure. *Zoo Biology*, **22**, 323–34.

Hu, J. (1990). *Research and Progress in the Biology of the Giant Panda*, Chengdu: Sichuan Publishing House of Science and Technology, pp. 316–21.

Maple, T. L., Perkins, L. A., and Snyder, R. (1997). The role of environmental and social variables in the management of apes and pandas. In *Proceedings of the International Symposium on the Protection of the Giant Panda*, ed. A. Zhang and G. He. Chengdu: Sichuan Publishing House of Science and Technology, pp. 23–8.

Xie, Z. and Gipps, J. (2002). *The 2002 International Studbook for the Giant Panda (Ailuropoda melanoleuca)*. Beijing: Chinese Association of Zoological Gardens.
    (2003). *The 2003 International Studbook for the Giant Panda (Ailuropoda melanoleuca)*. Beijing: Chinese Association of Zoological Gardens.

20

Role and efficiency of artificial insemination and genome resource banking

JOGAYLE HOWARD, YAN HUANG, PENGYAN WANG, DESHENG LI,
GUIQUAN ZHANG, RONG HOU, ZHIHE ZHANG, BARBARA S. DURRANT,
REBECCA E. SPINDLER, HEMIN ZHANG, ANJU ZHANG, DAVID E. WILDT

INTRODUCTION

Historically, the breeding of giant pandas in *ex situ* programmes has been difficult due to behavioural incompatibility and interanimal aggression. Because some individuals fail to mate naturally, the potential loss of valuable genes is a major concern to effective genetic management (see Chapter 21). Consistently successful artificial insemination (AI) would allow incorporating genetically valuable males with behavioural or physical anomalies into the gene pool. This strategy becomes even more powerful when used in the context of a genome resource bank (GRB), an organised repository of cryopreserved biomaterials (tissue, blood, DNA and sperm) (see Chapter 7). The use of sperm cryopreservation and AI allows the movement of genes among zoos and breeding centres without needing to transfer animals, which is both stressful and costly.

'Assisted breeding' refers to the tools and techniques associated with helping a pair of animals propagate, from AI to embryo transfer to cloning, among others (Howard, 1999; Pukazhenthi & Wildt, 2004).

Giant Pandas: Biology, Veterinary Medicine and Management, ed. David E. Wildt, Anju Zhang, Hemin Zhang, Donald L. Janssen and Susie Ellis. Published by Cambridge University Press. © Cambridge University Press 2006.

With the exception of AI, there is not much need for most other assisted-breeding techniques for the giant panda. As will be demonstrated here, AI is quite adequate for dealing with most cases of infertility or with helping to maintain adequate gene diversity in the captive population. In fact, the major breeding facilities, especially the China Conservation and Research Centre for the Giant Panda (hereafter referred to as the Wolong Breeding Centre) and the Chengdu Research Base of Giant Panda Breeding, routinely use AI to increase pregnancy success. It is common practice at these centres to combine natural mating (with breeder males) with AI (using sperm from nonbreeders). Mate incompatibility and what is perceived as 'weak' sexual behaviour are the two most common reasons for using AI. However, the growing emphasis on genetic management (see Chapter 21) also elicits interest in assisted breeding to avoid the logistical challenges of moving pandas between facilities.

This chapter addresses four issues, the first being a comparison of the efficiency of AI only in the giant panda (without natural breeding) to the usual tactic of combining natural mating with AI (see Chapter 10). Second, we assess the efficacy of AI for enhancing reproduction in nonbreeding females who demonstrate weak behavioural oestrus or in nonbreeding males with compromised health. The third purpose is to determine the reproductive competency of semen cold stored at 4°C and used for multiple inseminations over consecutive days. Lastly, we evaluate the efficiency of AI with frozen–thawed spermatozoa using recent improvements in cryopreservation techniques (described in Chapter 7).

METHODS

Semen processing for assisted breeding

Semen was collected by electroejaculation from nonbreeding males and assessed for ejaculate volume, sperm concentration, percentage sperm motility and sperm forward progression (on a scale of 0 to 5; 5 is best) using protocols described earlier (Huang *et al.*, 2000b; see also Chapter 7). For AI, ejaculates were diluted in either SFS (prepared fresh at the giant panda breeding facilities in China; described in Chapter 7) or TEST (Irvine Scientific, Santa Ana, CA) egg-yolk diluent containing 0% or 4 to 5% glycerol. These mixtures were then used for AI either while still fresh or after cold storage at 4°C (for up to 48 hours) (Olson et al., 2003a,b). For future AI, some aliquots were cryopreserved using the pellet or straw (0.25 ml) method of freezing (Huang et al., 2000a,b; see also Chapter 7).

Artificial insemination

Female giant pandas were monitored beginning mid January through May for natural oestrus. The time of AI was based on behavioural signs of oestrus (Kleiman et al., 1979; Bonney et al., 1982; Zeng et al., 1984) and urinary hormones (increased oestrogen) assessed by enzyme immuno-assay (see Chapter 8). Early indications of oestrus included increased vocalisation (particularly bleating), scent marking, restlessness and de-creased appetite. These changes were followed by more overt beha-viours including backwards walking and lordosis in the female (tail lifted and arched back) and strong interest in males. Vaginal cytology (increased cornified superficial cells) was assessed in selected females either during anaesthesia scheduled for AI or throughout the breeding season (Zeng et al., 1984; Durrant et al., 2002, 2003; see also Chapter 9). Longitudinal vaginal cytology profiles were mostly generated in USA institutions because pandas in China were not trained to accept vaginal swabs.

For AI, anaesthesia was induced in each female using injectable ketamine hydrochloride (Ketaset; Fort Dodge Laboratories, Inc., Fort Dodge, IA; ~5 mg kg^{-1} BW). The anaesthetised female was placed in a supine position, and a lubricated vaginal speculum (12 cm in length and 2 cm in diameter; Fig. 20.1) was inserted to visualise the cervix (Fig. 20.2). The insemination consisted of inserting a plastic or stain-less steel catheter into the external cervical os and advancing the catheter approximately 18 cm from the vulva. A total of 1 to 2 ml of fresh, cooled or frozen–thawed spermatozoa typically was used for each insemination, although a larger volume (3 to 6 ml) was used in some females. Females were monitored for pregnancy and birth of cubs (Fig. 20.3).

Historical data on AI success outside China were summarised for giant panda cubs born in Spain in 1982, in Japan on three occasions (1985, 1986 and 1988) and most recently in the USA in 1999 and 2005. Numbers were sufficient at the Wolong Breeding Centre to allow com-paring the efficiency of AI only with the combined use of natural

Figure 20.1. (a) A glass vaginal speculum and (b) plastic catheter used for transcervical artificial insemination in the giant panda. The speculum is used to expose the cervix, and then the insemination catheter is inserted into the external cervical os approximately 18 cm from the vulva.

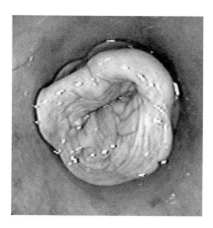

Figure 20.2. The external cervical os in a giant panda as viewed through a 12-cm long glass speculum.

breeding and AI (for breeding seasons 1998, 1999 and 2000). The reproductive competency of frozen–thawed spermatozoa was evaluated during the 2000, 2001, 2002 and 2003 breeding seasons at the Chengdu Research Base of Giant Panda Breeding.

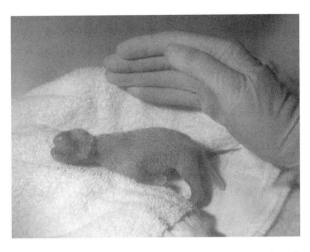

Figure 20.3. A giant panda neonate is approximately the size of the palm of a person's hand (an estimated mother-to-cub body size ratio of approximately 900 to 1).

RESULTS AND DISCUSSION

Reproductive success in the ex situ breeding programme

Giant pandas have been held *ex situ* in China since 1953 when a panda was taken from Guanxian County to Chengdu Zoo (Hu *et al.*, 1990). Two years later, Beijing Zoo also added giant pandas to its collection. The first *ex situ* giant panda birth, however, did not occur until 1963 when female studbook (SB) 25 (Li Li) gave birth to male cub SB 60 (Ming Ming) at Beijing Zoo (Kan & Shu-hua, 1964). In the late 1970s, both Beijing and Chengdu Zoos began investigating assisted reproduction, with the first successful AI and cub produced at Beijing Zoo in 1978 (Liu, 1979, 1981; Liu *et al.*, 1979). In this case, there were three males that did not breed naturally, so semen was collected for the first time by electroejaculation. Following fresh sperm insemination of four females, one giant panda (SB 132, Juan Juan) became pregnant and produced twins in September 1978, 15 years after the first birth by natural mating at this same institution. This was followed two years later by the first successful AI with frozen semen, which occurred in female SB 152 (Mei Mei) at Chengdu Zoo (Hu & Wei, 1990; Ye *et al.*, 1991; Zhang *et al.*, 1991).

Since these milestones, Chinese zoos and breeding facilities have made continuous progress in breeding giant pandas naturally and by AI. Animals have reproduced successfully in captive facilities in Beijing,

Kunming, Shanghai, Hangzhou, Chengdu, Chongqing, Fuzhou, Xi'an and at the Wolong Breeding Centre. To increase pregnancy success, it is now common practice (especially at the Chengdu and Wolong centres) to combine natural mating and AI, with the latter focused on using semen from males that normally fail to breed. Breeding facilities outside China have also used AI. Giant panda cubs have been born after AI of oestrual females at the Zoo de la Casa de Campo (Madrid, Spain), Ueno Zoological Gardens (Ueno, Japan), the Zoological Society of San Diego (San Diego, USA) (Hodges *et al.*, 1984; Moore *et al.*, 1984; Masui *et al.*, 1989; Durrant *et al.*, 2003) and the Smithsonian's National Zoological Park (Washington, DC). A compilation of the details associated with six successful inseminations conducted at these institutions is listed in Table 20.1. At Madrid Zoo, inseminations were conducted on the day of peak urinary oestrogen excretion (AI number 1), the day of oestrogen decline (AI number 2) and one day thereafter (AI number 3). Using a combination of fresh and frozen–thawed sperm, two cubs were born after a 160-day gestation. At Ueno Zoo, a single cub was produced from the same female in 1985, 1986 and 1988 following only one insemination each year with fresh semen inseminated on the day of oestrogen decline or 2 days later (see Table 20.1). The durations of gestation in this individual ranged from 102 to 121 days. At the Zoological Society of San Diego, a cub was born following three consecutive inseminations beginning on the day of urinary oestrogen decline (134-day gestation). At the National Zoological Park, a single cub was born following only one AI with fresh sperm inseminated after oestrogen decline (121-day gestation).

Efficiency of artificial insemination at the Wolong Breeding Centre

The Wolong Breeding Centre in the Wolong Nature Reserve was established in 1982, and currently achieves a high pregnancy success rate using the combined practice of natural mating and AI. From 1998 to 2000, seven females (4.5 to 10.5 years of age) were anaesthetised for transcervical AI without mating (Huang *et al.*, 2001, 2002). These females were selected for only AI (usually on two or three consecutive days) due to a young age (4.5 years; SB 432 and 434) or because of weak behavioural signs of oestrus (SB 444, 418, 446, 382 and 385) (Table 20.2). One particularly interesting case was a young female, SB 432, who was inseminated with only cold-stored semen from a wild-born, nonbreeding male with skin cancer (squamous cell carcinoma; see Chapter 4) (Table 20.3). Despite the chronic progression of the cancer, excellent quality semen

Table 20.1. Successful artificial inseminations (AI) in giant pandas outside China in Spain, Japan and the USA

Female SB no.	Age (years)	AI no.	Date of AI	Male SB no.	Semen type	Date of oestrogen decline	Motile sperm[a] (×10^6)	Gestation duration[b]; birth date	No. of cubs (sex SB no.)
Spain – Zoo de la Casa de Campo, Madrid[c]									
169	6.5	AI 1	27 Mar' 82	141	Fresh	28 Mar' 82	200	160 days; 4 Sep' 82	Two cubs (m. SB 249; f. SB 250)
		AI 2	28 Mar' 82	141	Fresh		200		
		AI 3	29 Mar' 82	141	Frozen		200		
Japan – Ueno Zoological Gardens[d]									
162	11.5	AI 1	9 Mar' 85	183	Fresh	9 Mar' 85	100	110 days; 27 Jun' 85	One cub (m. SB 293)
Japan – Ueno Zoological Gardens[d,e]									
162	12.5	AI 1	31 Jan' 86	183	Fresh	31 Jan' 86	120	121 days; 1 Jun' 86	One cub (f. SB 313)
Japan – Ueno Zoological Gardens[e]									
162	14.5	AI 1	15 Mar' 88	183	Fresh	13 Mar' 88	146	102 days; 23 Jun' 88	One cub (m. SB 345)
USA – Zoological Society of San Diego									
371	7.5	AI 1	9 Apr' 99	381	Fresh	9 Apr' 99	715	134 days; 21 Aug' 99	One cub (f. SB 487)
		AI 2	10 Apr' 99	381	Cold (for 24 hr)		651		
		AI 3	11 Apr' 99	381	Cold (for 48 hr)		635		
USA – Smithsonian's National Zoological Park									
473	7.0	AI 1	11 Mar' 05	458	Fresh	10 Mar' 05	520	121 days; 9 Jul' 05	One cub (m. SB 595)

[a] Total number of motile spermatozoa deposited transcervically and into the uterus.
[b] Duration of gestation was calculated as the first day of urinary oestrogen decline = Day 0.
[c] Data on female SB 169 from Hodges et al. (1984) and Moore et al. (1984). Fresh semen was collected from SB 141 located at the Zoological Society of London and shipped to Madrid for AI.
[d] Data on female SB 162 from Masui et al. (1989).
[e] Data on female SB 162 from K. Tenabe, pers. comm.

Table 20.2. *Results of artificial inseminations (AI) using fresh, cold-stored (4°C for 24 or 48 hours) or frozen–thawed semen in giant pandas at the Wolong Breeding Centre*[a]

Female SB no.	Age (years)	AI no.	Date of AI	Male SB no.	Semen type	Diluent-glycerol	Volume (ml)	Sperm motility (%)	Sperm progression	Motile sperm (×10^6)	Pregnant (gestation) (birth date)	No. of cubs (sex SB no.)
444	9.5	AI 1	9 Mar' 98	298	Fresh	SFS 0%	1.0	80	3.0	412	Yes	One cub
		AI 2	10 Mar' 98	298	Cold (24 hr)	SFS 0%	1.8	80	3.0	742	(135 days) (22 Jul' 98)	(f. SB 473)
418	9.5	AI 1	5 Apr' 98	298	Frozen – pellet	SFS 4%	1.0	—[data not available]—			No	–
		AI 2	6 Apr' 98	298	Fresh	SFS 0%	2.0	90	3.0	729		
446	10.5	AI 1	6 Apr' 98	298	Fresh	SFS 0%	2.5	90	3.0	912	Yes	One cub
		AI 2	7 Apr' 98	298	Cold (24 hr)	SFS 0%	2.0	80	2.5	648	(157 days) (11 Sep' 98)	(f. SB 477)
382	6.5	AI 1	14 Apr' 98	298	Fresh	SFS 0%	4.0	75	3.0	1062	Yes	Two cubs
		AI 2	15 Apr' 98	298	Cold (24 hr)	SFS 0%	6.0	70	2.5	1487	(110 days) (3 Aug' 98)	(f. SB 474;
		AI 3	16 Apr' 98	298	Frozen-pellet	SFS 4%	1.0	—[data not available]—				m. SB 475)

432	4.5	AI 1	4 Apr' 00	305	Cold (24 hr)	TEST 0%	3.0	65	2.0	1300	Yes (128 days) (10 Aug' 00)	One cub (f. SB 512)
		AI 2	5 Apr' 00	305	Cold (48 hr)	TEST 0%	3.0	58	2.0	580		
434	4.5	AI 1	5 Apr' 00	305	Cold (48 hr)	TEST 0%	3.0	58	2.0	580	No	–
		AI 2	6 Apr' 00	308	Frozen – pellet	SFS 4%	1.5	65	2.0	109		
385	7.5	AI 1	22 Apr' 00	357	Frozen – pellet	SFS 4%	1.5	73	2.0	168	No	–
		AI 2	23 Apr' 00	357	Frozen – pellet	SFS 4%	1.5	72	2.0	166		

[a] All females (except SB 432 and 434) demonstrated weak behavioural signs of oestrus and would not breed naturally. Females SB 432 and 434 were inseminated and not allowed to breed because of their young age (4.5 years) and first oestrus. Inseminant traits are represented as sperm motility (%), sperm progression (0 to 5; 5 is best) and the number of motile spermatozoa deposited intrauterine. Duration of gestation was calculated as day of the first AI (AI 1) = Day 0.

Table 20.3. *Relationship of urinary oestrogen excretion, behavioural signs of oestrus, vulva traits, vaginal superficial cells and timing of successful artificial insemination (AI) in giant panda SB 432 (4.5 years old) using cold-stored (4°C) semen from a nonbreeding, wild-born male (SB 305) with skin cancer (squamous cell carcinoma)*

Date	Urinary oestrogen (ng/ml Cr)	Behavioural signs of oestrus	Vulva traits	Vaginal superficial cells	Time of AI
24 Mar' 00	–	Begin scent-marking, chirping, some bleating			
26 Mar' 00	–	Increased bleating, decreased appetite	Redness of vulva		
27 Mar' 00	21.4	–			
28 Mar' 00	22.2	Masturbation, water play			
30 Mar' 00	34.5	–			
31 Mar' 00	86.8	–			
1 Apr' 00	101.0	–			
2 Apr' 00	148.4 (peak)	–			
3 Apr' 00	55.9	Peak oestrus, bleating, tail up, lordosis, backwards walking, masturbation, interest in males	Redness of vulva		
4 Apr' 00	36.0	Peak oestrus, bleating, tail up, lordosis, backwards walking, masturbation, interest in males	Redness of vulva	78%	AI 1, 48-hour post-peak oestrogen, semen cold stored for 24 hours
5 Apr' 00	22.0	–	–	88%	AI 2, 72-hour post-peak oestrogen, semen cold stored for 48 hours
6 Apr' 00	–	Some bleating, masturbation	Colour change to white		
5 Jun' 00	Male SB 305 dies from cancer				
10 Aug' 00	One female cub (SB 512) born after a 128-day gestation				

Details of inseminant traits are presented in Table 20.2.

was obtained and diluted in TEST egg-yolk diluent with no glycerol (one part semen to two parts diluent), then stored at 4°C for 24 and 48 hours until used for AI. During the same period (1998 to 2000), the combined practice of natural mating and AI was conducted in 18 breeding trials involving ten females (5.5 to 19.5 years of age) at the Wolong Breeding Centre (Table 20.4).

Excellent semen quality with high ratings of sperm motility, morphology and intact acrosomes was observed for all males used in these trials. Mean ($\pm$ SEM) ejaculate traits in the four male sperm donors and six ejaculates used for the 'AI only' females were: ejaculate volume 3.3 ± 0.5 ml; sperm concentration $1429.8 \pm 235.4 \times 10^6$ cells per ml; sperm motility $81.7 \pm 2.1\%$; sperm forward progression (0–5; 5 is best) 3.1 ± 0.1; structurally normal sperm $79.3 \pm 9.2\%$; and normal acrosomes $98.0 \pm 1.0\%$.

For transcervical AI ($n = 15$) in these seven females, mean inseminate traits were: spermic volume inseminated 2.3 ± 0.4 ml; sperm motility $73.5 \pm 2.9\%$; sperm forward progression 2.5 ± 0.1; and total motile sperm inseminated per AI $684.2 \pm 118.2 \times 10^6$ cells.

Results revealed that transcervical AI using semen from non-breeding giant panda males was effective for producing pregnancies and live offspring (Fig. 20.4 Plate XXXIX). Cold storage of semen at 4°C maintained high sperm percentage motility, percentage live and forward progression for at least 48 hours *in vitro*. Four of seven (57.1%) giant pandas inseminated with fresh, cold-stored and/or frozen semen became pregnant and produced a total of five cubs after AI without natural mating (see Table 20.2). Mean gestation and litter size were 132.5 ± 9.7 days and 1.3 ± 0.3 cubs per litter, respectively. Three of the four pregnancies produced (in females SB 444, 446 and 382) resulted from the use of fresh semen on the first day of AI and cold-stored semen (4°C for 24 hours) from the same male on the second day of AI. And, in female SB 382, frozen–thawed semen was used on the third day. One cub produced from this strategy was female SB 473 (Mei Xiang), which was born in 1998 and currently resides at the Smithsonian's National Zoological Park in Washington, DC (see Fig. 20.4; Plate XXXIX). In turn, this female now has produced an offspring by AI (Table 20.1).

AI using only cold-stored semen from the nonbreeding male SB 305 with skin cancer was effective for producing a pregnancy in SB 432 (see Table 20.2 and 20.3). Because this male was wild born and had never reproduced, this was an especially important milestone in that AI allowed perpetuating valuable genes for the *ex situ* population. This

Table 20.4. Results of the combined use of natural breeding and artificial insemination (AI) in giant pandas at the Wolong Breeding Centre from 1998 to 2000 (18 AI trials; ten females)[a]

Female SB no.	Age (years)	Type of breeding	Date	Male SB no.	Pregnant (gestation)	Date of birth	No. of cubs (sex SB no.)	Sire (SB) of cub(s)[b]
397	14.5	Natural breeding	23 Mar' 98	308	Yes	13 Aug' 98	One cub	308
		AI – Frozen SFS 4%	24 Mar' 98	298	(143 days)		(f. SB 476)	
374	8.5	Natural breeding	5 Apr' 98	329	No	–	–	–
		Natural breeding	6 Apr' 98	308				
		AI – Fresh SFS 0%	7 Apr' 98	394				
230	19.5	AI – Fresh SFS 0%	18 Apr' 98	357	No	–	–	–
		Natural breeding	19 Apr' 98	308				
		Natural breeding	20 Apr' 98	308				
418	10.5	Natural breeding	2 Mar' 99	394	Yes	2 Aug' 99	Two cubs	394
		Natural breeding	3 Mar' 99	394	(153 days)		(f. SB 478;	
		AI – Fresh SFS 0%	4 Mar' 99	298			m. SB 479)	
374	9.5	Natural breeding	26 Mar' 99	308	No	–	–	–
		AI – Frozen SFS 4%	28 Mar' 99	298				
		AI – Frozen SFS 4%	29 Mar' 99	298				

397	15.5	Natural breeding	29 Mar' 99	308				
		Natural breeding	30 Mar' 99	329				
		AI – Frozen SFS 4%	31 Mar' 99	298	Yes (142 days)	18 Aug 99	Three cubs (m. SB 484, 485; f. SB 486)	308
414	8.5	Natural breeding	30 Mar' 99	394				
		AI – Frozen SFS 4%	1 Apr' 99	298	No	–	–	–
404	5.5	Natural breeding	7 Apr' 99	394				
		AI – Frozen SFS 4%	8 Apr' 99	298	No	–	–	–
446	11.5	Natural breeding	14 May 99	329; 308				Two sires
		Natural breeding	15 May 99	308				329
		AI – Frozen SFS 4%	16 May 99	298	Yes (134 days)	25 Sep 99	Two cubs (f. SB 495) (m. SB 496)	308
444	10.5	Natural breeding	5 Jun' 99	308				
		Natural breeding	6 Jun' 99	308				
		AI – Fresh SFS 0%	7 Jun' 99	329	Yes (92 days)	5 Sep 99	One cub (m. SB 492)	308
374	10.5	Natural breeding	20 Mar' 00	308				
		AI – Frozen SFS 4%	21 Mar' 00	308	Yes (151 days)	18 Aug 00	Two cubs (f. SB 516; stillbirth SB 517)	308
404	6.5	Natural breeding	9 Apr' 00	394				
		Natural breeding	10 Apr' 00	394				
		AI – Fresh SFS 0%	11 Apr' 00	357	Yes (134 days)	21 Aug' 00	One cub (m. SB 518)	394

Table 20.4. (cont.)

Female SB no.	Age (years)	Type of breeding	Date	Male SB no.	Pregnant (gestation)	Date of birth	No. of cubs (sex SB no.)	Sire (SB) of cub(s)[b]
414	9.5	Natural breeding	8 Apr' 00	394	Yes (122 days)	8 Aug' 00	One cub (f. SB 511)	394
		Natural breeding	9 Apr' 00	394				
		Natural breeding	10 Apr' 00	399; 394				
		AI – Fresh SFS 0%	11 Apr' 00	357				
382	8.5	Natural breeding	24 Mar' 00	394	Yes (139 days)	10 Aug' 00	Two cubs (m. SB 513; f. SB 514)	394
		Natural breeding	25 Mar' 00	394				
		AI – Frozen SFS 4%	26 Mar' 00	305				
446	12.5	Natural breeding	13 May 00	329	Yes (136 days)	26 Sep' 00	One cub (m. SB 526)	329
		AI – Fresh SFS 0%	14 May 00	357				
		AI – Cooled 24 hours	15 May 00	357				
397	16.5	Natural breeding	1 Apr' 00	394	Yes (168 days)	16 Sep' 00	Two cubs (m. SB 524, 525)	394
		Natural breeding	2 Apr 00	308; 329				
		AI – Fresh SFS 0%	3 Apr' 00	305				

						6 Aug' 00	Two cubs (f. SB 509; m. SB 510)	329
418	11.5	Natural breeding	19 Apr' 00	329	Yes (109 days)			
		Natural breeding	20 Apr' 00	329				
		Natural breeding	21 Apr' 00	394; 308				
		AI – Fresh SFS 0%	22 Apr' 00	357				
358	6.5	Natural breeding	18 Apr' 00	308	No	–	–	–
		Natural breeding	19 Apr' 00	308				
		AI – Fresh SFS 0%	20 Apr' 00	357				

[a] All females demonstrated behavioural signs of oestrus and bred naturally. In all females (except SB 230), females bred naturally for one, two or three days before AI with fresh, cold-stored (4°C) or frozen–thawed semen. In SB 230, AI was conducted prior to natural breeding. Duration of gestation was calculated as first day of natural breeding = Day 0. [b] Sire of cub(s) was confirmed by molecular genetic analyses (see Chapter 10). All cubs were sired by males that mated females on the first day of oestrus.

Figure 20.4. Female giant panda cub SB 473 born after two transcervical artificial inseminations with fresh (AI number 1) and cold-stored (AI number 2) spermatozoa. (See also Plate XXXIX.)

pregnancy also demonstrated the reproductive competency of a 4.5-year-old female, while also revealing important information on the timing of AI since inseminations were performed 48 and 72 hours after the peak in urinary oestrogens (see Table 20.3). After combined natural mating and AI, females in 12 of 18 (66.7%) trials became pregnant and produced a total of 20 cubs (see Table 20.4). Interestingly, all the cubs were sired by the male that bred on the first day (as confirmed by molecular analyses; see Chapter 10), even when multiple males bred on two consecutive days (e.g. SB 397 in 1999 and 2000) (see Table 20.4). Furthermore, one set of twins with different sires was produced by female SB 446 after breeding with two different males on the first day of oestrus (see Table 20.4).

Overall, these results indicated that AI without natural mating was as effective for propagating giant pandas as natural mating combined with AI. Data also revealed that cold storage of semen at 4°C was a viable method for maintaining high-quality semen for AI over consecutive days.

Efficiency of artificial insemination with cryopreserved sperm at the Chengdu Research Base

The availability of a GRB and the efficient use of AI with frozen–thawed spermatozoa would facilitate genetic management of the *ex situ* giant panda population. Basic scholarly knowledge in gamete cryobiology is essential for allowing the consistent use of AI with frozen-thawed semen. There have been years of scientific research in China and the USA with the most recent efforts occurring during the Giant Panda Biomedical Survey. For details, see Chapter 7, which summarises the significant amount of new information generated about post-thaw sperm function, metabolism, membrane integrity, capacitation and the ability to undergo the acrosome reaction and decondensation in the presence of oocyte cytoplasm.

However, there have also been efforts to test the feasibility of producing giant panda offspring using AI with thawed spermatozoa. The Chengdu Research Base of Giant Panda Breeding, established in 1987, has had a long interest in sperm cryobiology beginning with its milestone cub birth in 1980 using thawed semen. From 2000 to 2003, and exploiting new information learned from the Biomedical Survey on sperm freezing, seven females (4.5 to 17.5 years of age) were serially anaesthetised in 14 insemination trials for a total of 55 transcervical inseminations without mating (Table 20.5). In each case, these females were demonstrating weak behavioural signs of oestrus and would not breed naturally. In 2000, two females (SB 297 and 425) were inseminated with a combination of cold-stored (4°C) and frozen–thawed sperm on different days of oestrus. Only frozen–thawed sperm were used in the remaining 12 trials.

For AI in all 14 trials, mean inseminate traits were: spermic volume inseminated 1.2 ± 0.1 ml; sperm motility $59.6 \pm 1.2\%$; sperm forward progression 3.1 ± 0.1; normal sperm $62.9 \pm 3.2\%$; and total motile sperm inseminated per AI $73.7 \pm 10.1 \times 10^6$. Overall, seven of the 14 (50.0%) insemination trials using cold-stored and/or frozen sperm resulted in a pregnancy, one of which resulted in abortion (see Table 20.5). Of the pregnancies going to term, the mean gestation and litter size were 165.7 ± 12.3 days and 1.2 ± 0.2 cubs per litter, respectively. For females receiving only thawed spermatozoa, six of the 12 (50.0%) became pregnant and produced a total of six cubs (including the aborted fetus) (see Table 20.5; Fig. 20.5; Plate XXXX). Together, these

Table 20.5. Results of artificial inseminations (AI) using cold-stored (4°C for 24 or 48 hours) or frozen–thawed semen in nonbreeding giant pandas at the Chengdu Research Base of Giant Panda Breeding (14 AI trials; seven females)[a]

Female SB no.	Age (years)	AI no.	Date of AI	Male SB no.	Semen type	Diluent-glycerol	Volume (ml)	Sperm motility (%)	Sperm progression	Motile sperm (×10^6)	Pregnant (gestation) (birth date)	No. of cubs (sex SB no.)
407	5.5	AI 1	29 Mar' 00	287	Frozen–2 pellets	SFS 5%	1.0	40	–	–	No	–
		AI 2	30 Mar' 00	287	Frozen–2 pellets	SFS 5%	1.0	30	–	–		
		AI 3	31 Mar' 00	287	Frozen–2 pellets	SFS 5%	1.0	50	–	–		
		AI 4	1 Apr' 00	287	Frozen–6 pellets	SFS 5%	1.0	60	–	–		
		AI 5	2 Apr' 00	287	Frozen–2 pellets	SFS 5%	1.0	55	–	–		
		AI 6	6 Apr' 00	287	Frozen–2 pellets	SFS 5%	1.0	60	–	–		
297	14.5	AI 1	1 Apr' 00 am	386	Frozen–1 pellet	SFS 5%	1.0	50	3.0	31	Yes (163 days) (11 Sep' 00)	Two cubs (f. SB 522, 523)
		AI 2	1 Apr' 00 pm	386	Cold (for 24 hr)	Milk 0%	1.0	70	–	274		
		AI 3	2 Apr' 00	386	Cold (for 48 hr)	Milk 0%	0.5	50	–	98		
425	4.5	AI 1	6 Apr' 00	386	Frozen–1 straw	SFS 5%	1.0	60	3.0	44	No	–
		AI 2	7 Apr' 00	386	Cold (for 0 hour)	Milk 0%	1.5	70	–	218		
		AI 3	8 Apr' 00	386	Cold (for 24 hour)	Milk 0%	1.0	65	–	135		
		AI 4	9 Apr' 00	386	Cold (for 48 hour)	Milk 0%	1.5	55	–	410		
		AI 5	10 Apr' 00	386	Frozen–2 straws	SFS 5%	2.0	55	–	77		

ID	Age	AI	Date	No.	Sample	Conc.	Vol.	Motility	Count	Motility2	Pregnant	Gestation	Cub
403	6.5	AI 1	23 Apr' 00	386	Frozen–2 pellets	SFS 5%	1.0	45	–	28	No		–
		AI 2	24 Apr' 00	386	Frozen–1 straw	SFS 5%	1.0	60	3.5	44			
		AI 3	25 Apr' 00	386	Frozen–1 straw	SFS 5%	1.0	55	–	41			
		AI 4	3 May 00	386	Frozen–1 straw	SFS 5%	1.0	60	–	44			
		AI 5	4 May 00	386	Frozen–2 pellets	SFS 5%	1.0	40	–	27			
		AI 6	5 May 00	386	Frozen–2 pellets	SFS 5%	1.0	40	–	27			
408	5.5	AI 1	21 May 00	287	Frozen–1 straw	SFS 5%	1.0	60	3.0	34	Yes	(108 days)	One cub (f., SB 521)
		AI 2	22 May 00	287	Frozen–1 straw	SFS 5%	1.0	60	–	34		(6 Sep' 00)	
		AI 3	23 May 00	287	Frozen–1 straw	SFS 5%	1.0	60	–	34			
		AI 4	24 May 00 am	287	Frozen–1 straw	SFS 5%	1.0	60	–	34			
		AI 5	24 May 00 pm	287	Frozen–1 straw	SFS 5%	1.0	60	–	34			
		AI 6	25 May 00	287	Frozen–2 straws	SFS 5%	2.0	60	–	68			
425	5.5	AI 1	19 Mar' 01	287	Frozen–3 straws	SFS 5%	2.0	65	3.0	129	Yes	(187 days)	One cub (m. SB 536)
		AI 2	20 Mar' 01	287	Frozen–2 straws	SFS 5%	2.0	65	3.0	86		(22 Sep' 01)	
		AI 3	21 Mar' 01 am	287	Frozen–2 straws	SFS 5%	2.0	60	3.0	79			
		AI 4	21 Mar' 01 pm	287	Frozen–2 straws	SFS 5%	2.0	60	3.0	68			
407	6.5	AI 1	31 Mar' 01 am	287	Frozen–1 straw	SFS 5%	1.0	60	3.0	34	No		–
		AI 2	31 Mar' 01 pm	287	Frozen–2 straws	SFS 5%	2.0	65	3.0	68			
		AI 3	1 Apr' 01	287	Frozen–1 straw	SFS 5%	1.0	65	3.0	34			
		AI 4	2 Apr' 01	287	Frozen–2 straws	SFS 5%	1.2	65	3.5	86			

Table 20.5. (cont.)

Female SB no.	Age (years)	AI no.	Date of AI	Male SB no.	Semen type	Diluent-glycerol	Volume (ml)	Sperm motility (%)	Sperm progression	Motile sperm (×10^6)	Pregnant (gestation) (birth date)	No. of cubs (sex SB no.)
407	7.5	AI 1	20 Mar' 02	390	Frozen–1 straw	SFS 5%	1.0	60	3.5	57	Yes (155 days) (22 Aug' 02)	One cub (m. SB 553)
		AI 2	21 Mar' 02	390	Frozen–1 straw	SFS 5%	1.0	65	3.5	62		
		AI 3	22 Mar' 02	390	Frozen–1 straw	SFS 5%	1.0	70	3.5	67		
		AI 4	23 Mar' 02	342	Frozen–1 straw	SFS 5%	1.0	60	3.5	32		
401	8.5	AI 1	28 Mar' 02	390	Frozen–1 straw	SFS 5%	1.0	65	3.5	62	No	–
		AI 2	29 Mar' 02	386	Frozen–1 straw	SFS 5%	1.0	55	3.0	47		
453	4.5	AI 1	13 Apr' 02	390	Frozen–1 straw	SFS 5%	1.0	55	3.0	20	No	–
		AI 2	14 Apr' 02	390	Frozen–1 straw	SFS 5%	1.0	60	3.0	22		
		AI 3	15 Apr' 02	386	Frozen–1 straw	SFS 5%	1.0	55	3.0	47		
425	7.5	AI 1	9 Mar' 03	386	Frozen–2 straws	SFS 5%	2.0	50	3.0	59	Yes (198 days) (23 Sep' 03)	One cub (m. SB 577)
		AI 2	10 Mar' 03	386	Frozen–1 straw	SFS 5%	1.0	60	3.0	50		
		AI 3	11 Mar' 03	386	Frozen–1 straw	SFS 5%	1.0	60	3.0	50		

297	17.5	AI 1	17 Mar' 03	377	Frozen–2 straws	SFS 5%	2.0	70	3.5	72	Yes		One cub
		AI 2	19 Mar' 03	377	Frozen–2 straws	SFS 5%	2.0	80	3.5	83	(183 days)		(f. SB 576)
		AI 3	20 Mar' 03	377	Frozen–2 straws	SFS 5%	1.0	80	3.0	83	(16 Sep' 03)	No	–
407	8.5	AI 1	18 Mar' 03	342	Frozen–1 straw	SFS 5%	1.0	70	3.0	36			
		AI 2	19 Mar' 03	342	Frozen–1 straw	SFS 5%	1.0	60	3.0	30			
		AI 3	20 Mar' 03	342	Frozen–1 straw	SFS 5%	1.0	60	3.0	30			
453	5.5	AI 1	8 Apr' 03	377	Frozen–2 straws	SFS 5%	2.0	70	3.0	72	Yes		Aborted
		AI 2	9 Apr' 03	377	Frozen–2 straws	SFS 5%	2.0	70	3.0	72			
		AI 3	10 Apr' 03	377	Frozen–2 straws	SFS 5%	1.0	70	3.0	72			

[a] All females demonstrated weak behavioural signs of oestrus and would not breed naturally. Duration of gestation was calculated from the day of the first AI (= Day 0).

Figure 20.5. Female giant panda cub SB 576 born after three transcervical artificial inseminations with only frozen–thawed spermatozoa. (See also Plate XXXX.)

results demonstrated the reproductive competency of frozen–thawed spermatozoa deposited transcervically in giant pandas experiencing 'weak' oestrus.

PRIORITIES FOR THE FUTURE

This is the first examination of the efficacy of AI in the context of a significant number of giant pandas. The data are compelling in that they illustrate a feasible strategy for producing living, healthy offspring when natural breeding fails. We have determined that AI is particularly valuable:

1.   for female pandas that are not in strong oestrus;
2.   when the pair is so aggressive that injury is likely;
3.   when an animal is young and behaviourally inexperienced; or
4.   for a genetically valuable sperm donor that is critically ill.

However, our evaluation also provides encouragement for the utility of AI with cooled or frozen semen to help meet one of the highest priorities of *ex situ* programmes, achieving a self-sustaining population through genetic management (see Chapter 21). Since maintaining adequate gene diversity requires either moving animals between breeding facilities or transporting germ plasm, there are some significant advantages for the latter approach. Certainly, it is less stressful on the giant panda (and probably the keeper) as well as less costly to transport sperm than the living individual. However, the movement of germ plasm is also attractive for bureaucratic reasons, especially since giant pandas are the property of competing federal agencies in China. Thus, shipping plastic straws containing sperm causes fewer concerns than moving living, successful breeders.

In the context of assisted breeding, we suggest that there are two high priorities for future study.

Continue testing and improving artificial insemination efficiency with fresh and thawed spermatozoa

In Chapter 7, we reviewed those biological areas in basic gamete function that require more investigation, including in sperm cryobiology. However, more effort also needs to be directed at the practical application of the technology, especially identifying the optimal day for AI. Whether there is a single 'ideal' day (or time) or if multiple days are required is critical information for minimising the number of anaesthesia episodes and avoiding wastage of precious spermatozoa. For example, our initial analyses here seemed to suggest that pregnancy success was higher if sperm were deposited on the day coincident with declining excretory oestrogen concentrations. However, solid comparative data are required, including a more systematic and detailed correlation of ideal AI time with behaviours and vaginal cytology. Identifying the best day for AI would be possible by inseminating a female on each day of oestrus with sperm from a different male and then determining parentage using already established molecular techniques (see Chapter 10). Likewise, to conserve germ plasm, we need to know the minimum number of spermatozoa necessary to achieve successful conception. Although the giant panda produces mega-numbers of spermatozoa, understanding how many are needed in a typical AI will allow more efficient use (or long-term storage) of germ cells.

Developing genome banking strategies and resources to facilitate genetic management

Because we now know that cryopreserved sperm and AI can result in a high incidence of pregnancy success (of about 50%) in the giant panda, it is time to use transported semen more seriously to meet genetic management goals. However, the prerequisite step toward the wide scale use of frozen semen is the establishment of a formal GRB. The advantages of such banks have been described in detail (Wildt, 1997; Wildt *et al.*, 1997). In addition to providing germ plasm to avoid the stress and cost of transporting living animals, a GRB:

1. helps ensure that more animals are successful breeders (through AI);
2. allows sperm to be used over generations rather than the lifetime of the male;
3. saves space in that fewer males need to be maintained at a breeding site;
4. preserves existing gene diversity from unforeseen catastrophes, e.g. disease epidemic or fire.

However, any effective GRB is only as good as a written protocol that is adhered to by all the institutions interested in sharing germ plasm. In previous literature, such practices have been referred to as a GRB Action Plan (Wildt, 1997), a document that details the justification for such an effort as well as information on:

1. genetic and population management goals;
2. accessibility of existing animals for banking;
3. amounts of spermatozoa to cryopreserve from targeted males;
4. technical aspects, including standard methods for sperm collection, evaluation, frozen storage, thawing and AI;
5. ownership of resulting offspring.

Within each of these is a host of issues that range from detailed record-keeping to ensuring security of the repository to the source of funds for maintaining the bank and using the stored materials. Many required practices are simply common sense, such as making sure that frozen samples are stored in at least two sites (two liquid nitrogen tanks in separate buildings or even different institutional facilities). In general, it has also been agreed that, to be effective and worthwhile, all holders of giant pandas must benefit. Therefore, a top priority is ensuring that sperm be shared on the basis of collective decision-making by managers who are adhering to genetic and demographic targets. Ballou *et al.*, in Chapter 21, describe a scenario being explored by the Chinese that would also involve a 'global management' plan – one where frozen semen would be shipped between giant panda facilities throughout the world. Thus, a high priority is to begin to implement this plan on a practical scale, using stored giant panda sperm (when necessary) to maintain gene diversity and to continue to help grow the population.

Finally, we also see a future where giant pandas in nature, all of which are producing surplus germ plasm, become an important component of a GRB. The opportunistic banking of wild, occasionally captured males would provide yet more insurance for saving those truly wild genes while creating a resource for the occasional infusion of genetic variation into the *ex situ* population.

ACKNOWLEDGEMENTS

We thank staff of the China Conservation and Research Centre for the Giant Panda and the Chengdu Research Base of Giant Panda Breeding for sharing and assisting in summarising data.

REFERENCES

Bonney, R. C., Wood, D. J. and Kleiman, D. G. (1982). Endocrine correlates of behavioural oestrus in the female giant panda (*Ailuropoda melanoleuca*) and associated hormonal changes in the male. *Journal of Reproduction and Fertility*, **64**, 209–15.

Durrant, B., Czekala, N., Olson, M. *et al.* (2002). Papanicolaou staining of exfoliated vaginal epithelial cells facilitates the prediction of ovulation in the giant panda. *Theriogenology*, **57**, 1855–64.

Durrant, B. S., Olson, M. A., Amodeo, D. *et al.* (2003). Vaginal cytology and vulvar swelling as indicators of impending estrus and ovulation in the giant panda (*Ailuropoda melanoleuca*). *Zoo Biology*, **22**, 313–21.

Hodges, J. K., Bevan, D. J., Celma, M. *et al.* (1984). Aspects of reproductive endocrinology of the female giant panda (*Ailuropoda melanoleuca*) in captivity with special reference to the detection of ovulation and pregnancy. *Journal of Zoology (London)*, **203**, 253–67.

Howard, J. G. (1999). Assisted reproductive techniques in nondomestic carnivores. In *Zoo and Wild Animal Medicine. Current Therapy IV*, ed. M. E. Fowler and R. E. Miller. Philadelphia, PA: W. B. Saunders Co., pp. 449–57.

Hu, J. and Wei, F. (1990). Development and progress of breeding and rearing giant pandas in captivity within China. In *Research and Progress in Biology of the Giant Panda*, ed. J. Hu. Chengdu: Sichuan Publishing House of Science and Technology, pp. 322–5.

Hu, J., Liu, T. and He, G. (1990). *Giant Pandas with Graceful Bearing.* Chengdu: Sichuan Publishing House of Science and Technology.

Huang, Y., Li, D. S., Du, J. *et al.* (2000a). Cryopreservation experiment on sperm of the giant panda. *Chinese Journal of Veterinary Medicine*, **26**, 13–14.

Huang, Y., Li, D. S., Zhang, H. M. *et al.* (2000b). Electroejaculation and semen cryopreservation in the giant panda. *Journal of Sichuan Teachers College (Natural Science)*, **21**, 238–43.

Huang, Y., Wang, P. Y., Zhang, H. M. *et al.* (2001). Efficiency of artificial insemination in giant pandas at the Wolong Breeding Center. *Journal of Andrology Supplement*, abstract 118.

Huang, Y., Wang, P., Zhang, G. *et al.* (2002). Use of artificial insemination to enhance propagation of giant pandas at the Wolong Breeding Center. In *Proceeding of the Second International Symposium on Assisted Reproductive Technology for the Conservation and Genetic Management of Wildlife*, ed. N. Loskutoff. Omaha, NE: Omaha's Henry Doorly Zoo, pp. 172–9.

Kan, O. and Shu-hua, T. (1964). In the Peking Zoo – the first baby giant panda. *Animal Kingdom*, **57**, 44–6.

Kleiman, D. G., Karesh, W. B. and Chu, P. R. (1979). Behavioural changes associated with oestrus in the giant panda (*Ailuropoda melanoleuca*) with comments on female proceptive behaviour. *International Zoo Yearbook*, **19**, 217–23.

Liu, W. (1979). The test of giant panda artificial insemination. *Chinese Science Report*, **9**, 415.

Liu, W. (1981). A note on the artificial insemination of the giant panda. *Acta Veterinaria et Zootechnica Sinica*, **12**, 73–6.

Liu, W. X., Ye, J. Q., Li, C. Z. and Liao, G. X. (1979). Artificial insemination experiment on the giant panda. *The Chinese Zoo Annual*, **2**, 20–4.

Masui, M., Hiramatsu, H., Nose, N. *et al.* (1989). Successful artificial insemination in the giant panda (*Ailuropoda melanoleuca*) at Ueno Zoo. *Zoo Biology*, **8**, 17–26.

Moore, H. D. M., Bush, M., Celma, M. *et al.* (1984). Artificial insemination in the giant panda (*Ailuropoda melanoleuca*). *Journal of Zoology (London)*, **203**, 269–78.

Olson, M. A., Huang, Y., Li, D. *et al.* (2003a). Assessment of motility, acrosomal integrity and viability of giant panda (*Ailuropoda melanoleuca*) sperm following short-term storage at 4°C. *Zoo Biology*, **22**, 529–44.

Olson, M. A., Huang, Y., Li, D., Zhang, H. and Durrant, B. (2003b). Comparison of storage techniques for giant panda sperm. *Zoo Biology*, **22**, 335–45.

Pukazhenthi, B. S. and Wildt, D. E. (2004). Which reproductive technologies are most relevant to studying, managing and conserving wildlife? *Reproduction, Fertility and Development*, **16**, 33–46.

Wildt, D. E. (1997). Genome resource banking: impact on biotic conservation and society. In *Reproductive Tissue Banking: Scientific Principles*, ed. A. M. Karow and J. K. Critser. San Diego, CA: Academic Press, pp. 399–439.

Wildt, D. E., Rall, W. F., Critser, J. K., Monfort, S. L. and Seal, U. S. (1997). Genome resource banks: 'living collections' for biodiversity conservation. *BioScience*, **47**, 689–98.

Ye, Z. Y., He, G. X., Zhang, A. J. *et al.* (1991). Studies on the artificial pollination method of the giant panda. *Journal of Sichuan University (Natural Science)*, **28**, 50–3.

Zeng, G. Q., Meng, Z. B., Jiang, G. T., He, G. X. and Xu, Q. M. (1984). The relationship between estrogen concentration and estrous behaviors during the estrous cycle of the giant panda. *Acta Veterinaria et Zootechnica Sinica*, **30**, 324–30.

Zhang, A. J., Ye, Z. Y., He, G. X. *et al.* (1991). Studies on the conception effect of frozen semen in the giant panda. *Journal of Sichuan University (Natural Science)*, **28**, 54–9.

21

Analysis of demographic and genetic trends for developing a captive breeding masterplan for the giant panda

JONATHAN D. BALLOU, PHILIP S. MILLER, ZHONG XIE, RONGPING WEI,
HEMIN ZHANG, ANJU ZHANG, SHIQIANG HUANG, SHAN SUN,
VICTOR A. DAVID, STEPHEN J. O'BRIEN, KATHY TRAYLOR-HOLZER,
ULYSSES S. SEAL, DAVID E. WILDT

INTRODUCTION

The foundation of any managed breeding programme for animals living in captivity is a studbook. This is the chronological listing of animals in the historical captive population detailing birth and death dates, gender, parentage, locations, transfers and local identification numbers (Glatston, 1986). Analyses of these data provide critical information on past trends in population size, age-specific reproductive and survival rates, age structure, numbers of founders, degree of inbreeding, loss of genetic diversity and other measures useful for evaluating temporal changes in a captive population. This information then becomes the basis for making management recommendations to enhance the demographic and genetic security of the captive population (Ballou & Foose, 1996). Demographic security is needed to ensure that an adequate number of breeding-aged animals are available to reproduce at the rates needed to grow or maintain the population at its desired size. Genetic diversity is required for the population to remain healthy and to adapt to changing environments (i.e. experience natural selection).

Giant Pandas: Biology, Veterinary Medicine and Management, ed. David E. Wildt, Anju Zhang, Hemin Zhang, Donald L. Janssen and Susie Ellis. Published by Cambridge University Press. © Cambridge University Press 2006.

The 2001 International Studbook for the Giant Panda contains detailed life history information on 542 giant pandas that have lived in zoos around the world (Xie & Gipps, 2001). The first entry, giant panda Studbook (SB) Number 1, is Su Lin, a wild-caught female who arrived at Brookfield Zoo on 2 February 1937 (see Chapter 1). A quick scan of the studbook leaves one with the impression that the captive population's dynamics are dominated by entry and subsequent death of wild-caught animals without sustainable reproduction. However, this situation has changed. Since 1990, the population has experienced increased reproduction with decreased mortality, the result being a substantial expansion in population size. The outlook is cautiously optimistic.

Partially in response to the expanding population and partly due to recent molecular analyses to clarify paternity (see Chapter 10), the first genetic management workshop for giant pandas was held in Chengdu, China, in January 2002. The aims of this meeting were to:

1.   discuss broad objectives for the *ex situ* breeding programme, including setting population size objectives;
2.   brainstorm an organisational structure for a cooperative breeding programme;
3.   begin making breeding recommendations for the worldwide population based on demographic and genetic needs.

While these concepts were new for China, models could be shared from experiences of the largely successful Species Survival Plan (SSP) of the American Zoo and Aquarium Association or AZA (Foose, 1989; Hutchins & Wiese, 1991). The SSP concept was implemented in 1981 as a cooperative captive population management programme for selected taxa in USA zoos and aquaria. SSPs were initially developed to maintain healthy, self-sustaining populations that were genetically diverse and demographically stable. However, over time they have evolved to become more comprehensive and conservation-oriented, encompassing diverse research, public education, fund raising, field project and reintroduction activities.

In this chapter we present results of analyses used to develop the first ever set of global *ex situ* breeding recommendations for the giant panda. The basis of our technical calculations was the studbook, which provided the demographic and genetic data for developing projections and determining relationships of potential mates. We examine changing demographic trends over the past ten years while evaluating the population's genetic status using pedigree analysis. Our primary objective is to evaluate the population's self-sufficiency: does it have the demographic momentum and genetic foundation to create a long-term, self-sustaining *ex situ* population that can serve as a healthy and viable back-up for giant pandas living in nature? Population viability analyses are used to help answer this question.

GENERAL METHODS

All analyses presented here were based on *The 2001 International Studbook for the Giant Panda* (Xie & Gipps, 2001), which is maintained using the software SPARKS (Single Population Analysis and Record Keeping System) (ISIS, 1994). A major change to the studbook over the earlier 1999 edition has been the resolution of many of the uncertain paternities. To optimise pregnancy success, female pandas are often naturally mated plus artificially inseminated with sperm from one or more males, a practice leading to questionable paternity of resulting cubs. Molecular genetic analyses by David *et al.* (see Chapter 10) have resolved many of these uncertainties, making the 2001 edition of the studbook (as well as more recent editions) much more valuable for assessing population viability.

Despite new data, some information remains as estimates, for example, the age at which wild-caught pandas enter the captive population. Often, ages are estimated on the basis of dental wear and overall physical condition. For our analysis, we assumed that wild-caught pandas without estimated birth dates entered the population as young adults unless noted otherwise.

Details of the specific methods used for the demographic and genetic analyses are provided in the appropriate sections below. PM2000 Version 1.17 software (Pollak *et al.*, 2002) was used for many of these analyses whereas VORTEX Version 8.32 software (Miller & Lacy, 1999) was used to evaluate population viability. SAS version 8.03 software (Statistical Analysis System, SAS, 2001) was used for all statistical testing.

RESULTS

Changing demographics in the *ex situ* giant panda population

Demography refers to the characteristics of a population's age structure, growth rate and vital statistics, such as reproductive and survival rates. Perhaps the most telling feature of the changing demography of the global *ex situ* giant panda population is its growth, specifically an increase in total number of individuals living in zoos and breeding centres (Fig. 21.1). However, how much of this 6% annual growth rate is due to adding pandas from nature (i.e. catching more pandas for zoos) versus improved ability to produce captive-born young? Until 1989, virtually all growth was due to importing wild-caught (i.e. born in the wild or wild-born) individuals. In the late 1980s, however, the trend began to change, with a significant increase in captive-born offspring. Since 1990, the latter component of the population has expanded about 12% per year while there has been a 5% annual decline in the number of wild-born or -caught pandas (see Fig. 21.1). Thus, the population is in transition from a 'sink' (one whose survival relies on wild-caught animals) to a 'source' (one that produces captive-born animals without the need for capturing animals from nature).

An increasing proportion of captive-born animals is indicative of improved overall reproduction and/or survival. We next examined how reproductive and survival success has changed over time. We first focused on fecundity, defined as the number of births in a year divided by the number of reproductive-age giant pandas. This was calculated separately for males (six to 29 years of age) and females (five to 20 years) and then averaged. Results showed that fecundity had increased exponentially over the last ten years at about 10% per year ($p = 0.004$;

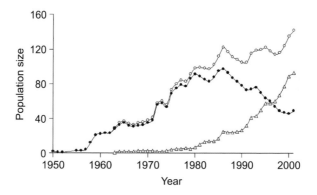

Figure 21.1. Growth in the captive population since 1950 showing captive-born versus wild-caught components. ○, total; ●, wild-caught; △, captive-born.

$r^2 = 0.59$, using a weighted exponential regression; Fig. 21.2). While fecundity was only about 10% in 1990, it had risen to 40% by 2001.

This increased fecundity was not due to increased litter size. Since 1995, average litter size has remained at 1.52 with 50% of births producing litters of one cub, 48% producing two and 2% producing three cubs. Rather, the fecundity surge has been due to a 6% annual increase in the proportion of animals breeding ($r^2 = 0.49$; $p = 0.01$, using a weighted exponential regression), a rise that has been almost identical for males and females. Based on these trends, 23% of the females can be predicted to breed on an average basis (with a standard deviation of $\pm$ 6%). Thus, there is good news – an impressive rise in fecundity – but also not-so-good news in that still only one in four adult giant pandas reproduces. This dependence upon so few individuals to do all the breeding has significant implications on population genetic viability (see below). In particular, the number of contributing captive-born males is exceptionally low, mostly because of sexual incompatibilities with females (see Chapter 5).

It is important to distinguish between reproductive success in wild-caught versus captive-born giant pandas. Historically, most of the reproduction has occurred in the former (Fig. 21.3). With gradually fewer wild-caught pandas in the population, reproduction will probably decrease unless captive-born pandas can increase their reproductive contributions. The proportion of births from the latter group is increasing.

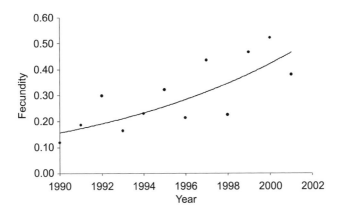

Figure 21.2. Increase in fecundity (number of births per number of breeding-age males and females, averaged) over time. The trend indicates a 9.9% per year increase in reproduction since 1990.

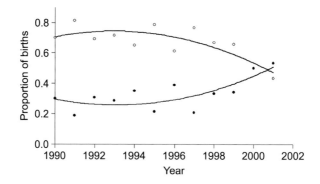

Figure 21.3. Proportion of births each year being produced by captive-born or wild-caught parents. ○, wild-caught parents; ●, captive-born parents.

In fact, in 2001 for the first time more offspring were produced from captive-born parents (one or both captive born) than wild-born counterparts (one or both parents wild-born; see Fig. 21.3). However, this finding does not consider the different number of breeding-age individuals in the two groups. On a 'per-panda basis', average fecundity (number of births to parents in a group per number of breeding-age animals in a group, averaged over both sexes) does not differ between wild-born and captive-born parents (paired Student's t-test; d.f $= 10$; $t = 1.796$; $p = 0.20$; Fig. 21.4). Although fecundity has increased in both groups since 1990, the improvement has only been statistically significant in the wild-born giant pandas (see Fig. 21.4). Overall, it seems that captive-born giant pandas are capable of reproducing at the rate of their wild-born counterparts, although substantial year-to-year variation remains.

Figure 21.5 illustrates the age-specific fecundity patterns from all reproductive events since 1990. Fecundity rates are calculated separately for males and females. Each birth attributes 0.5 of that birth to each parent so that, for example, female m_x = number of births to females at age x multiplied by 0.5 divided by the number of females of age x. This functionally assumes an equal sex ratio at birth and attributes all reproductive events to both the sire and dam. The points have been smoothed with the Brass function (Gage, 2001). Since data on opportunity to breed are not recorded in the studbook, these distributions represent observed fecundity patterns given historical management practices and do not portray maximum potential fecundity. Additionally, wild-born fecundity data are approximate because the

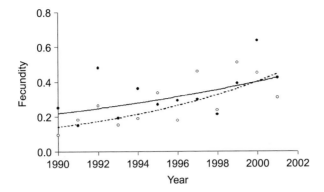

Figure 21.4. Fecundity of captive-born (●,——) and wild-caught (○,-----) giant pandas. Exponential regression is significant for wild-caught ($p = 0.005$; $r^2 = 0.56$), but not for captive-born ($p = 0.065$; $r^2 = 0.30$) individuals.

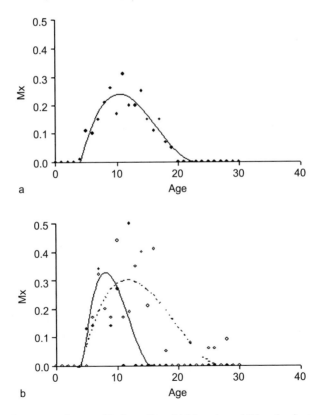

Figure 21.5. Age-specific fecundity of (a) female and (b) male giant pandas (solid line, captive males; dashed line, wild-caught males). Distributions have been smoothed with the Brass function (Gage, 2001).

ages of most of these animals were estimated upon their entry into the *ex situ* population.

Because they were almost identical, wild-caught and captive-born female fecundities were combined (see Fig. 21.5a). Results demonstrated that females begin breeding at five years of age and peak at ten to 11 years; no female over the age of 19 has given birth. The earliest age of reproduction has been six years for captive-born and five years for wild-born males (see Fig. 21.5b). Wild-born male fecundity extends to older age classes than captive-born counterparts because the latter cohort is still relatively young (not having been in captivity long enough to breed at older ages). Wild-born males have successfully mated and produced offspring at 28 years.

Survival

Survivorships (l_x), or the proportion of animals surviving from birth to age class x, were calculated using all animals existing in the population from 1 January 1990 to 1 February 2002 (the study window). Animals alive in the population in January 1990 as well as wild-caught animals entering the population during the window were treated as left-censored animals on the date and age they entered the study window. Escapes and animals alive at the end of the window were right-censored at the date and age they left the population (Hosmer & Lemeshow, 1999). Ages of animals at events (entry, exit and death) were recorded in days, and survivorship was calculated using Kaplan–Meier product limit estimators with the SAS Proc PHREG (SAS, 2001). Analyses (including wild-caught animals) were based on estimated dates of birth (as described above). Giant pandas of unknown sex were alternately included as males and females; all mortality of unknown-sex animals occurred before one year of age.

To test for changes in survival rates over time in different animal groupings, the study window was further divided into a pre-1995 window (ending on 31 December 1995) and a post-1995 window (starting 1 January 1996). Survival rates were calculated for males and females and for captive-born and wild-caught groups for both pre- and post-1995 windows. Statistical comparisons used proportional-hazards likelihood ratio tests (Proc PHREG; SAS, 2001).

Overall, there were no differences in survival success between males and females ($p = 0.25$), the pre- and post-1995 time periods ($p = 0.27$) or captive-born and wild-caught animals (comparing adult

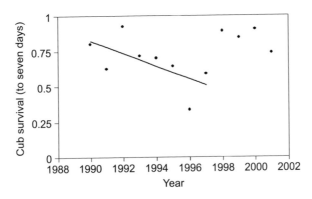

Figure 21.6. Cub survival (to seven days) per year. The declining annual trend was reversed in 1998.

age only; $p = 0.32$). In contrast, a detailed analysis of early survival rates over time found substantial improvements in cub survival to seven days of age (Fig. 21.6). From 1990 through 1997, cub survival decreased sharply, averaging less than 50% by 1997. Since 1998, however, survival has ranged from 73 to 90%, an improvement attributable to several factors. For instance, in 1998 and 1999 there was a new focus on the rearing of newborn cubs. When twins were produced, one cub was placed with the mother and the other in the nursery. The cubs were then switched frequently so that both could benefit from mother-rearing (see Chapter 13). This differed from previous protocols in which one cub remained with the mother while the other was relegated strictly to nursery-rearing. A new milk formula was also developed at this time and used to supplement twins or orphans (see Chapter 13).

Because there were no significant differences in survival rates over time for adults, overall pre- and post-1995 rates from captive-born and wild-caught animals were combined to allow comparing survivorship curves for male versus female giant pandas (Fig. 21.7). Although there was no gender difference ($p > 0.05$), the female survival curves tended to be higher than males because of a lower mortality for females that were one to two years of age.

Life table projections

To determine population viability and to help develop objectives for optimal population size, it is necessary to predict likely population

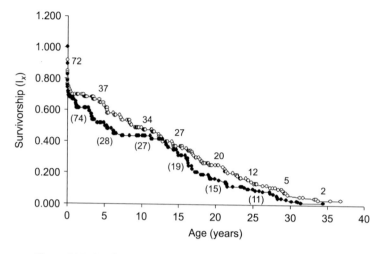

Figure 21.7. Survivorship for male and female giant pandas based on data since 1990. The number of animals at risk every five years is shown above the line for females and below for males. ●, male; ○, female.

growth rates. With the age-specific survival (l_x) and fecundity rates (m_x) estimated above, this now becomes possible (Caughley, 1977). The annual growth rate (λ) can be determined by using the Euler equation:

$$\Sigma\lambda^{-x}l_x m_x = 1.0$$

where the sum is over all age classes (Caughley, 1977). The λ cannot be determined directly and is usually calculated using iterative computer software (e.g. PM2000; Pollak *et al.*, 2002).

For the giant panda, the λ estimated from the studbook life table is 1.028. This value means that, if a stable population of giant pandas was to consistently reproduce and survive at the rates indicated in Figures 21.5 (smoothed fecundity) and 21.7 (survivorship), it would grow at 2.8% per year. This rate is lower than that actually observed over the last ten years (6% per year) because it only includes growth due to reproduction and survival. In reality, the actual population has also experienced growth due to immigration of wild-caught animals. The 1.028 annual growth rate, if applied to this population, would result in a captive population of 206 pandas by 2016 with no further influx of wild-caught animals.

Population genetics

Captive breeding programmes strive to retain genetic diversity and avoid inbreeding to promote population viability in captivity while maximising the potential for animals to adapt to nature if ever re-introduced (Ballou & Foose, 1996). Achieving this goal also minimises genetic deterioration (i.e. inbreeding depression, loss of genetic variation and adaptations to the captive environment) that can potentially occur in *ex situ* populations (Frankham *et al.*, 1986). Genetic management of captive populations is usually based on pedigree analysis (Lacy *et al.*, 1995). The pedigree provides information on the degree of inbreeding, the amount of genetic diversity retained relative to the founders, the level of relatedness among the living population and genetic substructuring (i.e. differences among various regions, institutions and the like). However, information on parentage may frequently be missing due to poor record-keeping, husbandry practices that make it difficult to identify parentage (e.g. species kept in mixed-sex groups) or multiple inseminations by several males. Any or all of these factors complicate pedigree-based management strategies.

Pedigree and genetic analyses are critical for the captive giant panda population because of the extensive use of artificial insemination (AI), especially using semen from multiple males to inseminate a single female (see Chapters 10 and 20). Of the 261 captive-born animals listed in the 2001 studbook (Xie & Gipps, 2001). 103 (39.5%) are listed with multiple potential sires. Recently, molecular genetics has been used to resolve the majority of these pedigree uncertainties (see Chapter 10). Yet, 17.5% of the current captive gene pool still remains unresolved. Although 40 individuals have some level of uncertainty in their pedigree, all can be resolved by determining the paternity of 17 individuals, many of which are individuals born in 2001 whose biosamples are awaiting genetic evaluation.

For the purposes of the present analysis, we made paternity assumptions for these 17 individuals based on other available information. For example, in some pairings, AI was used one or more times after a natural mating. In cases where paternity was later verified using molecular genetics, the male participating in the natural matings was always confirmed as the true sire (in contrast to the AI donor) (see Chapter 10). Whenever possible, it is important to use modern genetic technologies to resolve issues of uncertain paternity. Meanwhile, most pedigree analyses will continue to rely on some level of assumed parentage in the population.

For our analysis, two giant panda groups were excluded: poor reproductive candidates, which were senescent or which exhibited severe behavioural or medical problems; and animals at institutions that were unavailable for coordinated captive breeding. These 18 animals were identified at the January 2002 Population Management Workshop in Chengdu (SB 133, 161, 199, 203, 204, 214, 217, 230, 264, 288, 290, 300, 307, 358, 365, 416, 447 and 499). With these pandas excluded, 124 individuals remained in the genetically analysed population.

The modern *ex situ* giant panda population stands out from many other captive wildlife populations because of the significant number of wild-caught individuals that have entered, and continue to enter, the programme. Although giant pandas are no longer prospectively 'harvested' for zoo collections, animals are occasionally 'rescued' often as a result of nutrition or health problems or panda–human conflicts (see Chapter 2). These individuals then become opportunistically in-corporated into captive collections. Of the 542 animals listed in the current studbook, 281 (51.8%) have been wild-caught, whereas of the current living population of 142 individuals, 49 (34.5%) are wild-caught. The result is that the population has a strong genetic foundation given that these wild-caught individuals do eventually contribute their genes to subsequent generations. Table 21.1 provides the current genetic summary of the population.

Forty-one wild-caught individuals (called 'founders') already have contributed genes to the population, and another 14 wild-caught animals are alive and of reproductive age, but have not produced offspring (these are called 'potential founders'). Due to the significant

Table 21.1. *Genetic status of the captive giant panda population as of 1 January 2002*

Number of founders	41
Potential additional founders	14
Proportional heterozygosity retained	97.0%
Potential proportional heterozygosity	98.9%
Founder genome equivalents	16.7
Average inbreeding coefficient	0.001
Average mean kinship	0.030

number of founders, 97% of the heterozygosity of the wild panda population has been retained, or captured, in the *ex situ* population. This is equivalent to the proportion of heterozygosity contained in a population founded by almost 17 completely unrelated individuals (i.e. 17 founder genome equivalents; Lacy, 1989). Under ideal genetic management, which includes successfully breeding all additional potential founders (Lacy, 1995), the proportion of heterozygosity retained could be increased to almost 99%. To date, inbreeding has been avoided, and there are only two inbred animals in the population (with inbreeding coefficients of 6.3%, i.e. their parents are first cousins). The level of kinship among all individuals (average mean kinship; see also 'Priorities for the future', p. 512) is only 3%. Overall, the captive population is genetically quite well off, much more so than other high-profile species in *ex situ* breeding programmes, for example, the black-footed ferret (Wisely *et al.*, 2002) and California condor (Ralls & Ballou, 2004).

Nonetheless, there is a serious challenge with significant genetic substructure from a pedigree perspective. Very few animals have been exchanged between breeding institutions, and each of the three primary breeding centres (at Beijing, Chengdu and the Wolong Nature Reserve) has founder lineages that are more- or- less unique to that institution. The Beijing and Chengdu facilities, each with representation from 12 founders, share representation from only three founders (SB 152, 174 and 202). The Chengdu and Wolong facilities, with 20 founders represented, share representation from only three founders (SB 231, 253 and 298). The Beijing and Wolong centres do not share any common founders.

This genetic fragmentation results in low levels of relatedness among institutions but high relatedness within institutions. Although the average kinship in the total population is 0.03, it is 0.08, 0.10 and 0.13 in the Wolong, Chengdu and Beijing centres, respectively. At the Beijing Zoo, animals are related on average more than half-sibs. The implication of high within-institution relatedness is that the number of genetically suitable breeding pairs at each institution is becoming limited. The proportion of all possible male/female pairings that involve related individuals (which would produce inbred offspring) is 60%, 56% and 33% at the Beijing, Chengdu and Wolong facilities, respectively. Limiting future pairings to within these institutions will only increase the level of relatedness, eventually resulting in inbreeding. Although data are not yet available on the effects of inbreeding in giant pandas, there is overwhelmingly strong scientific evidence in many other species that it is detrimental in small populations. In almost every species studied, inbreeding adversely affects population health, invariably increasing mortality and decreasing reproductive success (Ralls *et al.*, 1988; Lacy, 1997; Keller & Waller, 2002).

An additional risk associated with this genetic fragmentation is the potential loss in genetic diversity given a catastrophic loss of animals at any one of the breeding centres, e.g. due to disease, natural disaster or other unforeseen calamity. With many founder lineages present in only one facility, loss of individuals at that facility could cause the extinction of those lineages. If offspring were transferred among centres (and other zoos), founder representation would be distributed, decreasing the risk of being lost.

In summary, while the overall genetic status of the captive population is good, gene flow among breeding centres is needed in the near future to avoid inbreeding and, thus, inbreeding depression. We next examine the potential effect of inbreeding depression in unmanaged, fragmented populations versus the benefits of a managed cooperative breeding programme.

Impact of inbreeding in managed versus unmanaged populations

To examine the influence of inbreeding depression on the probability of extinction, we modelled the *ex situ* giant panda population using the population viability analysis (PVA) software VORTEX (Miller & Lacy, 1999) and SIMPOP (Lacy & Ballou, 2002). Demographic and genetic parameters used for VORTEX, derived from the demographic and genetic analyses of the studbook, are presented in Table 21.2. The scenarios modelled were:

1. A small isolated population with an initial population size set at 15 individuals. This scenario was used to model the long-term effects of inbreeding in a single, small, closed institution (with no immigration or new founders) of a size similar to the current Beijing Zoo collection.
2. A larger isolated population with an initial population size set at 45 individuals. This example was developed to resemble a closed population of size similar to the current collections of the larger

Table 21.2. *Parameters used in VORTEX and SIMPOP population viability analyses*[a]

PVA parameter	Value used
Proportion of females breeding	0.30[b]
Maximum breeding age	20
Age (years) for first reproduction (males/females)	5/5
Male mortality	
Age 0	0.330
Age 1	0.053
Age 2	0.023
Age 3	0.086
Age 4	0.019
Adult	0.073
Female mortality	
Age 0	0.282
Age 1	0.014
Age 2	0.014
Age 3	0.018
Age 4	0.063
Adult	0.065[c]
Litter size	
% litters with 1 cub	50%
% litters with 2 cubs	48%
% litters with 3 cubs	2%
Inbreeding depression	3.14 lethal equivalents with 50% of these being lethal recessives[d]
Catastrophe	Arbitrary: 2% chance occurring each year; 50% reduction in survival when occurs
Male mating	In unmanaged populations, 20% of males available; in globally managed population, all males available with breeders and matings determined by mean kinship and relatedness of mates using SIMPOP
Carrying capacity	Set at initial population sizes for each scenario

[a] Unless noted, all parameters were derived from analysing data from the International Giant Panda Studbook. Inclusion of these parameters in VORTEX and SIMPOP resulted in a population with a deterministic growth rate of $\lambda = 1.027$, comparable to the observed life table growth rate of 1.028; [b] Studbook results showed 21.8% of adult individuals breeding but improving over time;

significantly by establishing a cooperative captive breeding programme. It is worth noting that our results undoubtedly *underestimated* the effects of population isolation on extinction probability. For example, in our model we only examined the detrimental effects of inbreeding depression on juvenile survival. However, inbreeding influences many other aspects of fitness, including adult survival and reproduction (Keller & Waller, 2002). If these variables had been factored in, the consequences of the isolated management scenario would have been even more severe.

The need for a global captive breeding programme

A global, cooperative captive breeding programme for giant pandas would mean that all holding institutions worldwide would work together to share animals (or germ plasm, as necessary) to maximise genetic heterozygosity, avoid inbreeding and distribute the species' valuable genes among multiple populations (Fig. 21.9). Such a partnership would ensure that the captive population remains viable, healthy and genetically 'insured' while fully capable of supporting species conservation *ex situ* and, if necessary, *in situ* and long into the future. Cooperative breeding programmes for some wildlife species have been

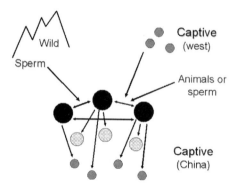

Figure 21.9. Metapopulation management of the captive giant panda population showing animal or gene flow among primary breeding centres (large, dark circles) and other captive institutions (medium and small circles) with the potential of future gene flow from the wild as well as animals held outside China (in the west).

in place for decades. However, unlike what is being proposed here, most exist only at a regional level, administered by regional zoo associations (e.g. the SSP coordinated by the AZA, or the European Endangered Species Programme coordinated by the European Association of Zoos and Aquaria; Princée, 2001). There are a few exceptions, including the global captive management programmes for the golden lion tamarin (Ballou *et al.*, 2002), which can serve as a model for the giant panda.

The global captive breeding programme for the giant panda would consist of multiple regional populations or subpopulations that would be managed as a metapopulation, i.e. an interacting set of populations (see Fig. 21.9). We foresee that the metapopulation would comprise a set of Chinese institutions, institutions outside China and the wild populations themselves. Transfer of animals or germ plasm among captive breeding institutions would maintain genetic diversity *within* institutions. Eventually (and if warranted) animal movements might also include the release of captive-born animals into vacant but suitable wild habitat, i.e. reintroductions. In theory, pandas living in nature might also occasionally be acquired to bolster the captive gene pool, but given the advances in reproductive technologies (see Chapter 20) it would make more sense to collect surplus germ plasm (i.e. sperm) from wild pandas for infusing new genes into captivity. Lastly, a metapopulation strategy could involve transferring animals between wild populations to maintain maximum genetic diversity in nature.

Most managers of captive wildlife populations have adopted a goal of maintaining 90% of the population's heterozygosity for the next 100 years (Soulé *et al.*, 1986). With the current level of retained heterozygosity at 97% and a realistic population growth rate of about 3% per year (see above), this would require a global captive population of about 340 giant pandas (estimated using PM2000 software; Pollak *et al.*, 2002). Under tighter genetic management, which could increase heterozygosity by successfully breeding wild-caught animals which have so far failed to breed, a smaller overall population would be needed. For example, if retention of 98% of the heterozygosity could be achieved in this manner, population size requirements would decrease to about 270 individuals. Similarly, a smaller population would be needed if the population growth rate was higher than 3%. For example, a doubling to 6% annually could allow the genetic target to be hit with an overall world population of 280 giant pandas.

Regardless of the various scenarios, it is apparent that to maintain the genetic fitness of the giant panda into the future will require space and resources for the management of about 300 individuals. Based on our knowledge of the panda's life table and current age structure, a 3% annual growth would initially require producing about 13 to 17 births per year and would need to increase over time, eventually (in about 30 years) requiring about 28 annual births to maintain the population at 300. This seems attainable, as the average number of births per year over the last ten years has been 13.3.

Demographic versus genetic priorities

Certainly, the population should not be constrained to grow at only 3%. The higher the growth rate, the more demographically and genetically secure the population becomes. However, growth in numbers should not be the only priority. Genetic management must also be considered. Decisions to maximise production now (e.g. by only mating the most willing breeders) could compromise the long-term genetic health of the population. As an extreme example, in a population of 45 giant pandas, the use of only one male as a breeder of most females would cause the inbreeding level to increase at more than twice the rate than if all males were used. The use of only a few highly successful breeders will also hasten the population becoming genetically adapted to the captive environment.

Captive breeding programmes include genetic management priorities to maintain high levels of gene diversity in the population, avoid inbreeding and minimise genetic adaptation to the captive environment (Ballou & Foose, 1996). This can be accomplished by implementing a breeding strategy that minimises mean kinship, as discussed above (Ballou & Lacy, 1995). A mean kinship value (the average kinship between that individual, i, and all individuals) would be calculated for every animal in the population using the equation:

$$mk_i = \Sigma k_{ij}/N$$

where k_{ij} is the kinship coefficient between individuals i and j . This is summed over all N individuals ($j = 1$ to N) in the genetically viable population and divided by N to give the average kinship of individual i to the entire population. (Note: usually relationships of individuals with wild-caught founders are not included in the calculation because managers are most interested in the genetic diversity already captured

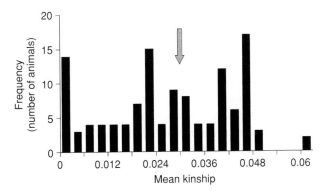

Figure 21.10. Distribution of mean kinship values in the giant panda population.

in the progeny of founders. Wild-caught animals that have not yet produced should not technically be counted as having contributed genetic material to the captive population; Lacy, 1995.)

Priority breeders are those with low *mk* values as they have fewer relatives in the population than animals with high *mk* values. By selecting breeders that minimise average mean kinship, the breeding programme maximises retention of heterozygosity and avoids inbreeding to the extent possible. Using a mean kinship strategy also tends to equalise the genetic contribution of founders in the population since descendants of under-represented founders will have low mean kinship and be preferentially bred. The distribution of *mk* values in the contemporary giant panda population is shown in Figure 21.10.

The 14 giant pandas with *mk* = 0 are those wild-caught animals that have not yet reproduced at the time of this writing. Priority should be placed on trying to breed these individuals before they become reproductively senescent; in the case of males, at the least multiple semen samples should be collected and used for AI and/or stored in the giant panda Genome Resource Bank (GRB; see Chapter 20). Similarly, the two giant pandas with the highest *mk* value (0.06) should be kept from contributing further to offspring production, unless their mean kinships decline relative to the rest of the population at some point in the future. If these males continue to breed, their genes will dominate the captive population's genetic pool, causing inbreeding to increase rapidly.

Captive breeding programmes often need to focus on demographic expansion during their early years when the population is extremely small and at a high risk of extinction. As the population grows, however, more emphasis can and should be placed on genetic management (Ralls & Ballou, 1992). The giant panda captive population is at a point where both increased population size and genetic management are needed. This can be accomplished by breeding all females while also attempting to minimise mean kinship. Females with low mean kinships should be placed in the best breeding situations and paired with males with similar low mean kinship values. However, the dilemma faced by this and other programmes is that some of the females and males with low mean kinships are also often the 'problem breeders', e.g. those with behavioural or medical challenges. Lack of reproductive success has led to their low mean kinship values. Thus genetic management becomes a balance of:

1.   placing breeding priority on those animals with low mean kinship, some of which may have questionable probability of breeding success, and;

2.   establishing a sufficient number of pairs with high probability of breeding success (but not necessarily high mk) to accomplish the desired number of offspring needed for targeted population growth.

Most captive breeding programmes are able to balance these concerns and achieve population growth while undergoing genetic management. Two examples for the reader to review include the black-footed ferret (Russell *et al.*, 1994) and the California condor (Ralls & Ballou, 2004). This same approach is needed for the *ex situ* population of giant pandas to ensure long-term genetic health and to establish and maintain a demographically secure and self-sustaining population.

PRIORITIES FOR THE FUTURE

The need for cooperation, coordination and integration

The International Studbook for the Giant Panda (e.g. Xie & Gipps, 2001) is the basis for evaluating demographic and genetic characteristics of the current captive population of this species. These data require updating annually before each reproductive season to allow the best breeding recommendations to be made. Studbook information should also be supplemented, when necessary, with molecular genetics data to resolve any uncertainties in paternity. Indeed, the *ex situ* population of giant pandas within China is growing rapidly at about 6% per year, and life-history characteristics have improved as captive-born animals account for more of the reproduction. Genetically, the population is healthy with many wild-caught founders represented – there is high genetic diversity and low levels of inbreeding. However, because there has been little gene flow between the primary breeding centres, populations within these institutions have significant genetic differences. Animals *within* institutions are becoming increasingly related, in part, because only a few males are producing most of the young. Such males (with high mean kinship values) need to be retired from the breeding programme. Otherwise, overbreeding of certain individuals (including females) will result in a population in the near future where it will be difficult to avoid pairing related individuals. A global cooperative genetic management programme that minimises mean kinship while ensuring exchanges of animals (or germ plasm) among institutions is key to a strong future for giant pandas living in captivity. The word 'global' indeed means world-wide cooperation that includes contributions from those giant pandas that have been provided on loan to western countries. Their offspring and/or germ plasm will also need to be widely dispersed according to genetic and demographic needs, including back into Chinese breeding programmes where, of course, the majority of animals live.

Finally, given these collective concerns, there is a need to develop a Cooperative Breeding Plan for the *ex situ* giant panda population, which is organised under a recognized 'management group'. During the 2002 Chengdu workshop, it was recommended that such a group be formed and include two co-coordinators (one from the Chinese Association of Zoological Gardens and one from China Wildlife and Conservation Association) as well as the current Giant Panda Studbook Keeper and representatives from all major breeding facilities. The purpose of this management group would be to develop and implement a formal Cooperative Breeding Plan to ensure that the captive population remains demographically and genetically viable based on annual breeding recommendations derived from yearly quantitative analyses. The management group should also develop a husbandry manual, monitor and provide advice to existing and potentially new breeding facilities, seek out funds to support high-priority research and training, and develop a website. When necessary, the management group should be supported by external experts (from China or abroad) in the areas of reproduction, nutrition, genetics, population biology, veterinary medicine, behaviour, pathology, conservation, education and genome resource banking. What is exciting is that most of these disciplinary components are already available and have started working together successfully.

ACKNOWLEDGEMENTS

The authors thank Friends of the National Zoo for financial support and all institutions for contributing data to *The International Studbook for the Giant Panda*.

REFERENCES

Ballou, J. D. and Foose, T.J. (1996). Demographic and genetic management of captive populations. In *Wild Mammals in Captivity*, ed. D. G. Kleiman, M. Allen, K. Thompson, S. Lumpkin and H. Harris. Chicago, IL: University of Chicago Press, pp. 263–83.

Ballou, J. D. and Lacy, R. C. (1995). Identifying genetically important individuals for management of genetic diversity in captive populations. In *Population Management for Survival and Recovery* ed. J. D. Ballou, M. Gilpin and T. Foose. New York, NY: Columbia University Press, pp. 76–111.

Ballou, J. D., Kleiman, D. G., Mallinson, J. J. C., Rylands, A. B., Valladares-Padua, C. and Leus, K. (2002). History, management and conservation role of the captive lion tamarin populations. In *The Conservation Program of the Lion Tamarins*, ed. D. G. Kleiman and A. B. Rylands, pp. 117–132. Washington, DC: Smithsonian Institution Press.

Caughley, G. (1977). *Analysis of Vertebrate Populations*. New York, NY: John Wiley & Sons.

Foose, T. J. (1989). Species survival plans: the role of captive propagation in conservation strategies. In *Conservation Biology and the Black-Footed Ferret*, ed. U. S. Seal, E. T. Thorne, M. A. Bogan and S. H. Anderson. New Haven, CT: Yale University Press, pp. 210–22.

Frankham, R., Hemmer, H., Ryder, O. A., *et al.* (1986). Selection in captive populations. *Zoo Biology*, **5**, 127–38.

Gage, T. B. (2001). Age-specific fecundity of mammalian populations: a test of three mathematical models. *Zoo Biology*, **20**, 487–99.

Glatston, A. R. (1986). Studbooks: the basis of breeding programs. *International Zoo Yearbook*, **25**, 162–7.

Hosmer, D. W. Jr., and Lemeshow, S. (1999). *Applied Survival Analysis*. New York, NY: John Wiley & Sons.

Hutchins, M. and Wiese, R. J. (1991). Beyond genetic and demographic management: the future of the Species Survival Plan and related AAZPA conservation efforts. *Zoo Biology*, **10**, 285–92.

ISIS. (1994). *SPARKS (Single Species Animal Record Keeping System)*. Apple Valley, MN: International Species Information System.

Keller, L. F. and Waller, D. M. (2002). Inbreeding effects in wild populations. *Trends in Ecology and Evolution*, **17**, 230–41.

Lacy, R. C. (1989). Analysis of founder representation in pedigrees: founder equivalents and founder genome equivalents. *Zoo Biology*, **8**, 111–23.

(1995). Clarification of genetic terms and their use in the management of captive populations. *Zoo Biology*, **14**, 565–77.

(1997). Importance of genetic variation to the viability of mammalian populations. *Journal of Mammalogy*, **78**, 320–35.

Lacy, R. C. and Ballou, J. (2002). *Simpop: software to simulate genetic management in pedigreed populations*. Brookfield, IL: Chicago Zoological Society.

Lacy, R., Ballou, J., Starfield, A., Thompson, E. and Thomas, A. (1995). Pedigree analyses. In *Population Management for Survival and Recovery*, ed. J. D. Ballou, M. Gilpin and T. Foose. New York, NY: Columbia University Press, pp. 57–75.

Miller, P. S. and Lacy, R. C. (1999). *VORTEX: A Stochastic Simulation of the Extinction Process. Version 8 User's Manual*. Apple Valley, MN: IUCN–World Conservation Union/SSC Conservation Breeding Specialist Group.

Pollak, J., Lacy, R. C. and Ballou J. D. (2002) *PM2000: Population Management Software*. Ithaca, NY: Cornell University.

Princée, F. P. G. (2001). Genetic management of small animal populations. *Lutra*, **44**, 103–12.

Ralls, K. and Ballou, J. D. (1992). Managing genetic diversity in captive breeding and reintroduction programs. *Transactions of the 57th North American Wildlife and Natural Resources Conference*, 263–82.

(2004) Genetic management of California condors. *The Condor*, **106**, 215–28.

Ralls, K., Ballou, J. D. and Templeton, A. R. (1988). Estimates of lethal equivalents and the cost of inbreeding in mammals. *Conservation Biology*, **2**, 185–93.

Russell, W. C., Thorne, E. T., Oakleaf, R. and Ballou, J. D. (1994). The genetic basis of black-footed ferret reintroduction. *Conservation Biology*, **8**, 263–6.

SAS. (2001). *SAS/STAT User's Guide, Version 8.02*. Cary, NC: SAS Institute.

Soulé, M., Gilpin, M., Conway, W. and Foose, T. (1986). The millennium ark: how long a voyage, how many staterooms, how many passengers?, ed. K. Ralls and J. D. Ballou. (Proceedings of the Workshop on Genetic Management of Captive Populations.) *Zoo Biology*, **5**, 101–13.

Wisely, S., Buskirk, S. W., Fleming, M. A., McDonald, D. B. and Ostrander, E. A. (2002). Genetic diversity and fitness in black-footed ferrets before and during a bottleneck. *Journal of Heredity*, **93**, 231–7.

Xie, Z. and Gipps, J. (2001). *The 2001 International Studbook for the Giant Panda (Ailuropoda melanoleuca)*. Beijing: Chinese Association of Zoological Gardens.

22

Partnerships and capacity building for securing giant pandas *ex situ* and *in situ*: how zoos are contributing to conservation

DAVID E. WILDT, XIAOPING LU, MABEL LAM, ZHIHE ZHANG, SUSIE ELLIS

INTRODUCTION

The new information in this book is largely the product of a series of successful cross-cultural and biological experiments – that is, people with diverse backgrounds and skills working together over time to create scholarly information, which is already being used to enhance giant panda management. Much of the progress is the result of personal relationships that developed during the course of the Survey, which, in turn, provided some valuable lessons about working together in China. Among these is the importance of developing respectful, collegial partnerships. This does not mean a one-time meeting or research study but rather long-lasting relationships that are sustained over many years. This obviously requires substantial investments of time and money, and fierce commitments by all parties. In China, this also means the need for frequent face-to-face interaction.

Remarkably, all of this has transpired to benefit giant pandas, both *ex situ* and *in situ*. While this chapter briefly reviews why success occurred, its main purpose is to share new information about the larger impacts of these relationships. In particular, we examine how partnerships involving giant pandas are addressing one of China's most

Giant Pandas: Biology, Veterinary Medicine and Management, ed. David E. Wildt, Anju Zhang, Hemin Zhang, Donald L. Janssen and Susie Ellis. Published by Cambridge University Press. © Cambridge University Press 2006.

frequently identified needs – capacity building, thereby creating the next generation of skilled biologists and managers devoted to conserving Chinese wildlife and their habitats. Interestingly, zoos are a major force taking many of these steps forward.

WHY SUCCESS TO DATE?

There are three elements responsible for the significant increase in knowledge about giant panda biology, as follows.

The charisma and uniqueness of the species

As millions of visitors clamour to see giant pandas in zoos, scientists also are drawn to this special creature, but for different reasons. They are interested in its oddities of feature – a somewhat bear-like animal with fascinating morphological adaptations, including a functional opposable 'thumb', the ability to sit upright on its hindquarters for hours and its vocal bleats (rather than growls or roars). Taxonomists are intrigued by molecular data that prove a distinct relationship to the bear lineage, even though the giant panda's chromosomes (of which there are 21 pairs) more closely resemble that of the red panda (22 pairs) than most other ursids (37 pairs) (O'Brien, 1987). Nutritionists are mesmerised by its capacity to survive virtually on a grass (bamboo) but with the gastrointestinal tract of a nonruminant. Physiologists are perplexed by a reproductive strategy wherein the female is sexually receptive for less than one per cent of the year (one three-day period) followed by frequent twin production where one of the helpless neonates invariably dies due to maternal neglect. Most of all, many of us are intrigued by the question: can such a highly specialised species, which diverged from conventional bears 15 to 25 million years ago, survive in environments and contemporary times undergoing such radical change? All these mysteries attract significant scientific interest and expertise.

The development of relationships without borders

As recently as the mid 1990s, virtually all on-the-ground giant panda efforts were unlinked, especially in terms of collaboration and communication. The milieu was one of rivalry and secrecy rather than cooperation for a common good. Probably the most important contribution of the Conservation Breeding Specialist Group (CBSG) in China was in demonstrating the value of open communication. The 1996 masterplanning meeting (see Chapter 2) was a milestone in catalysing relationship building. The Chinese who faced the same challenges with giant pandas began talking and then working together, including extending more invitations to western scientists to collaborate. Institutional competition did not disappear, of course, but at least it existed in a healthier and more open atmosphere. CBSG's inherent philosophy of not swamping workshops with western 'experts' or telling the Chinese how to think or what to do was essential. Open working groups and frank discussions helped the participants bond and led to developing realistic expectations and recommendations for action. Then, however, everyone realised that the many identified priorities could not be addressed without significant internal and external partnerships. The resulting Biomedical Survey as a concept was a pledge to work together. Its successful completion demonstrated a remarkable commitment to circumvent institutional and disciplinary chauvinism to solve real problems. And all of this was carried out in a manner consistent with CBSG's emphasis on the power of science, through hypothesis testing, studying, learning and applying new knowledge to improve giant panda management, health and reproduction. This was possible only by the voluntary dismantling of many historical barriers followed by boldly working together across institutions, disciplines and cultures.

Money, specifically associated with giant panda loans to zoos outside China

The information in this book did not simply appear as the result of diligent scientific inquiry. It required significant financial investments by all parties. As described in Chapter 1, almost all giant pandas now living *ex situ* in zoos in the USA, Europe and Asia are linked to a loan process that provides animals in exchange for money, most of which is destined for supporting pandas in nature. Nowhere are the restrictions tighter than in the USA where zoos interested in exhibiting this species must comply with a strict US Fish and Wildlife Service import policy. In the USA, a host institution that displays a pair of giant pandas currently pays a fee of at least $1 million annually to one or two Chinese agency partners. The US Fish and Wildlife Service then works with this zoo and the appropriate Chinese agency to ensure that these funds are used to 'enhance' the survival of this species in the wild. This complex process includes identifying projects in China, which must then be approved by the Service and the contributing zoo. In reality, financial costs are much higher than $1 million annually because participating USA zoos devote substantial additional resources (people and funding) to implement their own research and training programmes. For example, in the case of the Smithsonian's National Zoological Park, these supplemental monies from 2001 to 2004 averaged $400000 annually. Interestingly, it is these latter funds being provided by all USA holding institutions (as well as supplementary support by Chinese partners) that have driven the increased understanding of giant panda biology – the information found in this book.

WHAT ZOOS DO FOR CONSERVATION

For years, zoos were consumers of wildlife, rather than institutions that contributed to conservation. People tend to love zoos, which often provide children with their first personal experiences with wild species. Because zoos can evoke 'nature', they offer a golden opportunity to contribute in diverse ways to conservation. However, it is too easy for zoos to claim erroneously a 'mission' of conservation. Miller *et al.* (2004) recently offered some useful conservation metrics for zoos, ranging from ensuring that 'conservation thought' defines organisational policy to generating funds for research that actually protects the environment, locally and internationally. In short, it is clear that educating the public – providing living dioramas of species along with strong conservation messages through signage and interactive opportunities – is admirable *but not enough*. And, in fact, many zoos have become more active players in the conservation world, including through political advocacy, scientific research, fundraising for *in situ* activities and the training of wildlife professionals (see review Hutchins *et al.*, 2003).

In the case of the giant panda, there are three major ways in which zoos are contributing to conservation, *ex situ* and *in situ*.

Fostering partnerships

Conservation science is like no other in that successful programmes must be based on extensive collaborations, in part because so many disciplines are required to deal with encountered biocomplexities.

These include not only biological factors but also cultural, social and economic intricacies associated with resolving most conservation challenges. Each of these problems is like a huge, messy jigsaw puzzle with many multifaceted pieces and no simple, quick solutions (Wildt *et al.*, 2003). Thus it is impossible for a single organisation or discipline (let alone an individual) to effectively and unilaterally address how to understand and protect any wild animal or wild place.

For the giant panda, all research and training programmes so far have been intimately and inextricably tied to organisational and personal relationships. These have occurred at three levels, the first being the early significant association between CBSG and Chinese federal agencies charged with panda management and protection (see Chapter 2). CBSG exists, in large part, because of donations from zoos worldwide that support its modest operating costs. Chinese authorities trusted CBSG because of its reputation as a neutral facilitating organisation – it had no agenda for pandas, only an interest in objectively responding to a call for advice from the Chinese government.

The second level has been between individual western zoos (wanting giant pandas for exhibit) and official partners within China. The latter specifically involves the China Wildlife and Conservation Association (CWCA, within the State Forestry Administration, SFA) and the Chinese Association of Zoological Gardens (CAZG, under the umbrella of the Ministry of Construction). Western zoo interests in pandas, combined with CWCA and CAZG awareness about financial needs within China, motivated these legal partnerships. The result then was the transfer of a pair of animals to each approved facility in return for substantial funding to be directed, mostly, at *in situ* conservation priorities.

Third, and in our opinion the most important, partnerships evolved from personal interactions among people devoted to giant panda science and related training. The result has been substantial cross-institutional and interdisciplinary studies (highlighted throughout this book) as well as a rapidly growing list of other publications throughout the scientific literature. We are frankly unsure why these social interactions progressed so quickly, including the formation of sincere friendships that continue today. Perhaps it has something to do with the natural camaraderie arising from the risks that like-minded people took in collecting biological data from numerous anaesthetised giant pandas, arguably one of the most precious species on the planet. It is also related to spending every waking moment together as a team, eating meals together and unwinding as a group in the evenings. These teams became like 'second families' – for example, sharing anecdotes, poking fun at each other's foibles, comparing family photographs and donning silly hats to celebrate birthdays. However, we suspect that it also has to do with a universal commitment among panda biologists who realise that the popularity of the species (and the related funding that it attracts) offers a unique opportunity to make a meaningful difference. In short, the experience brought out the best in everyone involved, both person-ally and professionally. All evidence so far indicates that these extensive collaborations have spawned enhanced information sharing, better com-munication and improved animal management, all of which has fostered even more interest in partnering.

Creating species-specific biological data: tools applicable to in situ conservation and a hedge (insurance) population

The *ex situ* studies conducted associated with the Biomedical Survey and by other investigators associated with this book could never have been accomplished in remote and uncontrollable field conditions. In fact, good hypothesis-driven studies to understand biological mechanisms about how each species 'works' behaviourally, physiologically and med-ically are almost impossible to do *in situ*. This is one reason that many of the 'life sciences' are mostly ignored in the field of conservation biology – it's difficult to collect and interpret data. Yes, certainly, efforts to monitor species' numbers and habitat quality are crucial but can we ultimately save a species in the absence of understanding its biology – its reproductive mechanisms, its sensitivity to stress or its vulnerability to diseases? Such issues cannot be addressed through conventional ecological studies but rather require controlled studies in a regulated environment (Wildt, 2004). For example, virtually all the new informa-tion in this book related to (for example) medical issues, male and female reproductive physiology/endocrinology, behaviour, nutrition, developmental biology and genetics could never have been collected from giant pandas living *in situ*.

The *ex situ* population provided a unique opportunity to create this scholarly knowledge, much of which could (and probably will) have usefulness in studying and protecting this species in nature. Table 22.1 lists disciplinary examples of how information and tools generated from the *ex situ* studies can be applicable to giant pandas in nature. These range from the many newly determined species metrics (e.g. useful for assessing age and health status of wild populations, including those in

Table 22.1. *Examples of how disciplines and tools developed in* ex situ *studies of giant pandas could have* in situ *application*

Discipline/tool	Application	Chapter (in this book)
Biomedical surveying	Multidisciplinary, teamwork approaches for collecting, recording and interpreting data on the health and reproductive status of wild individuals or populations	3
Endocrinology	Noninvasive measurement of adrenal function (stress) as a result of habitat disruption or other perturbations	8
Molecular biology	DNA extraction from faeces for noninvasive studies of population number, genetic variation, paternity and extent of home range in wild populations	10
Behaviour	Improved methods for assessing 'personality' traits, stress and the influence of maternal rearing on subsequent offspring reproductive success; application of panda scents to luring animals through landscape corridors or into unoccupied areas; ensuring that reintroduction candidates have normal adaptive and cognitive abilities	11, 12, 14
Veterinary medicine/ pathology	Availability of: anaesthetic details and safety levels; 'normative' data for animal size on basis of known age and for assessing health status or cause of death of individuals or populations; advanced tools (ultrasound, endoscopy) for collecting new medical data	15, 16, 17

Table 22.1. (cont.)

Discipline/tool	Application	Chapter (in this book)
Genome resource banking	Tools for opportunistic or planned collection of biomaterials (blood, tissue, sperm, DNA) for insurance, forensics and movement of genes (sperm) to maintain or infuse gene diversity	10, 20
Population biology	Extension of risk assessment software tools (VORTEX, SIMPOP); application of *ex situ* genetic management concepts and standards to metapopulation management of *in situ* populations; scientific data on when *ex situ* population size is sufficiently demographically stable to consider reintroduction	21

sudden decline) to expanding tools in risk assessment and molecular genetics (to understand population stability and to develop metapopulation management scenarios).

Given the progress occurring in China, it is only a matter of time before the *ex situ* giant panda population becomes self-sustaining (at about 300 individuals worldwide; see Chapter 21). The value of this collective group of animals as an 'insurance policy' for wild counterparts cannot be overemphasised. Although there are challenges to genetically managing giant pandas in captivity, this group retains 97% of the heterozygosity of the wild panda population because of the presence of many founders (see Chapter 21). Meanwhile, giant pandas living in the wild remain highly vulnerable to a long list of potential threats – from habitat fragmentation making some populations too small to be reproductively and genetically viable to unforeseen catastrophes ranging from bamboo die-offs to disease epidemics. As long as the *ex situ* population does not detract from more efforts to study giant pandas and their habitats *in situ*, those animals in captivity provide some comfort as a hedge against losses in gene diversity, populations or the entire species. Finally, although reintroduction currently appears premature (Mainka *et al.*, 2004), once demographically and genetically secure, the *ex situ* population could serve as the resource for releasable candidates.

Building capacity

In 1999, CBSG was invited by the SFA and the Sichuan Forestry Department to facilitate a workshop on 'Conservation Assessment and Research Techniques', which was held in the Wolong Nature Reserve. Attended by 65 participants, mostly Chinese, the workshop sought to identify the highest priorities that could benefit giant pandas (Yan *et al.*, 2000). As usual during such meetings, working groups formed to spend several days discussing issues, which in this case focused on monitoring, surveys, wild population dynamics/genetics, habitats, the needs of reserves and local communities, and the role of the captive panda population in supporting its wild counterpart. The resulting report provided a detailed blueprint for the future. In reading it, one is struck by a common theme that emerged from every working group. *The highest priority articulated by the Chinese participants was for more information sharing, specifically training.* This need was requested at multiple levels to achieve one ultimate purpose: to develop internal abilities to independently and efficiently study and conserve not only the giant panda but also all indigenous wildlife.

And, in recent years, American zoos (predominantly those holding giant pandas) have responded by providing capacity assistance. Table 22.2 lists the formal courses and course descriptions that have been provided by four USA panda-holding institutions – San Diego Zoo, Zoo Atlanta, the Smithsonian's National Zoo and Memphis Zoo – within China. *In situ* training opportunities have ranged from general conservation biology (Fig. 22.1) to forest health to geographical information system (GIS)/remote sensing for reserve managers to radiocollaring methods for black bears. Generally, courses involved equipment donations to participating institutions. For example, GIS courses were always accompanied by gifts of computers and software which were carefully distributed across the various panda reserves. For *ex situ* training, the emphasis has been on those practical tools that could enhance health and reproductive management of the captive panda population. Such courses ranged from methods and protocols for banking of biomaterials (e.g. sperm, tissue, blood products and DNA) to the design and enrichment of zoo environments for addressing

Table 22.2. Formal training activities for Chinese scientists/educators within China supported by USA zoos maintaining giant pandas

Course name (site of training)	Year	Total participants	Purpose	Sponsorship[a]
Conservation Education Department Development (Chengdu schools)	2000–03	350	Parent and teacher training to strengthen conservation education programmes for children	ZA, CAZG, CRB
Geographic Information Systems (GIS)/ Remote Sensing for Giant Panda Reserve Managers I (Wolong Nature Reserve)	2001	40	Creating computer maps, using maps to store and access bio-information, conducting field exercises, ground-truthing, remote sensing imagery; providing equipment and software to reserves	CBSG, SNZP, WWF, MZ, CZ, OP
Conservation and Wildlife Biology Management (Tangjiahe Nature Reserve)	2001, 2002	47	Basic and applied conservation biology philosophies and practical tactics across diverse disciplines from habitat assessment to veterinary care, including field exercises and student projects	SNZP, CWCA
Genome Resource Banking for Giant Panda Genetic Management and Conservation (Chengdu Research Base)	2001	29	Practical aspects of biomaterials, collection, storage, use and record-keeping, including for assisting in basis research and applied genetic management	SNZP, CWCA, CAZG, CRB
Environmental Enrichment (Ocean Park-Hong Kong)	2001	10	Basic concepts and methods for enriching captive animal environments and optimising well-being	ZSSD, CWCA

Table 22.2. (cont.)

Course name (site of training)	Year	Total participants	Purpose	Sponsorship[a]
Environmental Enrichment (China Conservation and Research Centre for the Giant Panda)	2001	40	Basic concepts and methods for enriching captive animal environments and optimising well-being	ZSSD, CWCA
Reptile and Amphibian Husbandry and Veterinary Training Workshop (Chengdu Zoo)	2002	30	Training in animal care, management and veterinary procedures for a herpetological collection	ZA, CAZG, MZ
Sino–USA Forest Health and Restoration (Guiyang, Guizhou Province)	2002	52	A forum for demonstration area staff and stakeholders to share information on progress and challenges in implementing forest health and restoration activities	MZ, USDA-FS, SFA
GIS Remote Sensing for Giant Panda Reserve Managers II and III (Tangjiahe Nature Reserve)	2002, 2004	57	Advanced training in computer maps, using maps to store and access bio-information, conducting field exercises, ground-truthing, remote sensing imagery, providing equipment and software for reserves and identifying data layers requiring improvement for better decision-making	SNZP, CWCA

Sino–USA Forest Health and Restoration Exchange Tours (Shaanxi, Guizhou, Yunnan, Jiangxi, Beijing)	2002, 2004	6	USA experts share management strategies with Chinese counterparts in multiple forest health and restoration demonstration areas	MZ, USDA–FS, SFA
Diagnostics and Clinical Pathology (Beijing Zoo)	2002	55	Diagnostic methods for infectious and toxicological diseases, research applications of clinical pathology, diseases of potential relevance to wildlife and resolution of clinical problems afflicting zoo species, including the giant panda	SNZP, CWCA, CAZG
Environmental Enrichment (Beijing Zoo)	2002	65	Illustrating value and methods of enriching *ex situ* environments to enhance the welfare, behaviour, health and reproduction of wildlife, including the giant panda	SNZP, CWCA, CAZG
Ex situ Genetic Management for the Giant Panda (Chengdu Research Base)	2002, 2003, 2004	15	Teaching the value of genetically managing small populations, including sharing tools useful to ensuring appropriate panda pairings to avoid inbreeding depression	SNZP, CWCA, CAZG, CRB, CBSG
Applied Environmental Education (Tangjiahe Nature Reserve)	2002	22	Training nature reserve personnel and local school teachers in how to use environmental education as a conservation tool to build support in local communities for biodiversity, conservation and ecotourism	SNZP, CWCA

Table 22.2. (cont.)

Course name (site of training)	Year	Total participants	Purpose	Sponsorship[a]
Designing a Conservation Education Summer Camp (Chengdu Research Base)	2002	20	Training personnel from Sichuan Provincial institutions (including schools) in designing conservation education camps for children	ZA, CAZG, CRB
Conservation Education Volunteer Training (Chengdu Research Base and Chengdu Zoo)	2002	150	Training student volunteers from four universities to assist with conservation education programmes	ZA, CAZG, CRB
Bear Ecology, Radiotracking and Conservation (Tangjiahe Nature Reserve)	2002, 2003, 2004	55	Using the black bear as a 'model' to build capacity in phylogenetics, carnivore capture, radiotelemetry, wild population monitoring; home range analysis; habitat selection, GIS analysis and ursid biology	SNZP, CWCA
Large Mammal Surveys (Tangjiahe Nature Reserve)	2003, 2004	36	Training in practical field methods for identifying and surveying mammals in Chinese nature reserves, assembling and analysing data for decision-makers	SNZP, CWCA
Conservation Education Instructor Workshop (Chengdu Research Base)	2004	32	Training panda facility personnel, local school teachers and university students to teach a newly developed conservation education curriculum and related camp	ZA, CAZG, CRB

Title	Year	No.	Description	Partners
Forest Certification in China: Latest Developments and Future Strategies (Zeijiang)	2004	60	Provide SFA officials with an introduction to forest certification and the importance of these standards in forest practices and management	MZ, FAO, SFA, USDA-FS
Veterinary Medical Evaluations of *ex situ* Giant Pandas in China (Chengdu Research Base)	2004	6	Medical evaluations of giant pandas with questionable health status. Sharing cutting-edge technologies in veterinary diagnostics, including ultrasound and gastrointestinal endoscopy	SNZP, ZA, CAZG, CRB
Zoo Masterplanning and Exhibit Design (Shanghai Zoo)	2004	120	Presented modern philosophies on the importance and value of zoo masterplanning and the essentials of exhibit design for optimal animal welfare, management, education and visitor enjoyment	AZA GPCF, CAZG, SNZP
Endocrine Techniques Workshop	2005	27	Training in concepts and laboratory techniques for noninvasive hormone monitoring	SNZP, CRB, CWCA, CAZG, OP, OHDZ

[a]AZA GPCF, American Zoo and Aquarium Association's Giant Panda Conservation Foundation; CAZG, Chinese Association of Zoological Gardens; CBSG, Conservation Breeding Specialist Group; CRB, Chengdu Research Base of Giant Panda Breeding; CWCA, China Wildlife Conservation Association; CZ, Columbus Zoo and Aquarium; FAO, Food and Agriculture Organisation of the United Nations; MZ, Memphis Zoo; OP, Ocean Park-Hong Kong; OHDZ, Omaha's Henry Doorly Zoo; SFA, State Forestry Administration; SNZP, Smithsonian's National Zoological Park; USDA-FS, United States Department of Agriculture–Forest Service; WWF, WorldWild Fund for Nature; ZA, Zoo Atlanta; ZSSD, Zoological Society of San Diego.

Figure 22.1. Participants in a Conservation Biology and Wildlife
Management Training course conducted at the Tangjiahe Nature Reserve
(photograph reproduced courtesy of Rudy Rudran).

animal welfare concerns. In most cases, more sophisticated training
was provided through advanced courses held in China or specialised
training provided to the Chinese locally or in the USA (see below).

When serving as principal investigators on research projects
conducted in China, Americans virtually always trained one or more
Chinese counterparts in scientific methods, data collection, interpretation and even manuscript preparation. Some projects were excellent
opportunities to study other Chinese fauna. For example, a course on
'Large Mammal Surveys' by the Smithsonian's National Zoo was
designed for the Chinese to test the efficacy of using camera traps to
count giant panda numbers in reserves. In reality, however, the study
allowed identifying and counting a diversity of species to determine
mammal densities for entire reserves. Furthermore, most courses did
not involve simply didactic-in-the-classroom activities but were also
highly field based and 'hands on' (Fig. 22.2). There often was an emphasis on the actual long-term collection and interpretation of habitat
and species data that were of direct interest to decision-makers, including reserve managers. Finally, not all courses were directed at scientists.
Two of the sponsoring zoos (Zoo Atlanta and the National Zoo) have
directed   substantial   environmental   education   training   towards

Figure 22.2. Participants in (a) the laboratory (learning GIS tools) followed by (b) a ground-truthing exercise in the field (photographs reproduced courtesy of William McShea and Melissa Songer).

teachers living in communities near or immediately adjacent to panda reserves or breeding facilities (Table 22.2). These projects also focused on actual local public awareness activities to highlight the panda's plight and the value of protecting species and habitats.

The overall impact of capacity building efforts to date has been impressive, and from both cultures. For example, the western faculty often stated "we learned as much, if not more, from the Chinese than they did from us", making this experience truly one of mutual information sharing. From a numbers perspective, 1324 Chinese professionals have benefited from these formal courses in China (see Table 22.2) with 104 experiencing course work or other specialised training in the USA. A substantial proportion of the within-China activities have been directed towards field biologists and managers. For example, 454 Chinese professionals benefited from courses led and sponsored by the Smithsonian's National Zoo. Of these, 56.6% (257 individuals) were involved directly in *in situ* conservation of giant pandas, other wildlife species and their habitats.

What is important is that these training opportunities have been specifically tailored to meet the explicit needs identified by the Chinese, who obviously are most informed about skill and knowledge gaps. Inevitably, the focus has been on highly practical courses – techniques and information that can be applied immediately. Post-course evaluations have been critical to improving offerings as well as designing new or advanced ones. Success has also probably been associated with decisions to conduct most formal courses within China. Generally, it is more economical to send a few faculty to China rather than large numbers of students to the west. Training within China also avoids distractions associated with travelling abroad while allowing the largest number (and most appropriate) people to participate, usually in the actual panda reserves that require study. This strategy allows a person to train in familiar environments, which, in turn, facilitates applying new knowledge and techniques more quickly to the immediate surroundings for which the trainee is responsible. All shared information is also provided in the local language with extensive use of translators, translated manuals and PowerPoint presentations. Efforts have also been made to identify 'star' students within courses who, if possible, can receive further nurturing to become future trainers (i.e. the concept of 'training trainers'). Most of all, success to date has been based on a strong and consistent presence in China – courses are not one-time efforts but rather have occurred with reliable frequency. The participating Chinese and westerners have become not only colleagues but also friends, which in itself has promoted more interest in working together for yet more capacity building. In essence, the provision of extensive training within China has been the ultimate tool for building trust and confidence between partners. In the end, it will enable the Chinese to be sufficiently independent in conservation capacity, taking on their most critical needs with the necessary scientific background and confidence to succeed.

PRIORITIES FOR THE FUTURE

Throughout this book, we have demonstrated the significant accomplishments and new knowledge in giant panda biology that have occurred in a few short years. These have been driven, in part, by scientific curiosity and available resources, mostly provided by zoos. Beyond the scholarship have been the tangibles, including a growing and healthier *ex situ* panda population. Most important has been the open sharing of information and expertise always directed towards the highest priority of Chinese authorities – in-country conservation capacity. In short, it is impossible to do too much training. While we may have a dream of training ourselves 'right out of a job', this is not achievable because the magnitude of the need in all regions (not just China) is too great. Thus a high priority is determining how to develop capacity-building programmes that are bigger, stronger and financially self-sustaining.

One cannot dismiss the impressive progress being made by USA zoos as part of, or as supplements to, their giant panda loan arrangements. For organisations planning to secure giant pandas for exhibit, it is imperative to plan, carry out and financially contribute to capacity building in areas identified as the highest priorities by Chinese partners. These cannot be efforts exclusively directed at *ex situ* facilities because we must always focus first on what is best for giant pandas *in situ*. Additionally, for USA zoos, panda loans are inextricably linked to conservation enhancement in the field. We also realise that not all zoos, including those with an interest in giant panda loans, have research staff capable of providing field training. In this case it is imperative that they subcontract such activities to organisations that do have such abilities, thereby building capacity through sponsorship.

Interestingly, although many research projects described in this book emanated from partnerships involving multiple USA zoos working together, most training was conducted by zoos working individually. This should, and is likely to, change because USA zoos are cooperating more closely now through the Giant Panda Conservation Foundation (under the umbrella of the American Zoo and Aquarium Association or AZA) and the AZA's Species Survival Plan (SSP). For example, the Giant Panda SSP programme now meets annually for several days with managers and scientists to discuss common interests, including planned activities in China. High on the agenda is how the respective holding institutions can work together more strategically and more effectively to share resources in the face of what seems to be ever-decreasing revenue, including what can be contributed to China programmes. Contrary to what might be expected, giant panda loans are not good business for USA zoos. US Fish and Wildlife Service policy prevents giant panda loans from being commercially beneficial. In 2004, the four USA zoos maintaining giant pandas collectively reported a net loss of more than $2 million annually (Giant Panda Conservation Foundation, 2004). This then generates a critical question: even though panda conservation 'wins', are such fiscal losses sustainable?

Interestingly, the answer is not necessarily connected to how much funding is needed by the Chinese authorities to ensure giant panda protection. This is because accessible monies from loans already outstrip available capacity to plan and implement conservation initiatives. Thus, this is yet more weighty justification for continuous capacity building to allow the effective spending of loan-producing dollars. An equally important priority is how zoos can work more effectively with Chinese authorities and (in the case of the USA) the US Fish and Wildlife Service to utilise loan dollars better and more rapidly . This requires the more rapid identification of *in situ* priorities on the part of

Chinese authorities and quicker assessment and/or approval of project lists by the Service. Since USA holding zoos have significant experience in scientific grant preparation and recognising quality proposals, these organisations should be used, whenever possible, to accelerate project selection and monitor project progress.

Meanwhile, zoos certainly have demonstrated their ability to support *in situ* conservation. However, there is a danger in assuming that these organisations can sustain such a financial burden. For this reason, there is wisdom in continuously seeking a reasonable balance between panda loan costs for zoos (which indeed are businesses), what can actually be done in China and making the entire process coordinated and self-sustaining. The latter certainly is achievable given that more zoos decide to become involved, especially with mitigated loan expenses. A few years ago when giant panda breeding was less reliable, the idea of more zoos extracting giant pandas from China for exhibits could have been objectionable. However, given the growth of the *ex situ* population, it is logical to believe that sufficient animals will be available for money-raising loan programmes, especially if the animals remain part of the worldwide genetic management activities (see Chapter 21). Such responsibilities should not be exclusive to North America but should be applicable to other zoos outside China. As outlined by Zhang *et al.* (in Chapter 19), there have been growing numbers of giant pandas moving to zoological parks in Europe and Asia, and each programme should contribute to species conservation in nature.

As they work within and outside China to better identify how to best use scarce resources, zoos will need the attention and cooperation of wildlife authorities. Zhang *et al.* (Chapter 19) and Ballou *et al.* (Chapter 21) have provided excellent within-China examples, for instance the need to facilitate the building of a centralised wildlife infectious disease laboratory and ensuring compliance with mating recommendations by assisting in the transfer of living animals or germ plasm (thereby avoid population inbreeding). In the USA, zoos could benefit by the US Fish and Wildlife Service being more flexible in its definition of what is considered '*in situ* enhancement'. Under current policy interpretation, most activities associated with the *ex situ* programmes are not considered enhancement. Clearly, the biological data and the numerous *in situ*-related spin-offs made throughout this book (as well as depicted in Tables 22.1 and 22.2) would disagree. Rather, we see one worldwide population of giant pandas – some of which live in nature and some in captivity – but all of which are linked through their biological and conservation value and irreplaceable species distinctiveness.

Finally, the giant panda is receiving this massive attention because of worldwide interest in it as a unique emblem of our planet's fragile biodiversity. The collective studies and stories in this book, we believe, give it status as a model – a flagship – for how people can work together scientifically across diverse cultures to learn and problem-solve together. We admit that the giant panda is something of an exceptional case; its high profile and the pressures within China to succeed certainly eased the way for our progress. We would expect it to be far more challenging to stimulate as much enthusiasm for less compelling species, many of which are as threatened or even more ecologically important than the giant panda. Therefore, our personal priority is to promote and, whenever possible, use this paradigm to ensure the continued scientific investigation of any number of the world's many threatened wildlife species, most of which now have not been as thoroughly studied as the giant panda. Regardless, partnerships and interdisciplinary studies are key to gathering the kinds of information needed to make wise and informed conservation management decisions.

ACKNOWLEDGEMENTS

The authors thank Ronald R. Swaisgood (Zoological Society of San Diego), Rebecca J. Snyder (Zoo Atlanta) and John Ouellette (Memphis Zoo) for sharing information on training activities.

REFERENCES

Giant Panda Conservation Foundation (2004). *USA American Zoo and Aquarium Association Giant Panda Initiative: Summary of Annual Financial Reports for the Years 2000–2003*, ed. D. Kelly and K. Heagney.

Hutchins, M., Smith, B. and Allard, R. (2003). In defense of zoos and aquariums: the ethical basis of keeping wild animals in captivity. *Journal of the American Veterinary Medical Association*, **7**, 958–66.

Mainka, S., Pan, W., Kleiman, D. and Lu, Z. (2004). Reintroduction of giant pandas. In *Giant Pandas: Biology and Conservation*, ed. D. Lindburg and K. Baragona. Berkeley, CA: University of California Press, pp. 246–9.

Miller, B., Conway, W., Reading, R. P. *et al.* (2004). Evaluating the conservation mission of zoos, aquariums, botanical gardens and natural history museums. *Conservation Biology*, **18**, 1–8.

O'Brien, S. J. (1987). The ancestry of the giant panda. *Scientific American*, November, 102–7.

Wildt, D. E. (2004). More meaningful wildlife research by prioritizing science, linking disciplines and building capacity. In *Experimental Approaches to Conservation Biology*, ed. M. S. Gordon and S. M. Bartol. Berkeley, CA: University of California Press, pp. 282–97.

Wildt, D. E., Ellis, S., Janssen, D. and Buff, J. (2003). Toward more effective reproductive science in conservation. In *Reproductive Sciences and Integrated Conservation*, ed. W. V. Holt, A. Pickard, J. C. Rodger and D. E. Wildt. Cambridge: Cambridge University Press, pp. 2–20.

Yan, X., Deng, X., Zhang, H. *et al.* (2000). *Giant Panda Conservation Assessment and Research Techniques Workshop, Final Report*. Apple Valley, MN: IUCN/SSC Conservation Breeding Specialist Group.

Index

Entries for figures appear in *italics*. Entries for tables appear in **bold**.

Color Plates

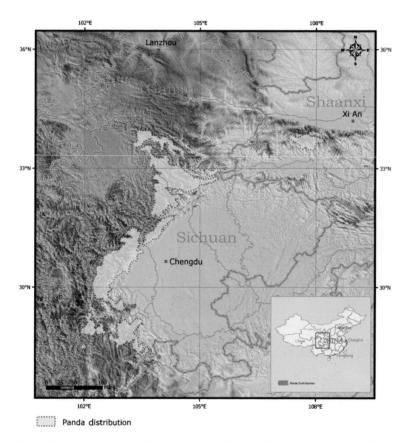

Plate I. Range map for remaining fragmented populations of giant pandas living in nature.

Plate II. Giant panda with Stunted Development Syndrome (SB 325) (photograph by D. Janssen).

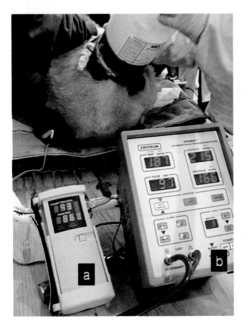

Plate III. Anaesthetised giant panda illustrating supplemental oxygen administration. (a) Portable pulse oximeter; (b) indirect blood pressure monitor.

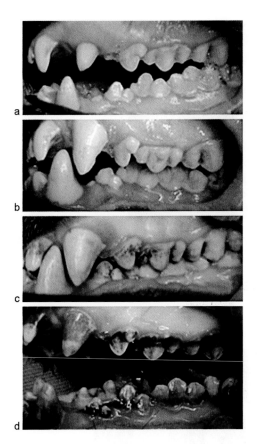

Plate IV. Lateral view of the dentition of four 18-month-old, juvenile giant pandas. (a) Male SB 461 and (b) female SB 452 are 'normal' compared to (c) female SB 453 and (d) male SB 454 who have enamel pitting and excessive teeth staining (along with chronic gastrointestinal disease and lower body weight for their ages).

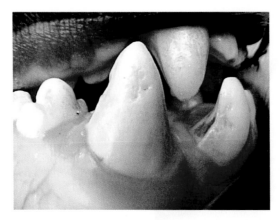

Plate V. Close-up of newly erupted canine teeth and lower third incisor of an 18-month-old giant panda male (SB 454). Defects in the enamel (including staining and pitting) can be observed.

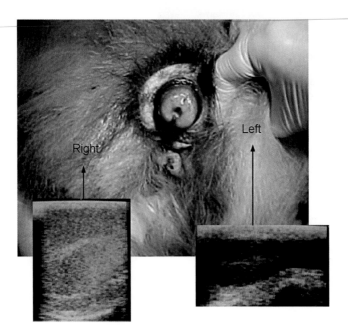

Plate VI. A 12.5-year-old male giant panda (SB 323) with a hypoplastic left testicle. The ultrasound images show the normal architecture and size of the right testis versus the hypoechoic condition of the smaller left testis.

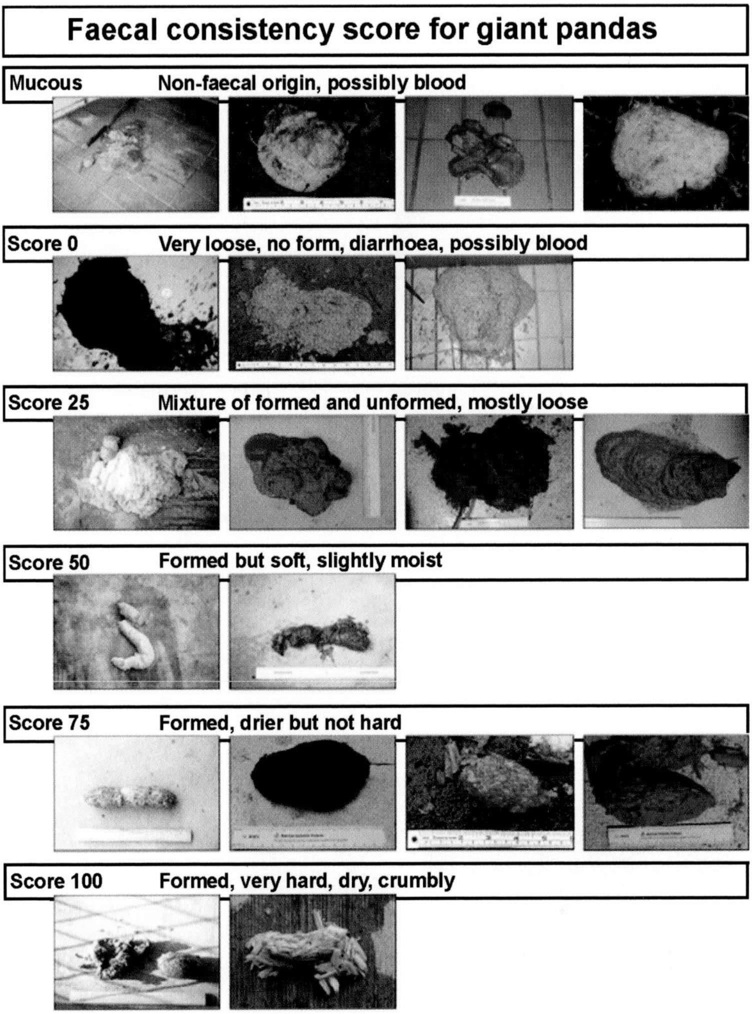

Plate VII. A standardised faecal grading system to routinely monitor and document faecal consistency (reproduced with permission from Edwards & Nickley, 2000).

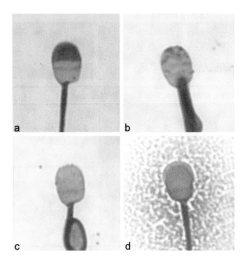

Plate VIII. Acrosomal morphology of the giant panda spermatozoon with a (a) normal apical ridge, (b) damaged apical ridge, (c) missing apical ridge or (d) loose acrosomal cap.

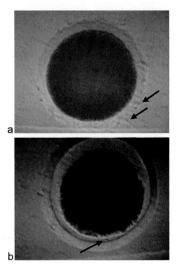

Plate IX. Sperm penetration of the zona pellucida of a (a) homologous giant panda and (b) heterologous cat salt-stored oocyte used to assess sperm function before and after cryopreservation.

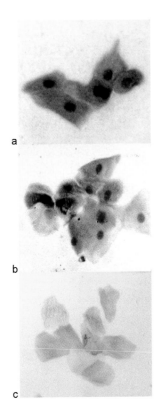

a

b

c

Plate X. Exfoliated vaginal
epithelial cells of the giant panda
stained with modified trichrome
Papanicolaou. (a) Nucleated
basophilic cells (×60);
(b) nucleated acidophilic cells
(×40); and (c) anucleated,
keratinised cells (×40).

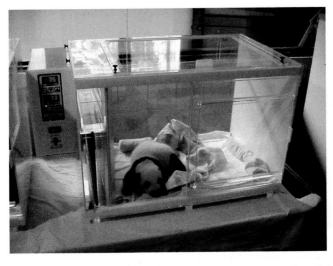

Plate XI. A controlled microenvironment (Animal Intensive Care Unit) used for
housing giant panda cubs from Day 0 to Day 50 (photograph by Mark S. Edwards).

Plate XII. Bean-bag support used to feed giant panda
cubs in an appropriate position (photograph by Mark S. Edwards).

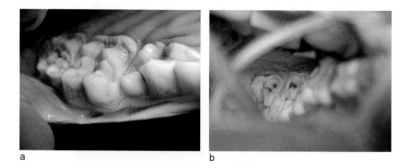

a   b

Plate XIII. View of the mandibular premolars and molars of (a) a 2.5-year-old giant
panda with minimal wear of the grinding surfaces and (b) a 26-year-old giant
panda (SB 381) with extreme wear of the grinding surfaces.

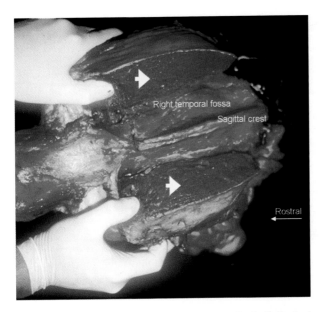

Plate XIV. Dorsal aspect of the adult giant panda skull displaying the large temporal muscles (arrows). The masticatory apparatus is massive relative to other carnivore species. The volume of the temporal fossa and the associated temporalis muscle are particularly large. Similar adaptations are seen in the red panda (a procyonid with a similar diet) and the hyaena (Davis, 1964). The relative mass of bone in the giant panda cranium and mandibles is also much greater than in other ursids.

Plate XV. Intestinal tract of the adult giant panda illustrating uniformity in diameter along its entire length. There is no discernible demarcation between the small and large intestine and no caecum. The extent of the tract is only four to five times the body length, which is more characteristic of a true carnivore than a species that survives on a fibrous bamboo diet.

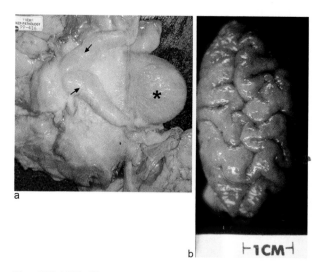

Plate XVI. (a) Bladder and ductus deferens of an adult giant panda. The glandular ductus deferens (arrows) opens into the neck of the bladder (*). (b) Ovary of an adult giant panda with its characteristic convoluted surface.

Plate XVII. The renculate kidney of the giant panda. (a) Whole kidney and (b) sagittal section. Each renculus has a thick cortex (1) and small medulla (2) with the papillae of each joining into a common minor calyx (3). In the renal fossa, the minor calyces unite into two major ones which, in turn, join to form the proximal end of the ureter. There is no renal pelvis in this species.

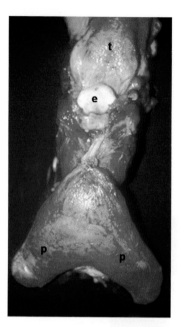

Plate XVIII. Dorsal view of the pharyngeal pouches of the adult giant panda. These paired sacs (p, shown here filled with water to facilitate viewing) open into the dorsal nasopharynx and may play a role in vocalisation. Some authors (Davis, 1964) describe the left sac as being much smaller than the right (15 mm vs. 130 mm in length), while others (Weissengruber *et al.*, 2001 and references therein) describe just a single medially located pouch. e, epiglottis; t, tongue.

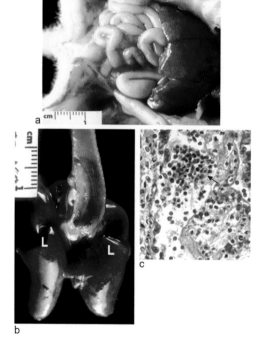

Plate XIX. (a) Gross necropsy (liver) of neonate with septicaemia and hepatitis caused by *Staphylococcus intermedius*. (b) Lungs (L) of a neonate that acquired pneumonia prenatally from a dam with an ascending urogenital tract infection that caused chorioamnionitis. (c) Microscopically the pulmonary alveolar spaces contained inflammatory cells (arrow). *Pseudomonas aeruginosa* was cultured from the lungs and *P. fluorescens* from heart blood.

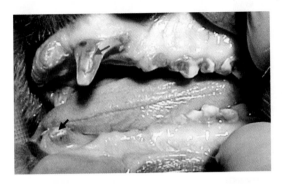

Plate XX. Giant pandas with Stunted Development Syndrome have dental abnormalities characterised by severely worn, stained teeth with dentine exposure (black arrow) and enamel hypo- or dysplasia (blue arrow).

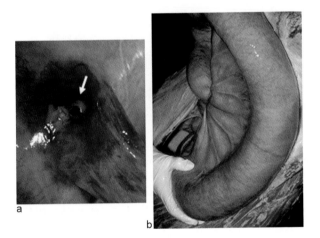

Plate XXI. (a) Chronic enteritis in a giant panda. Endoscopy demonstrating ulceration (arrow) of the intestine. (b) Megacolon found at necropsy.

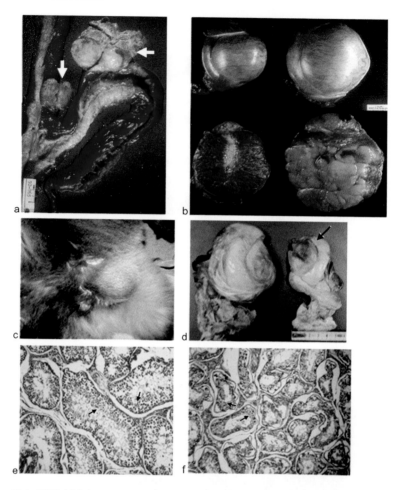

Plate XXII. (a) Leiomyoma (arrows) of the uterus in a 22-year-old giant panda. Leiomyomas are benign tumours of smooth muscle tissue and are common in Carnivora. (b) Seminoma (right) found incidentally in a 25-year-old individual. Both testes were surgically excised, and the neoplasm did not recur or metastasise. The top two items in the image depict the gross specimens and the bottom two their cut surfaces. (c) Gross and (d) excised testes of an adult giant panda (SB 323) with unilateral testicular hypoplasia (arrows). The photograph in (c) was taken during the CBSG Biomedical Survey when the panda was 13 years old; he died one year later. Histopathology of a normal (e) versus hypoplastic (f) testis demonstrating the absence of germ cells in the seminiferous tubules of the latter. Arrows indicate the germ cell layer.

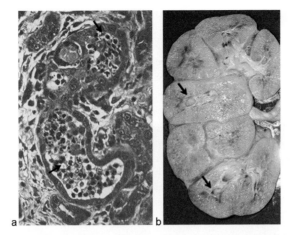

Plate XXIII. (a) Renal biopsy of a 14-year-old female giant panda with acute renal failure who was diagnosed with ascending coliform pyelonephritis. (Arrows designate neutrophils in the renal tubules.) (b) Renal cystic glomerular disease (arrows) in a 28-year-old male giant panda that died of chronic renal failure.

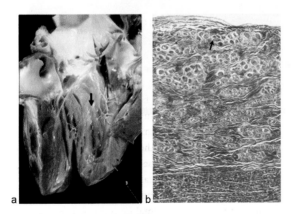

Plate XXIV. Endocardial fibrosis in a 22-year-old female giant panda that died suddenly of cardiac arrhythmia. (a) Gross necropsy demonstrating fibrosis of endocardial tissue (arrow). (b) Histopathology of endocardial tissue. Fibrosis disrupts the course of electrical signalling through conducting fibres (arrow).

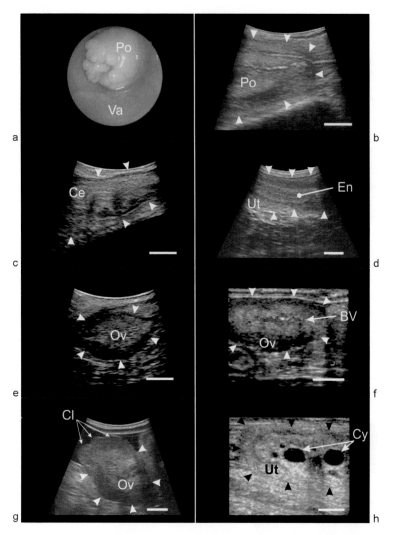

Plate XXV. Ultrasonographic images of an adult female giant panda at the Berlin Zoo. (a) Endoscopic image of the portio cervicalis (Po) and vagina (Va) during AI; (b) ultrasonographic image (TR-150, 7.5 MHz) of the portio (Po, bordered by arrow heads); (c) transrectal sonogram (7.5 MHz) of the cervix (Ce, arrow heads); (d) longitudinal aspect of the uterus (Ut, arrow heads) generated by transrectal ultrasound (TR-250, 7.5 MHz) showing the two layers of the endometrium (En); (e) sonographic image (TR-500, 7.5 MHz) of an inactive ovary (Ov, arrow heads) outside the breeding season; (f) colour-flow Doppler imaging (TR-500, 5.0–9.0 MHz) of the central blood vessels (BV, including veins and arteries) in the medulla of an inactive ovary (Ov, arrow heads); (g) sonogram of an ovary (Ov, arrow heads) containing at least three active corpora lutea (Cl); (h) degeneration of endometrial cysts (Cy) in the cranial part of the uterus (Ut, arrow heads) as a result of foetal resorption.

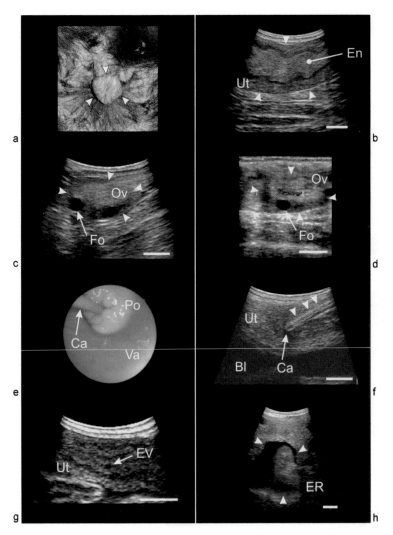

Plate XXVI. Ultrasound examination and AI of the female giant panda SB 378 at the Berlin Zoo. (a) Mildly swollen vulva (arrow heads) near oestrus; (b) transrectal sonogram (TR-500, 7.5 MHz) of the uterus (Ut, arrow heads) and endometrium (En) at the time of ovulation; (c) transrectal sonogram (TR-500, 7.5 MHz) of a Graafian follicle (Fo) on the ovary (Ov, arrow heads) shortly before ovulation; (d) colour-flow Doppler image (TR-500, 5.0–9.0 MHz) of the same ovary (Ov, arrow heads) showing the independent blood supply of the Graafian follicle (Fo); (e) endoscopic image of the transcervical passage of the AI catheter (Ca, 2 mm in diameter) showing the swollen portio (Po) containing opaque viscous mucus, and the mucosa of the vagina (Va); (f) sonographic verification (TR-250, 7.5 MHz) of

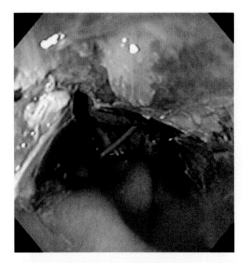

Plate XXVII. Ingesta obscuring visualisation and penetration of the pylorus.

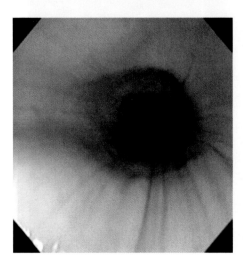

Plate XXVIII. Appearance of the oesophageal lumen between peristaltic waves.

Plate XXVI Caption (cont.)

the intrauterine position of the insemination catheter (Ca, arrow heads). Semen was deposited in the caudal part of the uterus (Ut), which was located dorsal to the urinary bladder (Bl); (g) sonographic evidence (TR-250, 7.5 MHz) of a nonimplanted embryonic vesicle (EV) in the enlarged uterus (Ut) one month after AI; (h) sonogram (TR-250, 7.5 MHz) showing the embryonic resorption site (ER, arrow heads) six months after AI.

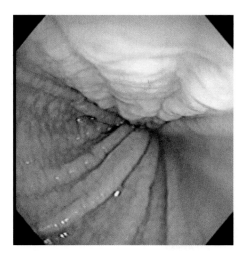

Plate XXIX. Appearance of the oesophageal lumen during a peristaltic wave.

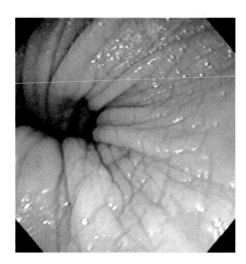

Plate XXX. Corrugated appearance of the oesophagus adjacent to the cardia.

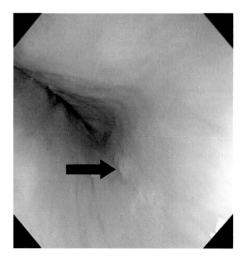

Plate XXXI. Focal lymphoid aggregates (arrow) in the oesophagus near the cardia.

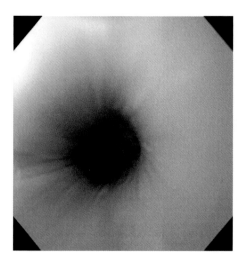

Plate XXXII. Smooth oesophageal mucosa observed throughout the oesophagus in a nine-month-old, juvenile giant panda.

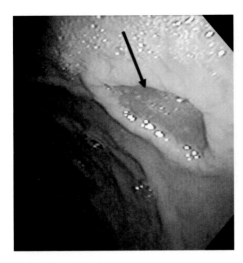

Plate XXXIII. Gastric ulceration (arrow) near the cardia in giant panda SB 454.

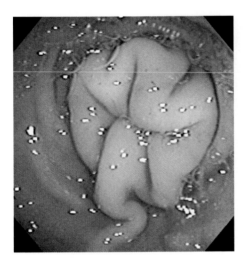

Plate XXXIV. Pylorus of an adult panda.

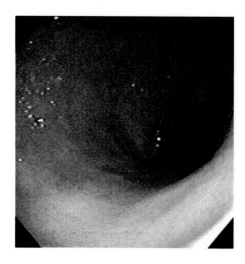

Plate XXXV. Grainy appearance of the duodenal mucosa.

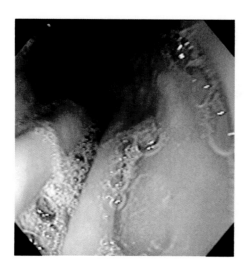

Plate XXXVI. Cavernous appearance of the duodenum with bilious mucoid fluid apparent.

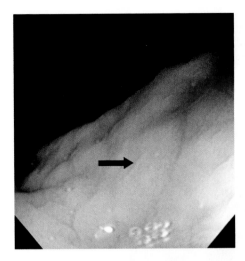

Plate XXXVII. Abnormal duodenal mucosa with patches of pale tissue (arrow).

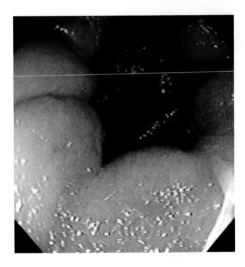

Plate XXXVIII. Typical colouration of the colonic mucosa.

Plate XXXIX. Female giant panda cub SB 473 born after two transcervical artificial inseminations with fresh (AI number 1) and cold-stored (AI number 2) spermatozoa.

Plate XXXX. Female giant panda cub SB 576 born after three transcervical artificial inseminations with only frozen–thawed spermatozoa.